AVIAN BIOLOGY
Volume III

CONTRIBUTORS

Ivan Assenmacher

B. K. Follett

Robert A. Hinde

Hideshi Kobayashi

B. Lofts

R. K. Murton

J. Schwartzkopff

Arnold J. Sillman

A. Tixier-Vidal

Masaru Wada

Bernice M. Wenzel

AVIAN BIOLOGY
Volume III

EDITED BY

DONALD S. FARNER

Department of Zoology
University of Washington
Seattle, Washington

JAMES R. KING

Department of Zoology
Washington State University
Pullman, Washington

TAXONOMIC EDITOR

KENNETH C. PARKES

Curator of Birds
Carnegie Museum
Pittsburgh, Pennsylvania

 1973

ACADEMIC PRESS New York and London
A Subsidiary of Harcourt Brace Jovanovich, Publishers

ACADEMIC PRESS, INC.
111 Fifth Avenue, New York, New York 10003

United Kingdom Edition published by
ACADEMIC PRESS, INC. (LONDON) LTD.
24/28 Oval Road, London NW1

Library of Congress Cataloging in Publication Data
Main entry under title:

Avian biology.

    Includes bibliographies.
    1.   Ornithology.   2.   Zoology–Ecology.
1.   Farner, Donald Stanley, DATE   ed.   II.   King,
James Roger, DATE   ed.
QL673.A9   598.2   79–178216
ISBN 0–12–249403–2

These volumes are dedicated to the memory of

A. J. "JOCK" MARSHALL

(1911–1967)

whose journey among men was too short by half

CONTENTS

Chapter 1.   Reproduction in Birds

B. Lofts and R. K. Murton

Chapter 2.   The Adenohypophysis

A. Tixier-Vidal and B. K. Follett

Chapter 7. Mechanoreception

J. Schwartzkopff

Chapter 8. Behavior

Robert A. Hinde

LIST OF CONTRIBUTORS

Numbers in parentheses indicate the pages on which the authors' contributions begin.

IVAN ASSENMACHER (183), Department of Animal Physiology, University of Montpellier II, Montpellier, France

B. K. FOLLETT (109), Department of Zoology, University College of North Wales, Bangor, Wales

ROBERT A. HINDE (479), M.R.C. Unit on the Development and Integration of Behaviour, Madingley, Cambridge, England

HIDESHI KOBAYASHI (287), Misaki Marine Biological Station, University of Tokyo, Misaki, Kanagawa-ken, Japan

B. LOFTS (1), Department of Zoology, University of Hong Kong, Hong Kong

R. K. MURTON, (1), Monk's Wood Experimental Station, Natural Environment Research Council, Huntingdon, England

J. SCHWARTZKOPFF (417), Ruhr-Universität Bochum, Lehrstuhl für Allgemeine Zoolgie, Bochum, West Germany

ARNOLD J. SILLMAN* (349), Jules Stein Eye Institute, UCLA School of Medicine, Los Angeles, California

A. TIXIER-VIDAL (109), Laboratoire de Biologie Moléculaire, Collège de France, Paris, France

MASARU WADA (287), Misaki Marine Biological Station, University of Tokyo, Misaki, Kanagawa-ken, Japan

BERNICE M. WENZEL (389), Departments of Physiology and Psychiatry, and Brain Research Institute, UCLA School of Medicine, Los Angeles, California

*Present address: Department of Biology, University of Pittsburgh, Pittsburgh, Pennsylvania.

PREFACE

The birds are the best-known of the large and adaptively diversified classes of animals. About 8600 living species are currently recognized, and it is unlikely that more than a handful of additional species will be discovered. Although much remains to be learned, the available knowledge of the distribution of living species is much more nearly complete than that for any other class of animals. It is noteworthy that the relatively advanced status of our knowledge of birds is attributable to a very substantial degree to a large group of dedicated and skilled amateur ornithologists.

Because of the abundance of basic empirical information on distribution, habitat requirements, life cycles, breeding habits, etc., it has been relatively easier to use birds instead of other animals in the study of the general aspects of ethology, ecology, population biology, evolutionary biology, physiological ecology, and other fields of biology of contemporary interest. Model systems based on birds have played a prominent role in the development of these fields. The function of this multivolume treatise in relation to the place of birds in biological science is therefore envisioned as twofold. We intend to present a reasonable assessment of selected aspects of avian biology for those for whom this field is their primary interest. But we view as equally important the contribution of these volumes to the broader fields of biology in which investigations using birds are of substantial significance.

Only slightly more than a decade has passed since the publication of A. J. Marshall's "Biology and Comparative Physiology of Birds," but progress in most of the fields included in this treatise has made most of the older chapters obsolete. Avian biology has shared in the so-called information explosion. The number of serial publications devoted mainly to avian biology has increased by about 20% per decade since

1940, and the spiral has been amplified by the parallel increase in page production and by the spread of publication into ancillary journals. By 1964, there were about 215 exclusively ornithological journals and about 245 additional serials publishing appreciable amounts of information on avian biology (P. A. Baldwin and D. E. Oehlerts, *Studies in Biological Literature and Communications, No. 4. The Status of Ornithological Literature, 1964.* Biological Abstracts, Inc., Philadelphia, 1964).

These stark numbers reflect only the quantitative acceleration in the output of information in recent times. The qualitative changes have been much more impressive. Avifaunas that were scarcely known except as lists of species a decade ago have become accessible to scientific inquiry as a consequence of improved transportation and facilities in many parts of the world. Improved or new instrumentation has allowed the development of new fields of study and has extended the scope of old ones. Examples that come readily to mind include the use of radar in visualizing migration, of telemetry in studying the physiology of flying birds, and of spectrography in analyzing bird sounds. The development of mathematical modeling, for instance in evolutionary biology and population ecology, has supplied new perspectives for old problems and has created a new arena for the examination of empirical data. All of these developments — social, practical, and theoretical — have profoundly affected many aspects of avian biology in the last decade. It is now time for another inventory of information, hypotheses, and new questions.

Marshall's "Biology and Comparative Physiology of Birds" was the first treatise in the English language that regarded ornithology as consisting of more than anatomy, taxonomy, oology, and life history. This viewpoint was in part a product of the times, but it also reflected Marshall's own holistic philosophy and his understanding that "life history" had come to include the whole spectrum of physiological, demographic, and behavioral adaptation. This treatise is a direct descendent of Marshall's initiative. We have attempted to preserve the view that ornithology belongs to anyone who studies birds, whether it be on the level of molecules, individuals, or populations. To emphasize our intentions we have called the work "Avian Biology."

It has been proclaimed by various oracles that sciences based on taxonomic units (such as insects, birds, or mammals) are obsolete, and that the forefront of biology is process-oriented rather than taxon-oriented. This narrow vision of biology derives from the hyperspecialization that characterizes so much of science today. It fails to notice that lateral synthesis as well as vertical analysis are inseparable partners in the search for biological principles. Avian biologists of both stripes have together contributed a disproportionately large share of the information and thought that have produced contemporary principles in zoogeography, systematics, ethology, demography, comparative physiology, and other fields too numerous to mention.

In part, this progress results from the attributes of birds themselves. They are active and visible during the daytime; they have diversified into virtually all major habitats and modes of life; they are small enough to be studied in useful numbers but not so small that observation is difficult; and, not least, they are esthetically attractive. In short, they are relatively easy to study. For this reason we find gathered beneath the rubric of avian biology an alliance of specialists and generalists who regard birds as the best natural vehicle for the exploration of process and pattern in the biological realm. It is an alliance that seems still to be increasing in vigor and scope.

In the early planning stages we established certain working rules that we have been able to follow with rather uneven success.

1. "Avian Biology" is the conceptual descendent of Marshall's earlier treatise, but is more than simply a revision of it. We have deleted some topics and added or extended others. Conspicuous among the deletions is avian embryology, a field that has expanded and specialized to the extent that a significant review of recent advances would be a treatise in itself.

2. Since we expect the volumes to be useful for reference purposes as well as for instruction of advanced students, we have asked authors to summarize established facts and principles as well as to review recent advances.

3. We have attempted to arrange a balanced account of avian biology as its exists at the beginning of the 1970's. We have not only retained chapters outlining modern concepts of structure and function in birds, as is traditional, but have also encouraged contributions representing a multidisciplinary approach and synthesis of new points of view. Several such chapters appear in this volume.

4. We have attempted to avoid a parochial view of avian biology by seeking diversity among authors with respect to nationality, age, and ornithological heritage. In this search we have benefited by advice from many colleagues to whom we are grateful.

5. As a corollary of the preceding point, we have not intentionally emphasized any single school of thought, nor have we sought to dictate the treatment given to controversial subjects. Our single concession to conceptual uniformity is in taxonomic usage, as explained by Kenneth Parkes in the Note on Taxonomy.

We began our work with a careful plan for a logical topical sequence through all volumes. Only its dim vestiges remain. For a number of reasons we have been obliged to sacrifice logical sequence and have given first priority to the maintenance of general quality, trusting that each reader would supply logical cohesion by selecting chapters that are germane to his individual interests.

DONALD S. FARNER
JAMES R. KING

NOTE ON TAXONOMY

Early in the planning stages of "Avian Biology" it became apparent to the editors that it would be desirable to have the manuscript read by a taxonomist, whose responsibility it would be to monitor uniformity of usage in classification and nomenclature. Other multiauthored compendia have been criticized by reviewers for use of obsolete scientific names and for lack of concordance from chapter to chapter. As neither of the editors is a taxonomist, they invited me to perform this service.

A brief discussion of the ground rules that we have tried to follow is in order. Insofar as possible, the classification of birds down to the family level follows that presented by Dr. Storer in Chapter 1, Volume I.

Within each chapter, the first mention of a species of wild bird includes both the scientific name and an English name, or the scientific name alone. If the same species is mentioned by English name later in the same chapter, the scientific name is usually omitted. Scientific names are also usually omitted for domesticated or laboratory birds. The reader may make the assumption throughout the treatise that, unless otherwise indicated, the following statements apply:

1. "The duck" or "domestic duck" refers to domesticated forms of *Anas platyrhynchos*.

2. "The goose" or "domestic goose" refers to domesticated forms of *Anser anser*.

3. "The pigeon" or "domesticated pigeon" or "homing pigeon" refers to domesticated forms of *Columba livia*.

4. "The turkey" or "domestic turkey" refers to domesticated forms of *Meleagris gallopavo*.

5. "The chicken" or "domestic fowl" refers to domesticated forms of *Gallus gallus;* these are often collectively called *"Gallus domesticus"* in biological literature.

6. "Japanese Quail" refers to laboratory strains of the genus *Coturnix,* the exact taxonomic status of which is uncertain. See Moreau and Wayre, *Ardea* **56**, 209–227, 1968.

7. "Canary" or "domesticated canary" refers to domesticated forms of *Serinus canarius.*

8. "Guinea Fowl" or "Guinea Hen" refers to domesticated forms of *Numida meleagris.*

9. "Ring Dove" refers to domesticated and laboratory strains of the genus *Streptopelia,* often and incorrectly given specific status as *S. "risoria."* Now thought to have descended from the African Collared Dove (*S. roseogrisea*), the Ring Dove of today *may* possibly be derived in part from *S. decaocto* of Eurasia; at the time of publication of Volume 3 of Peters' "Check-list of Birds of the World" (p. 92, 1937), *S. decaocto* was thought to be the direct ancestor of *"risoria."* See Goodwin, *Pigeons and Doves of the World* **129**, 1967.

As mentioned above, an effort has been made to achieve uniformity of usage, both of scientific and English names. In general, the scientific names are those used by the Peters "Check-list"; exceptions include those orders and families covered in the earliest volumes for which more recent classifications have become widely accepted (principally Anatidae, Falconiformes, and Scolopacidae). For those families not yet covered by the Peters' list, I have relied on several standard references. For the New World I have used principally Meyer de Schauensee's "The Species of Birds of South America and Their Distribution" (1966), supplemented by Eisenmann's "The Species of Middle American Birds" (*Trans. Linnaean Soc. New York* **7**, 1955). For Eurasia I have used principally Vaurie's "The Birds of the Palaearctic Fauna" (1959, 1965) and Ripley's "A Synopsis of the Birds of India and Pakistan" (1961). There is so much disagreement as to classification and nomenclature in recent checklists and handbooks of African birds that I have sometimes had to use my best judgment and to make an arbitrary choice. For names of birds confined to Australia, New Zealand, and other areas not covered by references cited above, I have been guided by recent regional checklists and by general usage in recent literature. English names have been standardized in the same way, using many of the same reference works. In both the United States and Great Britain, the limited size of the avifauna has given rise to some rather provincial English names; I have added appropriate (and often previously used) adjectives to these. Thus *Sturnus vulgaris* is "European Starling," not simply "Starling"; *Cardinalis cardinalis* is "North American Cardinal," not simply "Cardinal"; and *Ardea cinerea* is "Gray Heron," not simply "Heron."

Reliance on a standard reference, in this case Peters, has meant that certain species appear under scientific names quite different from those used in most of the ornithological literature. For example, the Zebra Finch, widely used as a laboratory species, was long known as *Taeniopygia castanotis*. In Volume 14 of the Peters' "Check-list" (pp. 357–358, 1968), *Taeniopygia* is considered a subgenus of *Poephila*, and *castanotis* a subspecies of *P. guttata*. Thus the species name of the Zebra Finch becomes *Poephila guttata*. In such cases, the more familiar name will usually be given parenthetically.

For the sake of consistency, scientific names used in Volume I will be used throughout "Avian Biology," even though these may differ from names used in standard reference works that would normally be followed, but which were published after the editing of Volume I had been completed.

Strict adherence to standard references also means that some birds will appear under names that, for either taxonomic or nomenclatorial reasons, would *not* be those chosen by either the chapter author or the taxonomic editor. As a taxonomist, I naturally hold some opinions that differ from those of the authors of the Peters' list and the other reference works used. I feel strongly, however, that a general text such as "Avian Biology" should not be used as a vehicle for taxonomic or nomenclatorial innovation, or for the furtherance of my personal opinions. I therefore apologize to those authors in whose chapters names have been altered for the sake of uniformity, and offer as solace the fact that I have had my objectivity strained several times by having to use names that do not reflect my own taxonomic judgment.

KENNETH C. PARKES

CONTENTS OF OTHER VOLUMES

Chapter 1

REPRODUCTION IN BIRDS

B. Lofts and R. K. Murton

I. Introduction

Reproduction is a cyclic phenomenon, and the majority of avian species are seasonal breeders whose physiological mechanisms regu-

1

lating gametogenesis are synchronized by environmental stimuli, ensuring that young are produced at the time of year best suited for their survival (Chapter 8, Volume I). Only in some domesticated species, or in those species inhabiting an environment showing little seasonal variation in food availability, has this ancestral cyclic pattern sometimes been lost (Lofts and Murton, 1968). Generally, the reproductive organs of most birds undergo a great annual variation in size and functional activity; the whole reproductive process from copulation through the fledging of the young is usually crowded into a few weeks, particularly in those birds that undergo long migrations to high-latitude breeding grounds.

The reproductive system of birds repeats the basic vertebrate pattern. Thus, the seasonal fluctuations in gonadal activity are produced by the environmental stimuli exerting their influence through a central nervous mechanism that leads to gonadotropin release from the adenohypophysis. This is mediated, as appears to be common throughout the vertebrates (Jørgensen and Larsen, 1967; Jørgensen, 1968), by the liberation of neurohumoral substances from the median eminence of the neurohypophysis, these being transported by portal vessels to the pars distalis. The neuroendocrine system, therefore, constitutes a finely integrated and balanced coordinating link between the organism and its environment. The system has evidently had a long evolutionary history, since neurosecretory cells have been identified in primitive coelenterates, platyhelminths, annelids, mollusks, and arthropods, as well as in all the vertebrate classes. The avian neuroendocrine system is discussed in considerable detail in later chapters, and a further elaboration is not necessary in this chapter. Therefore, we have restricted our dissertation to a consideration of the primary sexual organs, the accessory sexual structures, and the associated behavioral aspects that are concerned with the process of reproduction. We have also excluded any extensive consideration of the phenomenon of intersexuality, which is common in certain avian species. For details of this latter subject, the reader is directed to the excellent review by Taber (1964).

In common with those of all other vertebrate groups, the avian gonads are responsible both for the proliferation of gametes and also for the secretion of the steroid sex hormones that control the development and functional activity of the accessory sexual structures and secondary sexual characteristics. In mammals, these two functions are known to be regulated by the secretion of two distinct gonadotropic hormones from the pars distalis of the anterior pituitary gland, namely, the follicle stimulating hormone (FSH), which is primarily responsible for the regulation of gametogenetic activity of the germinal epithelium, and the luteinizing hormone (LH), which in the male regulates the secretory activity of the interstitial Leydig cells and in the female induces ovulation and the formation of ovarian corpora lutea. In birds, it has sometimes been suggested that the gonads may be regulated by only one type of gonadotropic secretion (Nalbandov, 1959, 1966; van Tienhoven, 1959), but pituitary cytology distinguishes two gonadotropic cell types as in the mammalian situation (Tixier-Vidal, 1963), and distinct FSH-like and LH-like fractions were earlier obtained by Fraps *et al.* (1947) from chicken pituitaries using a method previously applied to ovine glands. More recently, more purified samples of these two gonadotropic hormones have been separated from the chicken pituitary gland (Stockell-Hartree and Cunningham, 1969; Follett *et al.*, 1972; Scanes and Follett, 1972), and their specificity has been established by biological assay (Furr and Cunningham, 1970). There is also evidence of two anatomically distinct areas in the hypothalamus of the Japanese Quail responsible for the regulation of secretion of FSH and LH, respectively. Thus, electrolytic or radio frequency lesions in the anterior regions of the infundibular nuclear complex result in regression of the seminiferous tubules without any apparent depression of interstitial cell activity, whereas lesions placed in the medial, ventral, and posterior regions produce, in males, a regression of both gametogenesis and interstitial cell activity, and in females, a cessation of ovulation but no regression of the ovary or oviduct (Stetson, 1969, 1971, 1972a,b). Similarly, there is good evidence in cockerels, too, that the posterior infundibular region controls the release of an avian LH, whereas the release of FSH is associated with an anatomically separate region (Graber *et al.*, 1967).

The male and female gonads are derived from a pair of sexually undifferentiated primordia associated with the intermediate mesoderm (nephrotome). The primordial germ cells within these structures are derived from the embryonic splanchnopleur and migrate in the blood (Dubois, 1965) to be housed in these locations and become the presumptive germinal epithelium. They sink below the surface into the connective tissue (stroma). The left presumptive gonad receives a greater compliment of primordial germ cells than the right (Witschi, 1935), and thus establishes an asymmetrical gonadal development which generally persists throughout life. Initially, proliferation of the germinal epithelium in both presumptive gonads in either sex forms a potential testis (medullary tissue). In the female a second proliferation of cells gives rise to a cortex in the left gonad which then becomes the potential ovary. In some species, particularly the domestic hen, a few cortical cords may also sometimes be laid down in the embryonic right gonad as well (reviewed in Taber, 1964).

In males, the embryonic gonadal primordia develop into paired testes, but in the female of many species only the left organ develops into a functional ovary, and the right generally remains in an ambisexual state. When the left ovary is removed, or when it becomes non-functional due to some pathological condition, a compensatory development of the rudimentary gonad may take place under the influence of the increased circulation of gonadotropin. In the large majority of cases in which this occurs the rudiment develops into a testis or ovotestis. The experimental induction of this condition in chickens has demonstrated that age can be a modifying factor on the result; full spermatogenesis can develop in the resultant ovotestis if the operation is done at an early age, but if performed in older birds full spermatogenesis is rarely achieved, but ovulations can take place. Under natural conditions, many old domestic hens suffering from senile changes become masculinized, assuming the cock plumage and the capacity to crow, because the rudiment of medullary (testicular) tissue becomes functional. Such sex reversal is also relatively common and often spectacular in pheasant species in which, for example, a somber colored female Golden Pheasant (*Chrysolophus pictus*) can, as a result of such changes, assume the resplendent plumage of a male (see also Harrison, 1932). A high incidence of intersexual individuals may also occur in strains of the domestic pigeon (Riddle *et al.*, 1945), in which apparently a delay in the degeneration of the embryonic cortical tissue causes genetic males to develop with an intact right testis but a left testis in which the cortical component persists and differentiates into ovarian tissue (Lahr and Riddle, 1945).

Occasionally, adult birds are found with two functional ovaries, particularly among members of the Accipitridae, Falconidae, and Cathartidae (Domm, 1939), but even in these specimens it is sometimes unaccompanied by a corresponding development of the associated right oviduct. Functional right ovaries are also frequently found in pigeons (Brambell and Marrian, 1929) and in the Herring Gull *(Larus argentatus)* (Boss and Witschi, 1947).

Individuals without gonads are also sometimes found. Usually these lack both right and left Müllerian ducts and exhibit a general masculine appearance, thus resembling subjects experimentally castrated as embryos. Taber (1964) points out that in man gonadal agenesis or the Turner syndrome is accompanied by loss of one of the sex chromosomes, producing a neuter genotype resembling the female (genetic composition XO which resembles XX). Since males are the homogametic sex in birds, it may be significant that birds lacking gonads resemble the male.

II. The Testis

A. GENERAL STRUCTURE

The essential sex organs in the male bird are the paired testes which, unlike mammals in which they usually become housed in an extraabdominal sac, are located permanently in the body cavity just ventral to the anterior end of the kidneys. Each is attached by a short mesorchium to the dorsal body wall. They are supplied, together with their immediately contiguous ducts, by the spermatic arteries from the dorsal aorta; the testicular veins drain into the vena cava. Because of the asymmetry already established in the presumptive embryonic gonad, the left testis is commonly larger than the right, although in

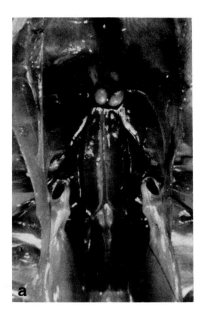

FIG. 1. Testes of European Tree Sparrow (*Passer montanus*) during (a) winter and (b) during the height of the breeding season. Note the tremendous seasonal increase in size and also the marked development of the vas deferens (arrowed). ×5.

the pigeon (Riddle, 1918) and in the turkey (Law and Kosin, 1958), the reverse is frequent.

Each testis is an ovoid, encapsulated body surrounded by a substantial fibrous coat, the tunica albuginea, with a fragile serous outer sheath, the tunica vaginalis. In seasonally breeding species, the testis is subject to great annual variations in size, sometimes by as much as 400- to 500-fold (Fig. 1). In the Japanese Quail, for example, testicular weights can be induced to increase from 8 mg to 3000 mg within 3 weeks when birds are artificially exposed to a stimulatory 20 hour photoperiod (Lofts et al., 1970). Such gross extremes are common in seasonally breeding species and imposes a great strain on the ensheathing tunics, which are replaced annually by a proliferation of new fibroblasts rebuilding a new capsule from beneath the old weakened covering (Marshall and Serventy, 1957). This upsurge of fibroblasts usually occurs during the postnuptial (regeneration) phase of the testicular cycle, and Marshall (1961a) has suggested that this may be automatically initiated by the collapse of the old tunic. Occasionally, in birds living in captivity or under abnormal climatic conditions, the testes may collapse, with the concomitant formation of a new tunic, even though reproduction has not occurred and the tubules are still packed with spermatozoa (Keast and Marshall, 1954; Marshall and Serventy, 1957). For some weeks after the postnuptial testicular regression, therefore, both the old and new tunics are seen in sectioned material (Fig. 2), and the testis appears to be surrounded by a double coat. This transient stage can sometimes be a useful parameter for distinguishing between a juvenile bird, in which the sexual organs have not yet undergone any expansion into a breeding condition, and an adult showing a postnuptial testicular collapse (Lofts et al., 1966).

Internally, each testis consists of a mass of convoluted seminiferous tubules, which in birds, unlike mammals, are anastomotic and not restricted by septa. They are lined by the germinal epithelium consisting of developing germ cells and nongerminal sustentacular or Sertoli cells. The latter have sometimes been described as forming a syncytium (Marshall, 1961a), but electron microscopy reveals them to be separate units whose cytoplasm becomes intimately wrapped around the germ cells, thus giving the appearance of a syncytium when viewed by the light microscope. Lofts (1968) has suggested that this surrounding of germ cells by folds of Sertoli cell might serve to maintain the integrity of a particular population of germ cells in a way analogous to the germinal cysts found in the testes of the anam-

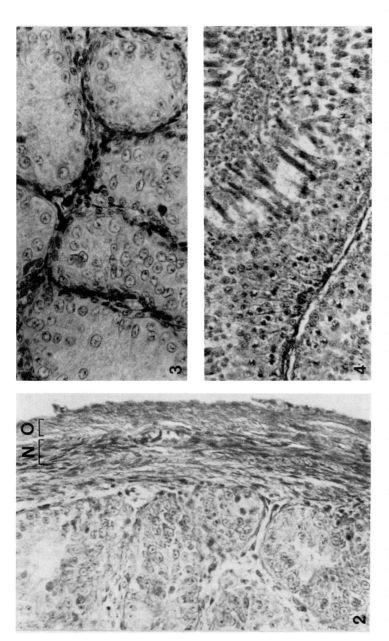

FIGS. 2–4. FIG. 2: The tunica albuginea of an adult pigeon (*Columba livia*) with regressed testis. The old worn outside tunic (O) is being replaced by an upsurge of fibroblasts laying down a new tunic (N) beneath the old. ×350. (From Lofts *et al.*, 1966a.) FIG. 3: Testis of House Sparrow (*Passer domesticus*) during December. The seminiferous tubules are regressed and contain only a peripheral ring of spermatogonia and Sertoli cells. ×500. FIG. 4: Testis of House Sparrow in April. The germinal epithelium is now several cells thick and all germinal stages are present, including bunched spermatozoa. The seminiferous tubules have expanded enormously and the interstitial tissue between adjacent tubules has become greatly compressed. ×350.

niotes. During the period of sexual quiescence, the germinal epithelium generally consists of a single layer of stem (type A) spermatogonia and Sertoli cells (Fig. 3). Because of the collapsed condition of the tubules, a lumen is sometimes not apparent during this stage because of the consequent compression of the Sertoli cells into the more confined area. However, with the advent of the breeding season, a recrudescence of mitotic activity in the stem spermatogonia causes the propagation of numerous germ cells which mature successively into secondary spermatogonia (type B), primary and secondary spermatocytes, spermatids, and spermatozoa, so that the germinal epithelium becomes several cells thick (Fig. 4). Each germinal stage arises in a coordinated fashion, so that the germinal epithelium is not a mass of independently developing germ cells, but is characterized by the different germinal stages being organized into well defined cellular associations. Thus, the sectioned gonad presents an orderly appearance, with the spermatogonia close to the basement membrane and the successive cell types appearing progressively toward the central lumen. There are three clearly defined phases in the rehabilitation of the germinal epithelium to a full breeding condition: (1) the period of spermatogonial multiplication, during which new spermatogonia are formed continuously, and some start maturing into primary spermatocytes; (2) the period of spermatocyte division, when the cells undergo their meiotic division producing secondary spermatocytes which mature into spermatids; (3) the period during which spermatids start elongating and start their transformation into the mature spermatozoa (Clermont, 1958).

In most single-brooded species, particularly those inhabiting temperate and high latitudes, once the gametogenetic resurgence begins, there is a rapid progression into the full breeding condition, and all seminiferous tubules appear uniform in cross section, with innumerable radiating bundles of spermatozoa. This is particularly evident in subarctic and arctic species in which the breeding season is particularly abbreviated and the birds migrate south before the onset of the harsh winter conditions. In continuously breeding forms, however, such as the domestic fowl and feral pigeon, and in multibrooded species in environments which remain equable for longer periods (e.g., *Columba palumbus*, *Zonotrichia leucophrys nuttalli*, and *Passer domesticus*), the breeding season may be protracted for 3–4 months, and the testes remain spermatogenetically active for a much longer period. In these forms, spermatozoa become released in waves throughout the more protracted breeding season. In mammals, it has been established that the spermatogenetic condition of the germinal epithelium is not synchronous throughout the length of any given tubule but shows a longitudinal progression in the form of a wave of development passing along the length of the tubule (Curtis, 1918; Cleland, 1951; Roosen-Runge and Barlow, 1953; Kramer, 1960). Since the particular stage of development reached in one segment of the tubule may differ from another segment further along, a transverse section of the testis will show an asynchronous pattern, with the germinal epithelium of adjacent sectioned tubules possibly in different stages of the epithelial cycle. Such an asynchronous pattern has sometimes been observed in birds, and Clermont (1958) distinguishes eight distinct stages in the epithelial cycle of the Pekin duck.

There is a tremendous variation in the morphological appearance of avian spermatozoa, but, like those of other vertebrates, they all consist essentially of the sperm head (with an apical cap above the acrosome containing the chromatic material), a midpiece, and a long propulsive tail. Brief discussions concerning the detailed anatomy and sperm motility may be found in Grigg and Hodge (1949), Bonadonna (1954), van Tienhoven (1961), and Bishop (1961). Once the spermatozoa are freed from the enfolding Sertoli cells, they migrate from the tubule lumina through the rete tubules and short vasa efferentia into the efferent ducts (vas deferens). These discharge into the expanded distal ends of the ducts, the seminal sacs, from which the spermatozoa are finally ejaculated into the urodeum (see Section IV).

The interstices between the convoluting seminiferous elements are packed with areolar connective tissue containing blood capillaries, lymph spaces, and the interstitial Leydig cells, which produce steroid sex hormone. Large numbers of melanoblasts may also be located in this tissue in some species, imparting a black or gray appearance to the organ. With the seasonal proliferation of the germinal epithelium and consequent expansion of the tubules, the interstitial tissue becomes dispersed and compressed into tight concentrations; fewer interstitial cells appear in testicular sections. Conversely, the interstitial tissue of the regressed testis of a wintering bird appears more conspicuous. These cells undergo well defined cyclic changes in their histophysiology (see Section II,B,2). They show the general characteristics of other steroid-producing tissues. Thus, histochemical techniques have established the presence of the Δ^5-3β-hydroxysteroid dehydrogenase (3β-HSDH) enzyme system in these cells (Arvy,

1962; Botte and Rosati, 1964; Woods and Domm, 1966). This enzyme catalyzes the conversion of Δ^5-3β-hydroxysteroids to Δ^5-3-ketosteroids (Samuels, 1960), which is known to be an important stage in the production of such steroids as progesterone and the androgenic steroid androstenedione. It has been found in all types of steroid hormone-producing tissues such as the adrenal, testis, ovary, and placenta (Deane and Rubin, 1965), and its presence in the Leydig cells provides strong evidence of their steroid biosynthetic capacity. Ultrastructurally, too, they have the general characteristics attributed to steroid-producing cells (Belt and Pease, 1956; Christensen and Gillim, 1969), in that, in the breeding season at least, they possess a well-developed agranular endoplasmic reticulum and numerous mitochondria with tubular cristae (Porte and Weniger, 1961). Generally, too, lipids are often found in their cytoplasm and react positively to tests for cholesterol.

The interstitial cells appear to be derived from fibroblast-like cells in the intertubular areas. In the young Japanese Quail, Nicholls and Graham (1972) have traced the evolution of the typical mature steroid-secreting Leydig cell from its fibroblast-like progenitor, by means of electron microscopy. It has been clearly established that quail subjected to short photoperiods from hatching exhibit no testicular growth or androgen secretion, but when exposed to a long photoperiod, rapid testicular development and androgen secretion take place (Follett and Farner, 1966a). When the interstitial tissue of young quail maintained under a 6 hour photoperiod are examined, three to four layers of spindle-shaped cells with elongate nuclei and prominent nucleoli are observed in the areas between adjacent tubules, with the more centrally located cells being somewhat less elongate but still basically fibroblastoid in their ultrastructural appearance (Nicholls and Graham, 1972). Generally, granular endoplasmic reticulum is found in these cells (Fig. 5), but when the birds are transferred to a highly stimulatory 20 hour photoperiod, the interstitial cells rapidly metamorphose into the typical secretory form; the cells and nuclei expand and develop the characteristic fine structure of steroid-secreting systems with the whole cytoplasmic area becoming filled with smooth endoplasmic reticulum in which are scattered rounded mitochondria with tubular cristae (Fig. 6). Nicholls and Graham have recorded the first discernible changes in this metamorphosis at about 3 days after the transfer to the increased photoperiod. Interestingly, Follett and colleagues (1972) have found that plasma levels of LH, measured by means of a radioimmunoassay technique, begin to increase about 4 days after such treatment, which is therefore in good agreement with the ultrastructural evidence.

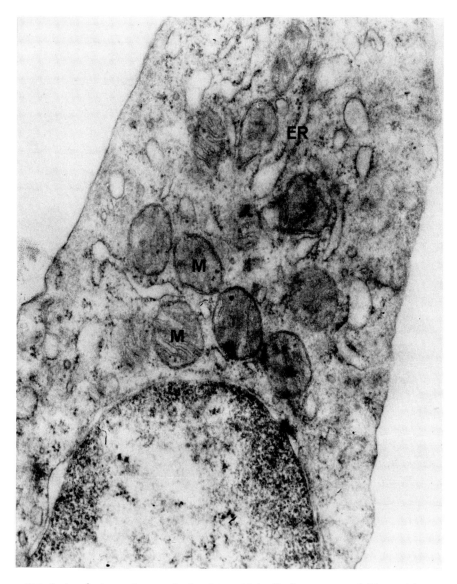

FIG. 5. An electron micrograph of an interstitial cell of a young quail (*Coturnix*) on a short photoperiod. It is basically fibroblastoid in shape with granular endoplasmic reticulum (ER) and mitochondria (M) with lamellar cristae. × 42,000.

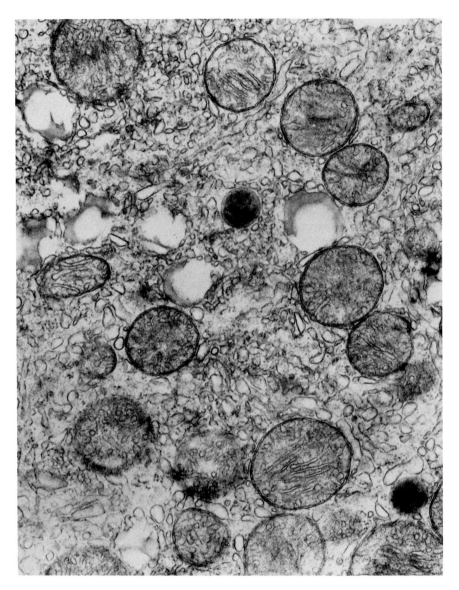

FIG. 6. An electron micrograph of an interstitial cell of a young quail (*Coturnix*) on a highly stimulatory photoperiod. The cell has now all the ultrastructural features of a steroid-producing tissue with smooth endoplasmic reticulum and many mitochondria with tubular cristae. ×42,000.

B. HISTOPHYSIOLOGY

1. General Histophysiology

Testicular activity is cyclic, and in some species there is evidence that these cyclic changes may, in part at least, be endogenous. Thus, the testes of young Budgerigars *(Melopsittacus undulatus)* (Vaugien, 1953), and Zebra Finches *(Poephila guttata castanotis)* (Marshall and Serventy, 1958), become spermatogenetically active and produce spermatozoa even when the birds are kept in almost total darkness. A similar endogenous rhythm has also been recorded in an equatorial African weaver finch *(Quelea quelea)* kept on an unvarying 12 hour photoperiod and thermostatically controlled temperature for 2.5 years (Fig. 7). Under these conditions, males show a seasonal testicular development and postnuptial regression similar to that observed in wild populations (Lofts, 1964). In the Pekin duck, too, the testes of birds kept for a number of years under conditions of either continuous darkness (Benoit *et al.*, 1956a), or continuous light (Benoit *et al.*, 1956b), similarly undergo a seasonal enlargement and regression.

Marshall (1951, 1955, 1961b) regards such an endogenous rhythm as the primary initiator of the seasonal reproductive periodicity. However, in many species such a rhythm is not apparent, and birds like the White-crowned Sparrow *(Zonotrichia leucophrys gambelii)*

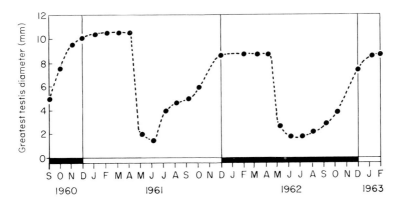

FIG. 7. Seasonal changes in the testicular size of the Red-billed Quelea (*Quelea quelea*) kept under an unchanging daily 12 hour photoperiod. (From Lofts, 1964.)

(Farner, 1959), Brambling *(Fringilla montifringilla)* (Lofts and Marshall, 1960), or Japanese Quail (Follett and Farner, 1966a) show no spermatogenetic recovery when maintained under winter light conditions and require the seasonal advent of a long photoperiod before a recrudescence of their gametogenetic activity occurs. Even those species possessing an apparently autonomous reproductive rhythm may still be influenced by environmental stimuli, and Marshall (1955) has likened the endogenous cycle to a cogwheel that is seasonally engaged by various environmental "teeth" that produce the final synchronization that ensures reproductive success.

Among various tropical species or species inhabiting a particularly equable environment with a continuous abundance of food, a precise annual breeding cycle may not occur (Miller, 1955; Marshall and Roberts, 1959). An equatorial population of *Zonotrichia capensis*, for example, has two complete cycles each year, and Miller (1959) considers this to be an expression of an endogenous 6 month rhythm which is only slightly affected by seasonal rainfall. Even in some temperate zone species, such as the populations of feral pigeons inhabiting our towns, the continuous supply of suitable food provided by a benevolent public and spillage in the docks has resulted in the evolution of a capacity for breeding throughout the year in a high proportion of the flocks (Lofts *et al.*, 1966a). Generally, however, although birds display widely diverse breeding patterns (Lofts and Murton, 1968), natural selection has favored the establishment of annual breeding cycles, particularly in the avifauna of the subarctic and temperate zones. Thus, testicular recovery usually begins in the spring, and the reproductive period is usually terminated by early summer whereupon the testes rapidly regress and remain small throughout the winter months.

The annual cycle can broadly be classified into three basic phases:

1. The regeneration (Marshall, 1961b), or preparatory (Wolfson, 1959a,b), phase comes immediately after reproduction in single-brooded species, and after the final ovulation of the season in multibrooded forms. Morphologically, it is marked by the rapid regression of the testes, which can be attributed to a changed and possible diminished gonadotropin release from the anterior pituitary (Benoit *et al.*, 1950; Greeley and Meyer, 1953; Lofts and Marshall, 1958; King *et al.*, 1966). In many species, this takes place at a time of year when the environmental photoperiod is still highly stimulatory, but the neuroendocrine apparatus no longer responds. The bird is said

to be in the refractory period (Bissonnette and Wadlund, 1932) and remains sexually quiescent even when artificially subjected to long photoperiods. Such postnuptial refractoriness is a feature of nearly all species so far submitted to controlled and critical photostimulation experiments, and it has been tacitly assumed or inferred that it is a phenomenon of universal occurrence among avian species (see Marshall, 1959). However, recent photoexperimentation on the Wood Pigeon (*Columba palumbus*) has shown this species, and probably others as well, to be without a photorefractory phase (Lofts *et al.*, 1967). The Wood Pigeon, like the Stock Dove (*Columba oenas*), is multibrooded with a breeding season extending into early autumn where agricultural grain harvests provide abundant food for rearing the later broods (Murton, 1958; Lofts *et al.*, 1966a). Reproductive activity ceases when autumn day lengths fall below a stimulatory level, and the bird then enters its regeneration phase. Spermatogenesis can, however, be immediately reactivated by artifically prolonging the photoperiod. A refractory period is not therefore a necessity for gonad rehabilitation as previously supposed (Marshall, 1961b).

The regeneration phase is characterized by a rehabilitation of both the interstitial tissue and seminiferous tubules (see below), and is a period of sexual quiescence during which the bird displays little or no sexual behavior. In many species, infiltration of the interstitial tissue by large numbers of leukocytes takes place during the early stages (Payne, 1969), and the replacement of the weakened testis tunic, and sometimes also the postnuptial molt, occurs during this phase. In the White-crowned Sparrow, for example, a rapid and intensive post-nuptial molt is inserted between the breeding period and the onset of autumnal migration (Morton *et al.*, 1969). However, the Wood Pigeon molts throughout the breeding season when the gonads are fully active, while the European Turtle-Dove (*Streptopelia turtur*) exhibits a partial molt immediately as it enters the refractory phase but defers the molt of its primaries until it reaches its African wintering grounds (Murton, 1968). The duration of the regeneration phase varies from species to species, and its termination is probably marked by the completion of the interstitial cell rehabilitation, which may sometimes manifest itself in the form of an autumnal resurgence of sexual behavior seen in many species (Morley, 1943; Marshall, 1952). In birds that have a refractory period, the end of the regeneration phase is also heralded by the restoration of photosensitivity.

2. The acceleration (Marshall, 1961b), or progressive (Burger,

1949; Wolfson, 1959a), phase succeeds the regeneration phase and is a period marked by the interstitial cells and seminiferous tubules responding to gonadotropic secretions from the adenohypophysis. The postnuptial molt is usually completed. A recrudescence of gametogenetic activity occurs under favorable environmental conditions but becomes retarded or stimulated by a variety of environmental inhibitors or accelerators that are mutually antagonistic (Marshall, 1959). Thus, the cycle hastens, slows, or sometimes stops altogether, depending upon the factors currently presented to it by the changing environment (Lofts and Murton, 1966). Temperature is perhaps the most important modifier of the testicular cycle during this stage, and its effect has been demonstrated experimentally by a number of investigators (Burger, 1948; Farner and Mewaldt, 1952; Engels and Jenner, 1956; Farner and Wilson, 1957). In addition, there are many field data that indicate a correlation between temperature and speed of testicular development (Marshall, 1949a; Lofts and Murton, 1966). Low temperatures will inhibit the cycle of most species, and an unusually cold spring will often nullify the accelerating effects of long sunny days. Among temperate zone birds, some species begin their acceleration phase in late summer or early autumn, but generally this becomes depressed or even halted by the onset of winter conditions until after the winter solstice. In some tropical or xerophilous species, the absence of rainfall may similarly retard the gametogenetic progress during this phase (Marshall, 1970), whereas among waterbirds and many others, a frequent inhibitor is the lack of a safe and traditional nesting site.

The acceleration phase is a period marked by increasingly intensive sexual activity and song, during which gametogenesis leading to the production of spermatozoa occurs. In some species, the winter feeding flocks begin to disintegrate as individuals start territorial selection and defensive displays. The phase varies enormously in duration, both interspecifically and intraspecifically. An interesting example of this is shown by the differences displayed by the resident British population of European Starlings (*Sturnus vulgaris*) and the continental starlings that migrate to Britain for the winter. Spermatogenetic activity begins in the British population in late September, but does not usually progress beyond the proliferation of spermatogonia and occasional primary spermatocytes until February, when a burst of activity populates the seminiferous tubules with secondary spermatocytes and later stages. Continental birds, on the other hand, show no spermatogonial division in autumn, and the first mitoses are

not seen until late December or early January, and primary spermatocytes not until early March. Thus, both populations respond to January and February photoperiods, though this marks a beginning of
spermatogenesis for continental birds but a resumption of the acceleration phase already started in the British population (Bullough,
1942; Lofts and Murton, 1968). Unlike continental birds, the British
population shows earlier and more intensive interstitial cell activity
in the autumn and winter, which is evident by the earlier changes in
bill coloration and development of the vas deferens. This early elevation of androgen titer is also responsible for autumnal sexual displays and even winter breeding in exceptionally mild winters, events
virtually absent in the life history of the continental population.

3. The culmination phase, during which actual ovulation and
insemination occur, may be regarded as a distinct component of the
annual cycle, since a species-specific requirement (Marshall, 1955)
is generally necessary before the cycle can culminate in oviposition.
By the end of the preceding phase, the male bird has reached a fully
reproductive condition with expanded testes containing seminiferous
tubules charged with masses of spermatozoa. Nevertheless, although
the internal physiology may be now wholly prepared for reproduction, many complicated behavioral factors may influence the final
culmination. The male generally assumes this reproductive state
before the female, and the final timing of oviposition then depends
on the female receiving the appropriate psychological stimuli, both
from her mate and the environment. The appropriate habitat and
and interpair displays are often essential to stimulate final oocyte
development, ovulation, and insemination. The action of stereotyped
behavior in causing specific hormone secretion is now well appreciated (Aschoff, 1955; Benoit, 1956; Lehrman, 1961, 1964; Immelmann,
1963; Hinde, 1967), and conditions of captivity, for example, have
been shown to inhibit gonadotropin secretion in the Pintail (*Anas
acuta*), thus preventing breeding (Phillips and van Tienhoven, 1960).

2. The Interstitial Cells

There is an extensive literature concerning the seasonal changes
observable in the avian interstitial tissue, but many of the early reports are contradictory, often being based on unsatisfactory and now
outmoded histological procedures. Because of their dispersal by
seasonal tubule expansion, the Leydig cells have sometimes been
stated to be absent from the gonad at certain times of the year (Bissonnette and Chapnick, 1930; Rowan and Batrawi, 1939), and an

inverse relationship between sexuality and interstitial (Leydig) cell activity has sometimes been claimed (Oslund, 1928). This, of course, is not true and the more recent histochemical and electron microscopic observations discussed in Section II,A have clearly established the close relationship between these cells and the androgen-dependent sexual structures. Furthermore, the selective destruction of the germinal epithelium of cockerels by roentgen radiation (Benoit, 1950), which leaves the interstitium and secondary sexual characters apparently unaffected, also indicate this tissue as the site of androgen production. Tumors of the interstitial cells result in an increased production of androgens and 17-ketosteroids (Sharma *et al.*, 1967).

In birds, as in the majority of seasonal vertebrates (Lofts, 1968), the interstitial cells undergo well defined seasonal secretory cycles involving a rhythmic accumulation and depletion of cholesterol-positive lipoidal material. The interstitial cells of young birds are generally heavily impregnated with such material (Lofts and Marshall, 1957). Then, at the approach of the sexual season and consequent buildup of spermatogenetic activity in the seminiferous elements, they become rapidly depleted of their lipids and cholesterol and become strongly fuchsinophilic (Benoit, 1927, 1929; Marshall, 1955). In a species such as the Northern Fulmar (*Fulmarus glacialis*), in which the young do not begin breeding until 7 years old, the interstitial cells of newly hatched birds are less lipoidal but become more heavily impregnated when the birds are just over 2 years old (Marshall, 1949b). In adults, the interstitial cells of the sexually quiescent winter gonad are generally small and often sparsely lipoidal (Fig. 8) with numerous fuchsinophilic elements that are more easily seen after dissolution of the lipids in wax-embedded material. During the acceleration phase, these cells rapidly increase in size and there is a buildup of the lipoidal inclusions, so that the interstitial tissue is seen to consist of compressed aggregations of heavily lipoidal and cholesterol-rich cells (Fig. 9). They also react positively to tests for 3β-HSDH. Then, as in juveniles, the lipoidal content rapidly diminishes (Fig. 10) at a time when androgen-dependent sexual displays (Murton *et al.*, 1969b) reach their maximum intensity. During this period, the cholesterol reaction also becomes weaker, and may disappear altogether, although 3β-HSDH activity remains strong (Fig. 11). The nuclei of Leydig cells also attain maximum diameter at this time, reflecting an increase in secretory activity (Muschke, 1953; Alfert *et al.*, 1955). In migratory waders (Charadriiformes), this depletion of interstitial lipids is often evident just before the birds leave their

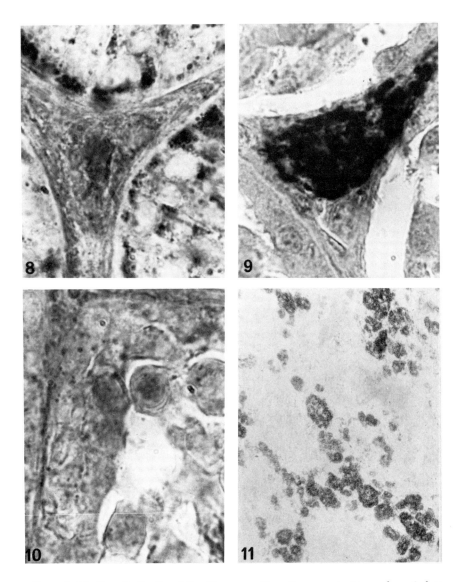

FIGS. 8–11. FIG. 8: Testis of the European Tree Sparrow in December. Gelatin section of the interstitial tissue stained with Sudan Black to show that the Leydig cells are only sparsely lipoidal, whereas there is a considerable amount of lipoidal material in the adjacent tubules. × 1000. FIG. 9: Testis of the European Tree Sparrow in February showing a considerable accumulation of cholesterol-rich lipoidal material in the interstitial Leydig cells. × 1000. FIG. 10: Testis of the European Tree Sparrow at the height of the breeding season. The interstitial tissue is compressed by the expanded tubules and the Leydig cells have become depleted of their lipids (cf. Fig. 9). × 1000. FIG. 11: Testis of pigeon during the breeding season showing the interstitial location of Δ^5-3β-hydroxysteroid dehydrogenase. × 130.

African wintering grounds on their north-bound migration (Lofts, 1962a).

With the advent of the regeneration phase, the now exhausted interstitial cells have reached the end of their secretory cycle, and it has been reported that they disintegrate so that ultimately their total number is drastically reduced (Marshall, 1949b, 1951, 1955; Jones, 1970). It may be that the massive invasion by leukocytes that takes place at this time clears the spent Leydig cells by phagocytic action. However, there is a need for further investigation by electron microscopy to confirm this point, and the possibility that the spent Leydig cells return to an inconspicuous fibroblast-like form should not be excluded. Concomitant with the atrophy of the exhausted generation, a new generation of juvenile interstitial cells begins to arise, presumably by their differentiation from fibroblast-like progenitors (Nicholls and Graham, 1972), and gradually begin to mature.

This seasonal replacement by new interstitial cells at the end of each breeding period is not unique to avian testicular cycles. A similar phenomenon has also been recorded in some snakes (Lofts et al., 1966b) and in the common frog (Rana temporaria) (Lofts, 1965). Marshall (1961b) suggests that the sequence is part of an endogenous rhythm that can occur even in the absence of any gonadotropic rhythm. Thus, even after complete removal of the adenohypophysis, Coombs and Marshall (1956) have reported that the interstitial cells of domestic cockerels still renew themselves and develop a new generation of Leydig cells with some lipoidal and cholesterol-positive material. It is doubtful, however, whether such cells would ever become secretory in the absence of gonadotropins.

The length of time necessary for interstitial cell rehabilitation probably varies from species to species, but there is evidence that in some birds at least it may be a fairly rapid process. Thus, Lofts and Marshall (1957) have recorded that the newly arisen interstitial cells of fifteen different migratory species were already beginning to manufacture small cholesterol-positive lipid droplets in their cytoplasm when they were autopsied at the time of their departure from Britain on their southward migration.

The cyclical waxing and waning of cellular lipids, which although in itself is insufficient to implicate unequivocally a steroid secreting role, is a useful index of the functional activity of the tissue. The lipids are both cholesterol-positive and strongly birefringent, reactions that are probably indicative of precursor material involved in androgen biosynthesis. The prenuptial buildup in birds often precedes the hypertrophy of the accessory sexual apparatus and behavioral activities thought to be dependent on androgen secretion. In young chicks, for example, the increase in concentration of birefringent material in the interstitial tissue is in close agreement with the gradual hypertrophy of the comb (Kumaran and Turner, 1949a), and in the House Sparrow (*Passer domesticus*) the level of 17α-hydroxylase activity in the interstitial tissue has been shown to increase rapidly between February and March (Fevold and Eik-Nes, 1962), a time when the interstitial cells are losing the lipoidal material accumulated earlier in January and February.

The sudden depletion of lipids and cholesterol at the height of the breeding activity has also been noted in a number of reptilian species, and it has been suggested that it is probably indicative of a rapid utilization of precursor material at a time of high androgen release (Lofts, 1968; Lofts and Bern, 1972).

The observations of Jones (1970) that the height of the epididymidal epithelium in the California Quail *(Lophortyx californicus)* attains its maximum thickness at the time when the interstitial tissue is showing this phenomenon are in agreement with such an interpretation. Certainly 3β-HSDH activity is high at this time (Fig. 11). In the Department of Zoology, University of Hong Kong, we have been attempting to correlate these histochemical events with the seasonal fluctuations in androgen production, measured by *in vitro* methods. For this, portions of testicular material have been taken at monthly intervals from the migratory Green-winged Teal *(Anas crecca)* throughout the year and incubated with radioactive pregnenolone as an added precursor. The biosynthetic activity of the tissue has been determined by standard chromatographic analysis of the steroids produced, and the percentage conversion of the added [16-^{3}H] pregnenolone to testosterone per unit weight of tissue has been calculated. Because of the seasonal dispersal of interstitial tissue during testicular expansion, fewer Leydig cells will be contained in the tissue from the sexually mature testis compared with the same weight of tissue taken from the regressed testes of winter birds. The conversion figures have therefore been multiplied by the mean testicular weight to give a truer picture of the seasonal pattern of steroid production by the whole gonad. Some preliminary results showing the seasonal variation are shown in Fig. 12. They support the hypothesis outlined above. Thus, the gradual elevation in androgen production from mid-February to early June coincided with the appearance of lipoidal inclusions in the interstitial tissue. Lipid depletion was observed in early July

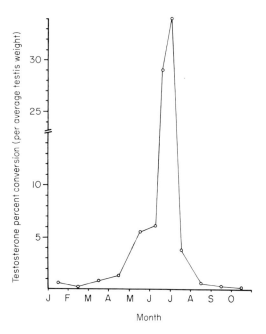

FIG. 12. The seasonal fluctuations in the *in vitro* production of testosterone from radioactive pregnenolone by the testis of the Green-winged Teal (*Anas crecca*). Testosterone production was high during the breeding season, but fell rapidly in August as the birds entered their refractory period. (From Chan and Lofts, 1973.)

at a time when androgen biosynthesis reached a peak, and the subsequent rapid decline marked the bird's entry into the postnuptial refractory period.

3. *The Sertoli Cells*

As well as the interstitial tissue, the Sertoli (sustentacular) cells of the tubule also undergo cyclic seasonal changes involving an accumulation and depletion of cholesterol-positive lipoidal material. Furthermore, ultrastructurally these cells show the fine structure normally associated with a steroid-producing tissue (Fig. 13) and also react positively to tests for 3β-HSDH (Arvy, 1962; Woods and Domm, 1966). The enzyme 17β-HSDH can also be demonstrated in the tubules of the European Tree Sparrow (*Passer montanus*) at the height of spermatogenetic activity (Fig. 14). An endocrine role is perhaps also suggested by the feminization effects seen in the plumage of cockerels suffering from Sertoli cell tumors (Siller, 1956).

Generally, when the germinal epithelium has advanced beyond the production of primary spermatocytes in the early acceleration

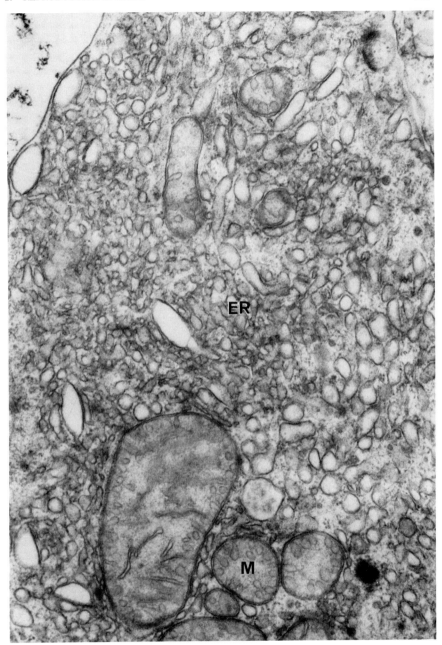

FIG. 13. An electron micrograph of the Sertoli cell of a photostimulated Japanese Quail. Note that it has all the characteristic ultrastructural features of a steroid-producing cell, namely, smooth endoplasmic reticulum (ER) and mitochondria (M) with tubular cristae. ×42,000.

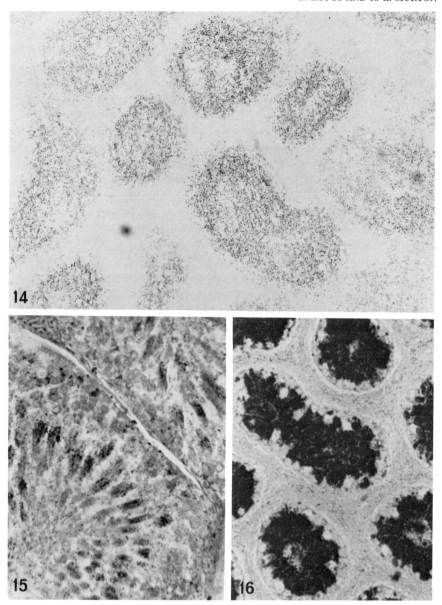

FIGS. 14–16. FIG. 14: Testis of the European Tree Sparrow during early acceleration phase showing a strong 17β-hydroxysteroid dehydrogenase reaction in the seminiferous tubules. ×150. FIG. 15: Testis of the European Tree Sparrow at height of breeding season. A scattering of fine lipid droplets have appeared in the area of the sperm bundles. ×350. FIG. 16: Testis of *Quelea* during the postnuptial period of testicular regression. The seminiferous tubules are filled with a dense amorphous mass of cholesterol-rich lipids. ×350.

phase, there is an absence of lipoidal material from the intratubular components. But as this phase draws toward its termination, and bunched spermatozoa are seen to be radially arranged around the tubule lumen, sections stained with sudan dyes show a light scattering of fine lipoidal granules concentrated in the region of the sperm heads (Fig. 15). These small droplets are a component of the residual bodies (Regaud, 1901) that form part of the spermatid protoplasm. The residual bodies subsequently become sloughed off from the spermatid during the end stages of spermateleosis and remain behind in the seminiferous tubules after evacuation of the spermatozoa. They are a common feature of the spermatogenetic process of probably all amniotic vertebrates (Lofts, 1968) and have been extensively studied in the rat, where they are known to be subsequently phagocytized by the Sertoli cell (Smith and Lacy, 1959; Lacy, 1960). In addition to their lipid inclusions, the residual bodies also contain a large amount of RNA, clusters of mitochondria and Golgi remnants (Smith and Lacy, 1959), and, in the rat at least, it has been suggested that they may be instrumental in maintaining the radial coordination of the associated germ cell population (Lacy, 1960; Lacy and Lofts, 1965).

In contrast to the slight scattering of intratubular lipids observed in the fully mature testis, a dramatic metamorphosis into a condition of heavily sudanophilic and strongly cholesterol-positive tubules takes place during the postnuptial regeneration phase. In sectioned material the seminiferous tubules appear to become filled with a dense amorphous mass of lipid and cholesterol, completely occluding the lumen (Fig. 16), but with the better resolution provided by the electron microscope, the lipid is generally seen to be within the Sertoli cell cytoplasm. While some of this lipid might be contributed by the lipoidal inclusions of the phagocytized residual bodies, much of it is probably produced by the Sertoli cell cytoplasm. Steroid dehydrogenase activity ceases to be demonstrable in these cells at this time. The tubules remain in this heavily lipoidal state for some time; then, concomitant with the recrudescence of spermatogenetic activity in the adjacent stem spermatogonia, the sudanophilic material rapidly disappears. The duration between the sudden accumulation in the Sertoli cell cytoplasm and the start of lipid depletion varies from species to species and may be as long as several months. For example, in the migratory Whimbrel (Numenius phaeopus), the seminiferous tubules have already become heavily lipoidal by the time the animals leave Britain and head south to their African wintering grounds. The gonads remain in this condition for about 5–6 months, and the Sertoli lipid clearance and spermatogenetic recovery begin just before the birds leave the contranuptial area and return north to breed. By the time they arrive in Britain, tubule lipids are absent and testes are reaching sexual maturity (Lofts, 1962a). In columbid species, such a massive postnuptial tubule steatogenesis does not occur, and the quantity of Sertoli lipid is very much less (Lofts *et al.*, 1967).

In view of the fine structure and histochemical evidence, the question arises as to whether this seasonal waxing and waning of cholesterol-rich lipoidal material in the Sertoli cells is indicative of a seasonal endocrine function as is apparent in the adjacent interstitial tissue. Tentative evidence for this has been provided by Lofts and Marshall (1959), who chromatographically analyzed testicular lipids extracted from birds with regressed gonads containing heavily lipoidal tubules but a lipid-free interstitium. The results showed the presence of progesterone, which also correlated with a positive progestogenic reaction in the blood subjected to a parallel bioassay. A similar analysis on birds with gonads with expanded nonlipoidal tubules but heavily lipoidal interstitial cells resulted in an absence of demonstrable progestogenic activity in the blood, and only androgenic steroids were identified in the testis extracts. By present-day standards, the above data would require more rigorous criteria for specific identification of the steroids, but so far as the authors are aware, no further confirmation of these data has yet been attempted. However, there is strong evidence in favor of this hypothesis in mammals (Christensen and Mason, 1965; Lacy *et al.*, 1969) and also in reptiles (Lofts, 1969b; Lofts and Choy, 1972), in which a separation of seminiferous tubules and interstitial tissue has been achieved by microdissection. By incubating seminiferous tubules with labeled precursors, their steroid biosynthetic capacity has been clearly established. Furthermore, in reptiles, Lofts (1972) has shown that testosterone is a major steroid being produced by this tissue, and that the production varies on a seasonal basis.

C. Control of Testis Function

1. Effects of Exogenous Hormones

a. Androgen. There is an extensive literature dealing with the effects of exogenous androgens on the functional activity of the vertebrate testis, and injections of androgenic steroids have been shown to have stimulatory effect in hypophysectomized and intact fishes (Lofts *et al.*, 1966), reptiles (Lofts and Chiu, 1968), and mammals (Albert,

1961; Boccabella, 1963; Harvey, 1963). In birds, administration of androgen has sometimes been reported to cause testicular regression (Bates *et al.*, 1937; Morato-Manaro *et al.*, 1938; Chu, 1940; Burger, 1944; Kumaran and Turner, 1949a,b; Kordon and Gogan, 1970), but has also been shown to induce testicular stimulation in adult Ring Doves (Lahr and Riddle, 1944), European Starlings (Burger, 1945), Red-billed Queleas (Lofts, 1962b), and House Sparrows (Pfeiffer, 1947). Furthermore, in the pigeon, testosterone administration maintains spermatogenetic activity and prevents testicular regression after hypophysectomy (Chu, 1940) and even induces a reestablishment of spermatogenesis in birds in which the gonads have been allowed to regress completely after removal of the pituitary gland (Chu and You, 1946). This is in contrast with the situation in the quail, in which Baylé *et al.* (1970) report that injections of testosterone propionate into hypophysectomized birds fail to prevent the degeneration of the germinal epithelium, although there is evidence that this process is retarded in androgen-treated birds. In *Quelea quelea*, a recovery of spermatogenetic activity can be produced by exogenous testosterone proprionate in the fully regressed gonads of regeneration phase birds (Lofts, 1962b), and although Pfeiffer (1947) has reported that spermatogonial multiplication is not augmented in the regressed testes of House Sparrows subjected to similar treatment, it has been our experience that such gonads can be stimulated when sparrows collected in November and maintained under an 8 hour photoperiod are given daily androgen injections. Furthermore, an elevation of endogenous androgens induced by injections of mammalian LH stimulating the interstitial cells has an even greater effect than that produced by exogenous androgens.

Much of the above controversy can probably be attributed to the fact that androgens have been administered to birds without cognizance being taken of the precise reproductive condition of the animal at the time of injection or implantation of the steroid. Thus, androgens given to birds just starting their seasonal testicular enlargement apparently retard the rate of spermatogenetic recrudescence (Burger, 1944; Pfeiffer, 1947; Lofts, 1962b), but if injections are delayed until spermatogenetic development has proceeded beyond the point of primary spermatocyte production, the same androgen treatment can accelerate gonadal enlargement (Pfeiffer, 1947; Kumaran and Turner, 1949b; Lofts, 1962b). In the sexually mature *Quelea quelea*, androgen therapy maintains the testes in breeding condition and prevents the postnuptial testicular collapse. The same is true of the European

Starling (Burger, 1944) and House Sparrow (Pfeiffer, 1947). In the pigeon, on the other hand, Chu (1940) has reported that daily injections of 2 mg of crystalline testosterone in sesame oil produces testicular atrophy in sexually mature birds. However, this observation may be somewhat speculative, since it was based on the results of only four specimens, and of these one killed after a week was normal and another killed after 30 days still contained spermatozoa. Furthermore, our own observations indicate that daily injection of testosterone propionate into feral pigeons does not cause a testicular regression but, on the contrary, appears to accelerate the maturation of spermatids into spermatozoa. This lends support to the suggestion

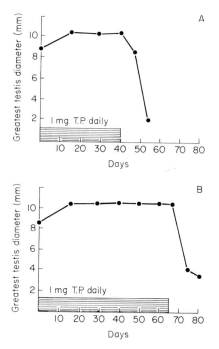

FIG. 17. These graphs show the effects on testis size of terminating daily androgen [1 mg per day of testosterone propionate (T.P.)] injections in the Red-billed Quelea (*Quelea quelea*). (A) This experimental bird was kept in full breeding condition by daily injections for 40 days. These were ended when control birds were in their refractory period. Rapid testicular regression took place soon after the termination of androgen therapy. (B) This experimental bird was maintained in full breeding condition even though the uninjected controls entered and passed through a full refractory phase. Again, termination of androgen therapy resulted in a rapid testicular regression, at a time when control birds were showing a spermatogenetic recovery. (From Lofts, 1962b.)

of Kumaran and Turner (1949b) that androgens facilitate the transformation of secondary spermatocytes in the cockerel, and is also in agreement with its known effects in lower vertebrates (Dodd, 1960; van Oordt and Basu, 1960; Lofts et al., 1966; Lofts and Chiu, 1968).

The stimulatory effects of androgens on the testes of hypophysectomized birds and on the completely regressed testes of intact wild birds is, in both cases, probably a direct effect on the seminiferous tubules, since Lofts (1962b) has shown that the termination of such therapy is immediately followed by a rapid regression (Fig. 17). However, although such evidence of a spermatokinetic effect of androgen has sometimes led to the suggestion that a seasonal resumption of interstitial cell activity may be responsible for the spring resurgence of gametogenesis, it does not explain the apparent contradiction that an inhibitory effect is produced during the early acceleration phase under similar experimental conditions. Lofts (1970) has suggested that there may be some antagonism between the exogenous androgen and gonadotropic hormones at a time when the germinal epithelium is particularly sensitive to FSH, but it is perhaps unwise to extrapolate from experiments using large nonphysiological doses that endogenous androgens have a direct interrelationship with the germinal epithelium. The recent observations by Murton et al., (1969a) on Greenfinches (Carduelis c. chloris) throws doubt on such an hypothesis. In an experiment devised to find the photoinducible phase (Pittendrigh, 1966), male Greenfinches with regressed testes were kept in groups under a 6 hour daily photoperiod with an additional 1 hour light pulse given as an interruption in the night. In each group, the night interruption occurred at a different time. The treatments are summarized in Fig. 18. After 3 weeks of such treatment, the birds were autopsied for testicular development and the plasma was assayed for LH levels by a radioimmunoassay technique (Wilde et al., 1967; Bagshaw et al., 1968). The results shown in Fig. 18 demonstrate that birds of group B had a high plasma titer of LH, which correlated with a highly stimulated interstitium with numerous Leydig cells heavily charged with lipids and swollen prominent nuclei, yet the seminiferous tubules were spermatogenetically inactive. This indicates that, in Greenfinches at least, high endogenous androgen levels produced by the interstitial cells do not stimulate a recrudescence of spermatogenetic activity in the sexually regressed testis, although the possibility of an influence on later germinal stages cannot be excluded. Furthermore, these data do not rule out the possibility that androgens from the Sertoli cells may also be involved.

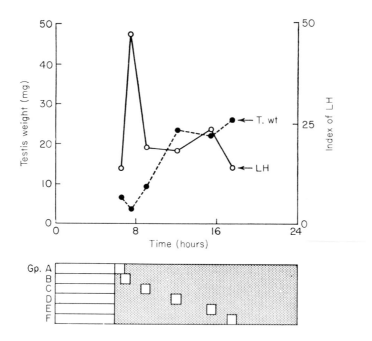

FIG. 18. The testicular weight (T. Wt) and plasma LH levels in groups of Green-finches (*Carduelis c. chloris*) subjected to asymmetric skeleton photoperiods. The lower diagram illustrates the light schedule each group of birds was held under. All were basically in a 6L:18D daily light regime, with an additional 1 hour light pulse given at night time. The time at which the birds experienced this latter differed in each group. The points on the graphs of testicular weight (●) and LH levels (○), represent the means of each particular group, and they show that a high peak of LH occurred in group (Gp.) B without any accompanying testicular development. In group F, on the other hand, the mean testicular weight was high, but plasma LH levels were comparatively low. (From Murton *et al.*, 1969a.)

b. Estrogen and Progesterone. There is scattered evidence that the male gonad may produce both estrogens and progesterone under both normal and pathological conditions. Chromatographic and bio-assay techniques have indicated the presence of progesterone in testicular extracts of a number of species (Lofts and Marshall, 1959; Höhn and Cheng, 1967; Höhn, 1970), and this is supported by the evidence of Fraps and his co-workers, who demonstrated the tes-ticular source of plasma progesterone or a closely allied substance in cockerels (Fraps *et al.*, 1949). Production of estrogen and proges-terone during the normal avian cycle is indicated, too, by the appear-ance of incubation patches in many species. These are produced by the synergistic action of these hormones with prolactin (Bailey, 1952; Steel and Hinde, 1963).

The suppressive effects of exogenous estrogen on the gonadotropic secretion in birds are well established (Parkes and Emmens, 1944; Lorenz, 1954; van Tienhoven, 1961), and estrogen administration to domestic fowl has been shown to produce testicular atrophy involving both seminiferous tubules and an inhibition of androgen secretion with a consequent decline of comb growth (Breneman, 1942, 1953).

There is a paucity of information regarding its effects on wild species, but the present authors have recently been investigating the consequence of daily injections of estradiol benzoate on testicular function in pigeons. We find that such injections produce a very characteristic histochemical response; spermatogenetic activity is suppressed, but there is no accompanying buildup of intratubular lipids (Fig. 19). In contrast, the interstitial cells become densely lipoidal and distended, with large rounded nuclei. In these cells the cytoplasm is heavily charged with an amorphous mass of sudanophilic material, and they appear as well-marked clusters (Fig. 20) interspersed throughout an extensive interstitium of small, spindle-shaped juvenile cells containing only a few small cytoplasmic lipid granules.

These results suggest a suppression of FSH leading to an atrophy of germinal stages beyond spermatogonia following estrogen administration. But, in view of the proliferation of juvenile interstitial cells and the massive buildup of sudanophilic lipids in the mature Leydig cells, an increased release of LH is indicated and would be in agreement with the known effects of estrogen in ovulating female mammals. Indeed, recent studies have confirmed that plasma levels of LH are markedly elevated following the treatment of male feral pigeons with estradiol benzoate (Murton, 1973). It seems likely, too, that there is a direct effect of estrogen on the interstitial cells causing an interference with the later stages of androgen biosynthesis resulting in an accumulation of precursor material, which would also lead to an atrophy of androgen-dependent secondary sexual characters such as comb growth. In the case of feral pigeons, new Leydig cells appear about the middle of the preincubation courtship cycle, and the displays typically occurring at this time are markedly enhanced by exogenous estrogens (Murton et al., 1969b).

Little is known regarding the effects of progesterone in male birds, and the available data are somewhat contradictory. Thus, an antigonadal effect has been recorded by some (Fox, 1955; Herrick and Adams, 1956), and a stimulation of testicular activity following pro-

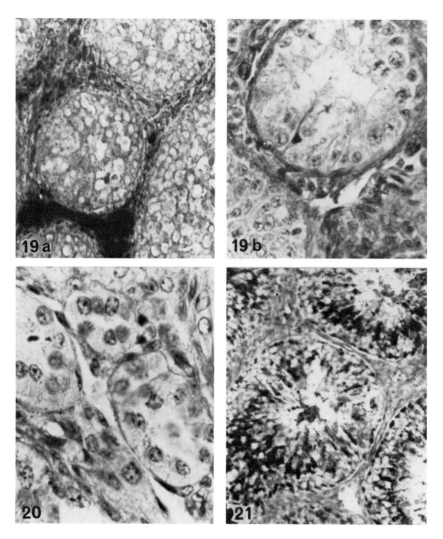

FIGS. 19–21. FIG. 19: (a) Testis of estrogen-treated pigeon. The interstitial Leydig cells have become densely lipoidal, but there is no lipoidal material in the adjacent regressed seminiferous tubules. Gelatin-embedded material stained with Sudan Black and carmalum. × 350. (From Lofts, 1970.) (b) Testis of progesterone-treated pigeon. Spermatogenesis has been completely suppressed and the tubules contain only a peripheral ring of spermatogonia. × 300. (From Lofts, 1970.) FIG. 20: A high-magnification photograph of the Leydig cells in an estrogen-treated pigeon. Note the hypertrophied clusters. × 1000. (From Lofts, 1970.) FIG. 21: A gelatin-embedded section of the testis of a progesterone-treated pigeon. Large quantities of cholesterol-positive lipids have accumulated in the Sertoli cells, but there are no densely lipoidal interstitial cells as noted after estrogen treatment (cf. Fig. 19a). × 350. (From Lofts, 1970.)

gesterone administration by others (Kar, 1949; Jones, 1969a). Farner, on the other hand, mentioned in discussion, that progesterone injections into *Zonotrichia leucophrys* produces a negative result (Farner and Follett, 1966). In pigeons, we have found a variable response to such injections, with some showing no effect on the testes, but others undergoing considerable testicular atrophy (Fig. 21). However, it is doubtful whether endogenous progesterone production has any antispermatogenetic effect, since Lehrman (1965) has good evidence that this hormone is released just prior to egg laying, at a time when there is no obvious testicular regression.

c. Prolactin. Prolactin is present in the pituitary gland of male birds (Bailey, 1952; Burrows and Byerly, 1936; Riddle and Bates, 1933) and has been known for many years to retard spermatogenesis and lead to testicular collapse in pigeons (Riddle and Bates, 1933), cockerels (Breneman, 1942; Bates *et al.*, 1937; Yamashima, 1952), and some passerine species (Lofts and Marshall, 1956; Meier and Dusseau, 1968; Meier, 1969). Apparently, exogenous prolactin does not induce a testicular regression in *Zonotrichia leucophrys gambelii* (Laws and Farner, 1960), *Carpodacus mexicanus* (Hamner, 1968), *Zonotrichia albicollis* (Meier and Dusseau, 1968), and *Lophortyx californicus* (Jones, 1969a), but it can inhibit photoperiodically induced gonad growth in *Z. l. pugetensis* (Bailey, 1950), *Z. l. nuttalli*, the Song Sparrow *Melospiza melodia* (Farner and Follett, 1966), the House Sparrow, and Wood Pigeon (Lofts and Murton, 1973).

The antigonadal effect of prolactin is suggested to be mediated by a suppression of FSH secretion (Bates *et al.*, 1937; Schooley, 1937; Payne, 1943; Nalbandov, 1945; Nalbandov *et al.*, 1945; Lofts and Marshall, 1956, 1958) and produces a buildup of intratubular lipids and cholesterol similar to the naturally occurring seasonal event (Lofts and Marshall, 1956). In *Ploceus philippinus*, feather regeneration, which is supposedly dependent on LH, is not affected by prolactin administration, but gonadal regression still takes place (Thapliyal and Saxena, 1964), again suggesting a suppression of FSH output. However, there is insufficient evidence to attribute a general antigonadal role to endogenous prolactin secretion under natural conditions. In multibrooded species, the seasonal testicular regression does not occur until after the completion of the last clutch, yet prolactin-induced brood patch formation and incubation behavior take place from the time the first clutch is produced. Furthermore, as in *Z. l. gambelii*, a reduction in gonad size is not produced in prolactin-injected pigeons, although the same treatment is capable of terminating spermatogenetic activity in the closely related Wood Pigeon and has an antigonadal effect in other strains (Lofts and Murton, 1973). In the California Quail, Jones (1969c) has recorded the highest pituitary prolactin content in males with testes in full breeding condition, and Gourdji (1970) has similarly recently reported that there is no antigonadal effect in domestic ducks, nor in Japanese Quail, and has further shown that an increase in hypophysial prolactin content under experimental photostimulation is accompanied by an increase in testicular growth. Prolactin does not inhibit the uptake of ^{32}P by the testes of cockerels *in vivo* (Stetson and Erickson, 1970). The assimilation of this material into the testis is stimulated by gonadotropins, and the measurement of its uptake into the testes of young cockerels is an established bioassay for gonadotropic hormones (Breneman *et al.*, 1962). Stetson and Erickson found that prolactin administration simultaneously with or up to 18 hours previous to the administration of either purified gonadotropins or pituitary extracts, not only had no inhibitory effect on ^{32}P uptake, but actually acted synergistically at high dose levels.

It seems likely that different levels of tolerance to prolactin secretion have arisen in different species, which would account for the variations in gonadal response. Certainly, there is no noticeable reduction in testis size in pigeons and doves when prolactin-dependent crop gland development and crop milk production is at its maximum, suggesting that columbid species must tolerate high levels of endogenous prolactin throughout the whole of their very protracted breeding seasons; there are, however, histological changes in the testes during the period of prolactin secretion. Höhn (1959, 1962) attributes a similar high testicular insensitivity to prolactin in the Brown-headed Cowbird *(Molothrus ater)* to an evolutionary adaptation to brood parasitism.

2. Pituitary–Gonad Axis (See also Chapter 2)

In birds, as in mammals, the hypothalamus exerts absolute control over gonadotropic secretion from the adenohypophysis, so that hypophysectomy, pituitary autotransplantation, or sectioning of the pituitary portal vessels produces a rapid testicular regression (e.g., Assenmacher, 1958; Benoit and Assenmacher, 1959; Ma and Nalbandov, 1963; Baylé and Assenmacher, 1967). The recent separation of avian FSH and LH fractions again shows a similarity to the mammalian situation in that testicular control is mediated by means of a dual hormonal regulation from the adenohypophysis. Lofts and Murton (1968) have proposed that the two gonadotropic hormones might not necessarily be secreted simultaneously, but rather, they may be secreted at different phases. Evidence in support of such a hypothesis appears to be provided by the Greenfinch experiments outlined earlier in this chapter and summarized in Fig. 18. In these birds, testicular growth (and, therefore, presumably FSH secretion) was most strongly induced in birds (group F) that were given a light pulse 17–18 hours after dawn, whereas the LH peak occurred in response to a light pulse between 7 and 8 hours after dawn. These results imply that there may be two separate hypothalamic centers controlling LH and FSH release, each perhaps receiving information from discrete rhythms of photoperiodic sensitivity, and the results of Stetson (1969, 1971, 1972a,b) outlined in Section I seem to provide anatomical evidence of such a mechanism.

Differences in the photoperiodic response mechanism are known to occur in birds, and this has in turn led to a variety of annual breeding patterns (Lofts and Murton, 1968). It has been suggested that this phenomenon, in terms of endogenous hormonal controlling mechanisms, must have its basis in differences in pattern in the release of the two gonadotropic hormones (Lofts *et al.*, 1970), and we have seen in the Greenfinch that a mechanism has evolved in which the photoinducible phase for LH is apparently entirely separate from that promoting FSH release. In *Coturnix*, Follett considers that LH and FSH may be secreted simultaneously at one circadian phase (Lofts *et al.*, 1970), and this may well also be true in the House Sparrow, where Murton *et al.* (1970a) have been able to show that there is a greater overlap of LH and FSH release. In this species, a marked peak of LH release in relation to a photoinducible phase early in the 24 hour cycle does not occur.

Since much of our knowledge about pituitary regulation of avian testicular function has, up to the present time, been derived from studying the effects produced by the administration of mammalian gonadotropins, there is a need for caution and further experimentation with avian preparations before firm conclusions are drawn. However, much of the evidence to date suggests that LH stimulates interstitial cell activity in basically the same way as in mammals. Thus, Leydig cell hyperplasia has been noted in LH-injected young cockerels and was seen to be accompanied by parallel comb growth (Taber, 1949), and in pigeons, similar LH treatment produces a proliferation in the numbers of recognizable Leydig cells and a buildup in their lipoidal inclusions and steroid dehydrogenase activity (Lofts and

Murton, 1973). Such results have recently been endorsed by the first preliminary experiments using purified avian LH. Follett *et al.* (1972) report that daily administration of 50 μg of avian LH for 10 days to sexually regressed young Japanese Quail left under non-stimulatory photoperiods stimulates Leydig cell development into typical secretory form with abundant smooth endoplasmic reticulum.

The fact that exogenous androgens cause a regression when administered to early acceleration phase birds (Lofts, 1962b) might indicate a negative feedback effect on the anterior pituitary, and the results of Kordon and Gogan (1970) showing that microimplants of testosterone within the ventromedial nucleus of the hypothalamus in ducks induces a testicular regression adds support to such a suggestion. The same investigators have also shown that the threshold of testosterone necessary to block light-stimulated gonad growth varies on a seasonal basis, being low at the time of testicular quiescence, increases when the gonads are stimulated, and decreases at the time of seasonal testicular regression. In *Spizella arborea* (Wilson, 1970) and Japanese Quail (Wada 1972; Stetson, 1972c) a similar inhibition of light-stimulated testicular growth is obtained by testosterone implants in the nucleus tuberis region. Evidence of a negative gonadal feedback mechanism is also suggested by the fact that the anterior pituitary of photostimulated castrated birds increase in weight more than the photostimulated intact controls (Benoit, 1961; Benoit *et al.*, 1950; Uemura, 1964), and that plasma levels of LH become up to five times higher in Japanese Quail subjected to such experimentation (Follett *et al.*, 1972). Furthermore, there is a significantly higher acid phosphatase content in the supraoptic and median eminence regions of photostimulated castrated *Zonotrichia leucophrys* than in the intact controls (Kobayashi and Farner, 1966). It would be informative to determine whether exogenous androgen administration would reduce the acid phosphatase content as it does plasma LH in quail.

Injections of purified preparations of mammalian FSH have a stimulatory effect on the germinal epithelium and can restore the spermatogenetic activity of a hypophysectomized bird (Lofts and Marshall, 1958) or an intact bird during the postnuptial regressive stage (Miller, 1949). Such treatment also produces a depletion of the intratubular lipids, and an inverse relationship seems to exist between the Sertoli cell lipid cycle and the activity of the germinal epithelium. Thus, a termination of spermatogenesis (and presumably FSH output), either naturally during the postnuptial refractory period or artificially as a consequence of hypophysectomy is generally accompanied by an increase in lipids and sterols in the Sertoli cell cytoplasm. Conversely, a recovery of gametogenetic activity during the acceleration phase, or by the exogenous administration of FSH, is always paralleled by a depletion of these substances. Such an association has also been noted in other vertebrate groups (Lofts, 1968), and in view of the steroidogenic capacity of this tissue, it has been suggested that the overall control of spermatogenesis by FSH might be mediated via a regulation of the secretory activity of the Sertoli cells (Lofts, 1968, 1972; Lofts and Bern, 1972). Such a mechanism would be more in line with the known actions of the other pituitary tropic hormones such as LH and ACTH, which are known to stimulate the secretory activity of other steroid-producing tissues. In the same way, FSH might thus stimulate the secretion of a steroid by the Sertoli cells, which, in turn, would have a spermatokinetic effect on the germinal epithelium. In view of the experimental evidence indicating that exogenous androgens can, under certain situations, stimulate spermatogenesis, an indirect stimulation of the germinal epithelium by FSH acting on the Sertoli cell would appear a more likely hypothesis than one that supposes that a protein hormone, on the one hand, and a steroid, on the other, can cause the same biological action, i.e., a stimulation of spermatogenesis. Evidence in support of such a mechanism is also beginning to appear in mammals (Lacy and Lofts, 1961, 1965; Lacy, 1967; Lacy et al., 1969; Lacy and Pettitt, 1969) and in lower vertebrates (Lofts, 1972).

3. The Refractory Period

The development of photorefractivity in those species that have photoperiodically controlled breeding cycles may be regarded as an adaptive mechanism preventing propagation at times when stimulatory day lengths still occur, but when other environmental needs for successful rearing of young are becoming scarce. In migratory species, it causes a premature termination of the breeding period so that the birds are able to undergo their postnuptial molt and autumnal hyperphagia before their contranuptial flight and while an adequate food supply is still available. The length of time before such birds regain their photosensitivity varies, but generally in many temperate zone species the animals remain in this sexually quiescent period for some 3–4 months. By the time they once again become photosensitive it is usually late autumn, and stimulatory light conditions are no longer present in their environment and will not be experienced until the following spring.

   The termination of photorefractivity has been shown to be advanced by experimentally subjecting birds to a period of short days (see review by Lofts and Murton, 1968), and it is only by exposure to a species-specific term of short photoperiods that these animals regain their capacity to respond to stimulatory day lengths. In some, such as *Junco hyemalis, Zonotrichia leucophrys, Z. albicollis, Sturnus vulgaris,* and *Anas platyrhynchos,* this requirement appears to be absolute, and refractory birds exposed to continuing long daily photoperiods remain sexually regressed. In others, on the other hand, the situation is less rigid, and even when experimentally exposed to continuous summer day lengths, they will spontaneously emerge from their photorefractory state. The domestic duck (Benoit *et al.,* 1956b) and *Quelea* (Lofts, 1962c) are two examples of such a situation. A circadian-based mechanism may be involved in determining the duration of photorefractivity. Thus, when groups of photorefractory House Sparrows are maintained for 35 days under daily photoperiods of 6 hours and an additional 1 hour light pulse given as an interruption of the dark period at 8, 12, or 16 hours from "dawn," respectively, only sparrows receiving a light interruption at 8 hours from "dawn" recover their photosensitivity and show gonadal stimulation when transferred to a stimulatory 16L–8D light schedule (Murton *et al.,* 1970b). These results indicate that the total daily quantity of light is unimportant, since it was the same in each group. Hamner (1968) similarly claims a circadian component timing the refractory period in *Carpodacus mexicanus.*

   The physiological basis for gonadal photorefractivity still remains enigmatic, and the hypothalamus, adenohypophysis, and the gonads themselves have all in turn been suggested as the possible site of the blockage. The early experiment by Benoit *et al.* (1950) demonstrating that a compensatory hypertrophy of the residual testis occurred in hemicastrated ducks when operated on during the non-refractory period, but not when the operation was performed during the refractory period, provided evidence suggesting a cessation of gonadotropin output during the refractory phase and indicated that the hypothalamo-pituitary axis and not the pituitary–testis axis as the primary site involved. Furthermore, Stetson and Erickson (1971) have evidence that the gonadotropic content of the pituitary of castrated White-crowned Sparrows shows a spontaneous decline at a time when intact birds enter their refractory phase, again suggesting a hypothalamopituitary block independent of any gonadal influence. Nevertheless, it is perhaps still premature to exclude completely any gonadal involvement during this period of photorefractivity. The histological condition of the interstitial tissue of a refractory bird is quite distinct from that of a winter bird that has regained its photosensitive potential but is still sexually regressed (Fig. 22), as has been observed in a number of species (Lofts and Murton, 1968). This condition can also be induced experimentally. For example, the interstitial tissue of wild Mallards *(Anas platyrhynchos)* caught during their refractory phase in August and maintained in a protracted refractory condition by exposure to artificial summer photoperiods to the end of December has been shown to be much more extensive

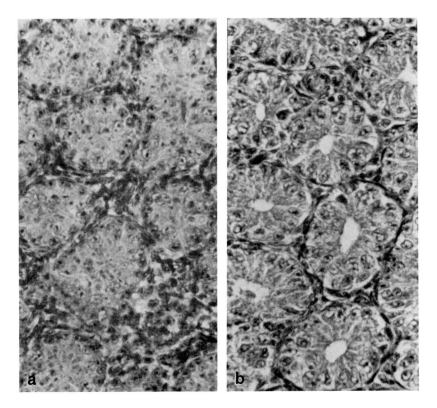

Fig. 22. (a) Testis of the Mallard *(Anas platyrhynchos)*. This specimen had been maintained in a photorefractory condition by exposure to a 16 hour photoperiod for 10 weeks. The seminiferous tubules are completely regressed and spermatogenetically inactive, but the interstitial tissue is extensive. ×400. (b) Testis of the Mallard in which the photorefractory condition has been terminated by exposure to an 8 hour photoperiod for 10 weeks. The interstitial tissue is much more condensed than in (a). ×400. (From Lofts, 1970.)

than that of birds caught at the same time but which have terminated their refractory state while being maintained under an 8 hour winter photoperiod. The testicular interstitial tissue of the latter appears to be much more lipoidal, although spermatogenetically both groups are identical (Lofts and Coombs, 1965). Hamner (1968) has recorded a similar accumulation of interstitial lipids at the termination of photorefractivity in *Carpodacus mexicanus*, and the same has been noted also in House Sparrows (see below). On the basis of this and other cytological data, Lofts and Murton (1968) have speculated that a release of LH might continue during the refractory phase and be a relevant factor. It is during this time that the differentiation of new Leydig cells occurs, and as has already been noted elsewhere, this rehabilitation of the interstitium with a subsequent release of androgen toward the end of the period is probably responsible for autumnal sexual activity in a variety of species.

Evidence implicating a more positive involvement of the testis in the underlying endocrinological mechanisms of the refractory period is also provided by the recent work of Murton *et al.* (1970b) on the manipulation of photorefractivity in *Passer domesticus* by circadian light regimes. Although each of three experimental groups of birds were given the same daily ration of light (6 hours plus an additional 1 hour light pulse given as an interruption of the dark period), these investigators found that after 35 days quite distinct cytological differences developed in the testis of each group. Thus, in the two groups that had experienced a light pulse at 12 hours and 16 hours, respectively, from the experimental dawn, the testes had acquired an expanded interstitium with numerous spindle-shaped cells, whereas in the group that had its night interruption 8 hours from dawn the interstitium was less extensive and the Leydig cells remained small but were more heavily lipoidal and the nuclei more rounded. Subsequent exposure to a daily 16 hour stimulatory photoperiod demonstrated that only the latter group had regained their photosensitivity. Thus, the possibility of a feedback effect from the testis during the refractory period cannot be excluded at this time, although a positive conclusion will be reached only when the circulating testosterone levels of the photorefractory bird are eventually measured.

III. The Ovary
A. General Structure

Reproduction in birds, unlike the other vertebrate groups, is exclusively oviparous. The left ovary, which becomes the dominant and functional primary sex organ in most species, is attached to the body wall by a short mesovarium in close apposition to the kidneys. It is supplied with blood by the ovarian artery, which is usually a branch of the left renolumbar artery but occasionally is a branch of the dorsal aorta (Nalbandov and James, 1949). The organ is drained by large veins that anastomose and converge into an anterior and posterior ovarian vein, both of which empty into the vena cava. Unlike the situation in the male, the functional morphology of the female gonad has been very little studied in wild species, and there is a paucity of information on the seasonal changes in ovarian histology. Most of our data have been derived almost exclusively from investigations of the domestic hen.

In seasonal breeding species, the ovary undergoes the same great variation in size (Fig. 23) that is evident in the testes, and in the European Starling *(Sturnus vulgaris)*, for example, ovarian weight can fluctuate from 8 mg at the time of sexual regression to a weight of over 1400 mg at the height of the breeding season. In the quiescent state, the ovary is a compact, often triangular, flattened organ with small follicles giving the surface a granular appearance. But as it develops into the breeding condition, the follicles become distended and bulge conspicuously from the surface of the ovary, which now assumes the appearance of a bunch of grapes. The largest follicles ultimately become suspended from the surface by a narrow isthmus of tissue, the pedicle (Fig. 24). The mature follicle is highly vascular and the surface has a conspicuous scarlike area, the stigma, which delineates the region where the follicle will eventually rupture during ovulation. The stigma appears macroscopically to be avascular, but it does, in fact, have small blood vessels traversing it (Nalbandov and James, 1949). In mammals, the ejection of the ovum during ovulation causes a rupture of the follicular capillaries resulting in the blood accumulating in the follicular cavity, but in birds the constriction of spiral vessels prevents such hemorrhage. In domestic species, such as the fowl, the ovary remains more or less in this well-developed state throughout the year (except during molting), but in the seasonally breeding species, the ovary undergoes a postnuptial regression into its quiescent state.

Histologically, the ovary follows the basic vertebrate pattern in consisting of an outer cortex containing the developing follicles surrounding a highly vascular medulla forming the ovarian stroma and composed primarily of connective tissue. At the periphery of the cortex is the germinal epithelium, which in the embryonic stage, starts its gametogenetic activity and proliferates a vast number of primary oocytes. By the time of hatching, the gonad is already endowed with

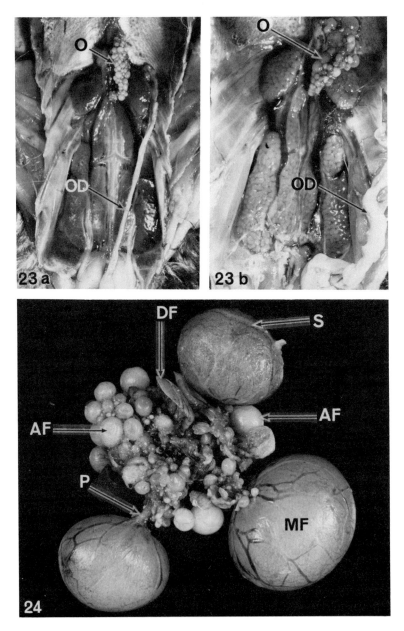

FIGS. 23–24. FIG. 23: Ovary (O) and oviduct (OD) of the European Tree Sparrow during (a) its sexually regressed winter condition and (b) during the breeding season. Note the tremendous development that takes place in the oviduct as a consequence of estrogen secretion from the developing ovary. × 4. FIG. 24: The ovary of a sexually mature chicken showing the pedicle (P), mature follicles (MF), atretic follicles (AF), and a discharged follicle (DF). The stigma (S) can also be seen near the top of one of the mature follicles. Natural size.

many thousands of oocytes (Hutt, 1949), and no further oocyte formation occurs in adult life. Follicular development proceeds from the endowment of oocytes propagated in the embryonic stage. Of these, only a very small percentage eventually mature fully, and the majority undergo follicular atresia while still in an early stage of maturation. Atretic follicles are, therefore, a prominent feature of the avian ovary.

Unlike the testes of seasonal birds, which in their regressed winter condition are generally gametogenetically inactive and show complete uniformity in histological condition, the ovary always contains large numbers of follicles in various stages of development (Fig. 25), as well as a spectrum of corpora atretica in various stages of atresia (Fig. 26). In its primary stage the follicle consists of a single layer of flattened granulosa cells ensheathing the oocyte, but as it matures, a multilayered thecal tissue forms around the outside and becomes highly vascularized. During this growth, the granulosa cells become separated by intercellular spaces filled with perivitelline substance (van Tienhoven, 1968), and villus-like elevations arise on the oocyte cell membrane to form the zona radiata. The oocyte completely fills the developing follicle, so that there is no antrum or other internal organization as in the mammalian follicle.

In the Rook *(Corvus frugilegus)*, which to date is the only seasonally breeding wild species in which the ovarian cytology and histochemistry has been studied in any detail, stromal fibroblasts become incorporated into the theca interna of the developing follicle and become rounded and glandular as maturation proceeds (Marshall and Coombs, 1957). With the approach of the breeding season, a proportion of the follicle population progress toward full maturity and eventually enter a phase of very rapid growth that results in conspicuous bulging from the surface. This culminating period of accelerated development is brought about mainly by the deposition of yolk, and Wyburn *et al.* (1965) have suggested that the ultrastructural elevations of the zona radiata may facilitate the passage of yolk materials into the egg at this time. The final yolk deposition can be very rapid and can, in the pigeon, produce an increase in the maximum follicular diameter from 5.5 mm to 20 mm within 8 days (Bartelmez, 1912). Sturkie (1954) reports a sixteenfold increase in ovum weight in chickens, in the same period before ovulation. After ovulation, the ruptured follicles rapidly regress, disintegrate, and become cleared by phagocytic action. No persistent corpora lutea of the mammalian pattern are formed, and in the chicken (van Tienhoven, 1959) and Rook (Marshall and Coombs, 1957) the postovulatory follicle becomes

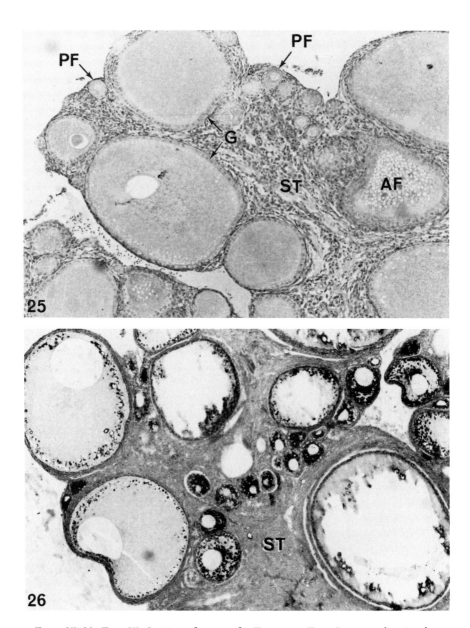

FIGS. 25–26. FIG. 25: Section of ovary of a European Tree Sparrow showing large numbers of follicles in various stages of development. Young, primary follicles (PF) occur near surface of ovary and consist of an oocyte surrounded by flattened granulosa cells. Granulosa layer (G) of the more enlarged follicle is much more hypertrophied. Also seen is a follicle becoming atretic and collapsing (AF). × 80. FIG. 26: Gelatin-embedded section of winter ovary of a European Tree Sparrow stained with Sudan Black and carmalum. A variety of follicles in various stages of lipoidal atresia are scattered throughout ovarian stroma (ST). × 80.

resorbed within a few days, but in Ring-necked Pheasants *(Phasianus colchicus)* (Kabat *et al.*, 1948) and in the Mallard they persist for as long as 3 months. In these species in which there is a more persistent postovulatory structure, it is derived mainly from the hypertrophic thecal cells (Chieffi and Botte, 1970).

The numerous follicles that undergo preovulatory degeneration may do so at different stages of maturity (Fig. 26). The first observable histochemical indication of this is usually a lipoidal zone surrounding the oocyte, which subsequently spreads so that the whole interior of the follicle becomes occluded with a densely sudanophilic mass of cholesterol-positive material. In the Rook, Marshall and Coombs (1957) report that this fatty atresia is accompanied by a proliferation of fibroblasts that invade the follicle and, for a time, remain as a distinguishable cluster of cells in the ovarian stroma. This phenomenon can also be observed in the Tree Sparrow (Fig. 27). These cellular masses subsequently lose their integrity with the disintegration of the follicle and become scattered singly, or in groups, in the stromal tissue (Fig. 28). Marshall and Coombs (1957) use the term "exfollicular gland cells" to distinguish these from the stromal interstitial cells that have developed from the ordinary connective tissue cells of the ovarian stroma and are regarded as being homologous with the interstitial Leydig cells of the testis. As in the latter, the ovarian interstitial cells possess the general ultrastructural characteristics of steroidogenic tissue and react positively to histochemical tests for 3β-HSDH, cholesterol, and sudanophilia. Chieffi and Botte (1970) also distinguish a pattern of atresia in which the yolk becomes extruded through the granulosa layer into the surrounding stromal tissue, where it becomes resorbed by phagocytic action. Once yolk resorption is completed only a few vacuolized cells, some capillaries, and strands of connective tissue remain to mark the site of the resorbed oocyte. In other cases, the oocyte may be invaded by cells derived from a hyperplastic activity of the granulosa tissue.

B. HISTOPHYSIOLOGY

1. General Histophysiology

Literature on the cyclic changes of the avian ovary is sparse, and very little is known beyond the follicular activity of domestic species. Data on the histophysiological changes in the ovary of the seasonally breeding bird throughout the year are almost nonexistent and have been reported, in any sort of detail, only in one species, the Rook

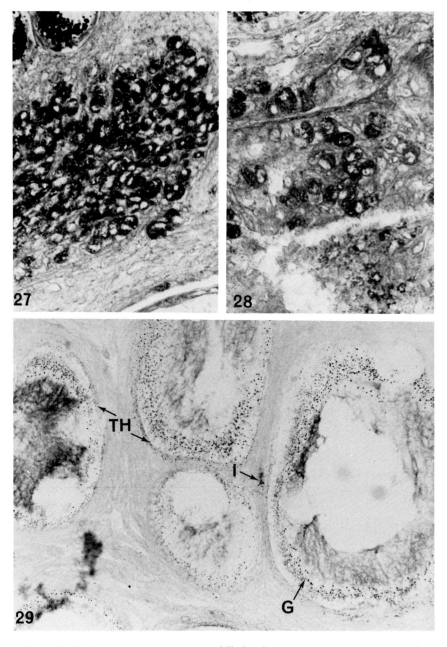

FIGS. 27–29. FIG. 27: An atretic ovarian follicle of a European Tree Sparrow that has become invaded by a mass of glandular and heavily lipoidal cells. × 500. FIG. 28: Glandular cells of an atretic follicle (Fig. 27) that have become scattered in the ovarian stroma. × 500. FIG. 29: Ovary of a European Tree Sparrow showing Δ^5-3β-hydroxy-steroid dehydrogenase reactions in the thecal area (TH), granulosa layer (G), and also in some interstitial cells (I) in the stroma. × 150.

(Corvus frugilegus) (Marshall and Coombs, 1957). In this bird, the follicles of the December ovary are too small to be observed on the surface macroscopically, but histologically, sectioned material shows some oocyte development to be taking place internally, and follicles up to 1 mm diameter occur in association with numerous smaller ones. Marshall and Coombs report that the largest follicles at this time of year have distended, sudanophilic, glandular cells incorporated in the theca interna, and that the granulosa layer of some of the biggest follicles disappear as such, as a proliferation of this tissue fills the structure with large lipid-free cells, which later on apparently become secretory. There is relatively little lipoidal atresia at this time of year.

In temperate-zone species, the advent of the spring weather is, as in the male, generally accompanied by a recrudescence of gonadal activity, and developing follicles become increasingly obvious, macroscopically, over the ovarian surface (Fig. 23b). In *Corvus frugilegus*, follicular recrudescence begins in late January and early February, and from then until the first ovulation in March the whole organ becomes studded with cholesterol-positive lipoidal structures, as increasing numbers of the smaller follicles start developing and become atretic. The stromal interstitial cells also become heavily charged with lipids at this stage and react positively to histochemical tests, both for cholesterol and 3β-HSDH. In the Rook, by March some follicles have enlarged enormously and a few enter the final vitellogenic stage and become ovulated. In a species such as the Band-tailed Pigeon *(Columba fasciata)*, on the other hand, no more than one such enlarged follicle is found in the breeding ovary, and this can be correlated with the fact that such species lay only a single egg per clutch (March and Sadleir, 1970). Generally, a species that lays several eggs per clutch will have an ovary containing several large follicles filled with yellow yolk.

After extrusion of the egg, the follicular wall collapses and becomes spotted with lipoidal material. In the hen, the granulosa cells become inflated with lipids and remain conspicuous for about 72 hours after ovulation (Wyburn *et al.*, 1966), but luteinization does not occur and no corpus luteum develops. In seasonal breeders the completion of the last clutch is succeeded by a rapid regression of the gonad, and individual follicles become increasingly difficult to distinguish by the naked eye, though internally great numbers of follicles continue to undergo a slow development followed by lipoidal atresia. According to Marshall and Coombs (1957), a differentiation of a new generation of interstitial tissue, a phenomenon analogous to the testicular interstitial cell rehabilitation during the regeneration phase, takes place in the Rook ovary during the autumn.

There is unequivocal evidence that the avian ovary, like its mammalian counterpart, is a source of androgenic, estrogenic, and also progestogenic secretions, even though a true corpus luteum is lacking. Chromatographic studies have clearly established that this organ, both in its embryonic stage (Cedard and Haffen, 1966; Weniger and Zeis, 1969) and in the adult bird (Layne et al., 1958; Ozon, 1965; Höhn and Cheng, 1967; Höhn, 1970; Furr, 1969a,b), has the capacity to produce these hormones, and androgenic, estrogenic, and progestational steroids have all been extracted from the plasma of the domestic fowl (O'Grady, 1968; O'Malley et al., 1968). The respective intraovarian sites of biosynthesis of these various steroids, however, is still somewhat speculative. We have already noted that the thecal and granulosa tissue of the preovulatory follicle, the postovulatory follicle, the atretic follicle, the interstitial cells of stromal origin, and also the exfollicular gland cells of Marshall and Coombs, at some stage all contain cholesterol-rich lipoidal inclusions and also have other glandular characteristics that might suggest every one of them to be possible loci of endocrine function, and this is confirmed by the fact that a positive reaction for 3β-HSDH activity has been reported for all these tissues (Botte, 1963; Chieffi and Botte, 1965; Woods and Domm, 1966; Narbaitz and de Robertis, 1968; Boucek and Savard, 1970; Arvy and Hadjiisky, 1970; Sayler et al., 1970) (see also Fig. 29). Furthermore, electron microscopy has shown that all these tissues at some stage possess the fine structure normally attributed to steroidogenic tissue.

2. The Interstitial Cells

Because of their apparent homology with the testicular Leydig cells, the ovarian interstitial cells arising from the stromal tissue are generally regarded to be the most likely source of androgenic secretion within the female gonad (Benoit, 1950; Taber, 1951; Marshall and Coombs, 1957). Techniques for displaying steroid dehydrogenase activity have indicated that in the hen the steroidogenic potential in this tissue develops early in the embryogenesis of the organ (Boucek et al., 1966) and not just in the post-hatching period, as had previously been supposed (Romanoff, 1960), and this has also been shown to be the case in Coturnix (Kannankeril and Domm, 1968; Sayler et al., 1970). In a seasonally breeding bird, such as the Rook, the cyclic development of these cells closely parallels the development of the sexual displays, becoming increasingly prolific and active as sexual activity heightens in the spring and autumn (Marshall and Coombs, 1957). In the hen, too, the 3β-HSDH activity in these cells increases in intensity as the animal enters its egg-laying period, and reduces in intensity as the follicular phase wanes. The work of Taber (1951) also strongly suggests that a stromal cell is responsible for androgen production. The comb of the domestic fowl is in both sexes under androgenic control and is sometimes used as a bioassay for androgenic analysis (reviewed by Dorfman and Shipley, 1956). Taber recorded a reappearance of foamy lipoidal interstitial cells concomitant with the decline of androgen production (comb regression) in hens after the cessation of gonadotropin stimulation. It seems likely that in the absence of gonadotropic stimulation and decline in androgen synthesis, the rapid accumulation of cholesterol-rich lipoidal precursor material in these cells resulted in them becoming more prominent. The more recent methods for the visualization of androgenic steroids by means of fluorescent antibody techniques confirm that the interstitial tissue is the primary locus of androgen secretion, though the thecal and granulosa layers also produce a weak reaction (Woods and Domm, 1966).

Boucek and Savard (1970) have recently investigated steroid formation in the hen's gonad at different stages of the ovarian cycle, by a chromatographic analysis of the bioconversion of [^{14}C]acetate into progesterone, androstenedione, testosterone, and estradiol-17β in an *in vitro* incubation. A parallel analysis of the distribution and intensity of 3β-HSDH and 17β-HSDH activity was also carried out to provide data on the intraovarian location of the sites of activity. It was discovered that whereas the same spectrum of radioactive steroids was manufactured, the relative ratios of one steroid to the other varied significantly with the state of the ovarian cycle. Thus, a relatively large proportion of androgenic steroids was synthesized by the gonad of the molting hen, whereas the steroid profile produced by the ovarian tissue of the laying hen showed equivalent amounts of progesterone, androgens, and estrogenic hormones. In the broody hen, chiefly progesterone and estrogen were synthesized with very little androgen. Significantly, the histochemical studies demonstrated that the high androgenic production by the molting hen coincided with an intense 17β-HSDH activity in the stromal cells, and also the thecal tissue, whereas only trace reactions were produced by these cells during the laying stage.

3. The Developing Follicle

In mammalian ovarian histophysiology, current opinion generally credits the thecal cells of the developing follicle with the secretion of estrogens, and the granulosa cells with the secretion of progesterone after they metamorphose into the granulosa lutein tissue of the corpus luteum (reviewed by Lofts and Bern, 1972). Support for this latter observation is also provided by the data showing that granulosa cells grown in tissue culture manufacture progesterone (Channing, 1966). In birds, the evidence is more conflicting, but the tremendous increase in the size of the oviduct which is coincident with sexual maturity is known to be an estrogen-dependent effect (Brant and Nalbandov, 1956) and indicates that estrogen secretion appears to be a consistent consequence of follicular development, as is the case in mammals. In a seasonally breeding species, the marked annual fluctuations in oviduct development closely parallel the activity of the ovary. Thus, in *Corvus frugilegus* the oviducal epithelial height in the sexually inactive winter condition is 8 μm, but cellular proliferation begins in early February as the vernal acceleration of follicular development starts, and the epithelial layer rapidly expands to a maximum height of about 40 μm by the end of the month. When follicular development diminishes, the oviducts become constricted again. Estrone, estradiol-17β, and estriol have all been found in avian ovarian tissue (Layne *et al.*, 1958; Ozon, 1965; Höhn and Cheng, 1967; Boucek and Savard, 1970), and the former two substances have also been identified in the plasma of the laying hen (O'Grady, 1968; O'Malley *et al.*, 1968). In addition to their vitally influential role in the seasonal preparation of the oviduct, estrogens have many effects on a wide variety of somatic and physiological processes in birds, and the reader is directed to the extensive reviews that have been published for details on these various aspects of estrogenic influence (Nalbandov, 1953; Sturkie, 1954; Parkes and Marshall, 1960; van Tienhoven, 1959, 1961; Marshall, 1961a).

The seasonal histochemical changes described by Marshall and Coombs (1957) in *Corvus frugilegus* seem to support the hypothesis that the avian thecal cells may similarly be the site of estrogen synthesis as they apparently are in the mammal. Thus, an accumulation of cholesterol-positive lipoidal material builds up in the glandular cells of the theca interna in the largest follicles during the winter months (Fig. 30a), and then rapidly becomes depleted as estrogen titers are reaching their peak and the mature follicles are undergoing their culminating vitellogenic phase prior to ovulation. By the time

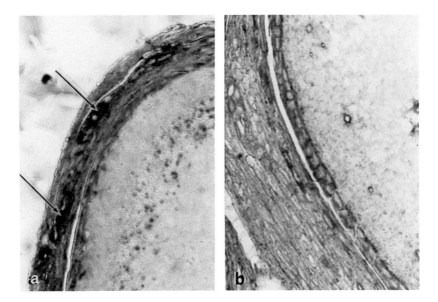

FIG. 30. (a) Heavily lipoidal glandular cells (arrowed) in the thecal layer of a large winter follicle in the ovary of a European Tree Sparrow. ×750. (b) Thecal layer of large follicle of the breeding ovary of a Tree Sparrow. The thecal gland cells have become depleted of their cholesterol-rich lipoidal content [cf. (a)]. Note the hypertrophy of granulosa layer. ×750.

yellow yolk appears in the oocytes, only slight traces of sudanophilic material remain in the cytoplasm of the greatly distended thecal cells (Fig. 30b). Vitellogenesis, and the consequent deposition of large quantities of yolk within the developing oocyte, is an estrogen-dependent phenomenon (for review, see Schjeide *et al.*, 1963), and the rapid depletion of the thecal cholesterol and lipids at this stage of the seasonal cycle may be indicative of a rapid utilization of precursor material at a time of high steroid synthesis, as has been suggested to be the case when a similar depletion occurs in the testicular Leydig cells (Marshall, 1961b). Certainly, the fine structure of the thecal cells and their positive response to tests for steroid dehydrogenase activity suggest a possible site of steroid synthesis, and the more recent demonstration that isolated thecal tissue from growing follicles of the hen's ovary has the capacity to convert cholesterol to estrogens (Botte *et al.*, 1966) provides strong support for the suggestion that this tissue may be the locus of estrogen production in the avian ovary. Sayler *et al.* (1970) have investigated the effects of photostimulation on the distribution of 3β-HSDH in the ovary of young Japanese Quail and have attempted to express changes in steroidogenesis by expressing the intensity of the 3β-HSDH histochemical reaction by means of an arbitrary scale based on the density of the formazan granulation (Wattenberg, 1958) deposited by the different steroidogenic tissues. In *Coturnix* that were raised from hatch to maturity under a stimulatory (16L:8D) light cycle, 3β-HSDH activity increased in all the ovarian steroid producing tissues, but from day 35 onward, the index of 3β-HSDH activity in the thecal tissue increased very rapidly from an average value of 1.3 to 4.3, whereas the enzyme activity in the granulosa tissue was less intense, rising from an average of 1.0 to 2.9 in the same period. Interestingly, the enzyme activity in the thecal tissue of birds kept under a short day length (8L:16D) was also much higher than that of the granulosa tissue.

According to Chieffi (1967) the granulosa cells are the probable site of estrogen biosynthesis, since 17β-HSDH is limited to this tissue (Chieffi and Botte, 1965, 1970). This enzyme catalyzes the transformation of testosterone into androstenedione and of estradiol into estrone. Arvy and Hadjiisky (1970) have also recorded a greater dehydrogenase activity in the granulosa cells both in the hen and in the quail. However, Marshall and Coombs (1957) point out that the granulosa cells in the seasonally breeding Rook are the only prominent ovarian cells that do not contain cholesterol during follicular development, and electron microscopy studies of the hen ovary have shown that granulosa cells start developing abundant agranular endoplasmic reticulum, mitochondria with tubular cristae, cholesterol, and sudanophilic granules, only at a time when the follicles are ready to ovulate, thus suggesting that their steroidogenic activity has been relatively slight up to this time.

Boucek and Savard (1970) have confirmed Chieffi's earlier observations of the localization of 17β-HSDH in the granulosa cells of the laying hen, but have also shown that in the molting bird the reverse situation develops and 17β-HSDH activity now becomes intense in thecal tissue but absent from granulosa cells, even though *in vitro* incubation and chromatographic analysis of the ovarian tissue show that there is a considerable bioconversion of radioactive acetate into estradiol-17β during this time. The same investigation has also shown that there was little or no progesterone production in these molting hens. In view of the absence of 17β-HSDH in granulosa tissue in these birds, and also the very marked reduction in the intensity of the 3β-HSDH as well, it seems not impossible that the granulosa cells might be involved in the synthesis of progesterone. The ovarian steroid profiles of the laying and broody hens both show considerable progesterone biosynthesis and, in both groups, steroid dehydrogenase activity occurs in the granulosa tissue, being particularly intense in the laying hen (Boucek and Savard, 1970). These observations are also supported by the data of Furr (1969a), which show that progesterone occurs in the blood of laying and broody hens, but is not detectable in the blood or ovaries of molting birds. Furthermore, the highest levels of progesterone are found in the follicles and only negligible amounts occur in the ovarian stroma (Furr, 1969b).

4. The Postovulatory Follicle

The postovulatory follicle has sometimes been suggested as the source of ovarian progestogenic steroids, and the ruptured follicles have been shown to contain progesterone (Furr, 1969b) and have the capacity of synthesizing steroids *in vitro* (Botte *et al.*, 1966). Furthermore, the experimental removal of recently ruptured follicles influences the retention time of eggs in the oviduct (Rothchild and Fraps, 1944a,b). Conner and Fraps (1954) have established that there appears to be a quantitative relationship between the amount of postovulatory tissue and oviposition, so that the greater the proportion removed, the greater the retardation in oviposition. It has also been established that when the next-to-last ruptured follicle is removed it has no effect (Rothchild and Fraps, 1944a), suggesting that whatever endocrine activity the postovulatory follicle possesses, it is rather short-lived, and this is also indicated by the rapid degeneration that generally occurs in most avian species.

The very transient endocrine phase of postovulatory follicles outlined by the experiments of Fraps and his colleagues is also supported by the electron microscopic studies of Wyburn and his co-workers (Wyburn *et al.*, 1966), who have recorded an extensively developed agranular endoplasmic reticulum in the granulosa cells of the recently ovulated follicle, together with an increase in sudanophilia and cholesterol content. There is also a very intense reaction to tests for 3β-HSDH (Botte, 1961, 1963; Chieffi and Botte, 1965), but this rapidly diminishes with the degeneration of the postovulatory structures. It seems likely that this very abbreviated postovulatory steroidogenic capacity in these cells is probably a hangover from the activity of the granulosa tissue during the period of high progesterone production immediately preceding the ovulatory process.

5. *The Atretic Follicle*

Cholesterol is abundant in the thecal tissue of the atretic follicles, and during the initial stages of atresia in the granulosa cells. In both layers, a positive 3β-HSDH activity has been reported (Woods and Domm, 1966; Chieffi and Botte, 1970). The latter investigators also report 17β-HSDH in the granulosa layer. Marshall and Coombs (1957) have suggested that the atretic follicles may be involved in progesterone production, since great numbers of follicles undergo such lipoidal atresia, and, in a seasonal breeder, this phenomenon builds up a considerable reservoir of cholesterol by the time of the seasonal ovulation period. However, Sayler *et al.* (1970) have established that when young *Coturnix* are reared to maturity under non-stimulatory light conditions (8L:16D), there is an increase in the numbers of atretic follicles, and when the 3β-HSDH reactions in the ovary are examined, sparse granulation is seen in the granulosa cells at the beginning of follicular regression. But in the later stages, this disappeared entirely, suggesting a loss of steroidogenic activity in developing follicles once atresis sets in. It is perhaps premature, however, to conclude that the atretia structures play no significant part in the endocrinology of the avian ovary, and it is evident that there is a need for much more information before the precise locus of any of the ovarian steroids can be established with any certainty.

6. *Vitellogenesis*

Attention has already been drawn to the fact that, in the days immediately preceding its ovulation, the avian follicle undergoes a very rapid expansion in size due to the storage of large quantities of yolk in the maturing oocyte (i.e., vitellogenesis). For example, in the chicken, the mass of yolk can increase from 100 mg to as much as 20 gm in the 7 days before ovulation (Romanoff and Romanoff, 1949). Although experiments using radioisotopes leave no doubt that protein synthesis can, and does, occur within the growing oocyte (Schjeide *et al.*, 1963), this accounts only for a minor portion of the yolk materials synthesized during this short time interval, and current evidence indicates that the greatest portion of the yolk lipids and proteins laid down in vitellogenesis arise elsewhere in the organism (Schechtman, 1956). Basically, the process can be divided into three successive stages, all of which are causally related. These are (1) the synthesis of the yolk proteins and lipids, (2) their transportation to the ovary and uptake by it, and (3) the incorporation of the serum proteins into the yolk elements.

In the avian oocyte, the yolk contains relatively more lipid (33%

wet weight) in relation to protein (16%) than is the case in many other vertebrate eggs. In fine structure, it is seen to consist of two major components, namely, the yellow and white yolk granules which appear to be freely suspended in the second component, the yolk fluid. The granules often have discrete limiting membranes and, internally, may contain osmophilic subgranules that are larger in the white granules than they are in the yellow ones (Bellairs, 1961). The yellow and white granules are laid down in alternating concentric layers that reflect a diurnal variation in carotenol deposition (Schjeide et al., 1963). Very few mitochondria are observable within the mature yolk mass, except during the early stages of yolk-granule formation (Schjeide and McCandless, 1962), nor is there much evidence of an endothelial reticulum. Much of the protein, together with some of the lipid, largely occurs in the granules (Schjeide and Urist, 1959), whereas the yolk fluid contains large quantities of lipid globules and a smaller proportion of proteins. Although there are slight differences in their detailed chemistry, the major yolk proteins consist, as in all other groups of oviparous vertebrates, of lipovitellins and phosvitin (Wallace, 1963a,b; Schjeide et al., 1963; Wallace et al., 1966; Jared and Wallace, 1968).

Evidence for the extraoocyte origin of most of these constituents stems from the earlier observations that phosphoprotein appears in the plasma of the laying hen (Laskowski, 1935, 1936; Roepke and Hughes, 1935). In view of its chemical similarity with the yolk protein vitellin, it was suggested that plasma phosphoprotein might be a precursor of the latter, which eventually became translocated into the egg (Roepke and Bushnell, 1936). The role of blood proteins in yolk formation was further emphasized by the demonstration, by immunological techniques, that even foreign proteins after being injected into the circulation could be detected in the yolk (Knight and Schechtman, 1954). Other changes in the constitution of the blood that have been observed to occur as a prelude to vitellogenesis are a very sharp elevation in the level of plasma calcium and also of lipids (Riddle and Reinhart, 1926; Laskowski, 1933; Lorenz et al., 1938).

More recently, Schjeide and co-workers (1963) have been able to show that, at the onset of laying, two new serum proteins appear in the hen which are chemically and immunologically similar to the three major egg yolk proteins—phosvitin and α- and β-lipovitellin. The same investigators also confirmed the transference of serum lipovitellin into the egg yolk by means of radioactive tracers. In this experiment serum lipovitellin was first extracted and labeled with ^{14}C, then injected into a laying hen. Within 10 hours, much of the radio-

active protein was lodged in the yolk proteins, indicating its transference from the circulation across the egg cell barriers into the yolk.
The mechanism of the transmission of such blood proteins into the
oocyte is far from clear, but it may be by micropinocytotic activity at
the oocyte surface (Press, 1959; Wyburn *et al.*, 1965). The follicular
epithelium certainly seems to play a significant part, and Patterson
et al. (1961) have clearly demonstrated a preferential transfer of serum
γ-globulins by the follicular tissue to developing ova and have found
that its concentration in the yolk is many times that in the serum.
Gonadotropins appear to be important for stimulating the mechanism
of this final transference of blood proteins into the oocyte (Phillips,
1959), and in some of the lower vertebrate groups, an increase in
serum protein transmission and yolk deposition has been shown to
result from FSH treatment, possibly by a stimulation of the process
of micropinocytosis.

The liver appears to be the extraoocyte locus of the synthesis of
yolk protein and lipid, and hepatectomy has been shown to prevent
the typical plasma changes associated with the onset of the vitellogenic phenomenon (Ranney and Chaikoff, 1951). Furthermore, radioactive phosphate injected into the wing vein of laying hens has been
shown to become incorporated first into a protein-bound component
in the liver which then becomes circulated in the plasma as a phosphoprotein, before finally being transferred into the oocyte and localized in the egg yolk (Flickinger and Rounds, 1956). Confirmation
is also provided by the demonstration that the *in vitro* incubation of
liver slices from laying hens can synthesize phosvitin, which is similar
to the yolk phosphoprotein (Heald and McLachlan, 1965), and Hawkins and Heald (1966) have similarly demonstrated the *in vitro* synthesis of triglycerides and established that it is much greater in laying
hens than in the liver of immature pullets. These triglycerides are
mainly transported to the oocyte as β-lipoproteins in the plasma
(Schjeide *et al.*, 1963), and eventually become concentrated within
the fluid yolk compartment of the egg as lipid globules. Although
β-lipoproteins occur in the plasma of immature as well as mature
birds, there is a pronounced lipemia at the onset of vitellogenesis
in the latter.

C. Control of Ovarian Function

1. Effects of Exogenous Hormones

The exogenous administration of different gonadotropic and gonadal hormones and their effects on the female gonad have been less

extensively studied than in males, and the little information that is available is generally based on observed changes at the morphological level, without any detailed records of the cytological and histochemical events that might have occurred. Most of our knowledge has been derived from studies on the domestic hen, with particular reference to the effects that such hormone treatment have on ovulation and oviposition. Information about the follicular phase is far less precise.

a. Androgen. The evidence for ovarian androgen synthesis, and the probable site of production, has already been discussed. The role of androgenic steroids in the regulation of secondary sexual structures and behavior is also well established (Witschi, 1961), but there is no good evidence that indicates that endogenous ovarian androgens have any significant role in the regulation of ovarian function. The available data are contradictory.

Chu and You (1946) have reported a stimulatory effect on the ovary of both immature and hypophysectomized adult pigeons after androgen administration, and testosterone stimulation of the ovary has also been recorded in *Passer domesticus* (Ringoen, 1943). In both these species, androgen injections greatly enhance follicular growth, and Ringoen has noted that the follicular epithelium of treated sparrows becomes stratified. Similarly, stratification of the follicular epithelium and a stimulation of follicular growth have also been noted in the hen ovary after testosterone propionate treatment (Breneman, 1955). An increase in the vacuolation of the cells and a greater ovarian sudanophilia was also noted in the latter investigation. On the basis of observing the effects of a variety of dose levels (0.1–100 μg) of testosterone propionate, Breneman (1955, 1956) has reported that androgen administration increases the height of the follicular epithelium, but a subsequent statistical analysis of his data has led van Tienhoven (1961) to conclude that no dose–response relationship existed and that the differences between controls and androgen-treated pullets could be accounted for by a sampling error. In the American Sparrow Hawk (*Falco sparverius*), large doses of testosterone propionate (140 mg over 30 days) fail to produce any gonadal changes, neither in the left nor right ovary (Nelson and Stabler, 1940).

b. Estrogen. Exogenous estrogens are known to delay ovulation in Ring Doves (Dunham and Riddle, 1942) and in domestic hens (Fraps, 1955), and have been reported to inhibit the seasonal development of ova in House Sparrows (Ringoen, 1940). Injections into hypophysectomized pigeons failed to induce any significant follicular maturation (Chu and You, 1946), and in 30-day-old pullets, estradiol administration did not cause any significant change in ovarian weight, although Breneman (1955, 1956) noted a general increase in sudanophilia and cholesterol content. In contrast to these observations, Phillips (1959) has reported a 32% increase in ovarian weight in 6-week-old pullets injected with diethylstilbestrol, and an even greater ovarian stimulation after similar injections into adult nonbreeding Black Ducks *(Anas rubripes)*. The same steroid can also stimulate ovarian development in wild Mallards and is said to act synergistically with gonadotropins to stimulate follicular development (Phillips and van Tienhoven, 1960). It may be that such contradictory results reflect varying capacities of different steroids to inhibit gonadotropic secretion (see also Section V).

Although to date there is insufficient evidence to attribute any direct effect of estrogenic hormones on gonadal function (i.e., with concomitant pituitary influence), there are ample data to show that such steroids have a vital indirect role by influencing vitellogenesis in the mature ovary. It is now reasonably well established that estrogen stimulates the liver to produce the protein and lipid precursors of yolk. Estrogen injections into female birds lead to a very rapid rise in the plasma levels of fatty acids and lipoproteins (Schjeide *et al.*, 1963; Heald and Rookledge, 1964), and Roos and Meyer (1961) report that estrogens also boost the serum level of lipoprotein lipase in pheasants. Phosphoproteins also appear in the circulation as a consequence of estrogen administration (Schjeide *et al.*, 1960, 1963), and, significantly, the same treatment can even induce similar changes involving the appearance of lipovitellin and phosvitin in the plasma of immature pullets and cockerels (Riddle, 1942; Ranney and Chaikoff, 1951; Schjeide and Urist, 1956; Mok *et al.*, 1961; Heald and McLachlan, 1964; Heald and Rookledge, 1964).

As might be expected in view of the above observations, estrogen injections also cause parallel changes to take place in the liver, and Schjeide *et al.* (1963) report that a very large increase in liver volume is produced in domestic fowl as a consequence of estrogen administration. In estrogen-treated birds, the liver tissue undergoes a hyperplasia, and the cells become greatly enlarged with a highly developed endoplasmic reticulum and an increased ribosomal and cytoplasmic protein content. The nuclear RNA content is also doubled in comparison with the average liver cell of nonestrogenized birds. Schjeide (1967) has established that lipid metabolism, and the release of such lipids into the plasma, is greatly stimulated in such birds. Although exogenous estrogen administration brings about a mobilization of yolk precursors, it does not necessarily result in an increase in the number of yolky follicles in the ovary, since gonadotropins are necessary for the transference of these substances from the plasma into the oocyte. Thus, in nonbreeding ducks injected with diethylstilbestrol, only those that were additionally treated with anterior pituitary extract developed large follicles with yellow yolk, whereas treatment with either estrogen or gonadotropin alone failed to stimulate large yolky follicles (Phillips, 1959). The same is true also of wild Mallards (Phillips and van Tienhoven, 1960).

c. Progesterone. In domestic fowl, progesterone administration inhibits egg laying if given early in the 24 hour ovulatory cycle, or if given at any time in large doses, and results in an increase in follicular atresia (Rothchild and Fraps, 1949; Adams, 1956; Juhn and Harris, 1956; van Tienhoven, 1961). A similar blockage of ovulation, together with a decrease in ovarian weight and follicular diameter, also occurs in progesterone-injected California Quail when given early in the daily ovulatory cycle (Jones, 1969a). In this latter species the oviduct weight of progesterone-treated females was also significantly smaller than in control birds, suggesting a decline in estrogen secretion presumably caused by an inhibition of FSH release. When, however, progesterone is implanted into immature pullets, it hastens follicular maturation and precipitates precocious egg production (Nalbandov, 1956), and its administration late in the egg-laying cycle of the adult causes premature ovulation, presumably by stimulating LH release (Fraps, 1970). The experiments of Ralph and Fraps (1960), which demonstrated the ovulation-inducing effects of small quantities of progesterone stereotactically injected into the hypothalamus, but not when injected into the pituitary, indicate that this ovulatory influence is probably mediated via the hypothalamo-hypophyseal axis. The investigations of Kappauf and van Tienhoven (1972) on the progesterone concentrations in peripheral plasma of laying hens have demonstrated that there is a significant rise in plasma concentrations at about 6 hours prior to ovulation.

d. Prolactin. Although prolactin was the first anterior pituitary hormone to be isolated in pure form, data on its effects on the ovary are much fewer than on the avian male gonad. As in the male, exogenous prolactin injections have sometimes been reported to have an anti-gonadal effect also in the female, and in the hen (Bates *et al.*, 1935), pigeon (Bates *et al.*, 1937), and White-crowned Sparrow (Bailey,

1950), treatment with this hormone inhibits follicular development with resultant atresia. In pigeons (Schooley, 1937), hens (Collias, 1950), House Sparrow (Vaugien, 1955), and Bank Swallow (*Riparia riparia*) (Peterson, 1955), the ovary has been reported to be regressed during the incubation period, and, in view of the good evidence showing that prolactin release is probably high during this time (Eisner, 1960), such observations are consistent with the experimental evidence showing an antigonadal effect with this hormone. However, in the Rook, the ovaries of incubating females possess many large follicles, and gonadal regression does not appear to take place until after the young have hatched (Marshall and Coombs, 1957). Furthermore, Jones (1969a) has reported that prolactin only interrupts, but does not stop egg laying in *Lophortyx californicus* and does not effect follicular diameters or ovarian and oviductal weights. Similarly, Juhn and Harris (1956) found that prolactin blocked the antigonadal effect of progesterone in domestic fowl and only temporarily interrupted egg laying when given alone. Thus, in females, too, there is at present insufficient evidence to attribute a general ovarian regulatory role to prolactin.

2. Pituitary–Gonad Axis (See also Chapter 2)

Ablation of the anterior pituitary gland produces a regression in the female gonad and extensive follicular atresia (Hill and Parkes, 1934; Opel and Nalbandov, 1958, 1961a,b; Nalbandov, 1959). This atresia can be delayed for about 5 days by injections of mammalian gonadotropins (Nalbandov, 1953) and for a longer period by avian pituitary material (Opel and Nalbandov, 1961a). Furthermore, ovarian activity and a recovery of ovary-dependent organs can be induced in hypophysectomized hens in which postoperative regression has been allowed to develop, by injection of a crude avian pituitary preparation (Mitchell, 1967a,b). In this investigation, Mitchell (1967a) allowed a 10 day postoperative period for complete ovarian and oviduct regression; he then injected different groups of hypophysectomized birds with varying doses of powdered acetone-dried material suspended in isotonic saline. The results are summarized in Fig. 31.

There is a very close correlation between ovarian function and gonadotropic concentration of the pituitary gland, both in domestic species (Breneman, 1955) and in a seasonally breeding bird such as the White-crowned Sparrow (King *et al.*, 1966), and a similar relationship between pituitary gonadotropin content and ovarian function has also been shown in *Coturnix*, where gonadotropin potency

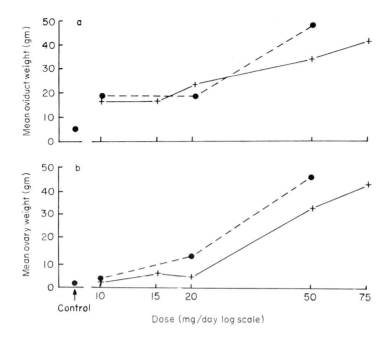

FIG. 31. Effects of chicken anterior pituitary extract on (a) oviduct weight and (b) ovarian weight of hypophysectomized hens, in which both organs had been allowed to become completely regressed before commencement of treatment; (●) treatment for 12 days; (+) treatment for 8 days. (From Mitchell, 1967a.)

and ovarian development undergo a parallel increase when birds are held under long daily photoperiods from the time of hatch (Tanaka *et al.*, 1965; Follett and Farner, 1966b).

Although female birds, like males, show a gonadotropic response to photostimulation, the ovary in most species can rarely be stimulated to full breeding maturity under experimental conditions. A block to growth occurs at the onset of vitellogenesis, presumably because other proximate factors are required (Polikarpova, 1940; Vaugien, 1948; Phillips and van Tienhoven, 1960; Lehrman, 1961; Farner *et al.*, 1966). Captivity itself is adequate to stop development, and captive female White-crowned Sparrows have pituitary glands whose gonadotropic potency is only about one-fourth that of reproductively active females in their natural habitat (King *et al.*, 1966). Similarly, a lack of gonadotropic potency has also been recorded in the pituitaries of captive wild Pintails *(Anas acuta)*, which failed to reproduce (Phillips and van Tienhoven, 1960). An exception appears to be domesticated strains of *Coturnix,* in which young females under long daily photoperiods can apparently develop to full reproductive condition (Follett and Farner, 1966a,b), a fact that enables the species to be utilized for commercial egg production.

As is also the case in male birds, practically all our knowledge on pituitary–gonad interrelationships is derived from studying the effects of exogenous purified mammalian gonadotropins or, as outlined above, by determining endogenous gonadotropic potencies of the anterior pituitary by assay methods that are not specific but generally measure the combined effects of FSH and LH. Investigations using purified avian FSH and LH have still to be done, and, therefore, caution must be exercised in extrapolating conclusions based on mammalian gonadotropins, particularly since there appear to be pharmacological and physiological differences between the avian and mammalian hormones. For example, follicular development cannot be stimulated in the ovaries of immature chicks by mammalian gonadotropins (Nalbandov and Card, 1946), yet crude avian pituitary extracts produce a marked stimulation (Das and Nalbandov, 1955). The lack of response to mammalian gonadotropins, however, alters with age, and in pullets nearing sexual maturity, exogenous hormone administration induces follicular development. Dessicated avian pituitary material is also more effective than purified mammalian gonadotropins in causing continued follicle development and ovulation in starved pullets (Morris and Nalbandov, 1961) and in stimulating precocious follicular development in the immature ovary (Taber *et al.,* 1958).

The differences between gonadotropins of avian origin and mammalian origin is further underlined by the fact that the partially purified avian hormones react poorly in mammalian assays (Herrick *et al.,* 1962; Furr and Cunningham, 1970; Licht and Stockell-Hartree, 1971). Interestingly, Licht and Stockell-Hartree (1971) have also tested avian FSH and LH in the lizard *Anolis carolinensis* and find that the relative potency is significantly greater than that estimated from standard rodent bioassays.

Bearing in mind the above reservations, there is good evidence that mammalian gonadotropins have considerable activity when injected into female birds, and ovarian stimulation as a consequence of such treatment has been demonstrated in a number of wild and domestic species (see reviews by Parkes and Marshall, 1960; van Tienhoven, 1961; Nalbandov, 1966). Furthermore, replacement therapy with mammalian gonadotropins is apparently effective in hypophysectomized pigeons (Chu and You, 1946). The evidence indicates that, as in mammals, FSH stimulates follicular development and estrogen secretion (as expressed by an increase in oviduct development) in a number of species (Bates *et al.*, 1935; Leblond, 1938; Witschi, 1955). Figure 32, for example, indicates the effect of daily injections of 0.1 mg/kg of mammalian FSH for 2 weeks into sexually regressed Green-winged Teal *(Anas crecca)* caught on their wintering grounds in Hong Kong. It can be seen that follicular development has been stimulated and there is a marked hypertrophy of the oviduct, indicating a stimulation of estrogenic secretion (Lofts, 1973). So far, the effects of injecting purified avian FSH have not been recorded, but a partially purified preparation injected into hypophysectomized hens maintained the ovary and oviduct weights at the control level when therapy commenced within an hour after the operation (Mitchell, 1970). There was also evidence that yolk deposition also continued after hypophysectomy in these birds. When hormone injections were delayed until 10–15 days after hypophysectomy, however, a resurgence of follicular development could not be induced at the dose level used (0.33, 0.5, and 0.75 mg/day). When avian FSII was tested on lizards, it stimulated a rapid development of the follicles and a significant increase in oviduct weight (Licht and Stockell-Hartree, 1971).

Nalbandov (1959) has pointed out that in birds, unlike most other vertebrate groups in which groups of follicles mature together and are destined to ovulate together, the growth and ovulation of follicles is continuous within the breeding season, and there are mechanisms that must prevent the maturation and ovulation of more than one egg at one time. Thus, the avian controlling mechanisms provide for the existence of a hierarchy of follicles of graded sizes, only one of which becomes ovulated at any one time and usually on any one day. There is considerable evidence that ovulation depends upon a cyclic release of LH, but ovulation will be induced only if the follicle has reached sufficient maturity to respond. Nalbandov (1961) has pointed out that the ovarian follicle is extremely resistant to the ovulatory action of exogenous LH until shortly before the expected release of endogenous LH, and in the domestic hen (Kappauf and Van Tienhoven, 1972) and also in the Japanese Quail (Follett *et al.*, 1972), there is good evidence that plasma LH levels display a significant peak some 6 to 8 hours prior to ovulation. The magnitude of the rise in plasma LH is much smaller than that found in mammals prior to ovulation, and may reflect an avian physiological adaptation, since a massive burst of LH might release more than one ovum from the ovary. Be-

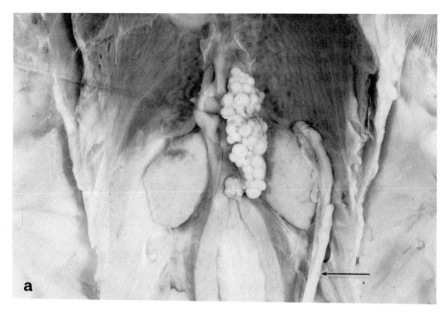

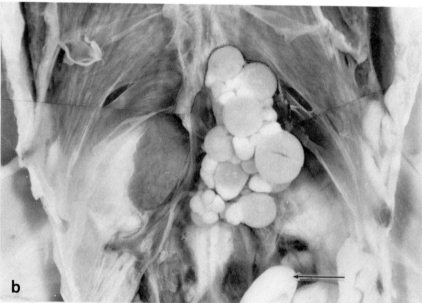

FIG. 32. (a) Ovary of Green-winged Teal (*Anas crecca*) in December. Note the small size of the follicles and the regressed condition of the oviduct (arrow). Actual size. (b) Ovary of December Teal given daily injections of 0.1 mg of mammalian FSH for 2 weeks. There has been a considerable stimulation of follicular development and the oviduct (arrow) is also hypertrophied. Actual size.

tween clutches the level of LH declines and becomes almost un-detectable (Heald *et al.*, 1968; Kappauf and van Tienhoven, 1972).

The release of LH from the anterior pituitary gland is under nervous control (see Chapter 4), and lesions placed in the ventral preoptic region of the hypothalamus block the ovulatory process (Ralph, 1959; Ralph and Fraps, 1959a; Egge and Chiasson, 1963; Novikov and Roudneva, 1964, 1969; Stetson, 1972b). Such lesions will also prevent progesterone-induced ovulations (Ralph and Fraps, 1959b), indicating that this steroid stimulates LH release via a hypothalamic route, and this is also indicated by the observation that electric stimulation of the preoptic area will induce a premature ovulation (Opel, 1964). There is an extensive literature dealing with how such a mechanism is involved in timing the egg-laying sequence of the domestic hen, and the reader is directed to the several excellent reviews for information about this topic (Nalbandov, 1959, 1962; Fraps, 1955, 1970; van Tienhoven, 1961).

IV. Accessory Sexual Organs

The accessory sexual organs are those parts of the reproductive system exclusive of the testis and ovary that are also vital to reproduction. These comprise the sperm ducts in the male and oviduct in the female, together with any special cloacal differentiation. The reproductive process may also be assisted by other differences between the sexes. For example, plumage color or pattern may differ, special plumes or wattles may become elaborated in one sex, bill color may change, or there can be voice differences. Such characters whereby the sexes differ are termed secondary sexual characters, the primary sexual characters (the gonads) obviously being excluded from this definition.

Two pairs of ducts, derived from primitive kidney ducts, develop in both sexes, these being the Müllerian and Wolffian ducts and each pair runs from the gonad to the urogenital sinus. In the male, the Wolffian ducts become the gonoducts draining the testes, and subsequently become the vasa deferentia and epididymides in the adult bird, while in the female they usually remain vestigial. In the female, the left Müllerian duct becomes the oviduct, while the right, and both these ducts in the male, remain vestigial; occasionally the ducts do not atrophy as appropriate. The occasional persistence of the right Müllerian duct in the female appears to be genetically controlled, since a high proportion of certain inbred strains of fowl exhibit two ducts instead of one (Morgan and Kohlmeyer, 1957).

A. The Male

Mature spermatozoa are released into the lumina of the spermatic tubules, which form the highly convoluted network of the testis. From here they are passed by a small number of tubules, the rete tubules, to the highly convoluted vasa efferentia (Fig. 33). Outside the breed-

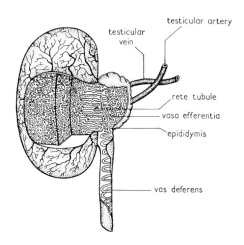

FIG. 33. Diagram of the internal network of the testis of the domestic fowl and its relationship with the efferent ducts. (From Marshall, 1961a.)

ing season, the rete tubules are inconspicuous, as they are sited in the tunica albuginea of the testis, but they do become enlarged and distinct as two to four tubes extending from the testis with the onset of reproductive activity (Bailey, 1953). In view of the fact that spermatozoa are rarely observed in the rete tubules, Bailey (1953) has concluded that their passage through this region is rapid and probably periodic. The spermatozoa next pass to thin tubules, the vasa efferentia, which also show some seasonal enlargement and which may become secretory. These leave the tunica albuginea and join to form a long coiled tube, the epididymis. This comprises a compact structure closely bound to the testis and lying adjacent to the kidney. With the onset of sexual activity, the epididymis hypertrophies (Fig. 34) to become a prominent white knob on the side of the testis. The cells of its epithelium are columnar and ciliated, and secrete a seminal fluid. Sperm are not stored in this organ but progress into a deferent duct, the vas deferens; two muscular tubes one from each of the paired epididymides therefore lead to the urodeum of the cloaca. These vasa

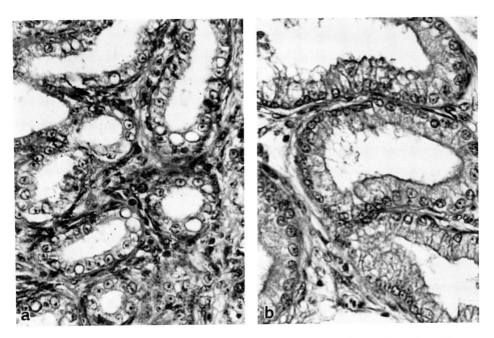

FIG. 34. (a) The epididymis of the European Tree Sparrow during December. The canals are collapsed and the lining epithelium is composed of cells that are cuboid in outline. ×500. (b) The epididymis during the breeding season. There has been a considerable enlargement of the canals, and the lining epithelium has hypertrophied so that the cells are now columnar in outline. ×500.

deferentia become enlarged, more convoluted, and distinct from the kidney during the breeding season, and before entering the cloaca are expanded to form a seminal sac. This increases thirty- to fortyfold in weight during the breeding season, becoming heavily charged with spermatozoa. A muscular sheath forces the posterior wall of the sperm sacs to expand into the cloaca as erectile papillae and doubtless facilitates the extrusion of sperm during copulation (see Nishiyama, 1955; Lake, 1957). However, these papillae are minute in most species, and during copulation the cloaca of each sex is everted and the papillae are brought into contact with the opening of the oviduct.

In some species part of the cloaca is modified as a penis-like structure. In waterfowl the penis is a vascularized sac which can be protruded by a muscle and retracted by a ligament. The ligament is bound spirally round the sac and gives the penis a twisted appearance. Sperm move along a spiral groove that passes from the cloacal papillae at the organ's base to its tip. Höhn (1960) has shown that the weight

of the penis of the Mallard *(Anas platyrhynchos)* increases sixfold
with the onset of the breeding season and that castration prevents
such development. Benoit (1936) had earlier noted that at puberty
the phallus became enlarged under the influence of testicular hormones, but outside the breeding season it did not regress again to its
juvenile dimensions but instead remained fairly constant in size.
Females of those species in which the male possess a "penis" have
a homologous but reduced structure termed the "clitoris." Their
cloacal opening is slightly modified to receive the intromissive organ
of the male.

Penis-like organs occur in tinamous (Tinamidae), curassows (Cra-
cidae), the ratite birds, and the ducks and geese (Anseriformes). Alone
among the Passeriformes, the Black Buffalo Weaver *(Bubalornis al-
birostris)* possesses an imperforate penis-like external organ. The
groups possessing these modified genitalia are diverse, and the
species appear to share no common feature that would indicate an
adaptive function for their special attributes. However, it may be significant that all the species concerned, except the ducks and geese,
habitually practice polygamy or polyandry. Thus, in tinamous, as in
nearly all other ratites, the male usually incubates the eggs and cares
for the young unaided by the female (Pearson, 1955). In the Slaty-
breasted Tinamou *(Crypturellus boucardi)*, the female lays an egg
for one male which incubates it and then, in turn, she moves on to
lay an egg for another male (Beebe, 1925). Conceivably, each male
must be ready to fertilize the female with a reduced opportunity for
precopulatory displays that would help synchronize the process, but
this view is speculative. The Ostrich *(Struthio camelus)*, Greater
Rhea *(Rhea americana)*, and other ratites, and such tinamous as *Nothocercus bonapartei* (Schafer, 1954) perform cooperative laying
whereby several females lay their eggs in the same nest that is then
attended solely by the male; in some cases the females may repeat
the performance for another male. Again selection may be favoring
more effective copulatory mechanisms in species that do not have
the benefit of a long period of intrapair courtship during which time
the endocrine state of the male can become adjusted (see below). The
ducks are also unusual in that pairing and copulation usually occur
on the water.

As already mentioned, seasonal changes in the size and degree of
development of the vasa deferentia, the epididymis, and the penis,
if present, parallel the seasonal changes observed in the primary sex
organs in those species showing cyclical activity; these structures

remain permanently developed in continuous breeding species such as the domestic fowl (Domm, 1939). This seasonal development apparently depends on androgenic hormones in the case of the vasa deferentia, epididymides (Witschi, 1945), and the seminal sacs (Riddle, 1927), and ablation of the testes will prevent the seasonal development of these organs in seasonally breeding species, whereas injections of testosterone propionate will induce their development into a fully breeding condition (Witschi, 1961). Some reservation is needed in that various pioneer experiments have not been repeated since purified hormones became available, and it is possible that early extracts were contaminated with gonadotropins; this reservation is occasioned by recent experiments into the factors inducing the black bill pigmentation of House Sparrows (see below). Moreover, in the case of the female canary, it seems that estrogenic hormones will induce oviduct development only in subjects kept on stimulatory photoperiods and not birds treated in winter (Hutchison, 1973). This suggests that the synergistic action of a gonadotropin is necessary, and it is likely that a parallel situation may apply in the case of the male. Critical experiments with hypophysectomized birds have not been performed.

Maximum development of the epididymides is noted at the beginning of the preincubation behavior cycle of the male pigeon, and the height of the columnar epithelial cells decreases up to the time of egg laying. If androgen-dependent, this change would indicate a declining androgen titer over the period leading up to egg laying, and for this there is other evidence (see Section V).

B. THE FEMALE

The right oviduct, like the right ovary, remains vestigial except in Raptores, and it is the left that becomes highly differentiated to satisfy the processes involved in producing the complex shelled and pigmented eggs. Even in species in which the incidence of two functionally developed ovaries is high, however, only one oviduct may develop. The developed oviduct is a tube attached to the body wall by dorsal and ventral ligaments and double folds of the peritoneum in which are situated a network of blood vessels; the oviduct is covered by the peritoneum. Its walls embody circular and longitudinal muscle fibers variously developed in different sections, and similarly, there is a variable development of a lining mucous membrane. In places, the membrane comprises longitudinal folds covered with a glandular and ciliated epithelium. Five anatomically distinguishable regions are discernible, namely, the infundibulum, magnum, isthmus, shell gland, and vagina. The extruded ovum is received from the ovary via the open anterior funnel-like infundibulum, which so surrounds the follicle that when the latter ruptures the ovum does not lay free in the body cavity. The ovum is also probably guided into the funnel by the ventral ligament. It is fertilized in the upper thinwalled neck of the oviduct, where the mucous membrane is manyfolded and before albumin is added to the yolk. In domestic fowls, fertile eggs can be laid 10–12 days following a single copulation, emphasizing the length of time that sperm can remain viable. However, fertility declines after about 5 days (Hammond, 1952), and it is a fact that wild birds normally copulate many times during the laying period. Polyspermy occurs as several spermatozoids enter the blastodics (12–25 in the pigeon), though only one unites with the female pronucleus. By the time the egg is laid, cell division causes the segmentation area or blastoderm to attain a diameter of nearly 5 mm in the domestic fowl.

The middle portion of the oviduct is the highly glandular magnum, where the white albumin is formed; this is deposited round the ovum in four layers to make a complex structure (see Romanoff and Romanoff, 1949). In the hen, this process takes from 3 to 4 hours to complete (Warren and Scott, 1935). The histological development of the oviduct has been studied in fowls (Richardson, 1935) and in a careful investigation of the domestic canary (Hutchison et al., 1968). While the ovaries comprise up to 1.5% of body weight of the canary, there is an approximately linear relationship between oviduct and ovary weights (both represented as percentages of total body weight), as might be suspected since oviduct development depends partly on ovarian hormones (see below). The formation of albumin coincides with the enlargement of one ovarian follicle more than the rest, and this also correlates with the completion of defeathering in the course of brood-patch development and intensive nest-building behavior. Relatively little oviduct development is apparent until nest-building activity begins; indeed only the division of epithelial cells of the magnum is to be noted prior to the collection of the first nest material. However, within the period of active nestbuilding, the oviduct increases markedly in length (decreasing again after egg laying), the epithelium divides and invaginates to form tubular glands, and these become distended and their cytoplasm charged with albumin granules; at the same time, the mucosa becomes very folded. The epithelium differentiates during this time into ciliated columnar cells and goblet cells. The histology of the magnum apparently remains uniform along its length at all stages of development.

After albumin has been deposited, the egg next passes into a less glandular and more muscular isthmus of the oviduct where the two shell membranes are secreted, these being comprised of felted protein fibers cemented together with albumin. The outermost membrane has a rough texture and so provides a key for the calcarious shell. This last, together with pigment, is added in the wider and expandable uterus, or shell gland. The shell is secreted in the form of calcium salts (about 97%) supported on an organic matrix of protein fibers. Any ground color is deposited during the final stages of shell formation, while blotches, streaks, or other surface markings are acquired after the shell has been completed. Because the egg is rotated on its long axis surface markings are often given a spiral configuration. The rotation is achieved by a powerful sphincter muscle, which when relaxed, allows the egg to pass into the cloaca and to be laid pointed end first. In the domestic fowl, about one-third of the eggs become reversed; the vagina bulges beyond the sphincter thereby turning the egg over, whereupon it is eventually laid blunt end first; whether this happens in other birds remains uncertain. It appears that this eventual oviposition is effected by a muscular contraction of the shell gland (uterus) regulated by the secretion of posterior pituitary hormones. Thus, a sudden decrease in the content of vasotocin in the neurohypophysis takes place in domestic hens (Tanaka and Nakajo, 1962a) and is correlated with a rise in plasma levels (Douglas and Sturkie, 1964; Sturkie and Lin, 1966), coincident with oviposition. Furthermore, a premature expulsion of the egg can be induced by exogenous oxytocin and vasopressin (Tanaka and Najajo, 1962b; Gilbert and Lake, 1963) and also vasotocin (Rzasa and Ewy, 1970), the latter being the most effective (Rzasa and Ewy, 1971).

The seasonal hypertrophy of the oviduct that occurs with sexual recrudescence (gonadotropins are implicated) is dependent on estrogen secretion by the ovary (Witschi and Fugo, 1940; Brant and Nalbandov, 1956; van Tienhoven, 1961) and estrogenic hormones have been shown to cause marked morphological changes in immature chick oviducts (Hertz et al., 1947) and to stimulate the differentiation of mature cell types (Kohler et al., 1969). But other hormones are involved in the full development of the tract, for progesterone and prolactin augment the increase in weight of the oviduct following estrogen injection in Ring Doves (Lehrman and Brody,

1957) and canaries (Hutchison *et al.*, 1967). In domestic fowls, estrogen alone will cause growth of the albumin-secreting glands and the secretion of ovalbumin. Daily injections (5 mg) of diethylstilbestrol into 4-day-old chicks causes a 300-fold increase in ovalbumin production within 15 days, and an increase from 5–10 mg to 1.5–3 gm in the oviduct weight is recorded after 3 weeks of such estrogen treatment (O'Malley *et al.*, 1967). Progesterone administration, on the other hand, appears to stimulate only the synthesis of the egg-white protein avidin (O'Malley, 1967; O'Malley *et al.*, 1967), and this has been shown to be effective, not only *in vivo* (Hertz *et al.*, 1943), but also *in vitro* when minced chicken oviduct tissue is incubated with progesterone (O'Malley, 1967). At the cellular level, O'Malley and his colleagues have demonstrated that estrogen treatment stimulates a large increase in nuclear transfer RNA and a definite but smaller increase in cytoplasmic transfer RNA (Dingman *et al.*, 1969).

C. Secondary Sexual Characters

Darwin appreciated that the physical differences between sexes in plumage and voice or their possession of various adornments were associated with courtship and that sexual selection depends on the success of certain individuals over those of the same sex in terms of producing surviving progeny. Because females normally choose their partner, and frequently rely on visual cues, isolating signal characters have evolved in many male birds to prevent confusion with, and hence hybridization between, closely related species. The avoidance of such interspecific competition for mates is to be distinguished from intraspecific competition for a mate, which, as Huxley (1938) recognized, leads to the maximum elaboration of display plumage in polygamous species.

Where a permanent plumage difference between the sexes imposes no other disadvantage, sexual dimorphism may have a genetic basis, so that hormones are unable to change the inherited sex type. Examples are provided by the House Sparrow *(Passer domesticus)* and European Bullfinch *(Pyrrhula pyrrhula).*

Witschi (1961) has provided examples of the endocrine basis of various secondary sexual plumage changes and acquisition of other nuptial adornments. Several of these variations have served for the bioassay of hormones. The development of the comb and wattles of newly hatched chicks, irrespective of sex, depends on androgen and can be stimulated by injections of exogenous testosterone. The pigmentation of the bill of birds is of two main kinds: (1) that produced by deposition of brown and black melanins in melanophores that become injected into those epidermal cells moving out from the area of proliferation and (2) yellow, red, and orange carotenoids that are not synthesized by the bird but are absorbed from the diet. The European Starling has a black bill during the contranuptial season, while both sexes acquire a yellow bill with onset of the breeding season. Injections of androgens induce the yellow color, but estrogen and progesterone do not; castration causes the loss of the yellow color (Witschi, 1961). In *Quelea quelea,* the bills of both sexes are red during the contranuptial season, but in females this color changes to a straw color during the breeding season and is an estrogen-dependent response.

The bill of the House Sparrow *(Passer domesticus)* turns black during the breeding season under the supposed influence of androgen; castration results in a loss of pigmentation in subjects with black bills (Witschi, 1961). Testosterone propionate, however, will not induce a black pigmentation in sparrows captured in winter and held under nonstimulatory day lengths, suggesting the implication of gonadotropins in the response (Fig. 35). Exogenous LH plus FSH (Fig. 35b) or androgen plus FSH (Fig. 35c) will induce a black bill pigmentation in subjects kept on short day lengths so that endogenous gonadotropin secretion is inhibited (Lofts *et al.,* 1973). Color change in the beak of the Paradise Whydah *(Vidua paradisaea)* apparently depends on LH and not gonadal hormones.

The assumption of secondary sexual plumage seems in many cases to depend on steroid hormones; for instance, castrated Ruffs fail to don the pectoral display plumes described above (van Oordt and Junge, 1936). Similarly, castrated male Black-headed Gulls *(Larus ridibundus)* fail to acquire the brown head typical of the breeding season (van Oordt and Junge, 1933). In phalaropes the female dons the more colorful breeding plumage, and in both Wilson's Phalarope *(Phalaropus tricolor)* and the Red-necked Phalarope *(Phalaropus lobatus),* it has been shown that the assumption of this plumage is dependent upon androgenic steroids. Thus, administration of testosterone propionate (but not estradiol or prolactin) to birds of either sex will stimulate the growth of feathers of the nuptial type in plucked areas (Johns, 1964). The reason that this colorful plumage is normally shown by the female and not the male can be attributed to the fact that there is a higher androgen secretion in the female (Höhn and Cheng, 1967; Höhn, 1970), the reverse of the situation in nearly all other species. In contrast to the mechanism in the preceding species,

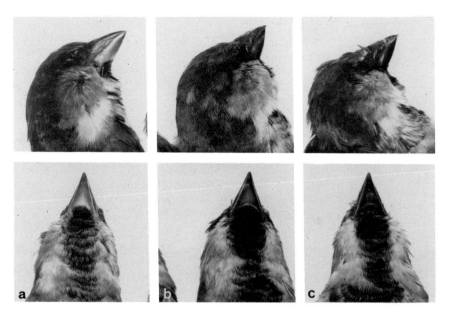

FIG. 35. (a) Pale yellow bill of the European House Sparrow treated with testosterone propionate during winter time and maintained on a 7 hour photoperiod. (b) Winter sparrow treated with exogenous LH plus FSH. The bill has turned black even though specimens were maintained under a 7 hour winter photoperiod. (c) Winter sparrow treated with testosterone propionate plus FSH. Again, the bill has turned black although androgen alone failed to stimulate the pigmentation [cf. (a)].

in which the acquisition of the nuptial plumage is apparently determined by the elaboration of gonadal androgens, the male-type plumage of the weaver *Euplectes orix* has long been known to depend on a gonadotropin, supposedly LH, and the species has long served as the bioassay animal for this hormone (Witschi, 1961). In this, and other weaver birds, a castrated male continues to develop a nuptial plumage. Females do not normally develop a gaudy plumage, but will do so after ovariectomy. In the natural condition ovarian estrogens inhibit potentiality of LH for stimulating nuptial plumage, and this is confirmed by the injection of estrogen into male weavers, which inhibits their assumption of the nuptial dress. Exogenous estrogens have a similar effect when injected into male fowls. Many of these experiments, however, were performed before purified hormones became available. Furthermore, because subjects were often held under photostimulatory day lengths, some reservation is needed in accepting previous conclusions about many of the supposed hormonal mechanisms governing secondary sexual characters.

D. INCUBATION PATCHES

In many avian species (but not all) the ventral apterium becomes defeathered and highly vascular and edematous (Fig. 36) before or during egg laying (Bailey, 1952). Hyperplasia of the epidermis also occurs, producing up to a sixfold increase in thickness (Selander and Yang, 1966). These so-called incubation patches, or brood patches, are in close contact with the eggs during incubation and supply the necessary warmth for development of the embryos. In addition, such a structure provides a sensitive surface in contact with the nest and eggs, which in the domestic canary at least, is a source of tactile stimuli influencing nest-building behavior (Hinde and Steel, 1962; Hinde, 1967; see also Section V). The sensitivity increases as the patch develops (Steel and Hinde, 1964).

Incubation patches usually develop in the females only (e.g., House Sparrow, White-crowned Sparrow), but in some species they may occur in both sexes (e.g., starlings and swallows). In the Red-necked Phalarope and Wilson's Phalarope, they develop only in the males, which in these species incubate the eggs, the female taking no part in this activity. Generally, there seems to be little correlation between the incubation habits of male birds and the possession of a brood patch, since the incubating males of some species, such as the House Sparrow (Selander and Yang, 1966) and Bushtit *(Psaltriparus minimus)* (Yapp, 1970), lack it, whereas the nonincubating cocks of others, such as the American genus of flycatchers *(Empidonax)*, possess it (Skutch, 1957; Yapp, 1970).

The chronology of the morphological changes that constitute the formation of an incubation patch appears to differ in various species. In the passerine species, defeathering occurs before egg laying (Bailey, 1952; Hinde, 1962; Selander and Kuich, 1963; Selander and Yang, 1966), but the timing of patch development in galliformes appears to be more retarded, so that defeathering becomes marked in Ring-necked Pheasants only after the laying of the fourth egg (Breitenbach *et al.*, 1965), and in California Quail after the tenth egg (Jones, 1969b). In *Zonotrichia leucophrys*, Bailey (1952) reports that defeathering from the ventral apterium is completed several days before oviposition, and increased vascularization and edema is not apparent until completion of this stage, but in the Red-winged Blackbird *(Agelaius phoeniceus)* (Selander and Kuich, 1963), Bank Swallow (Peterson, 1955), canary (Hinde, 1962; Steel and Hinde, 1963, 1964), and House Sparrow (Selander and Yang, 1966), epidermal hyperplasia, increased vascularization, and mild edema occur before defeathering is com-

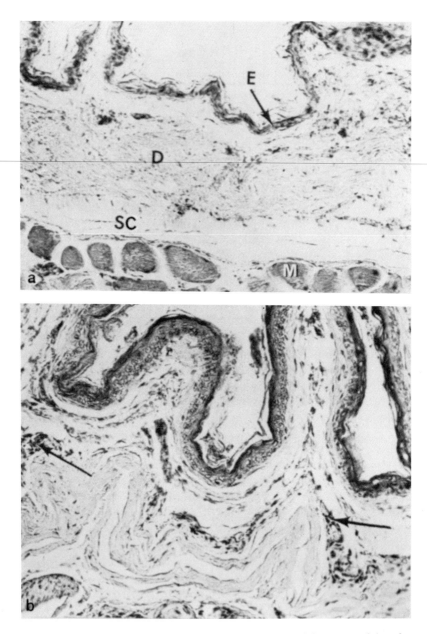

FIG. 36. (a) Ventral abdominal skin of a nonbreeding California Quail (*Lophortyx californicus*). Epidermis (E), dermis (D), subdermis (SC), subcutaneous muscle (M). × 290. (b) Ventral abdominal skin of incubating female showing hypertrophy and increased vascularization of the incubation patch. Note edema as evidenced by increased interstitial area in the dermis and leukocytes near blood vessels (arrows). × 290. (From Jones, 1960b.)

plete and, in some birds, begins weeks before egg laying. Thus, in both the European Starling (Lloyd, 1965) and House Sparrow (Selander and Yang, 1966) an appreciable increase in the numbers of dermal blood vessels is noticeable a month before oviposition. The increase in vessel diameter, however, occurs mainly during the egg-laying period.

The development of the incubation patch is under endocrine control and involves hormones of both gonadal and adenohypophyseal origin. Generally, it is formed under the influence of estrogen and prolactin, which are synergistic in their effects. In the passerines, exogenous estrogen administration has been shown to induce increased vascularization and defeathering in both intact (Bailey, 1952; Steel and Hinde, 1963; Selander and Yang, 1966) and ovariectomized (Steel and Hinde, 1964) birds, but similar treatment of hypophysectomized birds produces vascularization only (Bailey, 1952). When estradiol in combination with prolactin is given to hypophysectomized specimens, the complete patch is developed (Bailey, 1952; Hinde, 1967); in hypophysectomized *Zonotrichia* prolactin alone has no effect. Since the effect of estrogen on defeathering is augmented by prolactin, and the combination of these two hormones is necessary to produce full development in hypophysectomized birds, it seems likely that, in female passerines at least, this phenomenon is normally dependent upon the endogenous secretion of both these hormones. In California Quail, however, prolactin appears to play a more significant role than in the passerine mechanism, since Jones (1969b) reports that in this species injection of estradiol into reproductively active females fails to have any significant effect. Prolactin alone, on the other hand, stimulates epidermal hyperplasia, and when given in combination with estrogen produces full incubation patch development (i.e., defeathering, edema, vascularization, and dermal thickening). In this species, and also in the Bobwhite Quail (*Colinus virginianus*), prolactin-induced hyperplasia of the incubation patch can be stimulated not only *in vivo* but *in vitro* as well (Jones, 1969c). It seems that only the ventral skin is responsive to hormonal effects, since Jones *et al.* (1970) have shown that estrogen plus prolactin induces the ventral abdominal skin of the chicken to undergo hypervascularization and epidermal hyperplasia only if the skin is *in situ*. Ventral skin transplanted to the back exhibited epidermal hyperplasia but not hypervascularization. Dorsal skin was not responsive *in situ* nor when transplanted to the ventrum. Thus, hypervascularization of the chicken incubation patch is site-specific, but the epidermal hyperplasia is not. Furthermore, the data show that epidermal hyperplasia is not dependent on hypervascularization.

Progesterone has been found to augment the effects of estrogen in inducing defeathering in the female canary (Steel and Hinde, 1963) but is not essential for this process, since it can be induced by estrogen alone in ovariectomized birds (Steel and Hinde, 1964). When progesterone is given alone it has no effect on vascularization in canaries, House Sparrows, or California Quail, and in the last species it is also less effective as a synergist with estrogen (Jones, 1969b). In the canary, this hormone has been found to be necessary for the characteristic increase in sensitivity of the ventral skin during development of the incubation patch (Hinde et al., 1963; Hinde and Steel, 1964; Hutchison et al., 1967).

In *Phalaropus lobatus, P. tricolor,* and *Lophortyx californicus,* in which the males develop incubation patches, it has been shown that exogenous testosterone plus prolactin can induce incubation-patch development (Johns and Pfeiffer, 1963; Jones, 1969b). In the natural situation it seems likely that in such species, vascularization occurs when circulating androgen levels are high, and before defeathering of the patches is completed, the latter event being delayed until circulating prolactin levels increase, probably at the time of testicular regression. In the passerine species in which males usually do not incubate, exogenous androgen plus prolactin do not cause such feather loss in the cock, and in the parasitic Brown-headed Cowbird *(Molothrus ater),* a species that does not incubate its own eggs or develop a brood patch, the ventral skin of either sex is insensitive to estrogen–progesterone injections (Selander, 1960; Selander and Kuich, 1963).

In the Red-winged Blackbird, it has been shown that the experimentally induced patch (by exogenous estrogen or estrogen plus prolactin) differs histologically from those of wild incubating birds (Selander and Kuich, 1963), and in the House Sparrow, Selander and Yang (1966) have shown that the normal patch is more edematous than the hormonally induced ones, and it may be that the sensitivity to tactile stimulation that has been demonstrated by Hinde and his colleagues may be necessary for the complete development of the patch.

V. Breeding Behavior (Endocrinological Basis)

As already made clear in this chapter, gametogenesis depends on a changing pattern of hormone production culminating in the animal entering the phase of overt reproductive activity. At this stage, various behaviors become manifest to assist the process of breeding. These include the acquisition of a territory and its advertisement by song, the attraction and pairing with a mate, and the initiation of displays that lead to egg laying and nestling production. Increasing titers of androgenic or estrogenic hormones have long been assumed to be responsible for many sexual behaviors in male and female, respectively, but it is becoming increasingly apparent that the endocrine basis of reproductive behavior is complex and often highly specific (see review by Eisner, 1960). Research has been concentrated on various pigeon species and the canary, and while these are possibly reasonably representative of the class Aves, cognizance must be made of the possibility that other species have evolved special endocrine adaptations to suit their own peculiar ecological requirements.

Unpaired, yet sexually mature, male feral pigeons will usually advance with the bowing display toward other members of the same species (Fabricius and Jansson, 1963; Murton *et al.*, 1969b). Sometimes, an otherwise responsive male will fail to react to a receptive female but respond to other females, and it appears that the birds can recognize particular plumage morphs with which they are reluctant to pair. In free-living populations, this results in certain pairings of like individuals occurring with less than expected frequency, and the aversion appears to function as an out-breeding mechanism preventing the production of homozygous birds that have a reduced fertility; the gene concerned improves fertility and lengthens the breeding season if present in the heterozygous condition (Murton, 1973). In Ring Doves, much variation in the range of displays exhibited by males during the initial courtship phase could be attributed to their previous experience; an experienced bird might reduce or omit certain of the more aggressive postures (Hutchison, 1970a). Previous breeding experience by either male or female contributes to and improves the reproductive performance of Ring Doves under captive conditions (Lehrman and Wortis, 1967). Fundamental endocrine–behavioral interactions of this kind doubtless underlie the general fact that under field conditions older and more experienced birds reproduce more successfully.

There is good evidence that androgens mediate bowing behavior in both male Ring Doves (Erickson and Lehrman, 1964; Erickson *et al.*, 1967) and also feral pigeons (Murton *et al.*, 1969b), and they restore male-type behavior to castrates (Beach, 1948; Collias, 1950), increase social status and aggressiveness (Bennett, 1940), and induce malelike behavior in females (Leonard, 1939; Shoemaker, 1939). In the phalaropes, in which there appears to be a reversal of the normal situation and females display more aggressive patterns than the male, Höhn and Cheng (1967) have shown an unusually high ovarian testosterone content. Both Goodwin (1967) and Davies (1970) have shown that the bowing display differs from species to species, exhibiting a typical intensity in each. Davies (1970) who also examined F_1 hybrids of *Streptopelia* doves showed that these displayed with a characteristic bowing pattern that depended on inherited components of the display. In particular, the rate of bowing proved to be species-specific, while components of the bowing vocalization were similarly inherited in the same manner as morphological characters (Lade and Thorpe, 1964).

Bowing appears to function as a ritualized appeasement display; it combines crouched submissive postures (at termination of downward bow) with tail-fanning and head-lowering reminiscent of and probably derived from braking movements. However, the upright and advancing components of the display sequence embodying an almost "goose-stepping" gait represent aggressive components. The ambivalent nature of the display is indicated by the fact that if a bowing male catches up with the subject of his attention, he may (1) peck aggressively, (2) mount and attempt to copulate, or (3) turn and flee. Thus, the display involves causal factors for attacking, fleeing, and mating, and it is significant that different exogenous hormones injected into males at the beginning of their preincubation behavior cycle will alter the relative importance of these various display components (Murton *et al.*, 1969b). Exogenous FSH increases the incidence of driving, an aggressive display, and facilitates the appearance of sexual–aggressive components generally, while exogenous androgen facilitates a more natural sequence with bowing progressing to the more sexual manifestations of mounting and copulation. Hutchison's data (1970b) indicate that the chasing component of the display persisted in castrates, whereas bowing was eliminated, but the extent to which exogenous FSH will evoke displays in castrates has yet to be determined. It does seem probable that it may act synergistically with androgen. Moreover, on the basis of behavioral evidence, it is feasible that FSH is involved in the release of androgen (whether from the Leydig cells or Sertoli cells is uncertain; see Section II,B,3). Exogenous LH had little effect on behavioral sequences in this species, but it did increase the probability that attacking would follow bowing. In *Quelea quelea*, although androgen injections would not increase the social status of low rank birds kept in small groups, exogenous LH would (Crook and Butterfield, 1968).

Similarly, LH and not androgen seems to determine social status in *Sturnus vulgaris* (Davis, 1957, 1963; Mathewson, 1961).

Natural selection will favor males who can fertilize a potentially good mother, while females must ensure they have adequate protection from a mate and territory before accepting a male's advances. Initially the male's behavior is aggressive and centers on depositing his sperm in the female before another male can do so; hence copulation occurs early in the behavior sequence when aggressive behavior is paramount. Females will usually squat for a male only in a territory in which they also have become accepted, but this does not prevent males making sexual advances to all potential mates when on the feeding grounds and away from a territory. So, selection tends to advance the time of copulation for the male, retard it for the female, and results in a compromise balance. Having fertilized a female, a male must now ensure adequate protection for the forthcoming eggs and young; and to a large degree, the remaining needs of his courtship sequence are, so to speak, to mollify his partner and provide her with suitable nesting facilities. But it appears that the male is initially endowed with high titers of plasma androgen (and probably also of FSH) which would be detrimental to the progression of the cycle. In fact, it appears that feedback mechanisms become initiated in consequence of courtship and lead to a reduction in androgen levels.

It is known that implanted testosterone propionate in the preoptic region of the hypothalamus reestablishes copulatory behavior in capons so long as the necessary external stimuli are provided (Barfield, 1969), and such work is beginning to define the hormone-sensitive regions of the central nervous system which must underlie sexual behavior. This led Hutchison (1970b) to examine whether other sexual patterns of courtship preceding copulation might be linked to androgen-sensitive areas of the central nervous system, especially since there had been a suggestion that aggressive and copulatory activity might depend on separate androgen-dependent mechanisms in the central nervous system (Lisk, 1967). Hutchison, therefore, investigated the effects of intracerebral implants of crystalline testosterone propionate on castrated Ring Doves; he was careful to segregate his experimental subjects according to their experience and preparedness to perform the nest-demonstration display as a sequel to the bowing display. The type of courtship evoked and whether fragmentary or the complete sequence of chasing, bowing, and nest soliciting (nest demonstration) depended on the anatomical position of the implant. Only those males with anterior hypothalamic and preoptic implants showed the complete reestablishment of bowing, while implants in the area basalis and hypothalamicus posterior medialis resulted in maximum nest soliciting. Hutchison's data (1970b) are consistent with the hypothesis that the stimulatory effects resulting from posterior hypothalamic implants depended on the diffusion in low concentration of testosterone to an androgen-sensitive area in the preoptic and anterior hypothalamic area; i.e., mechanisms underlying chasing and bowing require higher hypothalamic concentrations of androgen than mechanisms underlying nest soliciting. This evidence does not support the view that different androgen centers control different behaviors (cf. Lisk, 1967; Erickson *et al.*, 1967). Nonetheless, separate sites are possibly involved if a wider range of behavior is considered. Electrical stimulation of the preoptic and anterior hypothalamic areas of the forebrain of pigeons induced an intensive bowing with all normal components, and this display could then become transformed into nest demonstrations via displacement preening (Åkerman, 1966a). Defense and escape behavior in the same species was induced by electrical stimulation of di- and telencephalic structures in the forebrain (Åkerman, 1966b).

Following castration, bowing and chasing are the first displays to be eliminated while nest demonstration persists for longer, suggesting that it depends on a lower threshold of androgen stimulation (Hutchison, 1970b). Indeed, Hutchison (1970b), using different sized hypothalamic testosterone implants, found that high-diffusion implants, delivering high concentrations, resulted in longer durations of bowing than smaller medium-diffusion implants, while low-diffusion implants induced nest soliciting in the virtual absence of chasing and complete absence of bowing. This at first suggests a simple and plausible explanation for the mechanism underlying the natural display sequence. But Murton *et al.* (1969b) also found that nest demonstration became fully established, albeit after a delay, in subjects maintained on high levels of exogenous testosterone treatment, so that these high dosages did not inhibit the onset of nest demonstration. Instead, these data suggest that a new mechanism becomes activated that can nullify the effects of exogenous androgen. The exogenous administration of estradiol monobenzoate immediately and consistently caused the appearance of nest-demonstration displays in intact feral pigeons (Murton *et al.*, 1969b) and also in Ring Doves (Vowles and Harwood, 1966), while hypothalamic implants similarly induced long durations of nest demonstration in castrated doves (Hutchison, 1970b). We do not know whether estrogen is involved in the normal cycle of the male, but the capacity to synthesize estrogenic steroids by avian testicular tissue is well established (Höhn and Cheng, 1967; Höhn, 1970).

There is other evidence that androgen levels may decrease during the preincubation display phase of the feral pigeon. Lofts *et al.* (1973b) incubated testicular and ovarian material taken from feral pigeons at different stages of the reproductive cycle with tritiated pregnenolone as the steroid precursor. The synthesizing ability of the gonads at different stages was then measured using thin-layer chromatography to separate the different steroids produced. During the initial period of bowing display, there was a peak in androgen synthesis (i.e., during the first 3 days of the cycle), but approximately 10 days later and just prior to egg laying, androgen synthesis was markedly reduced (Fig. 37). An increase in progesterone levels was also noted at this time. The capacity of the Leydig cell to synthesize a substance need not imply that the same substance is released from the cell, but it is almost certainly significant that quite separate evidence points to the existence of a high plasma titer of progesterone prior to egg laying in Ring Doves. Thus, males of this species that have previous breeding experience can be induced to incubate eggs by injections of progesterone (Lehrman, 1958a) or, in a smaller percentage of cases, following prolactin administration (Lehrman and Brody, 1961).

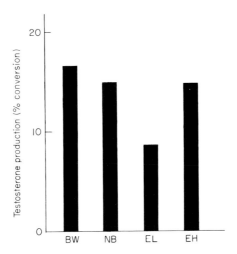

FIG. 37. The *in vitro* production of testosterone from radioactive pregnenolone by the pigeon testis during the bowing (BW), nest building (NB), egg laying (EL) and egg hatching (EH) periods of the reproductive behavior cycle. (From Lofts *et al.*, 1973b.)

All the same, when progesterone-injected and prolactin-injected doves were tested in bisexual pairs, more incubated than when tested alone, so the presence and behavior of a mate significantly influences hormone-induced incubation responses (Bruder and Lehrman, 1967; see also Lott and Brody, 1966; Lott et al., 1967).

Progesterone crystals chronically implanted into the brains of reproductively experienced Ring Doves induced incubation behavior if placed in the preoptic nuclei and lateral forebrain, while the sexual and aggressive components of male courtship were suppressed (Komisaruk, 1967). Male castrates, caused to exhibit courtship displays by treatment with testosterone, were selectively inhibited from bowing and induced to display nest soliciting alone when simultaneously injected with progesterone (Erickson et al., 1967). So it seems conceivable that rising progesterone levels could inhibit androgen effects leading to the suppression of sexual–aggressive displays and instead allow the emergence of pair-cementing displays, such as mutual allopreening (caressing) with the onset of nest building and egg laying. But we must still not neglect the possibility that an estrogenic phase occurs in the male pigeon during the middle part of the pre-incubation cycle, and certainly exogenous estrogen is more effective than exogenous progesterone in causing nest demonstration (Fig. 38).

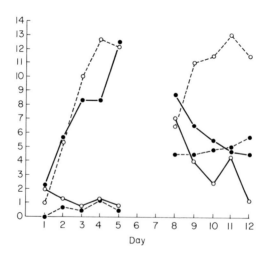

FIG. 38. Daily incidence of nest demonstration by paired male pigeons subjected to different hormone treatments. Birds were paired for 30 minutes per day beginning on day 1. The scores represent the mean number of 2 minute subperiods during which the behavior was recorded. Controls (●——●), estrogen (o---o), prolactin (o——o), progesterone (●---●). (From Murton et al., 1969b.)

A changed endocrine state during the preincubation cycle must depend either on exhaustion of a mechanism already engaged, which seems unlikely, or on the specific stimulation of neural–hormonal pathways elicited by specific behavioral events in the cycle. This second proposition is more likely in allowing for changes consequent to the natural or artificial disruption of the cycle. The preferential production of progesterone, for example, must depend initially on neural or hormone mechanisms initiated at higher levels of integration. We do not yet have the means to measure plasma FSH levels accurately and can only suspect that FSH titers decline during the preincubation behavior phase. The spermatogenetic sequence in feral pigeons shows that numerous spermatozoa are free in the tubule lumen at the start of the cycle and there exists a peak in the number of spermatids. By the middle phase, the spermatozoa lying free in the tubule lumina have disappeared, whereupon the existing spermatids mature into a new generation of spermatozoa at about the time of egg laying. Thereafter, there is a new wave of spermatogenetic activity, with numerous divisions of primary spermatocytes leading to a new production of spermatids during late incubation (Murton, 1973). Plasma LH titers have also been measured using a radioimmunoassay technique, and they also show a fluctuating titer during the preincubation period. Whatever the final answer proves to be, it seems highly probable that a changed pattern of gonadotropin secretion might be elicited by specific behaviors or other external stimuli and that a variable and probably synergistic balance of gonadotropin output would then result in a differential steroid production by the gonads.

Craig (1911) was the first to demonstrate that the caressing display of the male pigeon could facilitate ovulation in the female. Matthews (1939) later showed, by separating males and females with a glass plate, that the stimulus causing ovulation was visual and not tactile. Subsequently, Erickson and Lehrman (1964) found that castrated males separated from females by a glass plate were far less effective in stimulating oviduct development in the female than were intact birds. The elegant work of Lehrman and his colleagues has now well established that stimuli arising from association with a mate cause the birds to become interested in nest building, and that once in this state the stimuli produced by the presence of nest material facilitate the birds becoming interested in sitting on eggs (Lehrman, 1958b; Lehrman et al., 1961).

The behavioral acts involved in nest building by the male stimulate gonadotropin secretion (presumably FSH) in the female, causing gonadal hormone secretion from the ovary, which in turn leads to oviduct development. Thus, in both pigeons and canaries, nest-building activity is temporally associated with rapid oviduct development and ovulation followed a few days later by oviposition and the onset of incubation (Lehrman *et al.*, 1961; Warren and Hinde, 1961; Hinde, 1967). In both doves (Lehrman and Brody, 1957) and canaries (Steel and Hinde, 1963; Hutchison *et al.*, 1967), estrogen causes a moderate increase in oviduct weight, and whereas progesterone alone has no effect, it potentiates the effects of estrogen to produce maximum oviduct development. Prolactin can also augment the estrogen response, possibly by causing progesterone release. But while exogenous estrogens readily cause oviduct development (see, also, Brant and Nalbandov, 1956; van Tienhoven, 1961), very high doses are necessary to initiate actual nest building in doves (Lehrman, 1958a) and canaries (Warren and Hinde, 1959). Paired and photostimulated canaries in winter show nest-building behavior (Steel and Hinde, 1966a), while birds injected with PMS (pregnant mare serum) can lay without nest building (Steel and Hinde, 1966b). These observations support the likely situation that gonadotropin secretions potentiate nest-building behavior, which exogenous progesterone or prolactin certainly do not.

In Section IV,D, the development of incubation patches by the synergistic effects of estrogen and prolactin was discussed. In the female canary, incubation-patch development occurs coincidentally with nest-building and oviduct development (Hinde, 1967). Lehrman's studies (see 1965) make it clear for doves that exogenous prolactin, unlike progesterone, will not initiate incubation behavior but that prolactin can maintain such behavior once initiated (Lehrman and Brody, 1961, 1964). Moreover, judging from the development of the crop gland in pigeons, prolactin secretion is stimulated by the specific act of incubating eggs (see Riddle and Dykshorn, 1932; Patel, 1936; Lehrman, 1955, 1958b), though prolactin secretion can be induced in the male if he can see the incubating female and if he has previously associated freely with her during earlier courtship (Friedman and Lehrman, 1968). The appearance (artificially) of squabs once the female has laid eggs can also accelerate crop growth and, presumably, prolactin secretion (Hansen, 1966). The pituitary glands of domestic hens contain prolactin only so long as they are allowed to sit on eggs (Saeki and Tanabe, 1955).

To summarize, the preincubation reproductive cycle of the female appears to involve a period of increasing estrogen production, which it is tempting to attribute to increases in FSH production. Next follows a phase during which progesterone or prolactin secretion, or both, become involved, and the important endocrine problem needing to be resolved is whether increasing estrogen levels give rise to progesterone production which in turn stimulates prolactin secretion, or whether prolactin secretion is first stimulated leading in turn to progesterone production. Since Meites and Turner (1947) found no increase in pituitary prolactin level following progesterone injection, and since Lehrman's studies (1963) reveal no crop sac development in pigeons following progesterone administration, we may prefer the second alternative or acknowledge the possibility that nonmeasurable levels of prolactin were involved.

In the male feral pigeon, administration of exogenous estrogen causes (1) a regression of the testis tubules with only spermatogonia remaining present; (2) the encapsulation of the old Leydig cell generation, which becomes densely lipoidal; and (3) the appearance of a new Leydig cell generation, also heavily lipoidal (Section II,C,1). Coincidentally, plasma LH titers are markedly elevated, and it seems likely that the histological manifestations depend on the known elevation of LH coincident with a presumed suppression of FSH activity. In the male feral pigeon, LH titers increase during the end phase of the prelaying cycle and after the emergence of nest-soliciting displays (which are markedly enhanced by estrogenic hormone, see Fig. 38). While it has yet to be resolved whether estrogen is liberated during the male's prelaying cycle (though it is known that the testis can produce estrogen), it does seem reasonable to anticipate that rising estrogen production in the female suppresses FSH activity and stimulates LH production (the Holweg effect) and that such a change in gonadotropin balance provides the necessary conditions for progesterone (prolactin) secretion. It is of course known that there is a boost of LH production associated with ovulation in the domestic hen (Tanaka et al., 1966; see, also, Fraps, 1965). Analogous endocrine changes apparently occur in both male and female, particularly at the hypothalamic level, though the male cycle is advanced compared with that of the female. In the male, an initial FSH androgenic phase gives way to an LH and progesterone phase leading to incubation. In the female an FSH–estrogen phase is stimulated in consequence of the males' courtship, and this leads to oviduct development relatively late in the cycle; it is probably significant that in feral pigeons the female assumes the more aggressive and dominant role coincident with the nest-building phase, and she may even drive the male away at this stage of the cycle (Fabricius and Jansson, 1963; Murton et al.,

1969b). Significant developments in this field can be expected in the next few years.

REFERENCES

Adams, J. L. (1956). A comparison of different methods of progesterone administration to the fowl in affecting egg production and molt. *Poultry Sci.* **35**, 323–326.

Åkerman, B. (1966a). Behavioural effects of electrical stimulation in the forebrain of the pigeon. I. Reproductive behaviour. *Behaviour* **26**, 323–338.

Åkerman, B. (1966b). Behavioural effects of electrical stimulation in the forebrain of the pigeon. II. Protective behaviour. *Behaviour* **26**, 339–350.

Albert, A. (1961). The mammalian testis. In "Sex and Internal Secretions" (W. C. Young, ed.), 3rd ed., Vol. 1. Williams & Wilkins, Baltimore, Maryland.

Alfert, M., Bern, H. A., and Kahn, R. H. (1955). Hormonal influence on nuclear synthesis. 4. Karyometric and microspectrophotometric studies of rat thyroid nuclei in different functional states. *Acta Anat.* **22**, 185–205.

Arvy, L. (1962). Présence d'une activité steroïdo-3-β-ol-dehydrogénasique chez quelques Saurosidés. *C. R. Acad. Sci.* **255**, 1803–1804.

Arvy, L., and Hadjiisky, P. (1970). Histoenzymologie des activités deshydrogénasiques dans l'ovaire de *Coturnix coturnix* L. et de *Gallus gallus* L. (Phasianides, Galliformes). *C. R. Ass. Anat.* **54**, 50–62.

Aschoff, J. (1955). Jahresperiodik der Fortpflanzung beim Warmblütern. *Stud. Gen.* **8**, 742–776.

Assenmacher, I. (1958). Recherches sur le contrôle hypothalamique de la fonction gonadotrope préhypophysaire chez le canard. *Arch. Anat. Microsc. Morphol. Exp.* **47**, 448–572.

Bagshawe, K. D., Orr, A. H., and Godden, J. (1968). Cross-reaction in radioimmunoassay between human chorionic gonadotrophin and plasma from various species. *J. Endocrinol.* **42**, 513–518.

Bailey, R. E. (1950). Inhibition of light-induced gonad increase in White-crowned Sparrows. *Condor* **52**, 247–251.

Bailey, R. E. (1952). The incubation patch of passerine birds. *Condor* **54**, 121–136.

Bailey, R. E. (1953). Accessory reproductive organs of male fringillid birds. Seasonal variations and response to various sex hormones. *Anat. Rec.* **115**, 1–20.

Barfield, R. J. (1969). Activation of copulatory behaviour by androgen implanted into the preoptic area of the male fowl. *Horm. Behav.* **1**, 37–52.

Bartelmez, G. W. (1912). The bilaterality of the pigeon's egg. A study in egg organization from the first growth of the oocyte to the beginning of cleavage. *J. Morphol.* **23**, 269–328.

Bates, R. W., Lahr, E. L., and Riddle, O. (1935). The gross action of prolactin and follicle-stimulating hormone on the mature ovary and sex accessories of fowl. *Amer. J. Physiol.* **111**, 361–368.

Bates, R. W., Riddle, O., and Lahr, E. L. (1937). The mechanism of the anti-gonad action of prolactin in adult pigeons. *Amer. J. Physiol.* **119**, 610–614.

Baylé, J. D., and Assenmacher, I. (1967). Contrôle hypothalamo-hypophysaire du fonctionnement thyroïdien chez la Caille. *C. R. Acad. Sci.* **264**, 125–128.

Baylé, J. D., Kraus, M., and van Tienhoven, A. (1970). The effects of hypophysectomy and testosterone propionate on the testes of Japanese Quail *Coturnix coturnix japonica. J. Endocrinol.* **46**, 403–404.

Beach, F. A. (1948). "Hormones and Behaviour." Harper, New York.

Beebe, W. (1925). The variegated tinamou *Crypturus variegatus variegatus* (Gmelin). *Zoologica (New York)* **6**, 195–227.

Bellairs, R. (1961). The structure of the yolk of the hen's egg as studied by electron microscopy. I. The yolk of the unincubated egg. *J. Biophys. Biochem. Cytol.* **11**, 207–225.

Belt, W. D., and Pease, D. C. (1956). Mitochondrial structure in sites of steroid secretion. *J. Biophys. Biochem. Cytol.* **2**, 369–374.

Bennett, M. A. (1940). The social hierarchy in Ring Doves. II. The effect of treatment with testosterone propionate. *Ecology* **21**, 148–165.

Benoit, J. (1927). Quantité de parenchyme testiculaire et quantité d'hormone élaborée. Existe-t-il une "sécrétion de luxe" ou un "parenchyme de luxe." *C. R. Soc. Biol.* **97**, 790–793.

Benoit, J. (1929). Le déterminisme des caractères sexuels secondaires du coq domestique. *Arch. Zool. Exp. Gen.* **69**, 217–499.

Benoit, J. (1936). Stimulation par la lumière de l'activité sexuelle chez le Canard et la Cane domestiques. *Bull. Biol. Fr. Belg.* **70**, 487–533.

Benoit, J. (1950). Reproduction caractères sexuels et hormones. Déterminisme du cycle saisonnier. *In* "Traité de Zoologie" (P.-P. Grassé, ed.), Vol. 15, pp. 384–478. Masson, Paris.

Benoit, J. (1956). Etats physiologiques et instinct de reproduction chez des Oiseaux. *In* "L'instinct dans le comportement des animaux et de l'homme" (M. Autuori *et al.*, eds.), pp. 177–260. Masson, Paris.

Benoit, J. (1961). Opto-sexual reflex in the duck: Physiological and histochemical aspects. *Yale J. Biol. Med.* **34**, 97–116.

Benoit, J., and Assenmacher, I. (1959). The control by visible radiations of the gonadotropic activity of the duck hypophysis. *Recent Progr. Horm. Res.* **15**, 143–164.

Benoit, J., Assenmacher, L., and Walter, F. X. (1950). Réponses du mécanisme gonadostimulant á l'éclairement artificiel et de la préhypophyse aux castration bilatérale et unilatérale, chez le canard domestique mâle, au cours de la période de regression testiculaire saisonniere. *C. R. Soc. Biol.* **144**, 573–577.

Benoit, J., Assenmacher, I., and Brard, E. (1956a). Apparition et maintien de cycles sexuels non saisonniers chez le Canard domestique placé pendant plus de trois ans à l'obscurité totale. *J. Physiol. (Paris)* **48**, 388–391.

Benoit, J., Assenmacher, I., and Brard, E. (1956b). Etude de l'évolution testiculaire du Canard domestique soumis très jeune à un éclairement artificiel permanent pendant deux ans. *C. R. Acad. Sci.* **242**, 3113–3115.

Bishop, D. W. (1961). Biology of spermatozoa. *In* "Sex and Internal Secretions" (W. C. Young, ed.), 3rd ed., Vol. 2, pp. 707–796. Williams & Wilkins, Baltimore, Maryland.

Bissonnette, T. H., and Chapnick, H. M. (1930). Studies on the sexual cycle in birds. II. *Amer. J. Anat.* **45**, 307.

Bissonnette, T. H., and Wadlund, A. P. (1932). Duration of testis activity of *Sturnus vulgaris* in relation to type of illumination. *J. Exp. Biol.* **9**, 339–350.

Boccabella, A. V. (1963). Reinitiation and restoration of spermatogenesis with testosterone propionate and other hormones after a long term post-hypophysectomy regression period. *Endocrinology* **72**, 787–798.

Bonadonna, T. (1954). Observations on the submicroscopic structures of *Gallus gallus* spermatozoa. *Poultry Sci.* **33**, 1151–1158.

Boss, W. R., and Witschi, E. (1947). The permanent effects of early stilbestrol injections on the sex organs of the Herring Gull (*Larus argentatus*). *J. Exp. Zool.* **105**, 61–77.

Botte, V. (1961). Ricerche istologiche ed istochimiche sui follicoli post-ovulatori ed atresici di pollo. *Acta Med. Vet.* **7**, 359–380.

Botte, V. (1963). La localizzazione della steroide-3β-olo-deidrogenasi nell'ovaio di pollo. *Rend. Ist. Sci. Univ. Camerino* **4**, 205–209.

Botte, V., and Rosati, P. (1964). Il colesterolo e la Δ⁵-3β-idrossisteroide deidrogenasi nelle cellule interstiziali del testicolo di pollo trattato con estrogeni. *Acta Med. Vet.* **10**, 3–10.

Botte, V., Delrio, G., and Lupo di Prisco, C. (1966). Conversion of neutral steroid hormones "in vitro" by the theca cells of the growing ova and by the post-ovulatory follicles of the laying hen. *Exerpta Med. Found. Int. Congr. Ser.* **111**, Abstr. 738.

Boucek, R. J., and Savard, K. (1970). Steroid formation by the avian ovary *in vitro* (*Gallus domesticus*). *Gen. Comp. Endocrinol.* **15**, 6–11.

Boucek, R. J., Györi, E., and Alvarez, R. (1966). Steroid dehydrogenase reactions in developing chick adrenal and gonadal tissues. *Gen. Comp. Endocrinol.* **7**, 292–303.

Brambell, F. W. R., and Marrian, G. F. (1929). Sex reversal in a pigeon (*Columba livia*). *Proc. Roy. Soc., Ser. B* **104**, 459–470.

Brant, J. W. A., and Nalbandov, A. V. (1956). Role of sex hormones in albumen secretion by the oviduct of chickens. *Poultry Sci.* **35**, 692–700.

Breitenbach, R. P., Nagra, C. L., and Meyer, R. K. (1965). Studies of incubation and broody behavior in the pheasant (*Phasianus colchicus*). *Anim. Behav.* **13**, 143–148.

Breneman, W. R. (1942). Action of prolactin and estrone on weights of reproductive organs and viscera of the cockerel. *Endocrinology* **30**, 609–615.

Breneman, W. R. (1953). The effect of gonadal hormones alone and in combination with pregnant mare serum on the pituitary, gonad, and comb of White Leghorn cockerels. *Anat. Rec.* **117**, 533–534.

Breneman, W. R. (1955). Reproduction in birds: The female. *Mem. Soc. Endocrinol.* **4**, 94–110.

Breneman, W. R. (1956). Steroid hormones and the development of the reproductive system in the pullet. *Endocrinology* **48**, 262–271.

Breneman, W. R., Zeller, F. J., and Creek, R. O. (1962). Radioactive phosphorous uptake by chick testes as an endpoint for gonadotropin assay. *Endocrinology* **71**, 790–798.

Bruder, R. H., and Lehrman, D. S. (1967). Role of the mate in the elicitation of hormone induced incubation behavior in the Ring Dove. *J. Comp. Physiol. Psychol.* **63**, 382–384.

Bullough, W. S. (1942). The reproductive cycles of the British and Continental races of the Starling. *Phil. Trans. Roy. Soc. London, Ser. B* **231**, 165–246.

Burger, J. W. (1944). Testicular response to androgen in the light-stimulated Starling. *Endocrinology* **35**, 182–186.

Burger, J. W. (1945). Some effects of sex steroids on the gonads of the Starling. *Endocrinology* **37**, 77–82.

Burger, J. W. (1948). The relation of external temperature to spermatogenesis in the male Starling. *J. Exp. Zool.* **109**, 259–266.

Burger, J. W. (1949). A review of experimental investigations on seasonal reproduction in birds. *Wilson Bull.* **61**, 211–230.

Burrows, W. H., and Byerly, T. C. (1936). Studies of prolactin in the fowl pituitary. 1. Broody hens compared with laying hens and males. *Proc. Soc. Exp. Biol. Med.* **34**, 841–844.

Cédard, L., and Haffen, K. (1966). Transformations de la déhydroépiandrosterone par les gonades embryonnaires de Poulet, cultivées *in vitro*. *C. R. Acad. Sci.* **263**, 430.

Chan, K. M. B., and Lofts, B. (1973). Unpublished data.

Channing, C. P. (1966). Progesterone biosynthesis by equine granulosa cells grown in tissue culture. *Nature (London)* **210,** 1266.

Chieffi, G. (1967). Occurrence of steroids in gonads of non-mammalian vertebrates and sites of their biosynthesis. *Proc. Int. Congr. Horm. Steroids, 2nd, 1966* Int. Congr. Ser. No. 132, pp. 1047–1057.

Chieffi, G., and Botte, V. (1965). The distribution of some enzymes involved in the steroidogenesis of hen's ovary. *Experientia* **21,** 16–17.

Chieffi, G., and Botte, V. (1970). The problem of "luteogenesis" in non-mammalian vertebrates. *Boll. Zool. Agr. Bachicolt.* [2] **37,** 85–102.

Christensen, A. K., and Gillim, S. W. (1969). The correlation of fine structure and function in steroid-secreting cells with emphasis on those of the gonads. *In* "The Gonads" (K. W. McKerns, ed.), pp. 415–488. North-Holland Publ., Amsterdam.

Christensen, A. K., and Mason, N. R. (1965). Comparative ability of seminiferous tubules and interstitial tissue of rat testes to synthesize androgens from progesterone-4-^{14}C *in vitro. Endocrinology* **76,** 646–656.

Chu, J. P. (1940). The effects of estrone and testosterone and of pituitary extracts on the gonads of hypophysectomized pigeons. *J. Endocrinol.* **2,** 21–37.

Chu, J. P., and You, S. S. (1946). Gonad stimulation by androgens in hypophysectomized pigeons. *J. Endocrinol.* **4,** 431–435.

Cleland, K. W. (1951). The spermatogenetic cycle of the guinea pig. *Aust. J. Sci. Res., Ser. B* **4,** 344–369.

Clermont, Y. (1958). Structure de l'épithélium séminal et mode de renouvellement des spermatogonies chez le canard. *Arch. Anat. Microsc. Morphol. Exp.* **47,** 47–66.

Collias, N. (1950). Hormones and behaviour with special reference to birds and the mechanisms of hormone action. *In* "A Symposium on Steroid Hormones" (E. S. Gordon, ed.), pp. 277–329. Univ. of Wisconsin Press, Madison.

Conner, M. H., and Fraps, R. M. (1954). Premature oviposition following subtotal excision of the hen's ruptured follicle. *Poultry Sci.* **33,** 1051.

Coombs, C. J. F., and Marshall, A. J. (1956). The effects of hypophysectomy on the interstitial testis rhythm in birds and mammals. *J. Endocrinol.* **13,** 107–111.

Craig, W. (1911). Oviposition induced by the male in pigeons. *J. Morphol.* **22,** 299–305.

Crook, J. H., and Butterfield, P. A. (1968). Effects of testosterone propionate and luteinizing hormone on agonistic and nest building behaviour of *Quelea quelea. Anim. Behav.* **16,** 370–384.

Curtis, G. M. (1918). The morphology of the mammalian seminiferous tubule. *Amer. J. Anat.* **24,** 339–394.

Das, B. C., and Nalbandov, A. V. (1955). Responses of ovaries of immature chickens to avian and mammalian gonadotrophins. *Endocrinology* **57,** 705–710.

Davies, S. J. J. F. (1970). Patterns of inheritance in the bowing display and associated behaviour of some hybrid *Streptopelia* doves. *Behaviour* **36,** 187–214.

Davis, D. E. (1957). Aggressive behaviour in castrated Starlings. *Science* **126,** 253.

Davis, D. E. (1963). Hormonal control of aggressive behaviour. *Proc. Int. Ornithol. Congr., 13th, 1962* pp. 994–1003.

Dean, H. W., and Rubin, B. L. (1965). Identification and control of cells that synthesize steroid hormones in the adrenal glands, gonads and placentae of various mammalian species. *Arch. Anat. Microsc. Morphol. Exp.* **54,** 49–66.

Dingman, C. W., Aronow, A., Bunting, S. L., Peacock, A. C., and O'Malley, B. W. (1969). Changes in chick oviduct ribonucleic acid following hormonal stimulation. *Biochemistry* **8,** 489–495.

Dodd J. M. (1960). Gonadal and gonadotrophic hormones in lower vertebrates. *In*

"Marshall's Physiology of Reproduction" (A. S. Parkes, ed.), 3rd ed., Vol. 1, Part 2, pp. 417–582. Longmans, Green, New York.

Domm, L. V. (1939). Modifications in sex and secondary sexual characters in birds. *In* "Sex and Internal Secretions" (E. Allen, ed.), 2nd ed., pp. 227–327. Williams & Wilkins, Baltimore, Maryland.

Dorfman, R. I., and Shipley, R. A. (1956). "Androgens." Wiley, New York.

Douglas, D. S., and Sturkie, P. D. (1964). Plasma levels of antidiuretic hormone during oviposition in the hen. *Fed. Proc., Fed. Amer. Socs. Exp. Biol.* **23**, 150.

Dubois, R. (1965). La lignée germinale chez les reptiles et les oiseaus. *Année Biol.* **4**, 637–666.

Dunham, H. H., and Riddle, O. (1942). Effects of a series of steroids on ovulation and reproduction in pigeons. *Physiol. Zool.* **15**, 383–394.

Egge, A. S., and Chiasson, R. B. (1963). Endocrine effects of diencephalic lesions in the White Leghorn hen. *Gen. Comp. Endocrinol.* **3**, 346–361.

Eisner, E. (1960). The relationship of hormones to the reproductive behaviour of birds, referring especially to parental behaviour: A review. *Anim. Behav.* **8**, 155–179.

Engels, W. L., and Jenner, C. E. (1956). The effect of temperature on testicular recrudescence in juncos at different photoperiods. *Biol. Bull.* **110**, 129–137.

Erickson, C. J., and Lehrman, D. S. (1964). Effects of castration of male Ring Doves upon ovarian activity of females. *J. Comp. Physiol. Psychol.* **58**, 164–166.

Erickson, C. J., Bruder, R. H., Komisaruk, B. R., and Lehrman, D. S. (1967). Selective inhibition by progesterone of androgen-induced behaviour in male Ring Doves (*Streptopelia risoria*). *Endocrinology* **81**, 39–45.

Fabricius, E., and Jansson, A. M. (1963). Laboratory observations on the reproductive behaviour of the pigeon (*Columba livia*) during the pre-incubation phase of the breeding cycle. *Anim. Behav.* **11**, 534–547.

Farner, D. S. (1959). Photoperiodic control of annual gonadal cycles in birds. *In* "Photoperiodism and Related Phenomena in Plants and Animals," Publ. No. 55, pp. 717–750. Amer. Ass. Advance. Sci., Washington, D.C.

Farner, D. S., and Follett, B. K. (1966). Light and other environmental factors affecting avian reproduction. *J. Anim. Sci.* **25**, 90–115.

Farner, D. S., and Mewaldt, L. R. (1952). The relative roles of photoperiod and temperature in gonadal recrudescence in male *Zonotrichia leucophrys gambelii*. *Anat. Rec.* **113**, 612.

Farner, D. S., and Wilson, A. C. (1957). A quantitative examination of testicular growth in the White-crowned Sparrow. *Biol. Bull.* **113**, 254–267.

Farner, D. S., Follett, B. K., King, J. R., and Morton, M. L. (1966). A quantitative examination of ovarian growth in the White-crowned Sparrow. *Biol. Bull.* **130**, 67–75.

Fevold, H. R., and Eik-Nes, K. B. (1962). Progesterone metabolism by testicular tissue of the English Sparrow (*Passer domesticus*) during the annual reproductive cycle. *Gen. Comp. Endocrinol.* **2**, 506–515.

Flickinger, R., and Rounds, D. E. (1956). The maternal synthesis of egg yolk proteins as demonstrated by isotopic and serological means. *Biochim. Biophys. Acta* **22**, 38–42.

Follett, B. K., and Farner, D. S. (1966a). The effects of the daily photoperiod on gonadal growth, neurohypophysial hormone content, and neurosecretion in the hypothalamo-hypophyseal system of the Japanese Quail (*Coturnix coturnix japonica*). *Gen. Comp. Endocrinol.* **7**, 111–124.

Follett, B. K., and Farner, D. S. (1966b). Pituitary gonadotropins in the Japanese Quail

(*Coturnix coturnix japonica*) during photoperiodically induced gonadal growth. *Gen. Comp. Endocrinol.* **7**, 125–131.

Follett, B. K., Scanes, C. G., and Nicholls, T. J. (1972). The chemistry and physiology of the avian gonadotropins. *In* "Hormones Glycoproteiques Hypophysaires." Colloque Inserm, Paris, 1972. pp. 193–211.

Fox, T. W. (1955). Effects of progesterone on growth and sexual development in S. C. White Leghorns. *Poultry Sci.* **34**, 598–602.

Fraps, R. M. (1955). The varying effects of sex hormones in birds. *Mem. Soc. Endocrinol.* **4**, 205–218.

Fraps, R. M. (1965). Twenty-four hour periodicity in the mechanism of pituitary gonadotrophin release for follicle maturation and ovulation in the chicken. *Endocrinology* **77**, 5–18.

Fraps, R. M. (1970). Photoregulation in the ovulation cycle of the domestic fowl. *Colloq. Int. Cent. Nat. Rech. Sci.* **172**, 281–306.

Fraps, R. M., Fevold, H. L., and Neher, B. H. (1947). Ovulatory response of the hen to presumptive luteinizing and other fractions from fowl pituitary tissue. *Anat. Rec.* **99**, 571–572.

Fraps, R. M., Hooker, C. W., and Forbes, T. R. (1949). Progesterone in blood plasma of cocks and non-ovulating hens. *Science* **109**, 493.

Friedman, M., and Lehrman, D. S. (1968). Physiological conditions for the stimulation of prolactin secretion by external stimuli in the male Ring Dove. *Anim. Behav.* **16**, 233–237.

Furr, B. J. A. (1969a). Identification of steroids in the ovaries and plasma of laying hens and the site of production of progesterone in the ovary. *Gen. Comp. Endocrinol.* **13**, Abstr. 56.

Furr, B. J. A. (1969b). Progesterone concentrations in the blood and ovaries of laying, moulting and broody hens. *Gen. Comp. Endocrinol.* **13**, Abstr. 57.

Furr, B. J. A., and Cunningham, F. J. (1970). The biological assay of chicken pituitary gonadotrophins. *Brit. Poultry Sci.* **11**, 7–13.

Gilbert, A. B., and Lake, P. E. (1963). The effect of oxytocin and vasopressin on oviposition in the domestic hen. *J. Physiol. (London)* **169**, 52.

Goodwin, D. (1967). "Pigeons and Doves of the World," Publ. No. 663. British Museum (Natur. Hist.), London.

Gourdji, D. (1970). Prolactine et relations photo-sexuelles chez les Oiseaux. *Colloq. Int. Cent. Nat. Rech. Sci.* **172**, 233–258.

Graber, J. W., Frankel, A. I., and Nalbandov, A. V. (1967). Hypothalamic center influencing the release of LH in the cockerel. *Gen. Comp. Endocrinol.* **9**, 187–192.

Greeley, F., and Meyer, R. K. (1953). Seasonal variation in testis stimulating activity of male pheasant pituitary glands. *Auk* **70**, 350–358.

Grigg, G. W., and Hodge, A. J. (1949). Electron microscopic studies of spermatozoa. 1. The morphology of the spermatozoon of the common domestic fowl (*Gallus domesticus*). *Aust. J. Sci. Res., Ser. B* **2**, 271–286.

Hammond, J. (1952). Fertility. *In* "Marshall's Physiology of Reproduction," 3rd edition (A. S. Parkes, ed.), Vol. 2, pp. 648–740. Longmans, Green, New York.

Hamner, W. H. (1968). The photorefractory period of the House Finch. *Ecology* **49**, 211–227.

Hansen, E. W. (1966). Squab-induced crop growth in Ring Dove foster parents. *J. Comp. Physiol. Psychol.* **62**, 120–122.

Harrison, J. M. (1932). A series of nineteen pheasants (*Phasianus colchicus* L.) presenting analogous secondary sexual characteristics in association with changes in the ovaries. *Proc. Zool. Soc. London* 1, 193–203.

Harvey, S. C. (1963). The effect of testosterone propionate on germ cell number in the male rat. *Anat. Rec.* 145, 237.

Hawkins, R. A., and Heald, P. J. (1966). Lipid metabolism and the laying hen. IV. The synthesis of triglycerides by slices of avian liver *in vitro*. *Biochim. Biophys. Acta* 116, 41–45.

Heald, P. J., and McLachlan, P. M. (1964). The isolation of phosvitin from the plasma of the oestrogen-treated immature pullet. *Biochem. J.* 92, 51–55.

Heald, P. J., and McLachlan, P. M. (1965). The synthesis of phosvitin *in vitro* by slices of liver from the laying hen. *Biochem. J.* 94, 32–39.

Heald, P. J., and Rookledge, K. A. (1964). Effects of gonadal hormones, gonadotrophins and thyroxine on plasma FFA in the domestic fowl. *J. Endocrinol.* 30, 115–130.

Heald, P. J., Rookledge, K. A., Furnival, B. E., and Watts, G. D. (1968). Changes in luteinizing hormone content of the anterior pituitary of the domestic fowl during the interval between clutches. *J. Endocrinol.* 41, 197–201.

Herrick, R. B., and Adams, J. L. (1956). The effects of progesterone and diethylstilbestrol injected singly or in combination, on sexual libido, and the weight of the testes of single comb White Leghorn Cockerels. *Poultry Sci.* 35, 1269–1273.

Herrick, R. B., McGibbon, W. H., and McShan, W. H. (1962). Gonadotropic activity of chicken pituitary glands. *Endocrinology* 71, 487–491.

Hertz, R., Fraps, R. M., and Sebrell, W. H. (1943). Induction of avidin formation in the avian oviduct by stilbestrol plus progesterone. *Proc. Soc. Exp. Biol. Med.* 52, 142–144.

Hertz, R., Larsen, C. D., and Tullner, W. (1947). Inhibition of estrogen-induced tissue growth with progesterone. *J. Nat. Cancer Inst.* 8, 123–126.

Hill, R. T., and Parkes, A. S. (1934). Hypophysectomy of birds. III. Effects on gonads, accessory organs and head furnishings. *Proc. Roy. Soc., Ser. B* 116, 221–236.

Hinde, R. A. (1962). Temporal relations of brood patch development in domesticated canaries. *Ibis* 104, 90–97.

Hinde, R. A. (1967). Aspects of the control of avian reproductive development within the breeding season. *Proc. Int. Ornithol. Congr., 14th, 1966* pp. 135–153.

Hinde, R. A., and Steel, E. A. (1962). Selection of nest material by female canaries. *Anim. Behav.* 10, 67–75.

Hinde, R. A., and Steel, E. A. (1964). Effects of exogenous hormones on the tactile sensitivity of the canary brood patch. *J. Endocrinol.* 30, 355–360.

Hinde, R. A., Bell, R. Q., and Steel, E. A. (1963). Changes in sensitivity of the canary brood patch. *Anim. Behav.* 11, 553–560.

Höhn, E. O. (1959). Prolactin in the Cowbird's pituitary in relation to avian brood parasitism. *Nature (London)* 184, 2030.

Höhn, E. O. (1960). Seasonal changes in the Mallard's penis and their hormonal control. *Proc. Zool. Soc. London* 134, 547–558.

Höhn, E. O. (1962). A possible endocrine basis of brood parasitism. *Ibis* 104, 418–421.

Höhn, E. O. (1970). Gonadal hormone concentration in Northern Phalaropes in relation to nuptial plumage. *Can. J. Zool.* 48, 400–401.

Höhn, E. O., and Cheng, S. C. (1967). Gonadal hormones in Wilson's Phalarope (*Steganopus tricolor*) and other birds in relation to plumage and sex behaviour. *Gen. Comp. Endocrinol.* 8, 1–11.

Hutchison, J. B. (1970a). Differential effects of testosterone and oestradiol on male courtship in Barbary doves (*Streptopelia risoria*). *Anim. Behav.* 18, 41–51.

Hutchison, J. B. (1970b). Influence of gonadal hormones on the hypothalamic integration of courtship behaviour in the Barbary dove. *J. Reprod. Fert., Suppl.* **11**, 15–41.

Hutchison, J. B. (1973). Unpublished data.

Hutchison, R. E., Hinde, R. A., and Steel, E. A. (1967). The effects of oestrogen, progesterone and prolactin on brood patch formation in ovariectomized canaries. *J. Endocrinol.* **39**, 379–385.

Hutchison, R. E., Hinde, R. A., and Bendon, B. (1968). Oviduct development and its relation to other aspects of reproduction in domestic canaries. *J. Zool.* **155**, 87–102.

Hutt, F. B. (1949). "Genetics of the Fowl." McGraw-Hill, New York.

Huxley, J. S. (1938). The present standing of the theory of sexual selection. *In* "Evolution" (E. S. Goodrich and G. R. de Beer, eds.), pp. 11–42. Oxford Univ. Press (Clarendon), London and New York.

Immelmann, K. (1963). Tierische Jahresperiodik in ökologischer Sicht. *Zool. Jahrb., Abt. Syst.* **91**, 91–200.

Jared, D. W., and Wallace, R. A. (1968). Comparative chromatography of yolk proteins of teleosts. *Comp. Biochem. Physiol.* **24**, 437–443.

Johns, J. E. (1964). Testosterone-induced nuptial feathers in phalaropes. *Condor* **66**, 449–455.

Johns, J. E., and Pfeiffer, E. W. (1963). Testosterone-induced incubation patches of phalarope birds. *Science* **140**, 1225–1226.

Jones, R. E. (1969a). Effect of prolactin and progesterone on gonads of breeding California Quail. *Proc. Soc. Exp. Biol. Med.* **131**, 172–174.

Jones, R. E. (1969b). Hormonal control of incubation patch development in the California Quail *Lophortyx californicus. Gen. Comp. Endocrinol.* **13**, 1–13.

Jones, R. E. (1969c). Epidermal hyperplasia in the incubation patch of the California Quail, *Lophortyx californicus*, in relation to pituitary prolactin content. *Gen. Comp. Endocrinol.* **12**, 498–502.

Jones, R. E. (1970). Effects of season and gonadotropin on testicular interstitial cells of California Quail. *Auk* **87**, 729–737.

Jones, R. E., Kreider, J. W., and Criley, B. B. (1970). Incubation patch of the chicken: Response to hormones *in situ* and transplanted to a dorsal site. *Gen. Comp. Endocrinol.* **15**, 398–403.

Jørgensen, C. B. (1968). Central nervous control of adenohypophyseal functions. *In* "Perspectives in Endocrinology: Hormones in the Lives of Lower Vertebrates" (E. J. W. Barrington and C. B. Jørgensen, eds.), pp. 469–541. Academic Press, New York.

Jørgensen, C. B., and Larsen, L. O. (1967). Neuroendocrine mechanisms in lower vertebrates. *In* "Neuroendocrinology" (L. Martini and W. F. Ganong, eds.), Vol. 2, pp. 485–528. Academic Press, New York.

Juhn, M., and Harris, P. C. (1956). Responses in molt and lay of fowl to progesterone and gonadotrophins. *Proc. Soc. Exp. Biol. Med.* **98**, 669–672.

Kabat, C., Buss, I. O., and Meyer, R. K. (1948). The use of ovulated follicles in determining eggs laid by the Ring-necked Pheasant. *J. Wildl. Manage.* **12**, 399–416.

Kannankeril, J. V., and Domm, L. V. (1968). Development of the gonads in the female Japanese Quail. *Amer. J. Anat.* **123**, 131–146.

Kappauf, B., and van Tienhoven, A. (1972). Progesterone concentrations in peripheral plasma of laying hens in relation to the time of ovulation. *Endocrinology* **90**, 1350–1355.

Kar, A. B. (1949). Testicular changes in the juvenile pigeon due to progesterone treatment. *Endocrinology* **45**, 346–348.

Keast, J. A., and Marshall, A. J. (1954). The influence of drought and rainfall on reproduction in Australian desert birds. *Proc. Zool. Soc. London* **124**, 493–499.

King, J. R., Follett, B. K., Farner, D. S., and Morton, M. L. (1966). Annual gonadal cycles and pituitary gonadotropins in *Zonotrichia leucophrys gambelii*. *Condor* **68**, 476–487.

Knight, P. F., and Schechtman, A. M. (1954). The passage of heterologous serum proteins from the circulation into the ovum of the fowl. *J. Exp. Zool.* **127**, 271–304.

Kobayashi, H., and Farner, D. S. (1966). Evidence of a negative feedback on photoperiodically induced gonadal development in White-crowned Sparrow, *Zonotrichia leucophrys gambelii*. *Gen. Comp. Endocrinol.* **6**, 443–452.

Kohler, P. O., Grimley, P. M., and O'Malley, B. W. (1969). Estrogen-induced cytodifferentiation of the ovalbumin-secreting glands of the chick oviduct. *J. Cell Biol.* **40**, 8–27.

Komisaruk, B. R. (1967). Effects of local brain implants of progesterone on reproductive behaviour in Ring Doves. *J. Comp. Physiol. Psychol.* **64**, 219–224.

Kordon, C., and Gogan, F. (1970). Interaction du feed-back et de la photostimulation dans les régulations gonadotropes chez Mammifères et les Oiseaux. *Colloq. Int. Cent. Nat. Rech. Sci.* **172**, 325–350.

Kramer, M. F. (1960). "Spermatogenesis bij de Stier." Utrecht.

Kumaran, J. D. S., and Turner, C. W. (1949a). The endocrinology of spermatogenesis in birds. I. Effect of estrogen and androgen. *Poultry Sci.* **28**, 593–602.

Kumaran, J. D. S., and Turner, C. W. (1949b). The endocrinology of spermatogenesis in birds. II. Effect of androgens. *Poultry Sci.* **28**, 739–746.

Lacy, D. (1960). Light and electron microscopy and its uses in the study of factors influencing spermatogenesis in the rat. *J. Roy. Microsc. Soc.* [3] **79**, 209–225.

Lacy, D. (1967). The seminiferous tubule in mammals. *Endeavour* **26**, 101–108.

Lacy, D., and Lofts, B. (1961). The use of ionizing radiations and oestrogen treatment in the detection of hormone synthesis by the Sertoli cells. *J. Physiol. (London)* **161**, 23–24.

Lacy, D., and Lofts, B. (1965). Studies on the structure and function of the mammalian testis. I. Cytological and histochemical observations after continuous treatment with oestrogenic hormone and the effects of FSH and LH. *Proc. Roy. Soc., Ser. B* **162**, 188–197.

Lacy, D., and Pettitt, A. J. (1969). Transmission electron microscopy and the production of steroids by the Leydig and Sertoli cells of the human testis. *Micron* **1**, 15–33.

Lacy, D., Vinson, G., Collins, P., Bell, J., Fyson, P., Pudney, J., and Pettitt, A. J. (1969). The Sertoli cell and spermatogenesis in mammals. *In* "Progress in Endocrinology" (C. Gual, ed.), Int. Congr. Ser. No. 184, pp. 1019–1029. Exerpta Med. Found., Amsterdam.

Lade, B. I., and Thorpe, W. H. (1964). Dove songs as innately coded patterns of specific behaviour. *Nature (London)* **202**, 366–368.

Lahr, E. L., and Riddle, O. (1944). The action of steroid hormones on the male dove testis. *Endocrinology* **35**, 261–266.

Lahr, E. L., and Riddle, O. (1945). Intersexuality in male embryos of pigeons. *Anat. Rec.* **92**, 425–431.

Lake, P. E. (1957). The male reproductive tract of the fowl. *J. Anat.* **91**, 116–129.

Laskowski, M. (1933). Über den Calciumzustand im Blutplasma der Henne. *Biochem. Z.* **260**, 230–241.

Laskowski, M. (1935). Über die Phosphorverbindungen im Blutplasma der Legehenne. *Biochem. Z.* **279**, 293–300.

Laskowski, M. (1936). Vorkommen des Serumvitellins in Blute der Wirbeltiere. *Biochem. Z.* **284**, 318–325.

Law, G. R. J., and Kosin, I. L. (1958). Seasonal reproductive ability of male domestic turkeys as observed under two ambient temperatures. *Poultry Sci.* **37**, 1034–1047.

Laws, D. F., and Farner, D. S. (1960). Prolactin and the photoperiodic testicular response in White-crowned Sparrows, *Endocrinology* **67**, 279–281.

Layne, D. S., Common, R. H., Maw, W. A., and Fraps, R. M. (1958). Presence of estrone, estradiol and estriol in extracts of ovaries of laying hens. *Nature (London)* **181**, 351–352.

Leblond, C. P. (1938). Extra-hormonal factors in maternal behaviour. *Proc. Soc. Exp. Biol. Med.* **38**, 66–70.

Lehrman, D. S. (1955). The physiological basis of parental feeding behaviour in the Ring Dove (*Streptopelia risoria*). *Behaviour* **7**, 241–286.

Lehrman, D. S. (1958a). Effect of female sex hormones on incubation behaviour in the Ring Dove (*Streptopelia risoria*). *J. Comp. Physiol. Psycol.* **51**, 32–36.

Lehrman, D. S. (1958b). Induction of broodiness by participation in courtship and nest-building in the Ring Dove (*Streptopelia risoria*). *J. Comp. Physiol. Psychol.* **51**, 32–36.

Lehrman, D. S. (1961). Hormonal regulation of parental behaviour in birds and infra-human mammals. *In* "Sex and Internal Secretions" (W. C. Young, ed.), 3rd ed., pp. 1268–1382. Williams and Wilkins, Baltimore, Maryland.

Lehrman, D. S. (1963). On the initiation of incubation behaviour in doves. *Anim. Behav.* **11**, 433–438.

Lehrman, D. S. (1964). Control of behaviour cycles in reproduction. *In* "Social Behaviour and Organization among Vertebrates" (W. Etkin, ed.), pp. 143–166. Univ. of Chicago Press, Chicago.

Lehrman, D. S. (1965). Interaction between internal and external environments in the regulation of the reproductive cycle of the Ring Dove. *In* "Sex and Behavior" (F. A. Beach, ed.), pp. 355–380. Wiley, New York.

Lehrman, D. S., and Brody, P. N. (1957). Oviduct response to estrogen and progesterone in the Ring Dove (*Streptopelia risoria*). *Proc. Soc. Exp. Biol. Med.* **19**, 373–375.

Lehrman, D. S., and Brody, P. N. (1961). Does prolactin induce incubation behaviour in the Ring Dove? *J. Endocrinol.* **22**, 269–275.

Lehrman, D. S., and Brody, P. N. (1964). Effect of prolactin on established incubation behaviour in the Ring Dove. *J. Comp. Physiol. Psychol.* **57**, 161–165.

Lehrman, D. S., and Wortis, R. P. (1967). Breeding experience and breeding efficiency in the Ring Dove. *Anim. Behav.* **15**, 223–228.

Lehrman, D. S., Brody, P. N., and Wortis, R. P. (1961). The presence of the mate and of nesting material as stimuli for the development of incubation behaviour and for gonadotropin secretion in the Ring Dove (*Streptopelia risoria*). *Endocrinology* **68**, 507–516.

Leonard, S. L. (1939). Induction of singing in female canaries by injection of male hormone. *Proc. Soc. Exp. Biol. Med.* **41**, 229–239.

Licht, P., and Stockell Hartree, A. (1971). Actions of mammalian, avian and piscine gonadotrophins in the lizard. *J. Endocrinol.* **49**, 113–124.

Lisk, R. D. (1967). Neural localization for androgen activation of copulatory behaviour in the male rat. *Endocrinology* **80**, 754–761.

Lloyd, J. A. (1965). Seasonal development of the incubation patch in the Starling. *Condor* **67**, 67–72.

Lofts, B. (1962a). Cyclical changes in the interstitial and spermatogenetic tissue of migratory waders "wintering" in Africa. *Proc. Zool. Soc. London* **138**, 405–413.

Lofts, B. (1962b). The effects of exogenous androgen on the testicular cycle of the Weaver-finch *Quelea quelea*. *Gen. Comp. Endocrinol.* **2**, 394–406.

Lofts, B. (1962c). Photoperiodism and the refractory period of reproduction in an equatorial bird *Quelea quelea*. *Ibis* **104**, 407–414.

Lofts, B. (1964). Evidence of an autonomous reproductive rhythm in an equatorial bird (*Quelea quelea*). *Nature (London)* **201**, 523–524.

Lofts, B. (1965). Seasonal lipid changes and their possible significance in the testis of anura. *Proc. Int. Congr. Endocrinol., 2nd, 1964* Int. Congr. Ser. No. 83, pp. 100–105.

Lofts, B. (1968). Patterns of testicular activity. *In* "Perspectives in Endocrinology: Hormones in the Lives of Lower Vertebrates" (E. J. W. Barrington and C. B. Jørgensen, eds.), pp. 239–304. Academic Press, New York.

Lofts, B. (1969). Seasonal cycles in reptilian testes. *Gen. Comp. Endocrinol., Suppl.* **2**, 147–155.

Lofts, B. (1970). Cytology of the gonads and feed-back mechanisms with respect to photosexual relationships in male birds. *Colloq. Int. Cent. Nat. Rech. Sci.* **172**, 307–324.

Lofts, B. (1972). The Sertoli cell. *Gen. Comp. Endocrinol., Suppl.* **3**, 636–648.

Lofts, B. (1973). Unpublished observations.

Lofts, B., and Bern, H. A. (1972). The functional morphology of steroidogenic tissues. *In* "Steroids in Non-mammalian Vertebrates" (D. R. Idler, ed.), pp. 37–126. Academic Press, New York.

Lofts, B., and Chiu, K. W. (1968). Androgens and reptilian spermatogenesis. *Arch. Anat., Histol. Embryol.* **51**, 409–418.

Lofts, B., and Choy, L. Y. L. (1971). Steroid synthesis by the seminiferous tubules of the snake, *Naja naja. Gen. Comp. Endocrinol.* **17**, 588–591.

Lofts, B., and Coombs, C. J. F. (1965). Photoperiodism and the testicular refractory period in the Mallard. *J. Zool.* **146**, 44–54.

Lofts, B., and Marshall, A. J. (1956). The effects of prolactin administration on the internal rhythm of reproduction in male birds. *J. Endocrinol.* **13**, 101–106.

Lofts, B., and Marshall, A. J. (1957). The interstitial and spermatogenetic tissue of autumn migrants in Southern England. *Ibis* **99**, 621–627.

Lofts, B., and Marshall, A. J. (1958). An investigation of the refractory period of reproduction in male birds by means of exogenous prolactin and follicle stimulating hormone. *J. Endocrinol.* **17**, 91–98.

Lofts, B., and Marshall, A. J. (1959). The post-nuptial occurrence of progestins in the seminiferous tubules of birds. *J. Endocrinol.* **19**, 16–21.

Lofts, B., and Marshall, A. J. (1960). The experimental regulation of *Zugunruhe* and the sexual cycle in Brambling *Fringilla montifringilla. Ibis* **102**, 209–214.

Lofts, B., and Murton, R. K. (1966). The role of weather, food and biological factors in timing the sexual cycle of Woodpigeons. *Brit. Birds* **59**, 261–280.

Lofts, B., and Murton, R. K. (1968). Photoperiodic and physiological adaptations regulating avian breeding cycles and their ecological significance. *J. Zool.* **155**, 327–394.

Lofts, B., and Murton, R. K. (1973). Unpublished observations.

Lofts, B., Murton, R. K., and Westwood, N. J. (1966a). Gonad cycles and the evolution of breeding seasons in British Columbidae. *J. Zool.* **150**, 249–272.

Lofts, B., Pickford, G. E., and Atz, J. W. (1966b). Effects of methyl testosterone on the testes of a hypophysectomized cyprinodont fish, *Fundulus heteroclitus. Gen. Comp. Endocrinol.* **6**, 74–88.

Lofts, B., Phillips, J. G., and Tam, W. H. (1966c). Seasonal changes in the testis of the Cobra *Naja naja* (Linn.). *Gen. Comp. Endocrinol.* **6**, 466–475.

Lofts, B., Murton, R. K., and Westwood, N. J. (1967). Photo-responses of the Wood-pigeon *Columba palumbus* in relation to the breeding season. *Ibis* **109**, 338–351.

Lofts, B., Follett, B. K., and Murton, R. K. (1970). Temporal changes in the pituitary-gonadal axis. *Mem. Soc. Endocrinol.* **18**, 545–575.

Lofts, B., Murton, R. K., and Thearle, R. J. P. (1973a). The effects of testosterone propionate and gonadotrophins on the bill pigmentation and testes of the House Sparrow *(Passer domesticus)*. *Gen. Comp. Endocrinol.* (in press).

Lofts, B., Murton, R. K., and Yue, P. L. F. (1973b). Unpublished observations.

Lorenz, F. W. (1954). Effects of estrogens on domestic fowl and applications in the poultry industry. *Vitam. Horm. (New York)* **12**, 235–275.

Lorenz, F. W., Entemann, C., and Chaikoff, I. L. (1938). The influence of age, sex and ovarian activity on the blood lipids of the domestic fowl. *J. Biol. Chem.* **122**, 619–633.

Lott, D. F., and Brody, P. N. (1966). Support of ovulation in the Ring Dove by auditory and visual stimuli. *J. Comp. Physiol. Psychol.* **62**, 311–313.

Lott, D. F., Scholz, S. D., and Lehrman, D. S. (1967). Exteroceptive stimulation of the reproductive system of the female Ring Dove *(Streptopelia risoria)* by the mate and by the colony milieu. *Anim. Behav.* **15**, 433–437.

Ma, R. C. S., and Nalbandov, A. V. (1963). Hormonal activity of the autotransplanted adenohypophysis. *Advan. Neuroendocrinol., Proc. Symp., 1961* pp. 306–312.

March, G. L., and Sadlier, R. M. F. S. (1970). Studies on the Band-tailed Pigeon *(Columba fasciata)* in British Columbia. 1. Seasonal changes in gonadal development and crop gland activity. *Can. J. Zool.* **48**, 1353–1357.

Marshall, A. J. (1949a). Weather factors and spermatogenesis in birds. *Proc. Zool. Soc. London* **119**, 711–716.

Marshall, A. J. (1949b). On the function of the interstitium of the testis: The sexual cycle of a wild bird *(Fulmarus glacialis* L.). *Quart. J. Microsc. Sci.* **90**, 265–280.

Marshall, A. J. (1951). The refractory period of testis rhythm in birds and its possible bearing on breeding and migration. *Wilson Bull.* **63**, 238–261.

Marshall, A. J. (1952). The interstitial cycle in relation to autumn and winter sexual behaviour in birds. *Proc. Zool. Soc. London* **121**, 727–740.

Marshall, A. J. (1955). Reproduction in birds: The male. *Mem. Soc. Endocrinol.* **4**, 75–88.

Marshall, A. J. (1959). Internal and environmental control of breeding. *Ibis* **101**, 456–478.

Marshall, A. J. (1961a). Reproduction. *In* "Biology and Comparative Physiology of Birds" (A. J. Marshall, ed.), Vol. 2, pp. 169–213. Academic Press, New York.

Marshall, A. J. (1961b). Breeding seasons and migration. *In* "Biology and Comparative Physiology of Birds" (A. J. Marshall, ed.), Vol. 2, pp. 307–339. Academic Press, New York.

Marshall, A. J. (1970). Environmental factors other than light involved in the control of sexual cycles in birds and mammals. *Colloq. Int. Cent. Nat. Rech. Sci.* **172**, 38–48.

Marshall, A. J., and Coombs, C. J. F. (1957). The interaction of environmental internal and behavioural factors in the Rook *(Corvus f. frugilegus)* Linnaeus. *Proc. Zool. Soc. London* **128**, 545–589.

Marshall, A. J., and Roberts, J. D. (1959). The breeding biology of equatorial vertebrates: Reproduction of cormorants (Phalacrocoracidae) at lat. 0°20′N. *Proc. Zool. Soc. London* **132**, 617–625.

Marshall, A. J., and Serventy, D. L. (1957). On the post-nuptial rehabilitation of the avian testis tunic. *Emu* **57**, 59–63.

Marshall, A. J., and Serventy, D. L. (1958). The internal rhythm of reproduction in xerophilous birds under conditions of illumination and darkness. *J. Exp. Biol.* **35**, 666–670.

Mathewson, S. (1961). Gonadotrophic hormones affect aggressive behavior in Starlings. *Science* **134**, 1522–1523.

Matthews, L. H. (1939). Visual stimulation and ovulation in pigeons. *Proc. Roy. Soc., Ser. B* **126**, 557–560.

Meier, A. H. (1969). Antigonadal effects of prolactin in the White-throated Sparrow, *Zonotrichia albicollis. Gen. Comp. Endocrinol.* **13**, 222–225.

Meier, A. H., and Dusseau, J. W. (1968). Prolactin and the photoperiodic gonadal response in several avian species. *Physiol. Zool.* **41**, 95–103.

Meites, J., and Turner, C. W. (1947). Effects of sex hormones on pituitary lactogen and crop glands of common pigeons. *Proc. Soc. Exp. Biol. Med.* **64**, 465–468.

Miller, A. H. (1949). Potentiality for testicular recrudescence during the annual refractory period of the Golden-crowned Sparrow. *Science* **109**, 546.

Miller, A. H. (1955). Breeding cycles in a constant equatorial environment in Colombia, South America. *Proc. Int. Congr. Ornithol., 11th, 1954* pp. 495–503.

Miller, A. H. (1959). Reproductive cycles in an equatorial sparrow. *Proc. Nat. Acad. Sci. U.S.* **45**, 1095–1100.

Mitchell, M. E. (1967a). Stimulation of the ovary in hypophysectomized hens by an avian pituitary preparation. *J. Reprod. Fert.* **14**, 249–256.

Mitchell, M. E. (1967b). The effects of avian gonadotrophin precipitate on pituitary-deficient hens. *J. Reprod. Fert.* **14**, 257–263.

Mitchell, M. E. (1970). Treatment of hypophysectomized hens with partially purified avian FSH. *J. Reprod. Fert.* **22**, 233–241.

Mok, C. C., Martin, W. G., and Common, R. H. (1961). A comparison of phosvitins prepared from hen's serum and from hen's egg yolk. *Can. J. Biochem. Physiol.* **39**, 109–117.

Morato-Manaro, J., Albrieux, A. S., and Buno, W. (1938). Wirkung der Sexualhormone auf den Hahnenkamm. *Klin. Wochenscher.* **177**, 784–785.

Morgan, W., and Kohlmeyer, W. (1957). Hens with bilateral oviducts. *Nature (London)* **180**, 98.

Morley, A. (1943). Sexual behaviour in British birds from October to January. *Ibis* **85**, 132–158.

Morris, T. R., and Nalbandov, A. V. (1961). The induction of ovulation in starving pullets using mammalian and avian gonadotrophins. *Endocrinology* **68**, 637–646.

Morton, M. L., King, J. R., and Farner, D. S. (1969). Postnuptial and postjuvenile molt in White-crowned Sparrows in Central Alaska. *Condor* **71**, 376–385.

Murton, R. K. (1958). The breeding of Wood Pigeon populations. *Bird Study* **5**, 157–183.

Murton, R. K. (1965). "The Wood Pigeon." Collins, London.

Murton, R. K. (1968). Breeding, migration and survival of Turtle Doves. *Brit. Birds* **61**, 193–212.

Murton, R. K. (1973). Unpublished results.

Murton, R. K., Bagshawe, K. D., and Lofts, B. (1969a). The circadian basis of specific gonadotrophin release in relation to avian spermatogenesis. *J. Endocrinol.* **45**, 311–312.

Murton, R. K., Thearle, R. J. P., and Lofts, B. (1969b). The endocrine basis of breeding behaviour in the feral pigeon (*Columba livia*). I. Effects of exogenous hormones on the pre-incubation behaviour of intact males. *Anim. Behav.* **17**, 286–306.

Murton, R. K., Lofts, B., and Orr, A. H. (1970a). The significance of circadian based photosensitivity in the House Sparrow, *Passer domesticus. Ibis* **112**, 448–456.

Murton, R. K., Lofts, B., and Westwood, N. J. (1970b). Manipulation of photo-refractoriness in the House Sparrow *Passer domesticus* by circadian light regimes. *Gen. Comp. Endocrinol.* **14**, 107–113.

Muschke, H. E. (1953). Histometrische Untersuchungen am Rattenhoden nach Hypophysektomie und nach Choriongonadotropinzufuhr. *Endokrinologie* **3**, 281–294.

Nalbandov, A. V. (1945). A study of the effects of prolactin on broodiness and on cock testes. *Endocrinology* **36**, 251–258.

Nalbandov, A. V. (1953). Endocrine control of physiological functions. *Poultry Sci.* **32**, 88–103.

Nalbandov, A. V. (1956). Effect of progesterone on egg production. *Poultry Sci.* **35**, 1162.

Nalbandov, A. V. (1959). Neuroendocrine reflex mechanisms: Bird ovulation. *In* "Comparative Endocrinology" (A. Gorbman, ed.), pp. 161–173. Wiley, New York.

Nalbandov, A. V. (1961). Mechanisms controlling ovulation of avian and mammalian follicles. *In* "Control of Ovulation" (C. A. Villee, ed.), pp. 122–132. Pergamon, Oxford.

Nalbandov, A. V. (1962). The role of the endocrine system in the control of certain biological rhythms in birds. *Ann. N.Y. Acad. Sci.* **98**, 916–925.

Nalbandov, A. V. (1966). Hormonal activities of the pars distalis in reptiles and birds. *In* "The Pituitary Gland" (G. W. Harris and B. T. Donovan, eds.), pp. 295–316. Univ. of California Press, Berkeley.

Nalbandov, A. V., and Card, L. E. (1946). Effect of FSH and LH upon the ovaries of immature chicks and low-producing hens. *Endocrinology* **38**, 71–78.

Nalbandov, A. V., and James, M. F. (1949). The blood vascular system of the chicken ovary. *Amer. J. Anat.* **85**, 347–378.

Nalbandov, A. V., Hochhauser, M., and Dugas, M. (1945). A study of the effect of prolactin on broodiness and on cock testes. *Endocrinology* **36**, 251–258.

Narbaitz, R., and de Robertis, J. M. (1968). Post-natal evolution of steroidogenic cells in the chicken ovary. *Histochemie* **15**, 187–193.

Nelson, O. E., and Stabler, R. M. (1940). The effect of testosterone proprionate on the early development of the reproductive ducts in the female Sparrow Hawk (*Falco sparverius sparverius*). *J. Morphol.* **66**, 277–297.

Nicholls, T. J., and Graham, G. P. (1972). Observations on the ultrastructure and differentiation of Leydig cells in the testis of the Japanese Quail (*Coturnix coturnix japonica*). *Biol. Rep.* **6**, 179–192.

Nishiyama, H. (1955). Studies on the accessory reproductive organs in the cock. *J. Fac. Agr., Kyushu Univ.* **10**, 277–305.

Novikov, B. G., and Roudneva, L. M. (1964). Dependence of the ovarian function of ducks on the hypothalamus. *Zh. Obshch. Biol.* **25**, 390–393.

Novikov, B. G., and Roudneva, L. M. (1969). Particularités du contrôle hypothalamique du functionnement et du dévelopement des gonades chez les oiseaux. *Gen. Comp. Endocrinol.* **13**, 522.

O'Grady, J. E. (1968). The determination of oestradiol and oestrone in the plasma of the domestic fowl by a method involving the use of labelled derivatives. *Biochem. J.* **106**, 77–86.

O'Malley, B. W. (1967). *In vitro* hormonal induction of a specific protein (avidin) in chick oviduct. *Biochemistry* **6**, 2546–2551.

O'Malley, B. W., McGuire, W. L., and Korenman, S. G. (1967). Estrogen stimulation of synthesis of specific proteins and RNA polymerase activity in the immature chicken oviduct. *Biochim. Biophys. Acta* **145**, 204–207.

O'Malley, B. W., Kirschner, M. A., and Bardin, C. W. (1968). Estimation of plasma androgenic and progestational steroids in the laying hen. *Proc. Soc. Exp. Biol. Med.* **127,** 521–523.

Opel, H. (1964). Premature oviposition following operative interference with the brain of the chicken. *Endocrinology* **74,** 193–200.

Opel, H., and Nalbandov, A. V. (1958). A study of hormonal control of growth and ovulation of follicles in hypophysectomized hens. *Poultry Sci.* **37,** 1230–1231.

Opel, H., and Nalbandov, A. V. (1961a). Follicular growth and ovulation in hypophysectomized hens. *Endocrinology* **69,** 1016–1028.

Opel, H., and Nalbandov, A. V. (1961b). Ovulability of ovarian follicles in the hypophysectomized hen. *Endocrinology* **69,** 1029.

Oslund, R. (1928). Seasonal modifications in testes of vertebrates. *Quart. Rev. Biol.* **3,** 254–270.

Ozon, R. (1965). Detection of estrogen steroid hormones in the blood of adult chickens and of chick embryos. *C. R. Acad. Sci.* **261,** 5664–5666.

Parkes, A. S., and Emmens, C. W. (1944). Effects of androgens and estrogens on birds. *Vitam. Horm. (New York)* **2,** 361–408.

Parkes, A. S., and Marshall, A. J. (1960). The reproductive hormones in birds. *In* "Marshall's Physiology of Reproduction" (A. S. Parkes, ed.), 3rd ed., Vol. 1, Part 2, p. 583–706. Longmans, Green, New York.

Patel, M. D. (1936). The physiology of the formation of the pigeon's milk. *Physiol. Zool.* **9,** 129–152.

Patterson, R., Younger, J. S., Weigle, W. O., and Dixon, F. J. (1961). The metabolism of serum proteins in the hen and chick and secretion of serum proteins by the ovary of the hen. *J. Gen. Physiol.* **45,** 501–513.

Payne, F. (1943). The cytology of the anterior pituitary of broody fowls. *Anat. Rec.* **86,** 1–13.

Payne, R. B. (1969). Breeding seasons and reproductive physiology of Tricolored Blackbirds and Red-wing Blackbirds. *Univ. Calif., Berkeley, Publ. Zool.* **90,** 1–137.

Pearson, A. K. (1955). Natural history and breeding behaviour of the tinamou *Nothoprocta ornata. Auk* **72,** 113–127.

Peterson, A. J. (1955). The breeding cycle in the Bank Swallow. *Wilson Bull.* **67,** 235–286.

Pfeiffer, L. A. (1947). Gonadotrophic effects of exogenous sex hormones on the testes of sparrows. *Endocrinology* **41,** 92–104.

Phillips, R. E. (1959). Endocrine mechanisms of the failure of Pintails (*Anas acuta*) to reproduce in captivity. Ph.D. Thesis, Cornell University, Ithaca, New York.

Phillips, R. E., and van Tienhoven, A. (1960). Endocrine factors involved in the failure of Pintail Ducks *Anas acuta* to reproduce in captivity. *J. Endocrinol.* **21,** 253–261.

Pittendrigh, C. S. (1966). The circadian oscillation in *Drosophila pseudoobscura* pupae; a model for the photoperiodic clock. *Z. Pflanzenphysiol.* **54,** 275–307.

Polikarpova, E. (1940). Influence of external factors upon the development of the sexual gland of the sparrow. *Dokl. Akad. Nauk SSSR* **26,** 91–95.

Porte, A., and Weniger, J. P. (1961). Ultrastructure des cellules interstitielles du testicule d'embryon de poulet de 18 jours. *C. R. Soc. Biol.* **155,** 2181–2184.

Press, N. (1959). An electron microscopic study of a mechanism for the delivery of follicular cytoplasm to an avian egg. *Exp. Cell Res.* **18,** 194–196.

Ralph, C. L. (1959). Some effects of hypothalamic lesions on gonadotrophin release in the hen. *Anat. Rec.* **134,** 411–431.

Ralph, C. L., and Fraps, R. M. (1959a). Long-term effects of diencephalic lesions on the ovary of the hen. *Amer. J. Physiol.* **197,** 1279–1283.

Ralph, C. L., and Fraps, R. M. (1959b). Effect of hypothalamic lesions on progesterone-induced ovulation in the hen. *Endocrinology* **65,** 819–824.

Ralph, C. L., and Fraps, R. M. (1960). Induction of ovulation in the hen by injection of progesterone into the brain. *Endocrinology* **66,** 269–272.

Ranney, R. E., and Chaikoff, I. L. (1951). Effect of functional hepatectomy on estrogen induced lipaemia in the fowl. *Amer. J. Physiol.* **165,** 600–603.

Regaud, C. (1901). Etudes sur la structure des séminiferes et sur la spermatogenèse chez mamifères. *Arch. Anat. Microsc.* **4,** 101–123.

Richardson, K. C. (1935). The secretory phenomena in the oviduct of the fowl including the process of shell formation examined by micro-incineration technique. *Phil. Trans. Roy. Soc. London, Ser. B* **225,** 149–195.

Riddle, O. (1918). Further observations on the relative size and form of the right and left testes of pigeons in health and disease and as influenced by hybridity. *Anat. Rec.* **14,** 283–334.

Riddle, O. (1927). The cyclical growth of the vesicula seminalis in birds is hormone controlled. *Anat. Rec.* **37,** 1–11.

Riddle, O. (1942). Ovulation on estrogen production. *Endocrinology* **31,** 498–506.

Riddle, O., and Bates, R. W. (1933). Concerning anterior pituitary hormones. *Endocrinology* **17,** 689–698.

Riddle, O., and Dykshorn, S. W. (1932). Secretion of crop-milk in the castrated male pigeon. *Proc. Soc. Exp. Biol. Med.* **29,** 1213–1215.

Riddle, O., and Reinhart, W. H. (1926). Studies on the physiology of reproduction in birds. XXI. Blood calcium changes in the reproductive cycle. *Amer. J. Physiol.* **76,** 660–676.

Riddle, O., Hollander, W. F., and Schooley, J. P. (1945). A race of hermaphrodite-producing pigeons. *Anat. Rec.* **92,** 401–423.

Ringoen, A. R. (1940). The effect of theelin administration upon the reproductive system of the female English Sparrow *Passer domesticus* (Linnaeus). *J. Exp. Zool.* **83,** 379–383.

Ringoen, A. R. (1943). Effects of injections of testosterone propionate on the reproductive system of the female English Sparrow, *Passer domesticus (Linnaeus). J. Morphol.* **73,** 423–440.

Roepke, R. R., and Bushnell, L. D. (1936). A serological comparison of the phosphoprotein of the serum of the laying hen and the vitellin of the egg yolk. *J. Immunol.* **30,** 109–113.

Roepke, R. R., and Hughes, J. S. (1935). Phosphorus partition in the blood serum of laying hens. *J. Biol. Chem.* **108,** 79–88.

Romanoff, A. L. (1960). "The Avian Embryo. Structural and Functional Development." Macmillan, New York.

Romanoff, A. L., and Romanoff, A. J. (1949). "The Avian Egg." Wiley, New York.

Roos, T. B., and Meyer, R. K. (1961). Seasonal hormonally induced changes in plasma clearing factor and lipids in *Phasianus colchicus. Gen. Comp. Endocrinol.* **1,** 392–402.

Roosen-Runge, E. C., and Barlow, F. D. (1953). Quantitative studies on human spermatogenesis. I. Spermatogonia. *Amer. J. Anat.* **93,** 143–169.

Rothchild, I., and Fraps, R. M. (1944a). On the function of the ruptured ovarian follicle of the domestic fowl. *Proc. Soc. Exp. Biol. Med.* **56,** 79–82.

Rothchild, I., and Fraps, R. M. (1944b). Relation between light-dark rhythms and hour of lay of eggs experimentally retained in the hen. *Endocrinology* **35**, 355–362.

Rothchild, I., and Fraps, R. M. (1949). The induction of ovulating hormone release from the pituitary of the domestic hen by means of progesterone. *Endocrinology* **44**, 141–149.

Rowan, W., and Batrawi, A. M. (1939). Comments on the gonads of some European migrants collected in East Africa immediately before their spring departure. *Ibis* **3**, 58–65.

Rzasa, J., and Ewy, Z. (1970). Effect of vasotocin and oxytocin on oviposition in the hen. *J. Reprod. Fert.* **21**, 549–550.

Rzasa, J., and Ewy, Z. (1971). Effect of vasotocin and oxytocin on intrauterine pressure in the hen. *J. Reprod. Fert.* **25**, 115–116.

Saeki, Y., and Tanabe, Y. (1955). Changes in prolactin content of fowl pituitary during broody periods and some experiments on the induction of broodiness. *Poultry Sci.* **34**, 909–919.

Samuels, L. T. (1960). Metabolism of steroid hormones. *In* "Metabolic Pathways" (D. M. Greenberg, ed.), 2nd ed., Vol. 1, pp. 431–480. Academic Press, New York.

Sayler, A., Dowd, A. J., and Wolfson, A. (1970). Influence of photoperiod on the localization of Δ^5-3β-hydroxysteroid dehydrogenase in the ovaries of maturing Japanese Quail. *Gen. Comp. Endocrinol.* **15**, 20–30.

Scanes, C. G., and Follett, B. K. (1972). Fractionation and assay of chicken pituitary hormones. *Br. Poult. Sci.* **13**, 603–610.

Schäfer, E. (1954). Zur Biologie des Steisshuhnes *Nothocercus bonapartei*. *J. Ornithol.* **95**, 219–232.

Schechtman, A. M. (1956). Uptake and transfer of macromolecules with special reference to growth and development. *Int. Rev. Cytol.* **5**, 303–322.

Schjeide, O. A. (1967). Effects of estrogens on lipid metabolism in the chicken. *Progr. Biochem. Pharmacol.* **2**, 268–275.

Schjeide, O. A., and McCandless, R. (1962). On the formation of mitochondria. *Growth* **26**, 309–321.

Schjeide, O. A., and Urist, M. R. (1956). Proteins and calcium in serums of estrogen-treated roosters. *Science* **124**, 1242–1244.

Schjeide, O. A., and Urist, M. R. (1959). Proteins and calcium in egg yolk. *Exp. Cell. Res.* **17**, 84–94.

Schjeide, O. A., Binz, S., and Regan, N. (1960). Estrogen-induced serum protein synthesis in the liver of the chicken embryo. *Growth* **24**, 401–410.

Schjeide, O. A., Wilkens, M., McCandless, R., Munn, R., Peterson, M., and Carlsen, G. (1963). Liver synthesis, plasma transport and structural alterations accompanying passage of yolk proteins. *Amer. Zool.* **3**, 167–184.

Schooley, J. P. (1937). Pituitary cytology in pigeons. *Cold Spring Harbor Symp. Quant. Biol.* **5**, 165–179.

Selander, R. K. (1960). Failure of estrogen and prolactin treatment to induce brood patch formation in Brown-headed Cowbirds. *Condor* **62**, 65.

Selander, R. K., and Kuich, L. L. (1963). Hormonal control and development of the incubation patch in icterids, with notes on behavior in cowbirds. *Condor* **65**, 73–90.

Selander, R. K., and Yang, S. Y. (1966). The incubation patch of the House Sparrow, *Passer domesticus* Linaeus. *Gen. Comp. Endocrinol.* **6**, 325–333.

Sharma, D. C., Racz, E. A., Dorfman, R. I., and Schoen, E. J. (1967). A comparative study of the biosynthesis of testosterone by human testes and a virilizing interstitial cell tumour. *Acta Endocrinol. (Copenhagen)* **56**, 726–736.

Shoemaker, H. H. (1939). Effect of testosterone propionate on behaviour of the female canary. *Proc. Soc. Exp. Biol. Med.* **41**, 299–302.

Siller, W. G. (1956). A Sertoli cell tumour causing feminization in a brown leghorn capon. *J. Endocrinol.* **14**, 197–203.

Skutch, A. F. (1957). The incubation patterns of birds. *Ibis* **99**, 69–93.

Smith, K. B. V., and Lacy, D. (1959). Residual bodies of seminiferous tubules of the rat. *Nature (London)* **184**, 249–251.

Steel, E. A., and Hinde, R. A. (1963). Hormonal control of brood patch and oviduct development in domesticated canaries. *J. Endocrinol.* **26**, 11–24.

Steel, E. A., and Hinde, R. A. (1964). Effect of exogenous oestrogen on brood patch development of intact and ovariectomized canaries. *Nature (London)* **202**, 718–719.

Steel, E. A., and Hinde, R. A. (1966a). Effect of artificially increased day-length in winter on female domesticated canaries. *J. Zool.* **149**, 1–11.

Steel, E. A., and Hinde, R. A. (1966b). Effects of exogenous serum gonadotrophin (PMS) on aspects of reproductive development in female domesticated canaries. *J. Zool.* **149**, 12–30.

Stetson, M. H. (1969). Hypothalamic regulation of FSH and LH secretion in male and female Japanese Quail. *Amer. Zool.* **9**, 1078–1079.

Stetson, M. H. (1971). Control mechanisms in the avian hypothalamo-hypophyseal-gonadal axis. Ph.D. Thesis, University of Washington, Seattle.

Stetson, M. H. (1972a). Hypothalamic regulation of testicular function in Japanese Quail. *Z. Zellforsch. Mikrosk. Anat.* **130**, 389–410.

Stetson, M. H. (1972b). Hypothalamic regulation of gonadotropin release in female Japanese Quail. *Z. Zellforsch. Mikrosk. Anat.* **130**, 411–428.

Stetson, M. H. (1972c). Feedback regulation of testicular function in Japanese Quail: Testosterone implants in the hypothalamus and adenohypophysis. *Gen. Comp. Endocrinol.* **19**, 37–47.

Stetson, M. H., and Erickson, J. E. (1970). The antigonadal properties of prolactin in birds. Failure of prolactin to inhibit the uptake of ^{32}P by testes of cockerels *in vivo*. *Gen. Comp. Endocrinol.* **15**, 484–487.

Stetson, M. H., and Erickson, J. E. (1971). Endocrine effects of castration in White-crowned Sparrows. *Gen. Comp. Endocrinol.* **17**, 105–114.

Stockell-Hartree, A., and Cunningham, F. J. (1969). Purification of chicken pituitary follicle stimulating hormone and luteinizing hormone. *J. Endocrinol.* **43**, 609–616.

Sturkie, P. D. (1954). "Avian Physiology." Cornell Univ. Press (Comstock), Ithaca, New York.

Sturkie, P. D., and Lin, Y. C. (1966). Release of vasotocin and oviposition in the hen. *J. Endocrinol.* **35**, 325–326.

Taber, E. (1949). The source and effects of androgen in the male chick treated with gonadotrophins. *Amer. J. Anat.* **85**, 231–263.

Taber, E. (1951). Androgen secretion in the fowl. *Endocrinology* **48**, 6–16.

Taber, E. (1964). Intersexuality in birds. *In* "Intersexuality in Vertebrates including Man" (C. N. Armstrong and A. J. Marshall, eds.), pp. 285–310. Academic Press, New York.

Taber, E., Clayton, M., Knight, J., Gambrell, D., Flowers, J., and Ayers, C. (1958). Ovarian stimulation in the immature fowl by desiccated avian pituitaries. *Endocrinology* **62**, 84–89.

Tanaka, K., and Nakajo, S. (1962a). Participation of neurohypophysial hormone in oviposition in the hen. *Endocrinology* **70**, 453–458.

Tanaka, K., and Nakajo, S. (1962b). Oviposition-inducing activities of the chicken posterior pituitary extract, synthetic oxytocin and mammalian vasopressin. *World's Poultry Congr., Proc., 12th, 1962 Sect. B*, pp. 123–130.

Tanaka, K., Mather, F. B., Wilson, W. O., and McFarland, L. Z. (1965). Effect of photoperiods on early growth of gonads and on potency of gonadotropins of the anterior pituitary gland in *Coturnix. Poultry Sci.* **44**, 662–665.

Tanaka, K., Fujisawa, Y., and Yoshioka, S. (1966). Luteinizing hormone content in the pituitary of laying and nonlaying hens. *Poultry Sci.* **45**, 970–973.

Thapliyal, J. P., and Saxena, R. N. (1964). Absence of a refractory period in the Common Weaver bird. *Condor* **66**, 199.

Tixier-Vidal, A. (1963). Histophysiologie de l'adénohypophyse des oiseaux. *In* "Cytologie de l'adénohypophase" (J. Benoit and C. Da Lage, eds.), pp. 255–273, CNRS, Paris.

Uemura, H. (1964). Effects of gonadectomy and sex steroids on the acid phosphatase activity of the hypothalamo-hypophyseal system in the bird, *Emberiza rustica latifascia. Endocrinol. Jap.* **11**, 185–203.

van Oordt, G. J., and Jung, G. C. A. (1933). Der Einfluss der Kastration bei männlichen Lachmöwen (*Larus ridibundus* L.). *Wilhelm Roux' Arch. Entwicklungsmech. Organismen.* **128**, 165–180.

van Oordt, G. J., and Junge, G. C. A. (1936). Der Einfluss der Kastration auf männliche Kampfläufer (*Philomachus pugnax*). *Wilhelm Roux' Arch. Entwicklungsmech. Organismen.* **134**, 112–121.

van Oordt, P. G. W. J., and Basu, S. L. (1960). The effects of testosterone on the spermatogenesis of the common frog, *Rana temporaria. Acta Endocrinol. (Copenhagen)* **33**, 103–110.

van Tienhoven, A. (1959). Reproduction in the domestic fowl: Physiology of the female. *In* "Reproduction in Domestic Animals" (H. H. Coles and P. T. Cupps, eds.), 1st ed., Vol. 2, pp. 305–342. Academic Press, New York.

van Tienhoven, A. (1961). Endocrinology of reproduction in birds. *In* "Sex and Internal Secretions" (W. C. Young, ed.), 3rd ed., Vol. 2, pp. 1088–1169. Williams & Wilkins, Baltimore, Maryland.

van Tienhoven, A. (1968). "Reproductive Physiology of Vertebrates." Saunders, Philadelphia, Pennsylvania.

Vaugien, L. (1948). Recherches biologiques et experiméntales sur le cycle reproducteur et la mue des oiseaux passeriformes. *Bull. Biol. Fr. Belg.* **82**, 166–213.

Vaugien, L. (1953). Sur l'apparition de la maturité sexuelle des jeunes Perruches ondulées males soumises à diverses conditions d'eclairement: Le développement testiculaire est plus rapide dans l'obscurité complète. *Bull. Biol. Fr. Belg.* **87**, 274–286.

Vaugien, L. (1955). Sur les réactions ovariennes du Moineau domestique soumis, durant le repos sexuel, à des injections de gonadotrophine sérique de jument gravide. *Bull. Biol. Fr. Belg.* **89**, 1–15.

Vowles, D. M., and Harwood, D. (1966). The effect of exogenous hormones on aggressive and defensive behaviour in the Ring Dove (*Streptopelia risoria*). *J. Endocrinol.* **36**, 35–51.

Wada, M. (1972). Effect of hypothalamic implantation of testosterone on photostimulated testicular growth in Japanese Quail (*Coturnix coturnix japonica*). *Z. Zellforsch. Mikrosk. Anat.* **124**, 507–519.

Wallace, R. A. (1963a). Studies on amphibian yolk. III. A resolution of yolk platelet components. *Biochim. Biophys. Acta* **74**, 495–504.

Wallace, R. A. (1963b). Studies on amphibian yolk. IV. An analysis of the main body component of yolk platelets. *Biochim. Biophys. Acta* **74**, 505–518.

Wallace, R. A., Jared, D. W., and Eisen, A. Z. (1966). A general method for the isolation and purification of phosvitin from vertebrate eggs. *Can. J. Biochem.* **44**, 1647–1655.

Warren, D. C., and Scott, H. M. (1935). The time factor in egg production. *Poultry Sci.* **14**, 195.

Warren, R. P., and Hinde, R. A. (1959). The effect of oestrogen and progesterone on the nest-building of domesticated canaries. *Anim. Behav.* **7**, 209–213.

Warren, R. P., and Hinde, R. A. (1961). Roles of the male and the nest-cup in controlling the reproduction of female canaries. *Anim. Behav.* **9**, 64–67.

Wattenberg, L. W. (1958). Microscopic histochemical demonstration of steroid-3β-ol dehydrogenase in tissue sections. *J. Histochem. Cytochem.* **6**, 225–232.

Weniger, J. P., and Zeis, A. (1969). Formation d'oestrone et d'oestradiol radioactifs à partir d'acetate de Na-1-^{14}C par les ébauches gonadiques d'embryon de Poulet de 5 à 6 jours. *C. R. Acad. Sci.* **268**, 1306.

Wilde, C. E., Orr, A. H., and Bagshawe, K. D. (1967). A sensitive radio-immunoassay for human chorionic gonadotrophin and luteinizing hormone. *J. Endocrinol.* **37**, 23–35.

Wilson, F. E. (1970). The tubero–infundibular region of the hypothalamus: A focus of testosterone sensitivity in male Tree Sparrow (*Spizella arborea*). *In* "Aspect of neuroendocrinology" (W. Bergmann and B. Scharrer, eds.), pp. 274–286. Springer, Berlin.

Witschi, E. (1935). The origin of asymmetry in the reproductive system of birds. *Amer. J. Anat.* **56**, 119–141.

Witschi, E. (1945). Quantitative studies on the seasonal development of the deferent ducts in passerine birds. *J. Exp. Zool.* **100**, 549–564.

Witschi, E. (1955). Vertebrate gonadotrophins. *Mem. Soc. Endocrinol.* **4**, 149–165.

Witschi, E. (1961). Sex and secondary sexual characters. *In* "Biology and Comparative Physiology of Birds" (A. J. Marshall, ed.), Vol. 2, pp. 115–168. Academic Press, New York.

Witschi, E., and Fugo, N. W. (1940). Response of sex characters of the adult female Starling to synthetic hormones. *Proc. Soc. Exp. Biol. Med.* **45**, 10–14.

Wolfson, A. (1959a). The role of light and darkness in the regulation of the Spring migration and reproductive cycles in birds. *In* "Photoperiodism and Related Phenomena in Plants and Animals," Publ. No. 55, pp. 679–716. Amer. Ass. Advance. Sci., Washington, D.C.

Wolfson, A. (1959b). Role of light and darkness in the regulation of the refractory period in the gonadal and fat cycles of migratory birds. *Physiol. Zool.* **32**, 160–176.

Woods, J. E., and Domm, L. V. (1966). A histochemical identification of the androgen-producing cells in the gonads of the domestic fowl and albino rat. *Gen. Comp. Endocrinol.* **7**, 559–570.

Wyburn, G. M., Johnston, H. S., and Aitken, R. N. C. (1965). Specialised plasma membranes in the preovulatory follicle of the fowl. *Z. Zellforsch. Mikrosk. Anat.* **68**, 70–79.

Wyburn, G. M., Johnston, H. S., and Aitken, R. N. C. (1966). Fate of the granulosa cells in the hen's follicle. *Z. Zellforsch. Mikrosk. Anat.* **72**, 53–65.

Yamashima, Y. (1952). Notes on experimental brooding induced by prolactin injections in the domestic cock. *Annot. Zool. Jap.* **25**, 135–140.

Yapp, W. B. (1970). "The Life and Organization of Birds." Arnold, London.

Chapter 2

THE ADENOHYPOPHYSIS

A. Tixier-Vidal and B. K. Follett

I. Introduction

The aim of this chapter is to correlate the morphology of the pituitary cells with physiological analyses of pituitary function and
chemical studies on the pituitary hormones insofar as they have been
isolated. The past 15 years have seen great progress made into understanding these three facets of avian biology. In the field of morphology,
the development of cytochemistry and the introduction of the electron
microscope have permitted the easier identification of cell types and
improved assessment of their functional state. During the same
period there has been a rapid extension in our knowledge of avian
endocrinology, and especially of avian neuroendocrinology. Chemical and immunochemical studies on the adenohypophyseal hormones
have begun much more recently, but one can hope that in the near
future immunocytochemistry will aid in the identification of the cells
that produce various hormones. At present, these cellular localizations
can be established only on the basis of morphological data and must
still be considered tentative. Such morphological correlations are
nevertheless an important and necessary precedent for immuno-
cytological studies. Moreover, unlike cytology, the chemistry of the
avian pituitary hormones and the immunocytochemical identification
of the cell types will inevitably be limited to a small number of species.

II. Anatomy and Embryology

Most of the basic information on the anatomy and embryology of the
avian pituitary has been carefully and fully described by Wingstrand
(1951), and we shall refer to his monograph on a number of occasions.
Rather more attention will be paid, however, to recent studies concerned with cellular arrangement, pituitary vascularization, histo-
genesis, and the experimental analysis of pituitary development.

A. GENERAL ANATOMY

The gross anatomy of the pituitary is remarkably uniform among
birds. According to Wingstrand's nomenclature, the major part of the
adenohypophysis is the pars distalis, which lies ventral or rostroventral to the neurohypophysis, and which in some species is divided
externally into two parts by a transverse sulcus. The pars intermedia is
not present as a separate entity. The pars tuberalis consists of three
parts—the pars tuberalis proper, consisting of strings of cells cover-

ing part of the surface of the diencephalon; a portal zone, which consists of cells lodged between and along the portal vessels; and the pars tuberalis interna, which is intimately fused with the pars distalis. Depending on the species, these different parts may be developed unequally.

B. MICROSCOPIC ANATOMY

1. The Pars Distalis

The glandular cells are arranged in cords or in acini and are limited by a basement membrane which often appears infolded under the electron microscope. Between the cells lies an important plexus of blood capillaries, the fine structure of which has been described for the White-crowned Sparrow (*Zonotrichia leucophrys gambelii*) by Mikami *et al.* (1970). They are of the typically fenestrated type as seen also in the rat pituitary and in other endocrine organs. The capillary wall consists of a single layer of endothelial cells containing many diaphragm-covered fenestrations. On their perivascular faces the endothelial cells are lined by a basement membrane (Fig. 1). The majority of the glandular cells have direct contact with the pericapillary space, sometimes by means of narrow cytoplasmic extensions. The small chromophobic cells of the pituitary have no such relationship with the capillaries and lie in the center of the cell clusters. Under the light microscope, these appear to form aggregates of nuclei—the so-called *Kernhaufen* of German authors. Such figures are especially frequent in wild birds out of the breeding season.

Several histological features support the view that the pars distalis is divisible into a caudal lobe and a cephalic lobe. According to Rahn and Painter (1941) the most important difference between the lobes lies in the distribution of two types of acidophilic cells, the larger A1 cells being restricted to the caudal lobe, while the A2 cells, which are smaller and contain many fine granules, are characteristic of the cephalic lobe. Wingstrand (1951) confirmed this subdivision in a great number of species, and we shall revert to this important question later. The cellular arrangement in the two lobes is also frequently different, with the caudal region being characterized by numerous pseudo-acini and the cephalic by compact clusters or longitudinal cords of cells. However, it is important to note that in no species studied is there a true separation of the two lobes either by a connective tissue wall or by a distinct alignment of the cells, the two lobes invariably being mixed in the median region. Moreover, the trans-

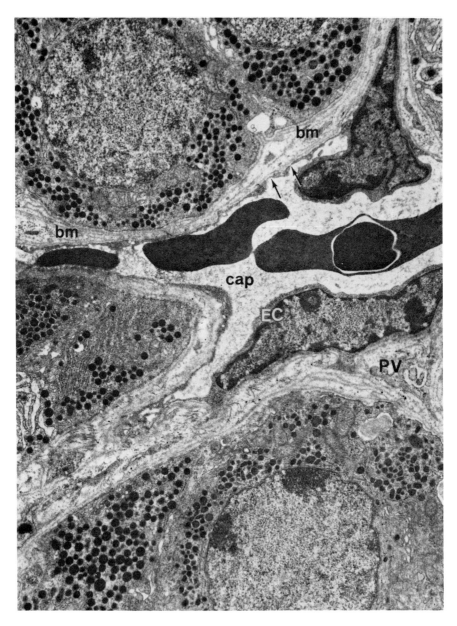

FIG. 1. An area of the caudal lobe of the pituitary gland of a pigeon showing the relationship between the glandular cells and the blood capillaries (cap). The endothelial cells (EC) containing diaphragm-covered fenestrations (arrow) are lined by a basement membrane (bm). The glandular acini are limited by another basement membrane. The space between the two basement membranes is occupied by collagen and perivascular cells (PV). × 9600.

verse sulcus of the pars distalis, whenever it exists, does not correspond with the histological subdivision.

A typical feature of the avian pars distalis is the presence in the adult of chromophobic and PAS (periodic acid–Schiff)-positive colloid droplets in the center of the acini. These formations are sometimes so numerous that the sections almost give the impression of being a thyroid gland. They have been described by several authors (e.g., Payne, 1942, 1946; Wingstrand, 1951), and their significance has often been discussed. More recent studies with the electron microscope show that the colloid has nothing to do with any pituitary secretion and is a fibrous material arising from the cytoplasm of the apical zones of several gland cells (Tixier-Vidal *et al.*, 1966).

2. *The Pars Tuberalis*

In the pars tuberalis proper, the flattened strings of cells are surrounded by a basement membrane (Mikami *et al.*, 1970). The preportal vessels course through these cell cords. They have the typical form of venous capillaries with a single layer of endothelial cells invested in a basement membrane. The junctions between these capillaries and the pars tuberalis cells are much less intimate than with the cells of the pars distalis. In the portal zone, the tuberal cells are arranged in long strings oriented with the blood vessels (see Mikami *et al.*, 1970). The cytoplasm of the glandular cells is frequently almost chromophobic, although after Bodian impregnation, some cells show a few argyrophilic granules. To our knowledge no work has been published on these cells since that of Wingstrand (1951).

C. VASCULARIZATION

As in mammals, the pars distalis is irrigated by a hypophysial portal system which arises as a primary plexus on the median eminence. This plexus is drained by portal vessels that enter the pars distalis in the dorsocentral region and then break up into a secondary plexus. Blood from the pars distalis is drained directly into the cavernous sinus which communicates with the carotid and jugular veins. Blood is supplied to the primary portal plexus by the infundibular arteries. This portal system was described almost simultaneously by Green (1951) in the chicken, by Benoit and Assenmacher (1951) and Assenmacher (1952) in the duck, and by Wingstrand (1951) in several other species. The crucial demonstration of blood flow from the median eminence to the pars distalis was first accomplished by Assenmacher (1958) in the duck.

Since these studies, further interest has centered on the differential distribution of the portal vessels between the lobes of the pars distalis, and the most thorough study by Vitums *et al.* (1964) in *Zonotrichia leucophrys gambelii* has established several points: (1) There are distinct anterior and posterior primary capillary plexus corresponding with the anterior and posterior divisions of the median eminence. (2) These plexus are drained by separate anterior and posterior groups of portal vessels. (3) The anterior vessels are distributed mainly in the cephalic lobe of the pars distalis, while the posterior vessels supply the caudal lobe. A point-to-point distribution of the portal vessels has also been described in less detail for fifteen other species of birds by Dominic and Singh (1969). The degree of vascular separation is species dependent, and anterior and posterior capillary plexus are not distinct in the Cattle Egret (*Bubulcus ibis*), the Indian Roller (*Coracias benghalensis*), and the White-breasted Kingfisher (*Halcyon smyrnensis*). In the Japanese Quail (Sharp and Follett, 1969), the portal vessels do not form separate groups beneath the median eminence before entering the pars distalis but pass directly as single vessels into the pituitary throughout its length. In most specimens, there is evidence for two portal plexus supplying the two lobes of the pars distalis. It must be emphasized that this distribution of portal vessels has not yet been shown to have any physiological significance.

D.   EMBRYOLOGY

1. *Morphogenesis*

The adenohypophysis is formed exclusively by Rathke's pouch. At an early stage of development (6 days in the domestic fowl) there is subdivision of the tissues, the proliferation of which result in the different parts of the adult adenohypophysis (Wingstrand, 1951; Assenmacher, 1951, 1952; Grignon, 1955, 1956; Thommes and Russo, 1959): (1) The aboral and oral lobes give rise, respectively, to the caudal and cephalic lobes. (2) The lateral lobes form the pars tuberalis. (3) The contact zone with the brain remains as a single layer of cells and corresponds morphologically with the pars intermedia. Later these cells lose their contact with the pars nervosa and are included in the pars distalis (cf. other vertebrates).

The vasculogenesis of the pituitary has been studied in the Pekin duck by Assenmacher (1951, 1952). The hypophyseal portal system is established simultaneously with the proliferation of the oral and aboral lobes of the anlage. The primary plexus on the diencephalon appears first and is drained by venules that penetrate the proliferating cellular cords of the gland and are later transformed into the capillaries of the secondary plexus. A similar situation occurs in the chick embryo (Grignon, 1956; Thommes and Russo, 1959).

2. Cellular Differentiation

The chronological appearance of the cell types has been studied in the chick embryo, but the results have varied with the methods and with the authors. Three kinds of staining methods have been used:

1. Tinctorial methods: Severinghaus (Rahn, 1939; Payne, 1942, 1946), Altmann's fuchsin (Payne, 1942, 1946), Azan technique (Wingstrand, 1951; Tixier-Vidal, 1954; Grignon, 1955), Romeis' cresazan (Grignon, 1955).

2. Silver staining according to Bodian (Wilson, 1949; Wingstrand, 1951).

3. Cytochemical techniques for glycoproteins [periodic acid–Schiff (PAS)] (Aronson, 1952; Tixier-Vidal, 1954; Grignon, 1955, 1956), Gomori's paraldehyde fuchsin (Tixier-Vidal, 1954).

The first morphological signs of cellular differentiation appear in the cephalic lobe where the first acidophils are generally found on day 10 or 11. The first PAS-positive cells occur either on day 6 (Aronson, 1952; Grignon, 1955) or on day 11 (Tixier-Vidal, 1954). In the caudal lobe, differentiated cells appear much later, with acidophils being seen on either day 15 (Wingstrand, 1951) or days 18–19 (Wilson, 1952; Tixier-Vidal, 1954), and the first PAS-positive cells occurring only after hatching. The appearance of acidophils and of PAS-positive cells within the cephalic lobe is closely associated with the more precocious establishment of the vascular pattern in this zone (Grignon, 1955, 1956; Thommes and Russo, 1959). Electron microscopy should provide a better estimate of when secretory activity starts in the embryo, and recently Guedenet et al. (1970) have identified the first secretory granules in the cephalic lobe on day 8.

Tentative correlations have been made in the embryonic chick between morphological and functional differentiation of the adenohypophysis. Melanotropic hormone (MSH) activity has been demonstrated in the 5-day embryo (Chen et al., 1940) and seems related to the appearance at the same time of the first strongly silver-positive cells (Wingstrand, 1951). By using a sensitive in vitro method, thyrotropic activity was first found on day 7 of development (Tixier-Vidal, 1956). This precedes the appearance of PAS-positive cells in the cephalic lobe. Later, however, there are clear parallels between the evolution in both numbers and cytology of the PAS-positive cells and changes in thyroid activity (Tixier-Vidal, 1954, 1958). Although there is some evidence for the presence in the embryonic chick of gonadotropic and corticotropic activities (Moszkowska, 1949, 1956; Case, 1952), no cellular localization for these activities has been suggested.

3. Experimental Embryology

Knowledge about the differentiation of the adenohypophysis has been greatly enhanced by the work of Ferrand and Le Douarin with the embryonic chick (Le Douarin *et al.*, 1967; Ferrand, 1969, 1970). Their main findings may be summarized as follows. Before the 25 somite stage, the infundibular floor of the forebrain is necessary to induce the formation and maintenance of Rathke's pouch. These results confirm the work of Stein (1929, 1933) and Hillemann (1943). After this stage, however, the adenohypophyseal anlage can develop normally in the absence of the hypothalamus. Thus, the 3-day embryonic adenohypophysis is maintained as a coelomic graft for 14 days and during this period undergoes normal histological development with acidophilic and PAS-positive cells appearing at the same time as *in situ*. Two acidophilic cell types are distinguishable, and thyrotropic activity appears under these conditions. This latter finding corroborates previous results (Tixier-Vidal, 1958) showing the independence of pituitary thyrotropic function from the hypothalamus in the chick embryo.

Ferrand's discoveries are of particular importance in view of the role previously assigned to pituitary vasculogenesis in the histological development of the gland. It now seems that during development the pituitary cells do not require specific substances carried from the infundibulum by the portal system. Some recent results of Ferrand (1971) indicate, moreover, that the vascularization exerts a favorable but only trophic effect; in chorioallantoic grafts on decapitated chick embryos, the new vascularization pattern stimulates the differentiation of PAS-positive cells in the caudal lobe.

III. Morphological Identification of Cell Types

Particular difficulties are encountered in identifying avian pituitary cell types because the cells are rather small and show only weak affinities for most dyes. As a first step, investigators used well-defined staining methods, such as those of Severinghaus, Bodian, Heidenhain,

Azan, and Romeis, and were able to distinguish three cell classes: (1) large and deeply staining acidophilic cells (A1) in the caudal lobe, (2) small and slightly staining acidophils (A2) localized at the rostral end of the cephalic lobe, and (3) basophilic cells that were generally small and difficult to stain and that tended to lie in the cephalic lobe (Rahn, 1939; Rahn and Painter, 1941; Payne, 1942, 1946; Wingstrand, 1951; Wilson, 1952). Rather more refined analyses became possible when glycoprotein-staining methods were introduced into pituitary cytology. This second phase of research was started by Legait (Legait and Legait, 1955, 1956), who identified PAS-positive and AF-positive cells in the hen pituitary. Thereafter, the methods were developed further by Mikami (1958) for the cockerel and by Herlant *et al.* (1960) and Tixier-Vidal *et al.* (1962) for the Pekin duck. The third phase of cytological research has employed the electron microscope. Such studies were first performed in the hen (Payne, 1965), duck (Tixier-Vidal, 1965), and pigeon (Tixier-Vidal and Assenmacher, 1966), and lately have been extended to include the chicken, White-crowned Sparrow, and Japanese Quail (Mikami, 1969; Mikami *et al.*, 1969; Dancasiu and Campeanu, 1970; Tixier-Vidal *et al.*, 1972).

A. TECHNOLOGY

In spite of the constant technical progress, it often remains difficult to correlate the results obtained by different authors in the various species. This seems to result from the many variations that occur both with the techniques and with the stainability of pituitaries of different species. Before outlining the present stage of our knowledge, it seems useful, therefore, to recall some of the more suitable staining methods.

1. Fixatives

The identification of the cell types requires excellent preservation of the cytoplasmic structures and especially of the secretory granules, and the choice of an appropriate fixative is of paramount importance. The most commonly used are Bouin Hollande sublimate according to Herlant (1960), Zenker-formol, formol calcium, and formol sublimate. Acid fixatives such as Bouin or Susa cause modifications in the affinities of the cells for the stains, in spite of giving a good general appearance to the tissue. For ultrastructural studies, osmic acid and glutaraldehyde give excellent results.

2. Tinctorial Methods

The classical trichrome methods have been used for a long time in

avian pituitary cytology, but more recently some specialized poly-
chromic methods have been introduced. One of the most commonly
used is the alizarin blue tetrachrome of Herlant (1960), which has the
advantage of differentiating five tinctorial types of cell — orange, rose,
purple, light blue, and very dark blue. This method gives the best
results if the tissues are fixed in Bouin-Hollande sublimate. Brookes'
method (1967) gives excellent distinction between the various types
of acidophils in birds. Matsuo has proposed a tetrachrome method
involving orange G, acid fuchsin, methyl green, and acid violet
(Matsuo et al., 1969). This technique stains the acidophils orange to
red to purple, the basophils green, and the amphophils violet or
purple; it invariably gives a deeply stained preparation. In the caudal
lobe, the results correlate well with those of Herlant's tetrachrome,
but in the cephalic lobe, the differentiation of the cell types is less
clear.

3. Cytochemical Methods

In contrast with the tinctorial methods, the cytochemical ones are
easier to standardize, since the underlying chemical reactions are
better understood. They offer more comparable results and have
come into widespread use.

a. Glycoproteins.   The PAS method plays a central role in pituitary
cytology, its importance resulting from the glycoprotein nature of the
gonadotropic and thyrotropic hormones (Section V). Often it is associated with acidic dyes such as orange G or naphthol yellow (Herlant
and Racadot, 1957), which allow for the simultaneous detection of
acidophils and glycoproteins. Matsuo frequently combines it with
methyl blue (Matsuo et al., 1969).

In combination with a cationic dye such as alcian blue, it offers a
useful means of distinguishing the acidic glycoprotein-containing
cells, and several techniques have been proposed, all involving
Alcian blue at low pH (Herlant, 1960; Purves, 1966). By using the
critical electrolyte concentration method of Scott and Dorling (1965),
it has been possible to confirm in duck and Japanese Quail that the
cells that retain Alcian blue are the richest in acidic groups (Tixier-
Vidal and Picart, 1971). The PAS methods may also be combined with
aldehyde thionin (Ezrin and Murray, 1963), but in our experience
the results have been less satisfactory than with Alcian blue–PAS
methods.

The technique of following PAS with lead hematoxylin (Mac-
Connail, 1947) is of particular value since it allows a deeply black-

staining cell to be distinguished from the rose-colored glycoprotein cells (Tixier-Vidal *et al.*, 1968). Pretreatment of the sections with methasol blue or luxol fast blue (Kluver and Barrera, 1953) before PAS can be useful in characterizing one glycoprotein cell type in the cephalic lobe (Herlant *et al.*, 1960; Tixier-Vidal, 1963).

Two methods (Rambourg, 1967; Thiery, 1967) have been used to detect glycoproteins with the electron microscope; both appear successful (Tixier-Vidal and Picart, 1968; Tixier-Vidal *et al.*, 1969; Guedenet *et al.*, 1970; Tixier-Vidal and Picart, 1971).

b. Gomori's Paraldehyde Fuchsin. The cytochemical significance of this dye is not clear. Although somewhat less selective than PAS, it generally stains glycoprotein-containing cells and in its various forms (Halmi, 1952, Gabe, 1953) has been applied to a number of species (fowl — Legait and Legait, 1955; Mikami, 1958; duck — Herlant *et al.*, 1960; Tixier-Vidal *et al.*, 1962; Red Bishop (*Euplectes orix*) — Gourdji, 1964; pigeon — Tixier-Vidal and Assenmacher, 1966; Ljunggren, 1969).

c. Amino Acids. Disulfide groups have been detected in the duck pituitary (Tixier-Vidal *et al.*, 1962) with the method of Barnett and Seligman (1954), the strongest reaction being seen in a single cell type in the caudal lobe. Two cell types in the cephalic lobe of the Japanese Quail (Tougard, 1971) show a high tryptophan content when stained by Glenner's rosindol method (Glenner and Lillie, 1957).

d. Biogenic Amines. In contrast with some mammals, no specific green or yellow fluorescence could be demonstrated by the Falck-Hillarp technique in the pituitary of the pigeon (Ljunggren, 1969). This means that the cells contain neither catecholamines nor serotonin, although some cell types, notably those staining with PAS and AF, are capable of storing noradrenaline if it is administered exogenously (Ljunggren, 1969). In the Japanese Quail, some cells in the cephalic lobe (PAS-positive, lead hematoxylin-positive) show an uptake of [^{3}H]serotonin *in vitro* (Tougard *et al.*, 1968).

e. Phospholipids. Phospholipids have been detected with Baker's acid hematin test in a caudal acidophil of the fowl pituitary (Mikami, 1958). In the Japanese Quail, the same technique gives the strongest reaction in two cell types in the cephalic lobe and one in the caudal lobe (Tougard, 1971).

f. Enzyme Cytochemistry. The localization of acid phosphatase has been studied with the electron microscope in duck and quail pituitaries (Tixier-Vidal and Picart, 1970, 1971). It appears that even in normal birds, lysosomes are very abundant in the pituitary cells. A similar observation has been made with the embryonic chick pituitary (Guedenet *et al.*, 1970). Acetylcholinesterase activity has been localized in PAS-positive cells of some bird pituitaries—White-crowned Sparrow (Haase and Farner, 1969, 1970), Brambling (*Fringilla montifringilla*) (Haase, 1971), House Sparrow (*Passer domesticus*), European Starling (*Sturnus vulgaris*), and the canary, but not in the Japanese Quail, chicken, or pigeon (E. Haase, personal communication).

4. *Ultrastructural Data—Their Importance as Indicators of the Functional State of Pituitary Cells*

Two kinds of criteria are useful in the study of pituitary cells, those that allow for the differentiation and identification of the cell types, and those that are indicators of the functional state of the cells. Information from the electron microscope falls into the latter category and is of particular importance in birds because the pituitary cells are so small and often difficult to stain. Some cytoplasmic structures, of course, such as the Golgi apparatus and the rough endoplasmic reticulum (RER) can be examined only with the electron microscope. When dealing with cellular function (Section IV) we shall attach some importance to ultrastructural data. The identification of cell types with the electron microscope is not easy, since only black and white pictures are obtainable. Two criteria are useful, however; the mean size and electron density of the secretory granules and the features of the RER or ergastoplasm.

B.   Tinctorial Affinities, Cytochemical Properties, and Ultrastructural Features

The tinctorial and cytochemical properties of the avian pituitary cells (Table I) have been studied thoroughly in only a few species— the domestic fowl, both male and female (Payne, 1942, 1943, 1946, 1955; Yasuda, 1953; Legait and Legait, 1955, 1956; Mikami, 1958, 1969; Perek *et al.*, 1957; Inoguchi and Sato, 1962), the pigeon (Schooley, 1937; Schooley and Riddle, 1938; Tixier-Vidal and Assenmacher, 1966; Ljunggren, 1969), the Pekin duck (Herlant *et al.*, 1960; Tixier-Vidal *et al.*, 1962; Tixier-Vidal, 1963), the Red Bishop (Gourdji, 1964), the quail (Tixier-Vidal *et al.*, 1968), and the White-crowned Sparrow (Matsuo *et al.*, 1969). In most, the affinities of the cells for the various stains are altered greatly by the physiological condition of the animal, and at first sight, therefore, the cell types seem to differ greatly among species. However, it is possible to find characteristic features for each cell type, provided they are studied at the time of maximum stainability. This invariably means comparing glands from birds in various physiological states. As in other vertebrates, some species are far better than others for morphological analyses, e.g., the Pekin duck.

A morphological analysis of the pituitary cell types necessitates some form of nomenclature. The International Committee (see van Oordt, 1964) recommended the use of a uniform and functional nomenclature, but this is still impossible in birds because of disagreement over the localization of several of the pituitary hormones. In this chapter, therefore, we shall continue to use Romeis' Greek-letter nomenclature as modified by Herlant. It is no better than any other system, but has been used widely for duck, pigeon, and quail pituitaries. Each cell type will be defined by its most typical tinctorial and cytochemical features. According to Herlant (1964), we shall distinguish two general classes of cells—those whose granules contain glycoproteins, and those whose granules contain simple proteins. Of the seven cell types recognized in the bird pituitary, only five can easily be assigned to these two classes. The remaining two cell types form a third class, characterized by having mixed affinities.

1. Proteinaceous Cells (Acidophils, Serous Cells)

a. α Cells (Caudal Acidophils). The α cells occur in the caudal lobe (Table I, Fig. 2). They have the same properties as the α cells of other vertebrate groups (see Herlant, 1964) and are selectively stained by all strongly acid dyes—orange G, erythrosine, naphthol yellow, and acid fuchsin. They contain disulfide groups due to a high concentration of cysteine and also contain phospholipids and lipoproteins. They are PAS-negative except after permanganate oxidation, when they appear pale rose due to the appearance of some hydroxyl groups. They are generally larger than any other avian cell type and show a strong polarity.

The α cells are among the types included in the classic A1 cells of Rahn and Painter. In fact, after staining with Heidenhain's azan, or with Matsuo's tetrachrome, it is possible to distinguish two types of A1 cells. With Herlant's tetrachrome and with PAS methods these cells can be distinguished as the acidophilic α cells and the γ glycoprotein cells (Fig. 2). α Cells are far less numerous in birds than in mammals.

TABLE I

SUMMARY OF THE MOST TYPICAL TINCTORIAL AND CYTOCHEMICAL PROPERTIES
OF THE AVIAN PITUITARY CELL TYPES

Classes			Tinctorial affinities		
	Herlant Nomenclature	Localization	Herlant tetrachrome[a]	Brooke method	Matsuo[a] method
Serous cells = acidophils	α caudal acidophils C.	caudal lobe	orange	yellow orange	orange
	η cephalic acidophils C.	cephalic lobe (rostro-ventral)	rose (erythrosinophil)	rose	?
Glycoprotein-containing cells	β cephalic PAS + C	cephalic lobe	purple to blue	blackish	violet
	δ Alcian blue + C.	cephalic lobe caudal lobe	pale blue	pale green	light green
	γ glycoprotein acidophils C.	caudal lobe	purple	brownish orange	purple
Lead–hematoxylin-positive cells	ϵ	cephalic lobe (center)	light purple to blue	green + rose	?
	κ	cephalic or both lobes	very deep blue	light green	?

TABLE I *(continued)*

| Cytochemical properties | | | | | | |
| Glycoproteins | | | | Phospholipids Baker | Amino acids | |
PAS hemalum orange G	Alcian blue PAS orange G	PAS lead hematoxylin	Aldehyde Fuchsin	hématéine or Luxol fast blue)	S−S	Tryptophan
yellow	pale yellow	−	+	+	+	−
yellow	very pale yellow	−	−	−	+	−
deep rose	deep rose	deep rose + light gray	violet in some species	+	−	+ (Japanese quail)
pale rose	pale blue	−	light to deep violet	−	−	−
brick (yellow + rose)	brick (yellow + rose)	gray or black in some species	− −	+	+	−
yellow brick	yellow brick	gray	−	−	−	−
brown	not distinguished	deep black	−	+ (in Japanese Quail)	−	+ (in Japanese Quail)

[a] Correlation between Matsuo's and Herlant's tetrachrome results have been made on sections of pituitaries of White-crowned Sparrow kindly communicated by Professor Donald S. Farner and Professor Arturs Vitums.

Under the electron microscope, α cells have been recognized in the duck (Tixier-Vidal, 1965) and in the White-crowned Sparrow (Mikami *et al.*, 1969), in which they seem to have the same features (Fig. 8). The granules are uniform in shape and size (250–300 nm), but not in density. The endoplasmic reticulum appears as a series of flattened sacs often dilated by large and irregular vacuoles.

   b. η Cells (Cephalic Acidophils, Erythrosinophilic Cells). These cells are characterized by their selective affinity for erythrosin in the tetrachrome stains of Herlant and of Brooke. They can also be distinguished from the α cells because they lie in the rostroventral region of the cephalic lobe, are irregular or polyhedric in shape, contain much less phospholipid, and are generally less easily stained. They seem to correspond with the A2 cells of Rahn and Painter (Fig. 3).

   Electron microscope observations of these cells have been made for the duck (Tixier-Vidal, 1965), the pigeon (Tixier-Vidal and Assenmacher, 1966), and the White-crowned Sparrow (Mikami *et al.*, 1969). They are characterized by their polymorphic secretory granules, the size and density of which vary with the species and with the functional state of the gland (Fig. 9). As with the α cells, the rough endoplasmic reticulum appears as flattened or dilated sacs.

2. Glycoprotein-Containing Cells

   a. β Cells (Cephalic PAS-Positive Cells). These cells of the cephalic lobe are the most strongly PAS-positive cells in the gland (Figs. 4 and 5). This is especially so in seasonally breeding species

---

FIGS. 2–7. FIG. 2, caudal lobe of the pituitary gland of a male Japanese Quail subjected to 8 short days (6L : 18D) after 40 long days (20L : 4D). Herlant tetrachrome: the γ cells are purple, the α cell red-orange, κ cell deep blue, and the δ cells are chromophobic or light blue. × 800. From Tixier-Vidal *et al.* (1968). *Z. Zellforsch. Mikrosk. Anat.* **92,** 610–635. © Springer-Verlag. Fig. 3, cephalic lobe of the pituitary gland of a lactating female pigeon sacrificed 4 days after hatching of the young. Herlant tetrachrome: the η cells, or prolactin cells, are erythrosinophilic and more or less degranulated. × 800. Fig. 4, cephalic lobe of the pituitary gland of a male Japanese Quail, sexually undeveloped. Alcian blue–PAS–orange G: the small deep rose cells represent β cells and are inactive; some very pale blue cells are also visible. × 800 (© Springer-Verlag, same as Fig. 2). Fig. 5, cephalic lobe of the pituitary gland of a male Japanese Quail, sexually active (subjected to 40 long days). Alcian blue–PAS–orange G: the β (gonadotropic cells) are deep rose and strongly stimulated; the δ cells are pale blue and dilated. × 800 (© Springer-Verlag, same as Fig. 2). Fig, 6, cephalic lobe of the pituitary gland of a male Japanese Quail, sexually undeveloped and treated with metapirone. Brooke's technique: cords of very active gray-rose ϵ cells (corticotropic). × 800 © Springer-Verlag, same as Fig. 2). Fig. 7, cephalic lobe of the pituitary gland of a male Japanese Quail subjected to 29 long days. PAS-lead hematoxylin: the κ cells are deeply black stained and the β cells are rose-gray. × 800 © Springer-Verlag, same as Fig. 2).

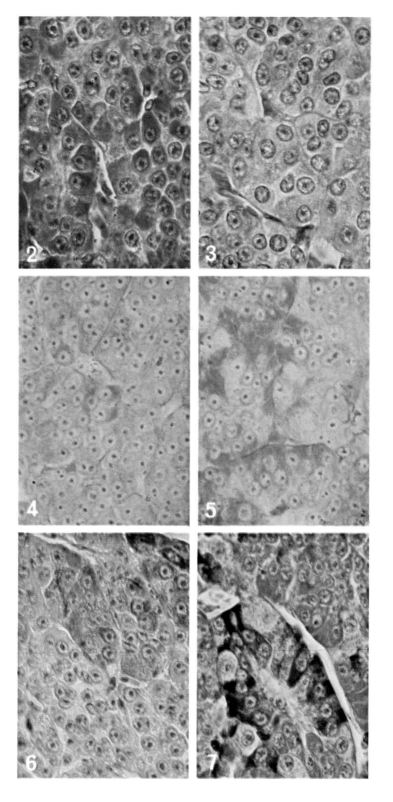

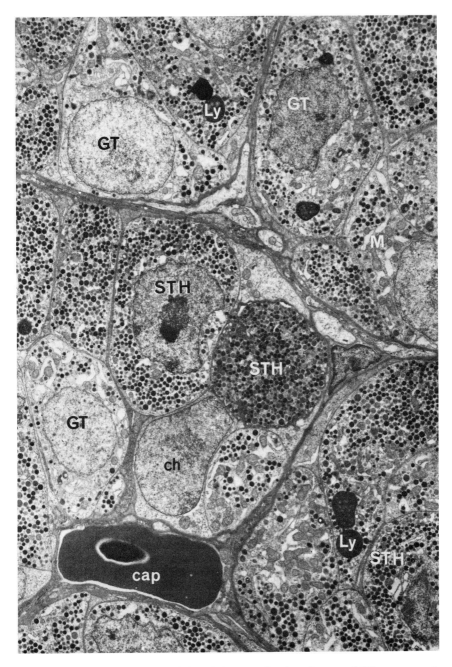

FIG. 8. An area of the caudal lobe of the pars distalis of a normal White-crowned Sparrow showing three types of cells—STH cells, gonadotropes (GT), and chromophobes (ch); Ly = lysome, M = mitochondria, cap = blood capillary. × 5000. (Courtesy of Professor S. Mikami.)

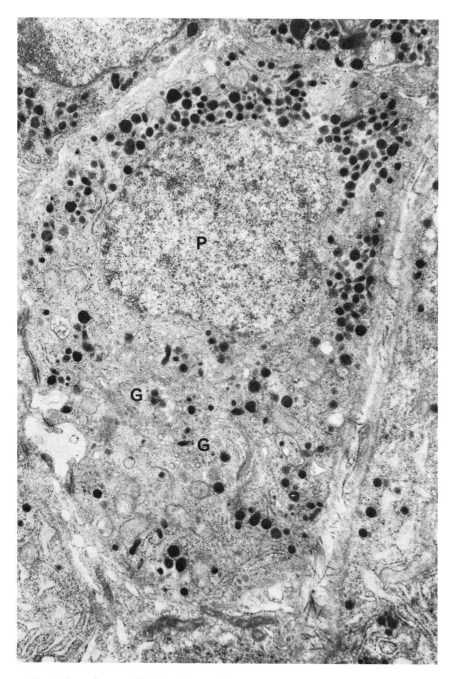

Fig. 9. A prolactin cell (P) in the cephalic lobe of the pituitary gland of a lactating female pigeon. It is characterized by small polymorphic secretory granules and shows a great development of the Golgi zone (G). × 15,600.

in which they are more easily distinguished than in domesticated birds such as the chicken and feral pigeon. They contain some phospholipid, and in the Japanese Quail, a high level of tryptophan. Typical acid dyes scarcely stain the β cells. With Herlant's tetrachrome, identification is sometimes difficult, since their affinities for aniline blue and alizarin blue depend upon the glycoprotein content, and they may be stained from purple to blue. The same difficulty occurs with the classic trichrome stains. Characteristically these cells are small and narrow with a strong polarity and could correspond well with the V cells described in the chicken by Mikami (1958). They belong to the basophil or amphophil class of pituitary cells.

At the ultrastructural level they have been described in the duck (Tixier-Vidal, 1965) and the Japanese Quail (Dancasiu and Campeanu, 1970; Tixier-Vidal *et al.*, 1972). Typically, they contain numerous round, uniformly dense granules of diameter 200–250 nm. The rough endoplasmic reticulum generally consists of scattered, flattened cisternae with no dilations (Fig. 10). These features are similar to the type A gonadotropic cell found in both pituitary lobes of the White-crowned Sparrow by Mikami *et al.* (1969).

b. δ Cells (Alcian Blue-Positive Cells). This cell type occurs in both lobes (duck, Japanese Quail, guinea fowl, pigeon, and Red Bishop), although it is often smaller in the caudal lobe. They are PAS-positive, though less so than the β cells, and after Alcian blue–PAS are selectively stained a blue color. In the duck, they stain selectively with AF, but in several other species the aldehyde–fuchsin is also taken up by β cells. Up to now, therefore, the best criterion for their identification seems their affinity for Alcian Blue (Fig. 5). Unfortunately, these cells are often chromophobic and in some species only acquire their characteristic affinities under limited conditions. Personal experience suggests that they appear best in the duck pituitary. With tinctorial methods, they stain with aniline blue and light green and belong, therefore, to the class of basophils. Their rounded shape with a central nucleus and a general paucity of granulations also helps to distinguish them from other cell types. In some species (Japanese Quail, ovariectomized hen and cockerel) they contain large PAS-positive droplets.

Under the electron microscope these cells are characterized by having a few small granules (50, 150, 200 nm) and by the great development of dilated and rounded vacuoles within the cisternae of the rough endoplasmic reticulum (Fig. 11). Glycoproteins are localized exclusively in the secretory granules (Tixier-Vidal and Picart, 1971), and this probably explains their low affinity for stains in the light

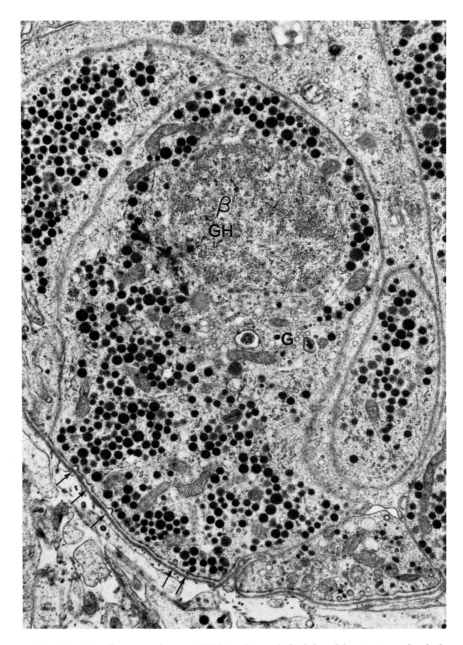

FIG. 10. A β cell, or gonadotrope (GH), in the cephalic lobe of the pituitary gland of a male Pekin duck castrated and photostimulated. It is characterized by round and dense secretory granules. This cell displays several signs of a strong activity. The Golgi zone (G) is extended. The diameter of the secretory granules varies between wide limits, and several pictures of exocytosis are seen (arrows). One notes that the polyribosomes are numerous but the cisternae of the RER are scarce and not vesiculated. ×9600.

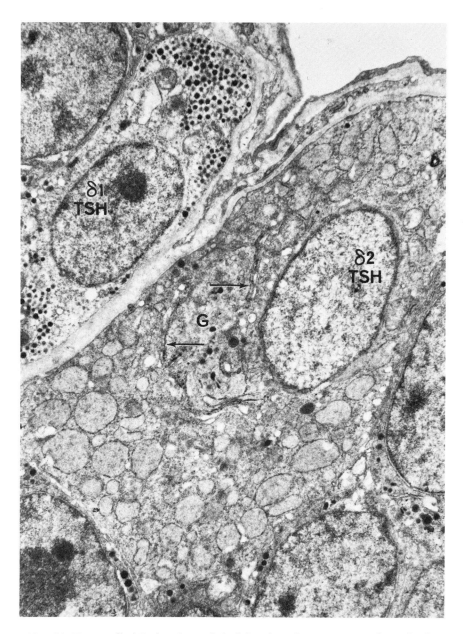

FIG. 11. Two δ cells (TSH) in the cephalic lobe of a male Japanese Quail sacrificed 14 days after bilateral thyroidectomy. Nonfrozen section treated for the localization of acid phosphatases. The two cells are characterized by small secretory granules. The δ1 is not stimulated. The δ2 is transformed in a thyroidectomy cell showing extremely scarce secretory granules, vesiculated cisternae of the RER, and a great enlargement of the Golgi zone (G). The acid phosphatase activity is localized to the inner saccules of the Golgi zone (arrows). × 12,000.

microscope. In the Japanese Quail the rounded cisternae contain large PAS-positive inclusions (Tixier-Vidal *et al.*, 1969).

c. γ Cells (Glycoprotein-Containing Acidophils). The γ cells, which occur in the caudal lobe, are a mixed cell type combining an affinity for PAS with some characteristics of acidophilic cells. The glycoprotein content changes considerably with the functional state of the gland, and such variation probably explains the variability in their reaction to the tetrachrome stains (Fig. 2). Sometimes they appear as acidophils, sometimes as cyanophils or amphophils. With the classic trichrome techniques, such as Heidenhain's azan, they can easily be confused with α cells. Also in common with the α cells, they contain S−S groups and phospholipids as well as having the same general shape. For these reasons they are considered together with the α cells to represent the A1 cell type of Rahn and Painter. In the majority of species they are somewhat more numerous than the α cells, although this does not apply to the Japanese Quail. In some species, particularly in the Red Bishop, they sometimes show a high affinity for lead hematoxylin (Gourdji, 1964).

γ Cells have been distinguished with the electron microscope in the duck and Japanese Quail (Tixier-Vidal, 1965; Dancasiu and Campeanu, 1970). They have dense, round secretory granules of diameter 300–400 nm. The rough endoplasmic reticulum is well developed and consists of flat cisternae lined with numerous ribosomes (Fig. 12). In the pigeon, the granules are polymorphic and the reticulum often dilated (Tixier-Vidal and Assenmacher, 1966).

3. Mixed Cell Types

a. ε Cells. These cells are evident only under conditions associated with a decrease in adrenocortical activity and are difficult to distinguish in normal birds. They are restricted to the center of the cephalic lobe. The cytoplasm usually has a flocculent appearance due to the small size of the secretory granules. They are slightly acidophilic and PAS-positive and have a distinct, although not strong, affinity for lead hematoxylin. With tetrachrome staining methods they react best to Brooke's technique, which stains them a gray-rose color set against a general green background (Fig. 6). These cells could well correspond with some of the "V cells" in the chicken (Mikami, 1958).

Ultrastructurally, they can be distinguished in the duck after metipirone treatment (Tixier-Vidal, 1965) and in the chicken and White-crowned Sparrow after adrenalectomy (Mikami, 1969; Mikami

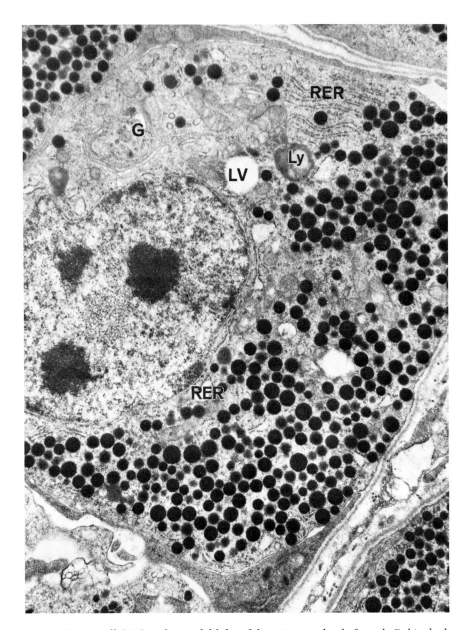

FIG. 12. A γ cell (LH) in the caudal lobe of the pituitary gland of a male Pekin duck castrated and photostimulated. The round secretory granules are larger than in the β cells. This cell shows also signs of high activity – the Golgi zone (G) is extended, the cisternae of the RER, either flat or dilated, are well developed; LV = lipoid vacuole, Ly = lysome. × 16,800.

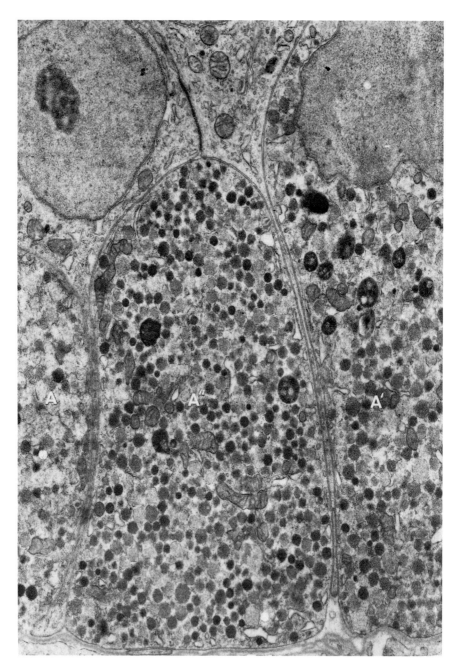

Fig. 13. An area of the cephalic lobe of the pituitary gland of a White-crowned Sparrow sacrificed 2 days after adrenalectomy and showing three adrenalectomy cells (A). The light cell (A) is fully developed adrenalectomy cell and A′ and A″ are transitional forms from ACTH cells. × 13,800. (Courtesy of Professor S. Mikami.)

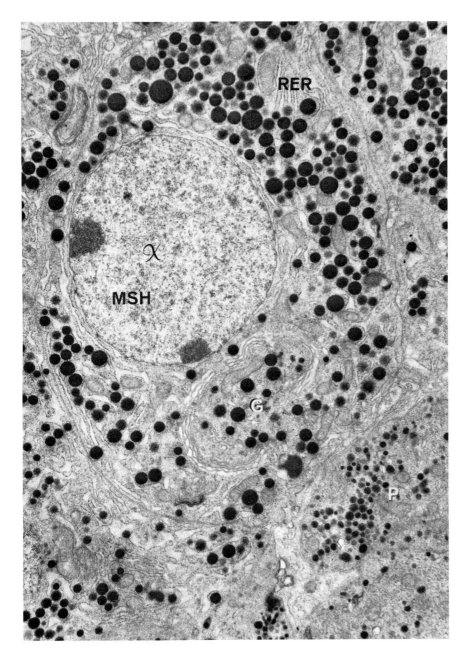

FIG. 14. A κ cell (MSH ?) in the cephalic lobe of the pituitary of a lactating female pigeon. It is characterized by large, round, and dense secretory granules. The Golgi zone (G) is dilated with condensing secretory granules and the cisternae of the RER are flattened; P is part of a prolactin cell. × 11,500.

et al., 1969), when the rough endoplasmic reticulum swells significantly. The secretory granules are dense and small (mean diameter 150 nm) in the duck and chicken. They are a little larger (220 nm)
in the White-crowned Sparrow and apparently show two distinct
phases of low and high electron density (Fig. 13).

b. *κ Cells (Lead–Hematoxylin Cells).* In contrast with the ε cell,
the κ cell is highly stainable, but it remains difficult to classify. With
Herlant's tetrachrome, the cells are dark blue and could thus be
considered as basophils. Indeed, they are often well stained with
toluidine blue at pH 4.5, sometimes with a slight metachromasia.
Kappa cells are also strongly stained by iron hematoxylin and lead
hematoxylin (Fig. 7). In the Japanese Quail, the cells contain high
levels of both phospholipid and tryptophan. Their low glycoprotein
content distinguishes them from β cells. There is some species variation in their localization within the gland. For example, in the Pekin
duck, they are scarce and restricted to the cephalic lobe, while in the
House Sparrow and Red Bishop they are more abundant but still
localized within the cephalic lobe. In the pigeon and Japanese Quail,
they occur throughout the gland.

Electron microscope observations of κ cells have been described
in the duck, pigeon, and Japanese Quail. In all, they contain round
dense granules, which are the largest in the gland (400–500 nm).
The endoplasmic reticulum is often well developed and appears as
flattened sacs lined with many ribosomes (Fig. 14).

In summary, this review confirms the pituitary cell pattern originally proposed by Rahn, Painter, Payne, and Wingstrand, but extends
it from 3 to 7 cell types. These cell types are not distributed evenly
through the gland, although the degree of segregation is not as pronounced as in some other vertebrates, notably the fishes. This becomes
clear if one compares serial sections cut either in the transverse or
sagittal plane. The mixed zone between the caudal and cephalic lobes
occupies a full third of the gland and, moreover, does not lie in a fixed
position but can vary depending on the physiological state of the bird
(see Tixier-Vidal *et al.*, 1968).

IV. Cytological Localization of the Pituitary Hormones

Two facts appear reasonably well established. First, histological
methods demonstrate the presence of seven cell types in the avian
pituitary. Second, the gland seems to secrete seven hormones — two

protein hormones [growth hormone (somatotropin, STH) and pro-
lactin]; three glycoprotein hormones [thyrotropin (TSH)] and two
gonadotropins (GTH) [luteinizing hormone (LH) and follicle-
stimulating hormone (FSH)]; and two polypeptide hormones [adreno-
corticotropin (ACTH) and melanotropic hormone (MSH)]. The aim
of the histophysiological study of pituitary cells is to correlate each
morphological cell type with one hormone. This presumes, of course,
that each cell type secretes one hormone, a hypothesis that is not altogether fully sustained.

Currently, two methods are available to establish the nature of the
hormone secreted by a particular cell type. The best is the immunocytochemical technique, which directly localizes the hormone within
its cell. As yet this method has not been applied to birds, but the
recent purification of some of the avian hormones suggests it should
be introduced fairly soon (see Section V). The alternative approach
is experimental and offers indirect evidence for the localization of a
particular hormone. Invariably this technique involves altering the
activity of the different pituitary target organs, thereby changing the
rate of secretion of a particular tropic hormone. Birds from a wide
number of physiological conditions must be studied, and the method
requires a preliminary knowledge of the pituitary cytology. The
method leads finally to tentative conclusions regarding the localization of the various tropic hormones.

A. THYROTROPIC ACTIVITY

1. Thyroidectomy

The thyrotropic cells have been identified following thyroidectomy
[chicken (Payne 1944; Mikami, 1958, 1969); Pekin duck (Tixier-
Vidal and Benoit, 1962); Japanese Quail (Tixier-Vidal et al., 1972;
White-crowned Sparrow (Mikami et al., 1969)] and treatment with
antithyroid drugs [chicken (Legait and Legait, 1956; Brown and
Knigge, 1958); duck (Tixier-Vidal et al., 1962; Tixier-Vidal, 1970:;
Red Bishop (Gourdji, 1964); Japanese Quail (Tixier-Vidal et al.,
1968)]. All authors agree that thyroidectomy strongly stimulates one
PAS-positive cell type which corresponds with the δ, or Alcian blue-
positive cell described earlier (Table I, Section III) and belongs to
the basophilic class of the avian pituitary. After thyroidectomy, the
cells are greatly hypertrophied and become hyalinized, less easily
stained, and often vacuolized. Electron microscope pictures of these
thyroidectomy cells are very similar in the chicken, White-crowned

Sparrow, and Japanese Quail (Mikami, 1969; Mikami *et al.*, 1969; Tixier-Vidal *et al.*, 1972), with the endoplasmic reticulum being greatly developed so as to occupy the greater part of the cytoplasm (Figs. 11 and 15). There is also a loss of secretory granules, the few remaining being very small (100 nm) and scattered along the cell membrane. The Golgi zone is as large as the nucleus and contains numerous small granules. In both the light and electron microscope it is possible to follow the progressive changes of the normal δ cell into a thyroidectomy cell.

Opinions differ over the localization of this thyrotropic cell type. In the chicken and White-crowned Sparrow, Mikami observed thyroidectomy cells only in the cephalic lobe and thus restricts TSH-producing cells to this zone of the pituitary. In the Japanese Quail, we also saw thyroidectomy cells in the cephalic lobe 13 and 40 days after the operation. Similarly, in the duck and the Red Bishop, after 1 month of treatment with thiourea, the δ cells were stimulated only in the cephalic lobe. But, in Japanese Quail, the same treatment led to an increase in δ cells in both lobes, and in ducks killed 14 months after thyroidectomy, there were hypertrophied δ cells in both lobes of the gland also. These conflicting results may be resolved if it is assumed that thyrotropic cells exist in both lobes, but that in the cephalic region they are more sensitive to modifications in the level of thyroidal hormones or more prone to become "thyroidectomy cells." Indeed, one must notice that small thyrotropic cells always persist in the cephalic lobe and that the thyroidectomy cell represents a maximum stage of stimulation. This hypothesis is not in conflict with Mikami's findings that TSH concentration in the normal chicken pituitary is higher in the cephalic lobe and that only in the cephalic lobe is there an increase in TSH content following thyroidectomy. Similarly, in the chick embryo, thyrotropic cells appear first at the rostral end of the cephalic lobe and only after hatching do they appear in the caudal lobe (Tixier-Vidal, 1954, 1958). TSH activity is exclusively found in the cephalic lobe of 1-day-old chick pituitaries (Brasch and Betz, 1971). The only other cells modified by thyroidectomy are the caudal acidophils, which undergo a marked degranulation and are reduced in volume and in number. This is considered a secondary phenomenon, the endocrine significance of which is not understood.

2. *Effect of Treatment with Thyroid Hormones*

After injection of thyroxine, the δ cells became highly chromophilic and stain deeply with Alcian blue. This effect occurs in both lobes of the gland (pigeon, Tixier-Vidal and Assenmacher, 1966; duck,

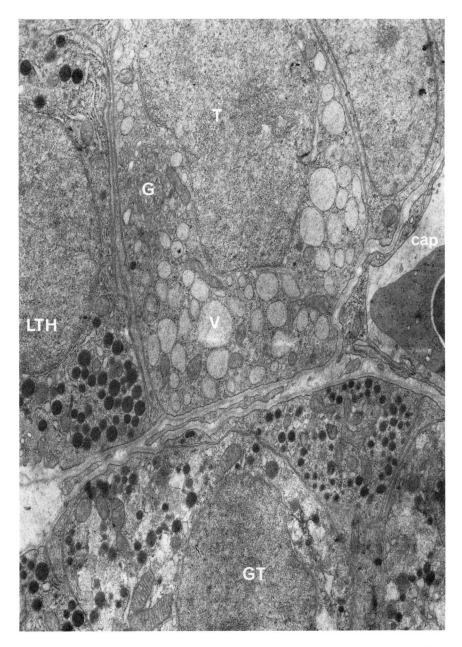

FIG. 15. An area of the cephalic lobe of a White-crowned Sparrow sacrificed 7 days after thyroidectomy, showing a thyroidectomy cell (T), prolactin cell (LTH), and gonadotrope (GT). The thyroidectomy cell contains enlarged vacuole throughout the cytoplasm and a well developed Golgi zone (G), cap is blood vessel. × 13,800. (Courtesy of Professor S. Mikami.)

Tixier-Vidal, 1970; Japanese Quail, Tixier-Vidal *et al.*, 1968). In thyroidectomized quail, injections of thyroxine induce regression of the thyrotropic cells and the appearance in the cytoplasm of numerous dense bodies with a high level of acid phosphatase (Tixier-Vidal *et al.*, 1972).

Several arguments, therefore, support the view that TSH is produced by the Alcian blue-positive, or δ, cells. These may be localized either in the cephalic or in both lobes, but the activity is greatest in the cephalic lobe. The properties of this cell are similar to the thyrotropic cell in other vertebrates (see Herlant, 1964).

B. GONADOTROPIC ACTIVITY

Information from three sources is generally used to identify the gonadotropin-producing cells: (1) the changes occurring during a natural or artificially induced breeding cycle, (2) the effects of castration, and (3) the effects of treatment with sex steroid hormones.

1. The Reproductive Cycle

The pronounced reproductive cycle of most wild birds, together with the ease of modifying this cycle experimentally, makes such species more attractive for cytological analysis than domesticated forms that are continually in full sexual activity. Nevertheless, in both the domestic chicken and the pigeon the pituitary undergoes cytological changes that have been studied relative to the ovarian cycle (Schooley, 1937; Legait and Legait, 1955, 1956; Payne, 1955; Perek *et al.*, 1957; Inoguchi and Sato, 1962; Ljunggren, 1969). The hormonal control of the ovary is complex, contributing to difficulties in interpretation. The clearest evidence concerns PAS-positive cells that display their greatest activity in laying hens and regress during broodiness. They seem to be involved in a gonadotropic function but may also have a role in TSH production. In addition, the caudal acidophils undergo modifications that could also be correlated with the ovarian cycle.

Changes in pituitary cytology through the reproductive cycle have been analyzed in the male of several species — the Pekin duck (Herlant *et al.*, 1960; Tixier-Vidal *et al.*, 1962; Tixier-Vidal, 1965), the Red Bishop (Gourdji, 1964), the Japanese Quail (Tixier-Vidal *et al.*, 1968), and the White-crowned Sparrow (Matsuo *et al.*, 1969; Haase and Farner, 1969). In the duck, Red Bishop, and Japanese Quail, the most striking changes concern the β or PAS-positive cells of the cephalic lobe (Figs. 4 and 5). This cell type is almost absent during sexual quiescence. Its appearance and numerical development is, however, closely tied both to the testicular cycle and to the gonadotropic content of the pituitary (see Section V,A,3). A second glycoprotein-containing cell, the γ cell, is also implicated in the sexual cycle. In the duck, the γ cells are highly condensed out of the breeding season but undergo degranulation and activation of the Golgi zone when testicular growth commences. Pronounced changes also occur in the γ cells of the Red Bishop and Japanese Quail during testicular development, and the cells become so degranulated as to be difficult to recognize with the light microscope. They reappear during testicular regression.

In the White-crowned Sparrow, Matsuo *et al.* (1960) have observed a good correlation between the reproductive cycle of both males and females and changes in the "light and deep basophils," which occur in both lobes. These PAS-positive cells contain acetylcholinesterase, and there is a parallel between the time course of AChE activity in the pars distalis and the growth of the testis (Haase and Farner, 1969).

The δ cells of the Pekin duck also show an annual cycle, but it differs in phase from that of the testis; they are in a resting state in March–April when the testis is undergoing maximum growth and when the thyroid is only weakly active (Tixier-Vidal *et al.*, 1962). In this species, therefore, we do not consider the δ cells as gonadotropic. In artificially photostimulated Japanese Quail, Red Bishop, and ducks, this dissociation between the β and δ cells is much less evident. There is, of course, positive evidence (Section I,V,A,1) that the δ cells in the Red Bishop and Japanese Quail have, as in the duck, a thyrotropic function. In the White-crowned Sparrow, the light basophils, which may be homologous to our δ cells, are also stimulated in phase with the testis, but it is not clear that these cells have a thyrotropic function.

2. Castration

a. Castration in Photosensitive Birds. The effects of castration are closely related to the photoperiod. In Japanese Quail, the pituitary cytology is not modified if the castrated birds are held on short daily photoperiods (Tixier-Vidal *et al.*, 1968), a fact corroborated by the small changes observed in both pituitary gonadotropin and plasma LH content (see Section V). The same appears true for the White-crowned Sparrow (Matsuo *et al.*, 1969) although in the electron microscope one cell type (type B gonadotropic) was found to be stimulated; but the AChE cells are not modified (Haase and Farner, 1970). In the castrated duck held on short days, the cephalic β cells are not modified (Tixier-Vidal and Benoit, 1962; Tixier-Vidal, 1968, 1970).

In contrast, there are marked cytological changes if the birds are killed after a period of photostimulation. Such a treatment also invariably leads to pituitary hypertrophy and to an increase in pituitary and plasma gonadotropins (see Section V). At the cytological level, two types of result have been reported—those of our own on the duck and Japanese Quail, and those of Matsuo and Mikami with the White-crowned Sparrow.

Three cell types are stimulated in the duck pituitary. Two of these— the β cell of the cephalic lobe (Fig. 10) and the γ cell of the caudal lobe (Fig. 12)—show an increased abundance and become hypertrophied and degranulated. The endoplasmic reticulum remains as flattened cisternae and there is no vacuolization of the cytoplasm. These effects agree well with the changes found during the annual reproductive cycle and confirm the view that the β and γ cells produce gonadotropic hormones. The third cell to be stimulated is the δ dell, previously identified as the TSH-producing cell. This cell also hypertrophies but in contrast with the other two types, also shows a high degree of vacuolization. Under the electron microscope, it appears rather similar to the thyroidectomy cell. One difference, however, is that the stimulated δ cells occur in both pituitary lobes (cf. short-term thyroidectomy). Castration seems, then, to stimulate thyrotropic activity, a finding supported by physiological experiments (Benoit and Aron, 1931; Tixier-Vidal and Assenmacher, 1961).

Much the same pattern of changes is observed in photostimulated castrated Japanese Quail, although the γ cells become difficult to interpret because of their low stainability. The δ cell of the castrated Japanese Quail (Fig. 16) differs from the thyroidectomy cell in two respects: (1) there is equal stimulation of the cell in both lobes, and (2) the cells contain numerous lysosomes. This suggests an intracellular degradation of the hormone and implies a lower rate of TSH release in the castrate than in the thyroidectomized bird. In fact, the stimulation of thyroid activity in the Japanese Quail following castration is much less pronounced than in the duck (Tixier-Vidal et al., 1972).

In the White-crowned Sparrow after castration and photostimulation, Mikami et al. (1969) described the stimulation of two cell types occurring in both lobes of the pituitary. One of them, the type A gonadotropic cell, hypertrophies and shows a well developed endoplasmic reticulum with dilated cavities and a large Golgi zone (Fig. 17A). The other cell type, type B gonadotropic cell, is characterized by

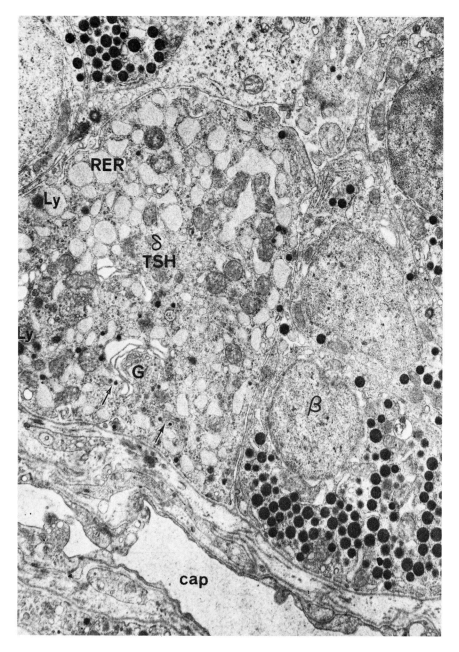

FIG. 16. A stimulated δ cell (TSH) in the cephalic lobe of a male Japanese Quail castrated and photostimulated. As in the thyroidectomy cell of Fig. 11, the RER cisternae are vesiculated and the secretory granules are small and scarce, but the Golgi zone (G) is less extended and several lysosomes are seen; Ly = lysomes, β = β gonadotrope, cap = blood capillary. × 12,000.

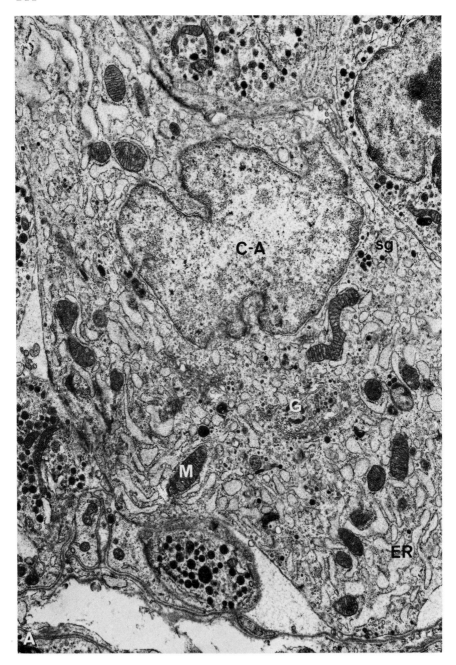

FIG. 17A. An area of the cephalic lobe of a White-crowned Sparrow castrated 65 days and sacrificed after photostimulation for 2 weeks. Type A castration cell. sg = secretory granules, M = mitochondrion, ER = endoplasmic reticulum, G = Golgi apparatus. (Courtesy of Professor S. Mikami.)

a vesiculated cytoplasm containing vacuoles without conspicuous limiting membranes and a few dense secretory granules (180–350 nm) (Fig. 17B). Curiously, this type B gonadotropic cell is stimulated in unilaterally gonadectomized and photostimulated birds and in 10-day castrates that have not been photostimulated. In photostimulated castrates of 14 and 65 days' treatment, the cell type is "rarely observed." The AChE cells are hypertrophied when castration is followed by photostimulation. They seem to represent one cell type only, occurring in both lobes (Haase and Farner, 1970).

If one tries to compare electron microscope pictures of Mikami *et al.* (1969) from the White-crowned Sparrow with ours from duck and Japanese Quail, some analogies do appear:

1. In normal sexually active males, type A gonadotropic cell is quite similar to our β cell.

2. In castrated photostimulated males, type A gonadotropic cell appears similar to our δ cell (Figs. 16 and 17).

3. In the three species, thyroidectomy cells show the same ultrastructural features (Figs. 11 and 15).

4. We can find no analogy between Mikami's type B gonadotropic cell and our γ cells (Figs. 12 and 17).

b. Castration in Chicken and Pigeon. Castration is the only method in these species of identifying the gonadotropic cells. In both it causes hypertrophy of the PAS- and AF-positive cells [pigeon (Schooley, 1937; Bhattacharyya and Sarkar, 1969); hen (Payne, 1942, 1947); cockerel (Mikami, 1958; Inoguchi and Sato, 1962)]. The castration cells are located in both lobes and Mikami (1969) considers that in each of them two cell types are involved as in the White-crowned Sparrow. In the pigeon, on the contrary, two gonadotropic cells are stimulated, the γ cells in the caudal lobe, and the β cells in the cephalic lobe; the δ cells, considered thyrotropic, are stimulated in both lobes (Bhattacharyya and Sarkar, 1969). An increase of the thyrotropic potency of the pituitary has been previously observed in the castrated pigeon (Schooley and Riddle, 1938).

There is no simple explanation for the differences between the two schools on this problem, although we feel that the explanation probably lies less in fact and more in interpretation. At the present moment, the two views can be summarized as follows:

1. The studies in the duck and Japanese Quail suggest that the gonadotropins are produced by the γ cells localized in the caudal lobe and the β cells in the cephalic lobe. Castration also stimulates the δ cells, suggesting a relationship between thyrotropic and gonadotropic functions.

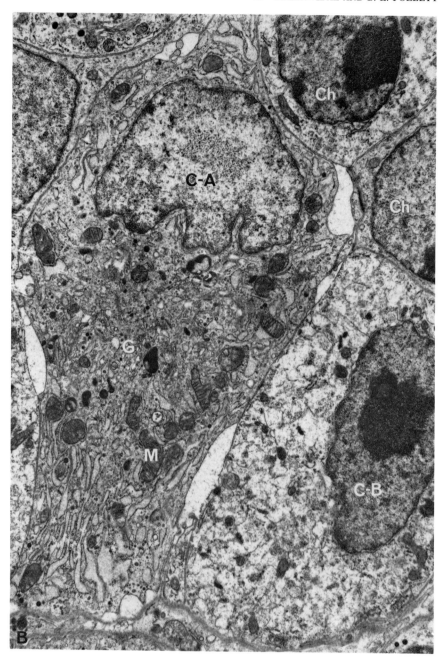

Fig. 17B. Same material as Fig. 17A. Two types of castration cells, A and B, are seen. Compare the type A gonadotrope with the stimulated δ cell (TSH) of a castrated Japanese Quail (Fig. 16). Ch = chromophobe, G = Golgi apparatus, M = mitochondrion. × 10,000. (Courtesy of Professor S. Mikami.)

2. In the White-crowned Sparrow and chicken, the gonadotropic activity is localized in two cell types occurring in both the cephalic and caudal lobes. The thyrotropic cells are small in size and number, suggesting that gonadotropic and thyrotropic functions are not interrelated.

3. Effects of Sex Steroid and of Thyroxine Injections

a. Testosterone.   The effects of testosterone propionate have been studied in the Pekin duck with both the light and the electron microscope. Both intact and castrated ducks have been used in combination with short and long days (Tixier-Vidal, 1968, 1970). Testosterone induced a storage of secretory granules in several types of pituitary cells, including the two gonadotropic cells, β and γ, and the thyrotropic δ cells. It partially inhibited the effects of castration plus photostimulation. These findings imply that testosterone has restrained the secretion of thyrotropin as well as the gonadotropins. There is good evidence from bio- and immunoassays (Section V) that gonadotropin secretion is blocked by testosterone, but the effects on TSH secretion are still unclear (see Chapter 3). Prolactin secretion is also inhibited by testosterone (Gourdji, 1967), and a storage of the secretory granules has been observed in the prolactin cells (Tixier-Vidal, 1970).

b. Thyroxine.   The effects of thyroxine injections have been studied in intact and castrated male Japanese Quail maintained on long days (Tixier-Vidal *et al.*, 1972). In both cases, thyroxine induced a storage of granules and a regression of both the β and δ cells, but the active dose is higher in the castrated than in the intact male. Numerous lysosomes appear in the two cell types.

In conclusion, the effects of testosterone and thyroxine do not aid the interpretation further. If one assumes that the three cell types — γ, β, and δ — each produce a different hormone, then one must conclude that they possess common receptors for testosterone and for thyroxine, and that the regulation of the pituitary–thyroid and pituitary–gonad axes are closely related. Another possibility may be that a common subunit exists in the avian thyrotropic and gonadotropic hormones, the synthesis of which is regulated by a common mechanism.

4. LH and FSH Localization

In the Pekin duck it is thought that the γ cells produce the LH (or ICSH) activity, while FSH arises from the β cells. Direct proof of this is unavailable, but in the duck, the two gonadotropic cells develop asynchronously during the annual cycle. The γ cells show a maximum in March but are regressed in April–May when the testes reach their greatest weight. Garnier (1971; Garnier and Attal, 1970) has shown that plasma testosterone reaches its peak in March. In October, there is another peak of plasma testosterone, although the testicular weight is minimal. At that time, the caudal lobe γ cells show some activity, while the β cells have almost disappeared. Castration at this time activates only the γ cells. The conclusion is that the γ cells are more closely related to testosterone production, and the β cells to testicular weight and development of the seminiferous tubules. The distribution of an LH activity in the caudal lobe and an FSH activity in the cephalic lobe has also been proposed for the day-old chick by Brasch and Betz (1971), who transplanted various regions of the pituitary onto the chorioallantoic membrane of partially decapitated chick embryos. Subsequently, the histology of the embryonic gonad was analyzed.

C. PROLACTIN

Prolactin does not have a target organ that can be ablated to stimulate hormone secretion; thus, the identification of the prolactin-producing cells largely rests on correlating cytological changes with the prolactin content of the pituitary. This content is raised during broodiness and care of the young in the hen (Burrows and Byerly, 1936; Riley and Fraps, 1942; Nakajo and Tanaka, 1956), the pigeon (Schooley, 1937), the California Gull (*Larus californicus*) (Bailey, 1952), the Ring-necked Pheasant (Breitenbach and Meyer, 1959) and the turkey (Cherms *et al.*, 1962). Nakajo and Tanaka (1956) have also shown that in nonbroody hens prolactin potency is greatest in the cephalic lobe, whereas the converse is true of broody birds. Prolactin content appears to follow an annual cycle concomitant with testicular development and is increased by experimental photostimulation (Pekin duck, Gourdji and Tixier-Vidal, 1966; White-crowned Sparrow, Meier and Farner, 1964; Japanese Quail, Gourdji, 1970).

Using the light microscope, the cytological changes occurring at broodiness have been studied in the pigeon (Schooley, 1937; Tixier-Vidal and Assenmacher, 1966) and the hen (Payne, 1955; Legait and Legait, 1955; Perek *et al.*, 1957; Inoguchi and Sato, 1962). In the latter, broodiness is accompanied by regression of the basophils and development of the cephalic ($\alpha 2$) and caudal ($\alpha 1$) acidophils. The former seem most closely related to broodiness. Broody or lactating pigeons show development of the cephalic acidophils that are typical erythrosinophils, as are the prolactin cells of several vertebrate classes (Herlant, 1964). The caudal acidophils showed no changes in the pigeon. Frequently, however, clumps of η cells are seen in the caudal lobe of broody or lactating pigeons, showing again how the two structures of the avian pituitary may become mixed when a strong stimulation is applied. The erythrosinophilic or η cells have been found in the cephalic lobe of several other species, including the duck (Herlant *et al.*, 1960; Tixier-Vidal *et al.*, 1962), the Red Bishop (Gourdji, 1964), the guinea fowl and *Streptopelia* sp. (Tixier-Vidal, 1963), the Japanese Quail (Tixier-Vidal *et al.*, 1968), and the White-crowned Sparrow (Matsuo *et al.*, 1969). In the duck and quail, their changes have been related to pituitary prolactin content, but this does not apply to the other species.

With the electron microscope, the prolactin cells have been identi-fied and related to broodiness in the turkey, hen, and pigeon (Cherms *et al.*, 1962; Payne, 1965; Tixier-Vidal and Assenmacher, 1966). They have also been identified in cultured duck pituitaries that synthesize and release prolactin at a constant rate (Tixier-Vidal and Gourdji, 1965). The prolactin cells of the broody turkey look like rat prolactin cells with large, round, secretory granules and numerous flat ergasto-plasmic cisternae. In broody hens and pigeons, they contain small polymorphic granules and have a much enlarged endoplasmic reti-culum (Fig. 9). Similar results were found with the duck pituitary, the granule size varying with the phase of cellular secretion. This cell type was stimulated by permanent light and degranulated after reserpine injections (Tixier-Vidal, 1965), which was consistent with the changes in pituitary prolactin content.

Mikami *et al.* (1969) tentatively identified a prolactin cell in *Zono-trichia leucophrys gambelii* (Fig. 15). It was mainly localized in the cephalic lobe and had characteristically larve oval or irregularly shaped granules. The cells were more numerous and active in the photostimulated bird.

D. CORTICOTROPIN

Corticotropic cells are most easily identified following adrenalec-tomy, but in birds this operation is difficult and often lethal. It was achieved by Mikami (1958), who was the first to assign ACTH secre-tion in the chicken to a cell type located in the cephalic lobe—the so-called V cells. These increased in both number and size and were arranged in palisade layers throughout the cephalic lobe. They con-tained faint granules that lost their PAS- and AF-positive character. The changes were maximal after 48 hours and were prevented by injections of DOCA. The same observations have been made in the adrenalectomized White-crowned Sparrow (Mikami *et al.*, 1969).

The ultrastructure of the pituitary adrenalectomy cells has been described for both species by Mikami (1969) (Fig. 13). Initially, the cells contained large secretory granules (220 nm) of both a low and a high electron density. Within 24–48 hours after adrenalectomy, the dense granules tended to disappear while the less dense granules became more numerous and lost their density. Finally, they appeared as vacuoles surrounded by the remnants of the granular membrane. In the well developed adrenalectomy cell, these vacuoles occupy the entire cytoplasm and the endoplasmic reticulum, and Golgi apparatus degenerates—an odd response in a highly stimulated cell.

Metapirone seems to act in birds as in other vertebrates by inhibiting corticosteroid synthesis, thus increasing ACTH secretion. Its effects on pituitary cytology are very clear in the duck and Japanese Quail (Tixier-Vidal and Assenmacher, 1963; Tixier-Vidal et al., 1968), with one cell type lying in the center of the cephalic lobe being stimulated and showing the characteristics of an adrenalectomy cell (Fig. 6). In our classification (Table I) this cell corresponds with the ϵ cell, being defined by its slight affinity for PAS and lead hematoxylin. It appears to be homologous to Mikami's V cell. Under the electron microscope, differences are apparent between ACTH cells in the duck treated with metapirone (Tixier-Vidal, 1965) and in the adrenalectomized chicken and sparrow (Mikami, 1969). The secretory granules in the duck are always dense, small (100–200 nm) and scarce, and lie along the cell membrane. The endoplasmic reticulum appears as parallel flat cisternae with small irregular vesicles or polyribosomes, and the Golgi zone is well developed.

E.   Melanotropin (MSH)

The bird pituitary lacks a pars intermedia, but MSH activity occurs in the pars distalis, apparently being somewhat greater in the cephalic lobe (Kleinholz and Rahn, 1939; Mialhe-Voloss and Benoit, 1954). The first evidence for the cell type responsible for this activity came from the duck, where a cell was found that stained strongly with Herlant's tetrachrome and with lead hematoxylin (Herlant et al., 1960; Tixier-Vidal et al., 1962). This cell type, the so-called κ cell, is restricted to the cephalic lobe. It bore affinities with a cell present in the fish pars intermedia (Stahl, 1958; Baker, 1963; Olivereau, 1964) and in the pars distalis of mammals (man and the pangolin), which also lack an intermediate lobe (Herlant, 1964). In all avian pituitaries so far examined with these two staining methods a κ cell has been found (see Section III) (Fig. 7).

The hypothesis that MSH derives from the κ cells has been recently examined further by Tougard (1972). There is a clear relationship between the number of κ cells and pituitary MSH activity in two breeds of duck (Pekin and Khaki Campbell), two pigeons (White Peacock and feral strains), and two strains of Japanese Quail (wild and unpigmented strains). In the ducks and pigeons, the number of κ cells was also correlated with feather pigmentation. While this evidence suggests MSH is produced by the κ cell, direct evidence is lacking. It is not helped by our ignorance of the role of MSH in birds. On several occasions we have reported strong changes in κ cell morphology (e.g., Tixier-Vidal, 1963; Tixier-Vidal and Assenmacher, 1966; Tixier-Vidal *et al.*, 1968), but these could not be correlated with a specific physiological event.

F. Growth Hormone or Somatotropic Cells

The caudal acidophils, or α cells, of the avian pituitary have the same properties as the α cell series in other vertebrates. In mammals, there is good evidence that the α cell secretes growth hormone (see Herlant, 1964), but in birds little is known of the existence or physiology of this hormone (see Section V). Some analogies, however, imply that the avian α cell might also be the source of growth hormone. They are numerous and active in young ducks, Japanese Quail, and pigeons and very much less obvious in older animals. Furthermore, they are degranulated and appear regressed in thyroidectomized birds (chicken, duck), but reappear following treatment with thyroxine. Similar changes occur in the rat, where the relation between thyroid activity and somatotropin secretion has been carefully investigated. The α cells of the duck are also strongly affected by other treatments such as castration, long days, and reserpine (Tixier-Vidal, 1963). Such variations are consistent with the intervention of growth hormone in a wide range of metabolic processes.

V. Chemistry and Physiology of the Hormones

A. The Gonadotropins (FSH and LH)

During the 1930s and early 1940s, many investigators showed gonadotropic activity in avian pituitary extracts (review, van Tienhoven, 1961), but whether this activity was associated with one or two hormones remained unclear. In fact, some results of Fraps *et al.* (1947) had suggested the existence of an LH and an FSH, but this was proven only when Stockell Hartree and Cunningham (1969) isolated two hormone fractions from chicken pituitaries, using as bioassays the standard mammalian techniques. These findings, which have greatly influenced this section, must lead to a reappraisal of much of the earlier work concerning the avian gonadotropins. The problem still remains, of course, as to whether physiologically the two hormones may act together in some form of gonadotropic complex (Nalbandov, 1961, 1966). This remains unresolved, although it seems unlikely, especially as Imaï and Nalbandov (1971) report differential ratios of FSH:LH activity in the plasma of domestic hens at different periods of the laying cycle.

1. Assay

Without specific assays, progress into understanding the nature, number, and physiology of the gonadotropins is made immeasurably more difficult, and it is therefore the more unfortunate that most of the assays used hitherto show one or more grave defects.

a. Bioassays in Mammals. The acceptable specific assays for mammalian LH and FSH are, respectively, the ovarian ascorbic acid depletion method (OAAD) in rats (Parlow, 1961) and the ovarian augmentation assay in rats or mice (Steelman and Pohley, 1953; Brown, 1955). This specificity seems also to apply to the two chicken gonadotropin fractions (Furr and Cunningham, 1970). Indeed, much of the basis for claiming the existence of chicken LH and FSH rests on the fact that of the two gonadotropin fractions isolated, one was active only in the Parlow assay, the other in the Steelman–Pohley assay. Both assay methods, however, are extremely insensitive to the avian hormones. Even with the OAAD assay it is usually impossible to assay the hormone present in a single pituitary gland of a hen if a factorial design (2+2) of standards and unknowns is used (Heald *et al.*, 1967). For estimates of plasma LH, a dose of 3 ml/rat had to be used by Nelson *et al.* (1965). The FSH assays are more insensitive, Imaï and Nalbandov (1971) using doses equivalent to two and four hen pituitaries in a (2+2) design.

While the FSH assay seems specific, problems have appeared recently with the OAAD assay because of its responsiveness to arginine vasotocin (Jackson and Nalbandov, 1969; Ishii *et al.*, 1970), one of the natural neurohypophyseal octapeptides in birds. Jackson and Nalbandov (1969) report that vasotocin also occurs in high concentration within the chicken adenohypophysis and that estimates of OAAD activity in simple pituitary extracts measure both vasotocin and LH. This appears to be a serious, although surmountable, problem since vasotocin can be destroyed by sodium thioglycolate treatment, or separated from LH by dialysis or chromatography on Sephadex. Nevertheless, the finding raises problems over the validity of published data on LH levels in the chicken pituitary (Nelson et al., 1965; Heald et al., 1967; Tanaka and Yoshioka, 1967). Plasma estimates may be more accurate, for although vasotocin levels increase at oviposition, the amounts present should not affect the assay. In summary, the OAAD method seems acceptable for avian LH only if vasotocin is removed from any extract.

b. Bioassays in Birds. The first valid assay depended upon the ability of gonadotropins to stimulate testicular growth in young domestic fowl chicks (Breneman, 1945; Breneman et al., 1959). A similar endpoint can be used in young turkey poults (Herrick et al., 1962) or in 14-day-old Japanese Quail reared under a nonstimulatory photoperiod (Tanaka et al., 1966a). With these methods, however, the gain in sensitivity over the mammalian assays is marginal, and the methods have been largely superseded by an assay that measures the testicular growth response in terms of uptake of radioactive phosphorus (Breneman et al., 1962; Follett and Farner, 1966).

As a bioassay the ^{32}P-uptake method is quite acceptable. The responses (Fig. 18) to mammalian gonadotropins (NIH-LH and FSH) remain stable over long periods, and adequate assay precision is obtainable with 6–8 chicks in each group and a typical factorial design. Again, the index of precision remains stable, as may be judged from three separate assay periods — in Pullman, where $\lambda = 0.154 \pm 0.006$ (SEM) (57) (Follett and Farner, 1966); in Leeds, $\lambda = 0.219 \pm 0.010$ (43) (Follett, 1970); and in Bangor, $\lambda = 0.228 \pm 0.014$ (42) (Scanes and Follett, 1972). The assay precision seems dependent on the strain of chick used. The method is sensitive to about 0.02 of a chicken pituitary (0.2 mg fresh weight) or to 0.2–0.4 of a pituitary from a Japanese Quail or White-crowned Sparrow (Follett and Farner, 1966; King et al., 1966). In practice, this means that single quail or sparrow pituitaries are assayable in a (2 + 1) design and that more complete potency estimates are possible with four or five pituitaries. Other advantages of the method include the ready availability, cheapness, and uniformity of 1-day-old chicks and the ability to process up to two hundred in a single assay. Full details are given in Follett and Farner (1966) and Follett (1970).

However, the ^{32}P-uptake technique has one major drawback that greatly limits its usefulness; it responds equally well to both chicken

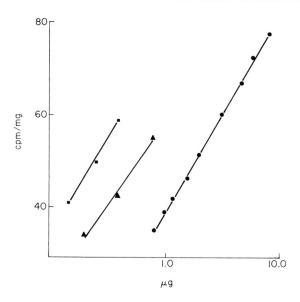

FIG. 18. Dose–response curves for gonadotrophins in the ^{32}P-uptake assay. The responses (abscissa) are expressed in terms of cpm/mg fresh testicular tissue. Doses are given in logarithmic units on the ordinate. The scale (micrograms) only applies to the most right-hand curve, which is for standard mammalian LH (NIH-LH-S16); this line is the composite derived from 43 assays. The left-hand curves, from single experiments, are drawn on an arbitrary scale and show dose–response lines, respectively, for quail and chicken pituitary extracts.

LH and FSH and must therefore be considered as a total gonadotropin bioassay (Furr and Cunningham, 1970; Follett *et al.*, 1972a). According to Burns (1969), the method can be made specific for avian LH by treating pituitary extracts with neuraminidase, which selectively destroys FSH. This conclusion was based on the effects of the enzyme on ovine FSH and on bovine LH. Unfortunately, our own results (C. K. Scanes, unpublished) suggest that both chicken gonadotropins are equally susceptible to neuroaminidase action. The assay is unaffected by arginine vasotocin. Its value has lain primarily in its sensitivity and thus its use in measuring the hormone content in small birds. It has also enabled the hypothalamic gonadotropin-releasing factors to be characterized and assayed (Follett, 1970), the OAAD method being unusable for this because of vasotocin contamination of the hypothalamic extracts. The weaver-finch assay for LH (Witschi, 1955) may be specific, but as yet it has not been tested with purified avian hormones.

c. *Immunoassays.* The last seven years have seen the introduction of radioimmunoassay methods for the measurement of many hormones, and assays are now available for human, ovine, bovine, simian, and rodent LH and FSH, as well as for human chorionic gonadotropin (reviews; Diczfalusy, 1969; Kirkham and Hunter, 1971). The advantages of immunoassay are a vast increase in sensitivity, allowing for the measurement of hormones in plasma, a more accurate estimate of potency, a stable and reproducible system, and the ability to measure many samples at once. Against this must be considered the problem of specificity. This includes both the specificity of an assay toward any single pituitary hormone, and also whether the system is measuring the same substance as the bioassay. Disparities, between bio- and immunoestimates of potency are quite common, and the problem remains central to all immunoassays.

A major technical problem is that most gonadotropin immunoassays have proved species specific; for example, antiovine LH does not cross react with chicken pituitary extracts when tested in an Ouchterlony plate, nor is it capable of blocking their biological activity (Moudgal and Li, 1961a; Bullock *et al.*, 1967). A weak cross reaction was observed, however, between chicken LH and antihuman LH in a hemagglutination inhibition system (Stockell Hartree and Cunningham, 1969). Bagshawe *et al.* (1968) have reported a stronger cross reaction in an HCG radioimmunoassay with plasma from laying hens, and this reaction was used to estimate the circulating "LH" levels in the Greenfinch (*Carduelis chloris*) and the House Sparrow (Murton *et al.*, 1969, 1970). No real evidence exists as to whether the assay is measuring avian LH, but the cross reactions are remarkable. Follett *et al.* (1972a) were unable to find an anti-HCG or antihuman LH serum that showed more than a trace ability to bind labeled chicken LH. The fact that different antisera to the same hormone show varied specificity is well established, and this is most probably the explanation for Bagshawe's results. Such cross reactions can be of great value, as exemplified by one antiovine LH serum produced by Niswender and Midgley that can be used to measure LH in sheep, rat, hamster, vole, and monkeys. In all cases, however, the assay has been validated with purified or semipurified hormones. The conclusions seem clear for avian physiologists; homologous assays using antisera and purified hormones from the species under study are the ideal; failing that an assay based on antibodies against the hormone from another species (heterologous assay) is acceptable if it can be validated. Both approaches became feasible when Stockell Hartree and Cunningham

(1969) purified chicken LH and FSH, and a homologous assay for avian LH has now been developed (Follett *et al.*, 1972a,b).

The assay was developed with LH extracted from broiler chicken pituitaries. The antisera produced in rabbits had the ability to inhibit the biological activity of avian but not ovine LH *in vivo*. More purified LH fractions were labeled with radioiodine (^{125}I). The label bound to the antibody (antiavian LH) was separated from the unbound or free label by precipitation with an antirabbit γ-globulin serum. When increasing amounts of unlabeled chicken LH were added to this system, the binding of the LH-^{125}I was progressively inhibited (Fig. 19), thus producing a standard curve. Chicken FSH appears to cross react only slightly in the assay (Fig. 19). It was also concluded after

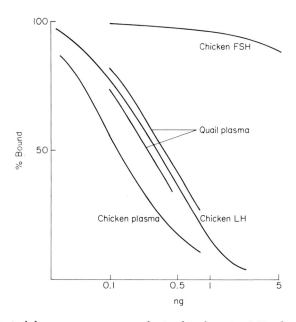

FIG. 19. Typical dose–response curves obtained in the avian LH radioimmunoassay. The response is expressed as percent bound — 100% being the amount of labeled LH bound to the antibody in the absence of unlabeled LH. Curves are shown for purified chicken LH and FSH in nanograms per tube. Plasma curves are on an arbitrary scale but range from 25 to 200 μl.

a series of physiological experiments that the assay was not measuring significantly the other glycoprotein hormone TSH (Follett *et al.*, 1972a). Pituitary extracts and plasma from both the chicken and the quail cross react fully in the assay and the immunoreactive LH (IR–

LH) level is measurable in the 2.5–200 μl of plasma (Fig. 19). Of more general interest is the ability of the assay to cross react with plasma and pituitary extracts of non-gallinaceous birds. It can measure LH in the Tree Sparrow (*Passer montanus*), the Brambling, the White-crowned Sparrow, the Pekin duck, the Green-winged Teal (*Anas crecca*), and the European Starling, the domestic pigeon, the Herring Gull (*Larus argentatus*) and the canary (Scanes *et al.*, 1972). If it can be validated for these species, the assay should prove of wider value to avian endocrinology. Even with its special problems, the immuno-assay appears superior to any available bioassay. The next few years will no doubt see assays developed for FSH and for the other tropic hormones of the adenohypophysis.

2. Chemistry

As yet, the avian gonadotropins have only been isolated from the domestic fowl, but it is encouraging that in this species they closely resemble mammalian LH and FSH in general chemistry and can therefore be purified by well-established methods. The following account summarizes the results of Stockell Hartree and Cunningham (1969) and of Scanes and Follett (1972).

The glycoprotein hormones (FSH, LH, and TSH) were extracted from broiler pituitaries (5–20,000) with a solution of ammonium acetate (6%) in 40% ethanol (pH 5.1). This is an efficient extraction procedure (>90%) which purifies the gonadotropins thirty- to sixty-fold. Growth hormone, prolactin, and ACTH remain in the insoluble pituitary residues. Further separation was effected on a column of carboxymethylcellulose. In dilute ammonium acetate solution (4 mM, pH 5.5) both LH and TSH are largely adsorbed into the cellulose network and form ionic bonds with the carboxylate groups. FSH remains unadsorbed and may be eluted directly. When the concentration of ammonium acetate is raised stepwise to 1 M (pH 5.5), LH and TSH come off together. This method is far from perfect and leaves some cross contamination of the gonadotropins. Subsequently, FSH was purified further by chromatography on columns of calcium phosphate, DEAE-cellulose, and Sephadex G-100. The material so produced has little LH and TSH activity.

A major problem has been the separation of LH from TSH, and as yet this has only been achieved by chromatography on DEAE-cellulose according to the method of Bates *et al.* (1968). The two hormones are very similar in their chemical properties, but under the right conditions, TSH can be bound preferentially to the column and a relatively TSH-free LH eluted. The LH fraction required further purification on columns of Amberlite IRC-50 and Sephadex G-100. Information on the amino acid composition of the LH is very similar to that of the mammalian LH's (C. K. Scanes, unpublished).

3. Physiology

The exact functions of avian FSH and LH have yet to be established, since the purified avian hormones have not been tested adequately in the hypophysectomized bird. However, experiments using mammalian gonadotropins suggest that in the male, FSH is responsible for growth of the seminiferous tubules, for the meiotic division of primary spermatocytes leading to secondary spermatocytes, and possibly for spermatogonial division (van Tienhoven, 1961; Nalbandov, 1966). LH is generally regarded as acting primarily on the Leydig cells to stimulate production of androgen (van Tienhoven, 1961; Nalbandov, 1966). For the attainment of full spermatogenesis, both hormones appear necessary; the reasons for this are unclear but may involve an intratesticular role for androgens in accelerating postspermatocyte development. In the female, the gonadotropins act to cause follicular growth as well as steroidogenesis, but the relative roles of FSH and LH within these processes have not been determined (van Tienhoven, 1961; Nalbandov, 1966; Mitchell, 1967). As in the mammals, LH is regarded as the hormone that induces ovulation (Nelson and Nalbandov, 1966; Fraps, 1970; Cunningham and Furr, 1972).

Preliminary experiments indicate that chicken LH, administered to quail, elicits a considerable testicular weight increase and causes differentiation of the Leydig cells (Follet et al., 1972b).

The pituitary gonadotropin content is dependent on the state of sexual maturity of the bird, and in seasonally breeding species, the level undergoes a cycle that reaches its maximum on the breeding grounds (Greeley and Meyer, 1953; King et al., 1966). Photorefractoriness leads to a precipitate decline in pituitary content. The stimulation by long daily photoperiods of gonadal growth also leads to an increase in pituitary gonadotropins (Assenmacher et al., 1962; Tanaka et al., 1965; Follett and Farner, 1966; Farner et al., 1966; Follett, 1970). Returning Japanese Quail to short daily photoperiods causes a decline in the pituitary content (Tixier-Vidal et al., 1968). Growth and the onset of sexual maturity in the chicken are likewise accompanied by an increase in pituitary gonadotropins (Breneman, 1955). In all these situations a high pituitary level is associated with an elevated level of gonadotropin secretion as assessed by gonadal growth.

These findings are generally confirmed when the circulating LH level in Japanese Quail is measured by radio-immunoassay (Nicholls *et al.*, 1973). Table II summarizes some results when Japanese Quail are transferred from short to long daily photoperiods. Two points are perhaps worth emphasizing: (1) There is detectable LH in unstimu-

TABLE II

EFFECT OF PHOTOSTIMULATION ON LUTEINIZING HORMONE (LH) IN
JAPANESE QUAIL PLASMA

Days on light schedule of 16L:8D[a]	LH (ng/ml) (mean ± S.E.M.)[b]	Combined testicular weight[c] (mg)
0	0.60 ± 0.06 (16)	11.6
1–2	0.61 ± 0.08 (6)	10.5
4–6	3.89 ± 0.57 (6)	75.8
8–10	4.78 ± 0.92 (6)	311.0
12–14	5.03 ± 0.41 (6)	753.0
18–22	4.15 ± 0.64 (6)	1857.0

[a] Birds were reared from hatch under a short day of 6L:18D. When the body weight was about 100 gm they were transferred to a stimulatory schedule of 16L:8D and killed thereafter at the times indicated.

[b] Plasma LH concentration is given in units (ng/ml) relative to a laboratory standard of pure avian LH (see Follett *et al.*, 1972a). The number of birds is shown in parentheses.

[c] Testicular weight is shown only to indicate the degree of sexual maturation attained; maximum testicular weight is about 2500 mg. The data are taken from Nicholls *et al.*, 1973.

lated young quail showing no testicular growth. Similarly, LH has been detected at low levels in the plasma of photorefractory White-crowned Sparrows; these observations suggest the hypothalamo–hypophyseal system may not be as inactive under these conditions as has often been assumed. (2) The plasma LH level is not correlated with the rate of testicular growth. Indeed, it remains relatively constant throughout the period of growth, although there is great individual variation. In the absence of data on the turnover of LH at different stages of sexual maturation, one cannot know whether the secretion rate remains stable or not over a month of photostimulation. There is no theoretical reason why the secretion rate needs to alter, but information on this point would be most valuable in understanding the dynamics of testicular growth and function in birds.

Under certain circumstances (rarely of natural occurrence) it is possible to separate gonadotropin synthesis in the pituitary from its release. The net result is that the pituitary level can be greatly elevated while testicular growth is either absent or at a low level (Farner and Follett, 1966; Follett *et al.*, 1967; Tixier-Vidal *et al.*, 1968).

Diurnal changes in pituitary gonadotropins have been measured in the Japanese Quail (Tanaka *et al.*, 1966b; Follett and Sharp, 1969; Bacon *et al.*, 1969), and in the White-crowned Sparrow (Stetson and Erickson, 1970). Some of the data are shown in Fig. 20, if only to

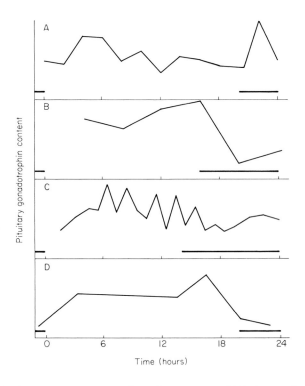

FIG. 20. Rhythms in pituitary gonadotrophin content of (A) *Zonotrichia leucophrys gambelii* (Stetson and Erickson, 1970); (B) Japanese Quail (Tanaka *et al.*, 1966b); (C) Japanese Quail (Bacon *et al.*, 1969); (D) Japanese Quail (Follett, 1971). The length of the dark period is shown by the black bars. The actual contents are not shown, since the data are not absolutely comparable; the scales were adjusted so that the extreme values filled the appropriate block.

emphasize their great disparity. Many of the experiments attempted to determine when the gonadotropic hormones are secreted under a stimulatory light schedule. This information could be important in establishing how the underlying circadian basis of avian photo-periodism is expressed endocrinologically (Follett and Sharp, 1969,

Follett, 1973). Unfortunately, the available results could support a variety of theories, and indeed, it seems unanswered yet whether there is a rhythm, or whether the changes may only reflect random fluctuations. The data are also subject to other problems, notably that they combine measurements of both LH and FSH, and that the "rhythms" might reflect storage rather than release of the hormone. Two further complications have arisen recently; the circulating LH level in a Japanese Quail represents only 0.5–1.5% of the stored pituitary content implying that quite large changes in plasma LH could occur with relatively minor alterations in the pituitary content. Second, Katongole *et al.* (1971) have shown in the bull that the plasma LH concentration cycles rapidly with five to ten peaks during a 24-hour period. The changes seem unrelated to daylight, feeding, or sleep. If this is a general phenomenon, it will make the detection of possible longer duration secretory rhythms a most difficult task. Daily rhythms in secretion must be distinguished from the type of data presented for the plasma LH levels in Greenfinches and House Sparrows (Murton *et al.*, 1969, 1970; Lofts *et al.*, 1970). These indicate in the Greenfinch that LH tends to be secreted more readily than FSH under an asymmetric skeleton photoperiod of 6L:0.5D:1L:16.5D, while the converse is true of the skeleton 6L:9D:1L:8D. The interpretation placed on these data is that LH and FSH may be secreted separately in time. The House Sparrow does not show the differential release.

As mentioned previously (Section IV,B), castration, when accompanied by photostimulation, leads to pituitary hypertrophy and to an increase in the gonadotropin content (e.g., Herrick *et al.*, 1962; Tixier-Vidal *et al.*, 1968; Follett, 1970). It also causes a fivefold increase in plasma LH in the Japanese Quail (Follett *et al.*, 1972a). If castrates are held on short daily photoperiods, these changes are not apparent (Tixier-Vidal *et al.*, 1968; Follett *et al.*, 1972a). Treatment of intact or photostimulated castrates with exogenous sex steroids causes a rapid fall in pituitary and plasma LH (Follett, 1970; Follett *et al.*, 1972a).

Virtually all studies of the female have concentrated on the egg-laying cycle of the domestic fowl, and estimates of pituitary and plasma LH (Nelson *et al.*, 1965; Heald *et al.*, 1967; Tanaka and Yoshioka, 1967) and FSH (Kamiyoshi and Tanaka, 1969; Imaï and Nalbandov, 1971) have attempted to establish the time at which these hormones are released in relation to ovulation. Neuroendocrine experiments have suggested that the release of LH (and FSH, Ferrando and Nalbandov, 1969) for ovulation occurs 6–8 hours before ovulation itself (Fraps, 1961, 1970), although some results of van Tienhoven *et al.* (1954) also suggest a burst of hormone release 14 hours prior to ovulation. In addition, one must expect that hormone secretion is associated with the maturation of follicles, a process that probably involves both hormones. The gonadotropin assays present a complicated picture with two to three peaks of both pituitary and plasma LH and FSH. Figure 21 is taken from Imaï and Nalbandov

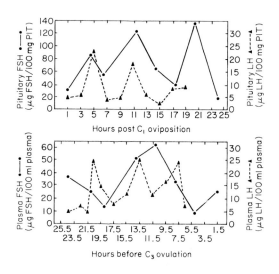

FIG. 21. FSH and LH levels in the pituitary and plasma of laying hens with respect to the times of oviposition and ovulation. LH levels are shown as dotted lines, FSH with a solid line. The data are drawn from Imaï and Nalbandov (1971), the LH figures having been published earlier by Norton *et al.* (1965).

(1971). Some difficulties of interpretation exist but the peak of plasma LH 8 hours prior to ovulation is assumed to cause that ovulation. The earlier peaks at 14 and 20 hours before ovulation are not fully understood. Imaï and Nalbandov (1971) interpret their data to suggest a minor release of FSH at about the time of ovulation that may be essential for rapid growth of the next follicle, and a major release at about 14–11 hours prior to ovulation. The latter may be necessary for maturation and ovulation of the next follicle. Cunningham and Furr (1972) have recently shown, using the LH radioimmunosassay, that only a single peak of LH occurs in the plasma, some 8 hours prior to ovulation.

B. Thyrotropin (TSH)

The dependence of the avian thyroid gland on the adenohypophysis together with its responsiveness to mammalian TSH, makes it certain that an avian TSH exists. However, relatively little is known of the hormone; rather surprisingly so, because the young chick is used extensively for TSH bioassay. What evidence there is suggests that TSH has a physiological action similar to that in mammals.

1. Assay

A great number of TSH assays have been developed and extensive reviews are available (Bates and Condliffe, 1966). The parameters most commonly used are radioactive iodine uptake or depletion from the thyroid gland, and the uptake of radioactive phosphorus by the gland. Either mammals or birds can be used as test animals. Since TSH shows some phylogenetic specificity (Dodd *et al.*, 1963), comments will be restricted to the methods using birds and that are likely to be of most value to avian physiologists.

a. ^{32}P-Uptake Assay. This technique is analogous to the gonadotropin-^{32}P assay and uses the incorporation of labeled phosphorus to measure the thyroidal hyperplasia following TSH injection. Methods are described in Lamberg (1955), Greenspan *et al.* (1956), and Scanes and Follett (1972). A typical schedule involves the 1-day-old chicks being killed 7 hours after TSH injection (subcutaneously) and 1 hour after a dose of sodium [^{32}P] phosphate. The thyroid glands are removed, weighed, and dried prior to counting. The method is sensitive to 2.5 mU of ovine TSH (NIH-TSH-S6). With a (2+2) assay design and six chicks at each dose level, the mean index of precision from sixteen assays was 0.192 ± 0.017 (SEM) (Scanes and Follett, 1972). Chicken and quail pituitary extracts give dose–response curves parallel to that of the mammalian TSH standard.

b. ^{131}I-Depletion Assay. An advantage of this method is that the thyroid glands need not be dissected. The best method (Bates and Cornfield, 1956; Bates and Condliffe, 1960, 1966) uses 1-day-old chicks from hens maintained on a low-iodine diet. The chicks are injected (subcutaneously) with 1–3 μCi ^{131}I, and the uptake is measured the following day with a probe counter. Those showing an uptake greater than 15% are then injected with TSH together with 8 μg of thyroxine and 500 μg of a goitrogen such as propylthiouracil to prevent recycling of any liberated iodide. One day later, the depletion is measured relative to the ^{131}I present 24 hours earlier. The injections of TSH, thyroxine, and goitrogen are then repeated, and depletion is measured after another 24 hours. Since each chick is its own control, the errors are much reduced and an index of precision of about 0.2 may be expected. The dose range is generally between 0.5 and 4 mU TSH.

c. ^{131}I-Uptake Assay. Five injections of TSH are given at 12 hourly intervals to young cockerel chicks. This is followed by an injection of ^{131}I, the chicks being killed 5 hours later (Shellabarger, 1954).

Radio-immunoassays for mammalian TSH are now available (see Kirkham and Hunter, 1971) but not yet for avian TSH. Antisera raised against impure chicken TSH fractions can block the biological activity of chicken TSH when tested in the ^{32}P-uptake assay (Follett et al., 1972a), but highly purified TSH suitable for radioiodination has yet to be prepared.

2. Identification

Pituitaries from a few avian species have been shown to contain TSH activity, viz., the chicken (Adams, 1946; Tixier-Vidal, 1956; Dodd et al., 1963), turkey (Adams, 1946), pigeon (Bates and Condliffe, 1960), duck (Adams, 1946), canary (Kobayashi, 1954), and the Japanese Quail (C. K. Scanes, unpublished observations). Circulating TSH has also been detected in the pigeon (D'Angelo and Gordon, 1950). Unfortunately, little is known either of the relative activity of avian TSH in the various bioassays or of the pituitary and plasma levels after physiological experimentation. Information is badly needed and should not prove too difficult to obtain with the assays now available. Chicken TSH has been partially purified (Scanes and Follett, 1972). Chemically, it is similar to mammalian TSH, being a glycoprotein of large molecular weight.

C. PROLACTIN

1. Assay

Although a number of prolactin assays have been developed (see Meites and Nicoll, 1966; Ensor and Ball, 1968) the most widely used continues to be that utilizing the crop-sac response in pigeons. This response to prolactin, which was originally demonstrated by Riddle et al. (1932, 1933), is remarkable; it apparently occurs only in birds of the order Columbiformes and represents a physiological adaptation akin to that of mammalian lactation (see Chapter 6, Volume II). Prolactin stimulates a proliferation of the entire mucosal epithelium of the crop-sac (see Dumont, 1955). Eventually the innermost layers of epithelial cells slough off into the crop-sac lumen and form a "milk." Mixed with food, this is regurgitated and fed to the young.

Many attempts have been made to quantify the response as an assay procedure. These have usually varied in the route and schedule of the hormone injections and in the methods used to determine the degree of crop-sac stimulation. In the systemic method of assay (Bates et al., 1963), prolactin is injected once daily for 4 or 7 days into adult pigeons and the crops weighed the day after the last injection. The technique is rather insensitive, the minimally effective dose being about 500 mU. Sensitivity is much increased (100- to 200-fold) if the hormone is applied locally to the crop-sac by an intradermal injection (e.g., Lyons, 1937; Grosvenor and Turner, 1958; Pasteels, 1966; Chadwick, 1966), and since the crop-sac on each side can be injected separately, each pigeon yields two observations. The drawback to this method has been to quantify the local responses, and most techniques have assessed crop-sac development on a subjective scale. This has been improved upon by Nicoll (1967), who devised an apparatus to stretch the pigeon hemicrops by a uniform amount, thereby allowing comparable disks of tissue to be cut from each crop. The disks are dried and weighed. Responses are measured as a percentage weight increase of the hormone-injected hemicrop over that of the saline-injected side. With this quantitative measure, Nicoll obtained a satisfactory log dose–response relationship and an improved index of precision. Ben-David (1967) has quantified the crop-sac assay by measuring tritiated thymidine incorporation after prolactin treatment.

A further point worth considering has recently emerged. There is a pronounced daily rhythm in the crop-sac response to prolactin (Meier et al., 1971), injections late in the daily photoperiod being two- to fivefold more effective than at earlier times in the day. This suggests that a strict regimen should be employed when assays are to be carried out routinely, and also that certain schedules give greater sensitivity.

Even with these improvements, however, the crop-sac response leaves much to be desired. The confidence limits on the estimate of potency are invariably rather wide, and the assay is too insensitive to measure routinely the circulating prolactin concentration. Two other approaches have therefore been explored. The first utilizes the obser-

vation that prolactin is readily separable from the other adenohypophyseal hormones by polyacrylamide gel electrophoresis. By staining
the prolactin band with aniline blue-black and measuring its density,
it has been possible to develop a quantitative assay (Nicoll *et al.*,
1969; Yanai and Nagasawa, 1969). The method is ten times more
precise than the crop-sac assay, but much less sensitive. Even so, it
can be used to detect the prolactin in single pigeon glands (Nicoll
et al., 1970). Radioimmunoassays have now been developed in a
number of mammalian species to measure prolactin. All are far more
sensitive than any bioassay but are not generally applicable because
of species specificity. Duck pituitary extracts do not cross-react in an
ovine prolactin radioimmunoassay (Bryant and Greenwood, 1968),
while semipurified chicken prolactin is ineffective in the equivalent
rat assay (K. Kortright, personal communication). It appears that a
homologous radioimmunoassay will be needed to measure avian
prolactin.

2. Physiology

The roles of prolactin are many, and the hormone has been implicated in a wide variety of processes, virtually all of which are associated directly or indirectly with reproduction. In many cases, prolactin
does not act alone but requires the presence of other endocrine secretions in order to produce a complete response. Such synergistic
effects, which are characteristic of prolactin action in other vertebrates,
are well seen in the development of the brood patch, fat deposition
and the induction of migratory restlessness. A detailed account of the
physiology of the hormone cannot be included in this chapter (reviews: van Tienhoven, 1961; Lehrman, 1961; Riddle, 1963; Chadwick,
1969), and comment will be restricted to those reports in which
pituitary prolactin content has been measured and related to some
event in the organism.

Many reports suggest that prolactin plays a part in the onset of
broodiness, acting both to induce behavioral changes and, with the
sex steroids, to cause hyperplasia of the brood patch (Lehrman,
1961; Riddle, 1963). In support of this, the level of pituitary prolactin
is 2.5–3 times greater in incubating than in nonincubating domestic
hens (Byerly and Burrows, 1936; Burrows and Byerly, 1936; Saeki and
Tanabe, 1955; Nakajo and Tanaka, 1956). Similarly, the level in Ring-
necked Pheasant hens (*Phasianus colchicus*) rises at the onset of
incubation to reach its greatest value 8–12 days later (Breitenbach
and Meyer, 1959). However, the prolactin content declines rapidly

during the latter stages of incubation and the early post-hatching period, the time at which birds display their most intense brooding activity. This implies that assays on pituitary levels may not truly reflect the periods of highest secretion. Jones (1969) has compared pituitary prolactin content with the degree of incubation patch development in the California Quail (*Lophortyx californicus*). In wild females, the level is much the same for birds taken out of the breeding season and during the stages of ovarian growth, egg laying, and early incubation (cf. pheasant). The content increases to reach its highest level while the females show broodiness. The phase of incubation patch development precedes this rise and occurs at the onset of incubation, and by the brooding stage the patch is regressing rapidly. These results suggest that prolactin secretion is greatest during early incubation and that this keeps the hormone content at a low level. As the birds finish incubation, the demand for prolactin for growth of the brood patch diminishes, and this may account for the rise in the pituitary level. Caged egg-laying females showing no brood patch development and incubation behavior have the highest prolactin levels of all. This may be an example in which prolactin synthesis occurs in the absence of significant release. Testicular growth is accompanied by a rise in prolactin content. In those males not forming an incubation patch, the prolactin level declined during testicular regression, but in those few that did form a patch, the content remained high. Here, then, is an example of a high level of prolactin secretion being correlated with a high pituitary content.

In the three species of the subfamily Phalaropodinae, only the male forms a brood patch and incubates the eggs. Again, prolactin is implicated (together with androgens); Höhn and Cheng (1965), using Wilson's Phalarope (*Phalaropus tricolor*), found more prolactin in the pituitaries of males than of females. An increase in pituitary prolactin content has also been reported in incubating California Gulls (*Larus californicus*) (Bailey, 1952) and turkeys (Cherms *et al.*, 1962).

Pituitary prolactin cells are activated if birds are photostimulated (see Section IV,C), and this is accompanied by an increase in the hormone content of the gland. In the male duck, the prolactin content shows a pronounced annual cycle that very much parallels the cycle in testicular growth, the two parameters reaching their maximum levels simultaneously (Assenmacher *et al.*, 1962; Gourdji and Tixier-Vidal, 1966; Gourdji, 1970). Exposure of ducks to continuous light in the autumn likewise increases the prolactin level and induces testicular growth. Two species of quail behave in a similar manner to the duck

(Japanese Quail, Gourdji, 1970; Alexander and Wolfson, 1970; California Quail, Jones, 1969). Some results of Gourdji (1970) suggest that the increase in prolactin content is a direct result of the increased day length and may be modified by changes in sex hormone secretion. Thus, castration of male ducks held on short days, or their treatment with exogenous testosterone, has no pronounced effects on pituitary prolactin content. Under conditions of continuous light, some effects were noted, with castration leading to a lower pituitary level. It is assumed that the high prolactin level found in these photostimulated birds reflects a high secretion rate, but little is known of the physiology of the hormone in these two species (see Gourdji, 1970). Not surprisingly, both fall into that group of birds in which prolactin does not suppress gonadotropin secretion.

A number of papers from Meier's laboratory have given a prominent role to prolactin in the regulation of fattening and of migratory restlessness in passerine species (see Meier et al., 1969). Most of the evidence derives from treatment with exogenous prolactin, and the fascinating discovery has been made that the time of administration of the hormone crucially affects the magnitude of the response. This could imply that under natural conditions the release of prolactin is more marked at one time of the day than another. Some evidence for this has been found in the White-throated Sparrow (Zonotrichia albicollis). During the period of vernal migration, the pituitary prolactin content shows a diurnal rhythm with a peak occurring 6 hours after dawn. Meier et al. (1969) interpret these data as indicating a release of prolactin during the afternoon. This is the same time of the day when exogenous prolactin is most effective in inducing fat deposition (Meier and Davis, 1967).

D. ADRENOCORTICOTROPIN (ACTH)

Indirect evidence for an avian ACTH has existed for many years, but surprisingly, it is only recently that its presence in the adenohypophysis has been established. The presence of ACTH in anterior and posterior pituitaries of the Pekin duck has been shown using the adrenal ascorbic acid depletion in the hypophysectomized rat (Mialhe-Voloss, 1955). Pituitary extracts from the chicken (De Roos and De Roos, 1964), duck (Stainer and Holmes, 1969), and pigeon (Péczely and Zoboray, 1967; Péczely et al., 1970) have all proved capable of stimulating corticosteroid production in vitro by adrenal tissues of both mammals and birds. An intravenous injection into chickens of a crude chicken pituitary extract also raises the adrenal venous plasma concentration of corticosterone within 30 minutes (Resko *et al.*, 1964). Rather more unequivocal evidence for chicken ACTH has come recently from Salem *et al.* (1970a,b). These authors used a well established ACTH bioassay based on adrenal ascorbic acid depletion (AAAD) in the hypophysectomized rat (see Evans *et al.*, 1966). In this system, chicken pituitary extracts gave a dose–response curve parallel with the purified mammalian standard, and the ACTH activity of a laying hen was estimated at 401.7 (270.0–597.7) mU per milligram of frozen tissue. This is somewhat higher than that reported by De Roos and De Roos (1964). The assay sensitivity is such that a complete (2 + 2) assay should be possible with 0.01 of a chicken pituitary, suggesting the AAAD method might be adequate for assays in single pituitaries of small birds.

Avian ACTH, like the mammalian hormone, is labile to boiling and to rapid adjustments in pH (Salem *et al.*, 1970b). This is in contrast to the ACTH-like activity that resides in the avian hypothalamus (Salem *et al.*, 1970a,b; Péczely and Zboray, 1967; Péczely *et al.*, 1970). The functional role of the hypothalamic ACTH-like substance is still a matter of speculation (see Chapter 3).

E. MELANOTROPIN (MSH)

Investigations on the melanotropin content of the bird pituitary have been directed toward the following: (1) appearance of this hormone during the development of the chick (Chen *et al.*, 1940; Rahn and Drager, 1941); (2) distribution of the hormone within the pars distalis of the chick (Kleinholz and Rahn, 1939) and the duck (Mialhe-Voloss and Benoit, 1954); (3) correlation between the pituitary level of MSH and the abundance of κ cells (Tougard, 1972; see Section IV,E). For these purposes, quantitative data have been obtained using different bioassays—the hypophysectomized lizard *Anolis* (Rahn and Drager, 1941), the hypophysectomized male frog (Mialhe-Voloss and Benoit, 1954), and Burger's method *in vitro* with the skin of *Anolis carolinensis* (Burger, 1961; Tougard, 1972). The specificity of these bioassays is not perfect, since they display some sensitivity to ACTH.

The MSH activity is probably the first to appear during the embryological life of the chick—on the 5th day. Its concentration in the pituitary increases very rapidly during the second half of the incubation but remains unchanged after hatching. Within the gland, the hormone is more concentrated in the cephalic lobe of the chick and the duck (thirty times more). In some species of duck and pigeon, there is a higher level of pituitary MSH in strains with colored feathers as compared to strains with white feathers. This finding (Tougard, 1972) suggests that MSH could intervene in the control of feather pigmentation. To our knowledge, this is the only indication as to a possible physiological role of this hormone in birds. As of now, the chemistry of avian MSH is not known.

F. Growth Hormone (Somatotropin, STH)

Pituitary extracts of chicken and turkey possess some growth-promoting activity since when injected into immature hypophysectomized rats, they cause widening of the tibial epiphyses, and increases in body weight and nitrogen retention (Solomon and Greep, 1959; Hazelwood and Hazelwood, 1961; Hirsch, 1961). In this assay, a rat pituitary contains about eight times as much activity as that of a chicken. This might reflect a low level of growth hormone in the chicken, but may equally well reflect a species difference in its structure and activity. A comparable phenomenon occurs with fish growth hormone, which is highly active in hypophysectomized fish but inactive in the rat (see Hoar, 1966). Chicken pituitary extracts do increase growth rate, nitrogen retention, and bone growth in hypophysectomized chicks, but the mammalian growth hormones of bovine and porcine origin have no effect even at large dose levels (Libby *et al.*, 1955; Glick, 1960; Nalbandov, 1966). Again, this may reflect a species specificity, since mammalian growth hormones alter plasma protein levels in the Budgerigar (*Melopsittacus undulatus*) (Rudolph and Pehrson, 1961), while pituitary tumors from Budgerigars cause growth in rats (Schlumberger and Rudolph, 1959). Immunologically, anti-bovine growth hormone shows no cross reaction with chicken pituitary extracts (Moudgal and Li, 1961b).

REFERENCES

Adams, A. E. (1946). Variations in the potency of thyrotropic hormone in animals. *Quart. Rev. Biol.* **21**, 1–32.
Alexander, B., and Wolfson, A. (1970). Prolactin and sexual maturation in the Japanese Quail, *Coturnix coturnix japonica. Poultry Sci.* **49**, 632–640.
Aronson, J. (1952). "Studies on the Cell Differentiation in the Anterior Pituitary of the Chick Embryo by Means of the PAS Reaction." Gleerup, Lund.
Assenmacher, I. (1951). Le développement embryologique du système porte hypophysaire chez le Canard domestique. *C. R. Acad. Sci.* **234**, 563–565.

Assenmacher, I. (1952). La vascularisation du complexe hypophysaire chez le Canard domestique. I. La vascularisation du complexe hypophysaire adulte. II. Le développement embryologique de l'appareil vasculaire hypophysaire. *Arch. Anat. Microsc. Morphol. Exp.* **41**, 69–152.

Assenmacher, I. (1958). Recherches sur le contrôle hypothalamique de la fonction gonadotrope préhypophysaire chez le Canard. *Arch. Anat. Microsc. Morphol. Exp.* **47**, 448–572.

Assenmacher, I. (1970). Importance de la liaison hypothalamo-hypophysaire dans le contrôle nerveus de la reproduction chez les Oiseaux. *Colloq. Int. Cent. Nat. Rech. Sci.* **172**, 167–192.

Assenmacher, I., Tixier-Vidal, A., and Boissin, J. (1962). Contenu en hormones gonadotropes et en prolactine de l'hypophyse du Canard soumis à un traitement lumineux ou réserpinique. *C. R. Soc. Biol.* **156**, 1555–1560.

Bacon, W., Cherms, F. L., MacDonald, G. J., and McShan, W. H. (1969). Gonadotropin potency variation in anterior pituitaries from sexually mature male quail (*Coturnix coturnix japonica*). *Poultry Sci.* **48**, 718–721.

Bagshawe, K. D., Orr, A. H., and Godden, J. (1968). Cross reaction in radioimmunoassay between human chorionic gonadotrophin and plasma from various species. *J. Endocrinol.* **42**, 513–518.

Bailey, R. E. (1952). The incubation patch of passerine birds. *Condor* **54**, No. 3, 121–136.

Baker, B. I. (1963). Effect of adaptation to black and white backgrounds on the teleost pituitary. *Nature (London)* **198**, 404.

Barrnett, R. J., and Seligman, A. M. (1954). Histochemical demonstration of sulfhydryl and disulfide groups of proteins. *J. Nat. Cancer Inst.* **14**, 769–803.

Bates, R. W., and Condliffe, P. G. (1960). Studies on the chemistry and bioassay of thyrotropins from bovine pituitaries, transplantable pituitary tumors of mice and blood plasma. *Recent Progr. Horm. Res.* **16**, 309–352.

Bates, R. W., and Condliffe, P. G. (1966). The physiology and chemistry of thyroid stimulating hormone. *In* "The Pituitary Gland" (G. W. Harris and B. T. Donovan, eds.), Vol. 1, pp. 374–410. Butterworth, London.

Bates, R. W., and Cornfield, J. (1957). An improved assay method for thyrotropin using depletion of I^{131} from the thyroid of day-old chicks. *Endocrinology* **60**, 225–238.

Bates, R. W., Garrison, M. M., and Cornfield, J. (1963). An improved bioassay for prolactin using adult pigeons. *Endocrinology* **73**, 217–223.

Bates, R. W., Garrison, M. M., Cooper, J. A., and Condliffe, P. G. (1968). Further studies on the purification of human thyrotropin. *Endocrinology* **83**, 721–730.

Ben David, M. (1967). A tentative bioassay for prolactin based on ^{3}H-methyl-thymidine uptake by the pigeon crop mucosal epithelium. *Proc. Soc. Exp. Biol. Med.* **125**, 705–708.

Benoit, J., and Aron, C. (1931). Influence de la castration sur le taux d'hormone préhypophysaire excito-sécrétrice de la thyroïde dans le milieu intérieur chez le Coq et le Canard. Notion d'un cycle saisonnier de l'activité préhypophysaire chez les Oiseaux. *C. R. Soc. Biol.* **108**, 786–788.

Benoit, J., and Assenmacher, I. (1951). Etude préliminaire de la vascularisation de l'appareil hypophysaire du Canard domestique. *Arch. Anat. Microsc. Morphol. Exp.* **40**, 27–45.

Bhattacharyya, T. K., and Sarkar, M. (1969). Adenohypophyseal cytology in normal and gonadectomised pigeons. *Acta Morphol.* **17**, 113–122.

Brasch, M., and Betz, T. W. (1971). The hormonal activities associated with the cephalic and caudal regions of the cockerel pars distalis. *Gen. Comp. Endocrinol.* **16,** 241–256.

Breitenbach, R. P., and Meyer, R. K. (1959). Pituitary prolactin levels in laying, incubating and brooding pheasants (*Phasianus colchicus*). *Proc. Soc. Exp. Biol. Med.* **101,** 16–19.

Breneman, W. R. (1945). The gonadotropic activity of the anterior pituitary of cockerels. *Endocrinology* **36,** 190–199.

Breneman, W. R. (1955). Reproduction in birds: The female. *Mem. Soc. Endocrinol.* **4,** 94–110.

Breneman, W. R., Zeller, F. J., and Beekman, B. E. (1959). Gonadotropin assay in chicks. *Poultry Sci.* **38,** 152–158.

Breneman, W. R., Zeller, F. J., and Creek, R. O. (1962). Radioactive phosphorus uptake by chick testes as an end-point for gonadotropin assay. *Endocrinology* **71,** 790–798.

Brookes, L. D. (1967). A stain for differentiating two types of acidophil in the pituitary. *Gen. Comp. Endocrinol.* **9,** Abstr. 22.

Brown, P. S. (1955). The assay of gonadotropin from urine of non-pregnant human subjects. *J. Endocrinol.* **13,** 59–64.

Brown, L. T., and Knigge, K. M. (1958). Cytology of the pars distalis in the pituitary gland of the chicken. *Anat. Rec.* **130,** 395.

Bryant, G. D., and Greenwood, F. C. (1968). Radioimmunoassay for ovine, caprine and bovine prolactin in plasma and tissue extracts. *Biochem. J.* **109,** 831–840.

Bullock, D. W., Mittal, K. K., and Nalbandov, A. V. (1961). Immunological and biological cross-reactivity of chicken and mammalian gonadotrophins. *Endocrinology* **80,** 1182–1184.

Burger, A. C. J. (1961). Occurrence of the three electrophoretic components with melanocyte-stimulating activity in extracts of single pituitary glands from ungulates. *Endocrinology* **68,** 698–703.

Burns, J. M. (1969). Luteinizing hormone bioassay based on uptake of radio-active phosphorus by the chick testes. *Comp. Biochem. Physiol.* **34,** 727–731.

Burrows, W. H., and Byerly, T. C. (1936). Studies of prolactin in the fowl pituitary. I. Broody hens compared with laying hens and males. *Proc. Soc. Exp. Biol. Med.* **34,** 841–844.

Byerly, T. C., and Burrows, W. H. (1936). Studies of prolactin in the fowl pituitary. II. Effects of genetic constitution with respect to broodiness on prolactin content. *Proc. Soc. Exp. Biol. Med.* **34,** 844–846.

Case, J. F. (1952). Relation entre les surrénales et l'antéhypophyse durant la vie embryonnaire. *Ann. N.Y. Acad. Sci.* **55,** No. 2, 147–158.

Chadwick, A. (1966). Prolactin-like activity in the pituitary gland of the frog. *J. Endocrinol.* **34,** 247–255.

Chadwick, A. (1969). Effects of prolactin in homoiothermic vertebrates. *Gen. Comp. Endocrinol., Suppl.* **2,** 63–68.

Chen, G., Oldham, F. K., and Geiling, E. M. K. (1940). Appearance of the melanophore expanding hormone of the pituitary gland in developing chick embryo. *Proc. Soc. Exp. Biol. Med.* **45,** 810–813.

Cherms, F. L., Herrick, R. B., McShan, W. H., and Hymer, W. C. (1962). Prolactin content of the anterior pituitary gland of turkey hens in different reproductive stages. *Endocrinology* **71,** 289–292.

Cunningham, F. J., and Furr, B. J. A. (1972). Plasma levels of luteinizing hormone and progesterone during the ovulatory cycle of the hen. *In* "Egg Formation and Production" (B. M. Freeman and P. E. Lake, eds.), pp. 51–64. British Poultry Science Ltd., Edinburgh.

Dancasiu, M., and Campeanu, L. (1970). Ultrastructure de l'adénohypophyse chez *Coturnix coturnix japonica*. *Rev. Roum. Endocrinol.* **7**, 129–133.

D'Angelo, S. A., and Gordon, A. S. (1950). The simultaneous detection of thyroid and thyrotrophic hormones in vertebrate sera. *Endocrinology* **46**, 39–54.

De Roos, R., and De Roos, C. C. (1964). Effects of mammalian corticotropin and chicken adenohypophysial extracts on steroidogenesis by chicken adrenal tissue *in vitro*. *Gen. Comp. Endocrinol.* **4**, 602–607.

Diczfalusy, E. (1969). Immunoassay of gonadotropins. *Acta Endocrinol. (Copenhagen), Suppl.* **142**.

Dodd, J. M., Ferguson, K. M., Dodd, M. H. I., and Hunter, R. B. (1963). The comparative biology of thyrotropin secretion. *In* "Thyrotropin" (S. C. Werner, ed.), pp. 3–38. Thomas, Springfield, Illinois.

Dominic, C. J., and Singh, R. M. (1969). Anterior and posterior groups of portal vessels in the avian pituitary. *Gen. Comp. Endocrinol.* **13**, 22–26.

Dumont, J. N. (1955). Prolactin-induced cytological changes in the mucosa of the pigeon crop during crop "milk" formation. *Z. Zellforsch. Mikrosk. Anat.* **68**, 755–782.

Ensor, D. M., and Ball, J. N. (1968). A bioassay for fish prolactin (paralactin). *Gen. Comp. Endocrinol.* **11**, 104–110.

Evans, H. M., Sparks, L. L., and Dixon, J. S. (1966). Chemistry of adrenocorticotrophin. *In* "The Pituitary Gland" (G. W. Harris and B. T. Donovan, eds.), Vol. 1, pp. 317–373. Butterworth, London.

Ezrin, C., and Murray, S. (1963). The cells of the human adenohypophysis in pregnancy, thyroid disease and adrenal cortical disorders. *In* "Cytologie de l'Adénohypophyse" (J. Benoit and C. De Lage, eds.), pp. 183–199. CNRS, Paris.

Farner, D. S., and Follett, B. K. (1966). Light and other environmental factors affecting avian reproduction. *J. Anim. Sci. Suppl.* **25**, 90–118.

Farner, D. S., Follett, B. K., King, J. R., and Morton, M. L. (1966). A quantitative examination of ovarian growth in the White-crowned sparrow. *Biol. Bull.* **130**, 67–75.

Ferrand, R. (1969). Influence inductrice exercée par le plancher encéphalique sur l'ébauche adénohypophysaire aux jeunes stades du développement de l'embryon de poulet. *C. R. Acad. Sci. Ser. D* **268**, 550–553.

Ferrand, R. (1970). Etude expérimentale de la différenciation de l'adénohypophyse chez l'embryon de poulet. *Année Biol.* **9**, 357–365.

Ferrand, R. (1971). Influence de la vascularisation sur la différenciation des types cellulaires de l'adénohypophyse des Oiseaux. *C. R. Soc. Biol.* **165**, 392–395.

Ferrando, G., and Nalbandov, A. V. (1969). Direct effect on the ovary of the adrenergic blocking drug dibenzyline. *Endocrinology* **85**, 38–42.

Follett, B. K. (1970). Gonadotropin-releasing activity in the quail hypothalamus. *Gen. Comp. Endocrinol.* **15**, 165–179.

Follett, B. K. (1973). Circadian rhythmicity and time measurement in avian photoperiodicity. *In* "The Environment and Reproduction in Mammals and Birds" (J. S. Perry and I. W. Rowlands, eds.). Blackwell, Oxford (*J. Reprod. Fertil. Suppl.* **19**, pp. 5–18.

Follett, B. K., and Farner, D. S. (1966). Pituitary gonadotropins in the Japanese Quail (*Coturnix coturnix japonica*) during photoperiodically induced gonadal growth. *Gen. Comp. Endocrinol.* **7**, 125–131.

Follett, B. K., and Sharp, P. J. (1969). Circadian rhythmicity in photoperiodically induced gonadotrophin release and gonadal growth in the Quail. *Nature (London)* **223**, 968–971.

Follett, B. K., Farner, D. S., and Morton, M. L. (1967). The effects of alternating long and short daily photoperiods on gonadal growth and pituitary gonadotropins in the White-crowned Sparrow *Zonotrichia leucophrys gambelii. Biol. Bull.* **133**, 330–342.

Follett, B. K., Scanes, C. G., and Cunningham, F. J. (1972a). A radioimmunoassay for avian luteinizing hormone. *J. Endocrinol.* **52**, 359–378.

Follett, B. K., Scanes, C. G., and Nicholls, T. J. (1972b). The chemistry and physiology of the avian gonadotropins. *In* "Hormones Glycoproteiques Hypophysaires," pp. 193–211. INSERM, Paris.

Fraps, R. M. (1961). Ovulation in the domestic fowl. *In* "Control of Ovulation" (C. A. Villee, ed.), pp. 133–162. Pergamon, Oxford.

Fraps, R. M. (1970). Photoregulation in the ovulation cycle of the domestic fowl. *Colloq. Int. Cent. Nat. Rech. Sci.* **172**, 281–306.

Fraps, R. M., Fevold, H. L., and Neher, B. H. (1947). Ovulatory response of the hen to presumptive luteinizing and other fractions from fowl anterior pituitary tissue. *Anat. Rec.* **99**, 571–572.

Furr, B. J. A., and Cunningham, F. J. (1970). The biological assay of chicken pituitary gonadotropins. *Brit. Poultry Sci.* **11**, 7–13.

Gabe, M. (1953). Sur quelques applications de la coloration par la fuchsine paraldéhydique. *Bull. Microsc. Appl.* **3**, 153–162.

Garnier, D. (1971). Variations de la testostérone du plasma périphérique chez le Canard Pékin au cours du cycle annuel. *C. R. Acad. Sci., Ser. D* **272**, 1665–1668.

Garnier, D., and Attal, J. (1970). Variations de la testostérone, du plasma testiculaire et des cellules interstitielles chez le Canard Pékin au cours du cycle annuel. *C. R. Acad. Sci., Ser. D* **270**, 2473–2475.

Glenner, G. G., and Lillie, R. D. (1957). The histochemical demonstration of indole derivatives by the post-coupled *p*-dimethyl-amino-benzylidene reaction. *J. Histochem. Cytochem.* **5**, 279–296.

Glick, B. (1960). The effect of bovine growth hormone, DCA, and cortisone on the weight of the bursa of Fabricius, adrenal glands, heart, body weight of young chickens. *Poultry Sci.* **39**, 1527.

Gourdji, D. (1964). La préhypophyse de l'Ignicolore mâle, *Pyromelana franciscana* au cours du cycle annuel. Thèse de Doctorat, 3e cycle, Paris.

Gourdji, D. (1967). Etude du déterminisme des variations du contenu hypophysaire en prolactine chez le Canard Pékin. Influence de la lumière permanente, de la castration et de la testostérone et de leurs interactions. *C. R. Acad. Sci., Ser. D* **264**, 1482–1484.

Gourdji, D. (1970). Prolactine et relations photosexuelles chez les Oiseaux. *Colloq. Int. Cent. Nat. Rech. Sci.* **172**, 233–258.

Gourdji, D., and Tixier-Vidal, A. (1966). Variations du contenu hypophysaire en prolactine chez le Canard Pékin mâle au cours du cycle sexuel et de la photostimulation expérimentale du testicule. *C. R. Acad. Sci., Ser. D* **262**, 1746–1749.

Greeley, F., and Meyer, R. K. (1953). Seasonal variation in the testis stimulating activity of male pheasant glands. *Auk* **70**, 350–358.

Green, J. D. (1951). The comparative anatomy of the hypophysis with special reference to its blood supply and innervation. *Amer. J. Anat.* **88**, 225–311.

Greenspan, F. S., Kriss, J. P., Moses, L. E., and Lew, W. (1956). An improved bioassay method for thyrotropic hormone using thyroid uptake of radiophosphorus. *Endocrinology* **58**, 767–776.

Grignon, G. (1955). Chronologie de la différenciation des éléments cellulaires du lobe distal de l'hypophyse chez l'embryon de Poule Rhode-Island. *C. R. Soc. Biol.* **149**, 1448.

Grignon, G. (1956). Développement du complexe hypothalamo-hypophysaire chez l'embryon de Poulet. Thesis, Université de Nancy, Société d'impression typographiques.

Grosvenor, C. E., and Turner, C. W. (1958). Assay of lactogenic hormone. *Endocrinology* **63**, 530–534.

Guedenet, J. C., Grignon, G., and Franco, N. (1970). Etude critique de la mise en évidence des cellules à grains glycoprotidiques de l'adénohypophyse chez le Poulet au cours de la vie embryonnaire et de la période post-natale. *Electron Microsc., Proc. Int. Congr.*, III, p. 565.

Haase, E. (1972). Effects of short days on the pituitary-gonadal axis of Bramblings *Fringilla montifringilla*. *Gen. Comp. Endocrinology* **18**, abst. 73, 594.

Haase, E., and Farner, S. D. (1969). Acetylcholinesterase in der pars distalis von *Zonotrichia leucophrys gambelii*. *Z. Zellforsch. Mikrosk. Anat.* **93**, 356–368.

Haase, E., and Farner, D. S. (1970). The function of the acetylcholinesterase cells of the pars distalis of the White-Crowned Sparrow, *Zonotrichia leucophrys gambelii*. *Acta Zool.* **51**, 99–106.

Halmi, N. S. (1952). 2 types of basophils in the rat pituitary: "Thyrotrophs" and "gonadotrophs" vs β and δ cells. *Endocrinology* **50**, 140 and 142.

Hazelwood, R. L., and Hazelwood, B. S. (1961). Effects of avian and rat pituitary extracts on tibial growth and blood composition. *Proc. Soc. Exp. Biol. Med.* **108**, 10–12.

Heald, P. J., Furnival, B. E., and Rookledge, K. A. (1967). Changes in the levels of luteinizing hormone in the pituitary of the domestic fowl during an ovulatory cycle. *J. Endocrinol.* **37**, 73–81.

Herlant, M. (1960). Etude critique de deux techniques nouvelles destinées à mettre en évidence les différentes catégories cellulaires présentes dans la glande pituitaire. *Bull. Microsc. Appl.* **10**, 37–44.

Herlant, M. (1964). The cells of the adenohypophysis and their functional significance. *Int. Rev. Cytol.* **17**, 299–381.

Herlant, M., Benoit, J., Tixier-Vidal, A., and Assenmacher, I. (1960). Modifications hypophysaires au cours du cycle annuel chez le Canard Pékin mâle. *C. R. Acad. Sci.* **250**, 2936–2938.

Herlant, M., and Racadot, J. (1957). Le lobe antérieur de l'hypophyse de la Chatte au cours de la gestation et de la lactation. *Arch. Biol.* **68**, 217–248.

Herrick, R. B., McGibbon, W. H., and McShan, W. H. (1962). Gonadotrophic activity of chicken pituitary glands. *Endocrinology* **71**, 488–491.

Hillemann, H. H. (1943). An experimental study of the development of the pituitary gland in the chick embryos. *J. Exp. Zool.* **93**, 347–373.

Hirsch, L. J. (1961). A study of the growth promoting principle in the domestic fowl. Ph.D. Thesis, University of Illinois, Urbana.

Hoar, W. S. (1966). Hormonal activities of the *pars distalis* in cyclostomes, fish and amphibia. *In* "The Pituitary Gland" (G. W. Harris and B. T. Donovan, eds.), Vol. 1, pp. 242–294. Butterworth, London.

Höhn, E. O., and Cheng, S. C. (1965). Prolactin and the incidence of brood patch formation and incubation behaviour of the two sexes in certain birds with special reference to Phalaropes. *Nature (London)* **208**, 197–198.

Imaï, K., and Nalbandov, A. V. (1971). Changes in FSH activity of anterior pituitary glands and of blood plasma during the laying cycle. *Endocrinology* **88**, 1465–1470.

Inoguchi, A., and Sato, Y. (1962). Acid- and aldehyde-fuchsinophile granules in so-called basophiles of the anterior pituitaries of capons and laying hens. *Arch. Histol. Jap.* **22**, No. 3, 273–280.

Ishii, S., Sarkar, A. K., and Kobayashi, H. (1970). Ovarian ascorbic acid depleting factor in pigeon median eminence extracts. *Gen. Comp. Endocrinol.* **14**, 461–466.

Jackson, G. L., and Nalbandov, A. V. (1969). A substance resembling arginine vasotocin in the anterior pituitary gland of the cockerel. *Endocrinology* **84**, 1218–1223.

Jones, R. E. (1969). Epidermal hyperplasia in the incubation patch of the California Quail, *Lophortyx californicus*, in relation to pituitary prolactin content. *Gen. Comp. Endocrinol.* **12**, 498–502.

Kamiyoshi, M., and Tanaka, K. (1969). Changes in pituitary FSH concentrations during an ovulatory cycle of the hen. *Poultry Sci.* **48**, 2025–2032.

Katongole, C. B., Naftolin, F., and Short, R. V. (1971). Relationship between blood levels of luteinizing hormone and testosterone in bulls, and the effects of sexual stimulation. *J. Endocrinol.* **50**, 457–466.

King, J. R., Follett, B. K., Farner, D. S., and Morton, M. L. (1966). Annual gonadal cycles and pituitary gonadotropins in *Zonotrichia leucophrys gambelii. Condor* **68**, 476–487.

Kirkham, K. E., and Hunter, W. M. eds. (1971). "Radioimmunoassay Methods." Churchhill, London.

Kleinholz, L. H., and Rahn, H. (1939). The distribution of intermedin in the pars anterior of the chicken pituitary. *Nat. Acad. Sci.* **3**, 145–147.

Kluver, H., and Barrera, E. (1953). A method for the combined staining of cells and fibers in the nervous system. *J. Neuropathol. Exp. Neurol.* **12**, 400–407.

Kobayashi, H. (1954). Thyrotrophin content in the pituitary body of the canaries receiving implants of sex steroids. *Annot. Zool. Jap.* **27**, 138–139.

Kobayashi, H., and Farner, D. S. (1966). Evidence of a negative feedback on photoperiodically induced gonadal development in the White-crowned Sparrow, *Zonotrichia leucophrys gambelii. Gen. Comp. Endocrinol.* **6**, 443–452.

Lamberg, B. A. (1955). Assay of thyrotrophin with radioactive indicators *Acta Endocrinol. (Copenhagen)* **18**, 405–420.

Le Douarin, N., Ferrand, R., and Le Douarin, G. (1967). La différenciation de l'ébauche épithéliale de l'hypophyse séparée du plancher encéphalique et placée dans des mésenchymes détérologues. *C. R. Acad. Sci.* **264**, 3027–3029.

Legait, H., and Legait, E. (1955). Modification de structure du lobe distal de l'hypophyse au cours de la couvaison chez la Poule Rhode Island. Essai d'interprétation de la valeur des deux types principaux de cellules cyanophiles. *C. R. Ass. Anat.,* **84**, 188–199.

Legait, H., and Legait, E. (1956). Nouvelles recherches sur les modifications de structure du lobe distal de l'hypophyse au cours de divers états physiologiques et expérimentaux chez la Poule Rhode Island. *C. R. Ass. Anat.* **91**, 902–908.

Lehrman, D. S. (1961). Gonadal hormones and parental behavior in birds and infra-human mammals. *In* "Sex and Internal Secretions" (W. C. Young, ed.), Vol. 2, 3rd ed., pp. 1268–1382. Williams & Wilkins, Baltimore, Maryland.

Libby, D. A., Meites, J., and Schaible, J. (1955). Growth hormone effects in chickens. *Poultry Sci.* **34**, 1329.

Ljunggren, L. (1969). "Studies on Seasonal Activity in Pigeons with Aspects on the Role of Biogenic Monoamines in the Endocrine Organs." Gleerup, Lund.

Lofts, B., Follett, B. K., and Murton, R. K. (1970). Temporal changes in the pituitary gonadal axis. Mem. Soc. Endocrinol. 18, 545–577.

Lyons, W. R. (1937). Preparation and assay of mammotrophic hormone. Proc. Soc. Exp. Biol. Med. 35, 645–648.

MacConnail, M. A. (1947). The staining of the central nervous system with lead-hematoxylin. J. Anat. 81, 371–372.

Matsuo, S., Vitums, A., King, J., and Farner, D. S. (1969). Light microscope studies of the cytology of the adenohypophysis of the White-crowned Sparrow, Zonotrichia leucophrys gambelii. Z. Zellforsch. Mikrosk. Anat. 95, 143–176.

Meier, A. H., and Davis, K. B. (1967). Diurnal variations of the fattening response to prolactin in the White-throated Sparrow, Zonotrichia albicollis. Gen. Comp. Endocrinol. 8, 110–114.

Meier, A. H., and Farner, D. S. (1964). A possible endocrine basis for premigratory fattening in the White-crowned Sparrow, Zonotrichia leucophrys gambelii. Gen. Comp. Endocrinol. 4, 584–595.

Meier, A. H., Burns, J. I., and Dusseau, J. W. (1969). Seasonal variations in the diurnal rhythm of the pituitary prolactin content in the White-throated Sparrow, Zonotrichia albicollis. Gen. Comp. Endocrinol. 12, 282–289.

Meier, A. H., Burns, J. T., Davis, K. B., and John, T. M. (1971). Circadian variations in sensitivity of the pigeon crop-sac to prolactin. J. Interdisc. Cycle Res. 2, 161–172.

Meites, J., and Nicoll, C. S. (1966). Adenohypophysis: Prolactin. Annu. Rev. Physiol. 28, 57–88.

Mialhe-Voloss, C. (1955). Activité corticotrope des lobes antérieur et postérieur de l'hypophyse chez le Rat et le Canard. J. Physiol. (Paris) 47, 251–254.

Mialhe-Voloss, C., and Benoit, J. (1954). L'intermédine dans l'hypophyse et l'hypothalamus du Canard. C. R. Soc. Biol. 148, 56–59.

Mikami, S. (1958). The cytological significance of regional patterns in the adenohypophysis of the fowl. J. Fac. Agr., Iwate Univ. 3, No. 4, 473–545.

Mikami, S. (1969). Morphological studies of the avian adenohypophysis related to its function. Gunma Symp. Endocrinol. [Proc.] 6, 151–170.

Mikami, S., Vitums, A., and Farner, D. S. (1969). Electron microscopic studies on the adenohypophysis of the White-crowned Sparrow, Zonotrichia leucophrys gambelii. Z. Zellforsch. Mikrosk. Anat. 97, 1–29.

Mikami, S., Oksche, A., Farner, D. S., and Vitums, A. (1970). Fine structure of the vessels of the hypophysial portal system of the White-crowned Sparrow, Zonotrichia leucophrys gambelii. Z. Zellforsch. Mikrosk. Anat. 106, 155–174.

Mitchell, M. E. (1961). Stimulation of the ovary in hypophysectomized hens by an avian pituitary preparation. J. Reprod. Fert. 14, 249–256.

Moszkowska, A. (1949). Pouvoir corticotrope et gonadotrope de l'hypophyse embryonnaire du Poulet. C. R. Soc. Biol. 143, 1322.

Moszkowska, A. (1956). Activité gonadotrope de l'antéhypophyse d'embryon de Poulet. Arch. Anat. Microsc. Morphol. Exp. 45, No. 1, 65–76.

Moudgal, N. R., and Li, C. H. (1961a). An immunochemical study of sheep pituitary interstitial cell-stimulating hormone. Arch. Biochem. Biophys. 95, 93–98.

Moudgal, N. R., and Li, C. H. (1961b). Immunochemical studies of bovine and ovine pituitary growth hormone. Arch. Biochem. Biophys. 93, 122–127.

Murton, R. K., Bagshawe, K. D., and Lofts, B. (1969). The circadian basis of specific gonadotropin release in relation to avian spermatogenesis. J. Endocrinol. 45, 311–312.

Murton, R. K., Lofts, B., and Orr, A. H. (1970). The significance of circadian based photosensitivity on the House Sparrow *Passer domesticus. Ibis* **112,** 448–456.

Nakajo, S., and Tanaka, K. (1956). Prolactin potency of the cephalic and the caudal lobe of the anterior pituitary in relation to broodiness in the domestic fowl. *Poultry Sci.* **35,** 989–994.

Nalbandov, A. V. (1961). The gonadotropic complex. *In* "Human Pituitary Gonadotropins" (A. Albert, ed.), pp. 339–342. Thomas, Springfield, Illinois.

Nalbandov, A. V. (1966). Hormonal activity of the pars distalis in reptiles and birds. *In* "The Pituitary Gland" (G. W. Harris and B. T. Donovan, eds.), Vol. 1, pp. 295–316. Butterworth, London.

Nelson, D. M., and Nalbandov, A. V. (1966). Hormone control of ovulation. *In* "Physiology of the Domestic Fowl" (C. Horton Smith and E. C. Amoroso, eds.), pp. 3–10. Oliver & Boyd, Edinburgh.

Nelson, D. M., Norton, H. W., and Nalbandov, A. V. (1965). Changes in hypophysial and plasma LH levels during the laying cycle of the hen. *Endocrinology* **77,** 889–896.

Nicholls, T. J., Scanes, C. G., and Follett, B. K. (1973). Pituitary and plasma LH in Japanese Quail during photoperiodically induced gonadal growth and regression. *Gen. Comp. Endocrinol.* in press.

Nicoll, C. S. (1967). Bioassay of prolactin. Analysis of the pigeon crop-sac response to local prolactin injection by an objective and quantitative method. *Endocrinology* **80,** 641–655.

Nicoll, C. S., Parsons, J. A., Fiorindo, R. P., and Nichols, C. W. (1969). Estimation of prolactin and growth hormone levels by polyacrylamide disc electrophoresis. *J. Endocrinol.* **45,** 183–196.

Nicoll, C. S., Fiorindo, R. P., McKennee, C. T., and Parsons, J. A. (1970). Assay of hypothalamic factors which regulate prolactin secretion. *In* "Hypophysiotropic Hormones of the Hypothalamus: Assay and Chemistry" (J. Meites, ed.), pp. 115–150. Williams & Wilkins, Baltimore, Maryland.

Olivereau, M. (1964). L'hématoxyline au plomb permet-elle l'identification des cellules corticotropes de l'hypophyse des téléostéens. *Z. Zellforsch. Mikrosk. Anat.* **63,** 496–505.

Parlow, A. F. (1961). Bio-assay of pituitary luteinizing hormone by depletion of ovarian ascorbic acid. *In* "Human Pituitary Gonadotrophins" (A. Albert, ed.), pp. 300–310. Thomas, Springfield, Illinois.

Pasteels, J. L. (1963). Recherches morphologiques et expérimentales sur la sécrétion de prolactine. *Arch. Biol.* **74,** 439–553.

Payne, F. (1942). The cytology of the anterior pituitary of the fowl. *Biol. Bull.* **82,** 79–11.

Payne, F. (1943). The cytology of the anterior pituitary of broody fowls. *Anat. Rec.* **86,** 1–13.

Payne, F. (1944). Anterior pituitary thyroid relationships in the fowl. *Anat. Rec.* **88,** 337–350.

Payne, F. (1946). The cellular picture in the anterior pituitary of normal fowls from embryo to old age. *Anat. Rec.* **96,** 77–91.

Payne, F. (1947). Effects of gonadal removal on the anterior pituitary of the fowl from 10 days to 6 years. *Anat. Rec.* **97,** 507–518.

Payne, F. (1955). Acidophilic granules in the gonadotrophic secreting basophiles of laying hens. *Anat. Rec.* **122,** 49–56.

Payne, F. (1965). Some observations on the anterior pituitary of the domestic fowl with the aid of the electron microscope. *J. Morphol.* **117**, 185.

Péczely, P., and Zboray, G. (1967). CRF and ACTH activity in the median eminence of the pigeon. *Acta Physiol.* **32**, 229–239.

Péczely, P., Baylé, J. D., Boissin, J., and Assenmacher, I. (1970). Activités corticotrope et "CRF" dans l'éminence médiane, et activité corticotrope de greffes hypophysaires chez le Pigeon. *C. R. Acad. Sci.* **270**, 3264–3267.

Perek, M., Ekstein, B., and Sobel, H. (1957). Histological observations on the anterior lobe of pituitary gland in moulting and laying hens. *Poultry Sci.* **36**, 954–958.

Purves, H. D. (1966). Cytology of the adenohypophysis. *In* "The Pituitary Gland" (G. W. Harris, and B. T. Donovan, eds.), Vol. 3, pp. 147–232. Butterworth, London.

Rahn, H. (1939). The development of the chick pituitary with special reference to the cellular differentiation of the pars buccalis. *J. Morphol.* **64**, 483–517.

Rahn, H., and Drager, G. (1941). Quantitative assay of the melanophore dispersing hormone during the development of the chicken pituitary. *Endocrinology* **29**, 725–730.

Rahn, H., and Painter, B. I. (1941). A comparative histology on the bird pituitary. *Anat. Rec.* **79**, 297–311.

Rambourg, A. (1967). Détection des glycoprotéines en microscopie électronique: Coloration de la surface cellulaire et de l'appareil de Golgi par un mélange acide chromique-phosphotungstique. *C. R. Acad. Sci., Ser. D* **265**, 1426–1428.

Resko, J. A., Norton, H. W., and Nalbandov, A. V. (1964). Endocrine control of the adrenal in chickens. *Endocrinology* **75**, 192–200.

Riddle, O. (1963). Prolactin in vertebrate function and organization. *J. Nat. Cancer Inst.* **31**, 1039–1110.

Riddle, O., Bates, R. W., and Dykshorn, S. W. (1932). A new hormone of the anterior pituitary. *Proc. Soc. Exp. Biol. Med.* **29**, 1211–1212.

Riddle, O., Bates, R. W., and Dykshorn, S. W. (1933). The preparation, identification and assay of prolactin, a hormone of anterior pituitary. *Amer. J. Physiol.* **105**, 191–216.

Riley, G. M., and Fraps, R. M. (1942). Relationship of gonad-stimulating activity of female domestic fowl anterior pituitaries to reproductive condition. *Endocrinology* **30**, 537–541.

Rudolph, H. J., and Pehrson, N. C. (1961). Growth hormone effect on the blood plasma proteins in the parakeet. *Endocrinology* **69**, 661.

Saeki, Y., and Tanabe, Y. (1955). Changes in prolactin content of fowl pituitary during broody periods and some experiments on the induction of broodiness. *Poultry Sci.* **34**, 909–916.

Salem, M. H. M., Norton, H. W., and Nalbandov, A. V. (1970a). A study of ACTH and CRF in chickens, *Gen. Comp. Endocrinol.* **14**, 270–280.

Salem, M. H. M., Norton, H. W., and Nalbandov, A. V. (1970b). The role of vasotocin and of CRF in ACTH release in the chicken. *Gen. Comp. Endocrinol.* **14**, 281–289.

Scanes, C. G., Goos, H. J. T., and Follett, B. K. (1972). Cross-reaction in a chicken LH radioimmunoassay with plasma and pituitary extracts from various species. *Gen. Comp. Endocrinol.* **19**, 596–600.

Scanes, C. K., and Follett, B. K. (1972). Fractionation and assay of chicken pituitary hormones. *Brit. Poultry Sci.* **13**, 603–610.

Schlumberger, H. C., and Rudolph, H. J. (1959). Growth promoting effect of a transplantable pituitary tumor in parakeets. *Endocrinology* **65**, 373.

Schooley, J. J. (1937). Pituitary cytology in pigeons. *Cold Spring Harbor Symp. Quant. Biol.* **5**, 165–179.

Schooley, J. P., and Riddle, O. (1938). The morphological basis of pituitary function in pigeons. *Amer. J. Anat.* **62**, 313–350.

Scott, J. E., and Dorling, J. (1965). Differential staining of acid glycosaminoglycans (mucopolysaccharides) by alcian blue in salt solutions. *Histochemie* **5**, 221–233.

Sharp, P. J., and Follett, B. K. (1969). The blood supply to the pituitary and basal hypopthalamus in the Japanese Quail (*Coturnix coturnix japonica*). *J. Anat.* **104**, 227–232.

Shellabarger, C. J. (1954). Detection of thyroid stimulating hormone by I^{131} uptake in chicks. *J. Appl. Physiol.* **6**, 721.

Solomon, J., and Greep, R. O. (1959). The growth hormone content of several vertebrate pituitaries. *Endocrinology* **65**, 334–335.

Stahl, A. (1958). Sur la présence d'une dualité cellulaire au niveau du lobe inter-médiaire de l'hypophyse de certains poissons. *C. R. Soc. Biol.* **152**, 1562–1565.

Stainer, I. M., and Holmes, W. N. (1969). Some evidence for the presence of a cortico-trophin releasing factor (CRF) in the duck (*Anas platyrhynchos*). *Gen. Comp. Endocrinol.* **12**, 350–359.

Steelman, S., and Pohley, F. H. (1953). Assay of the follicle stimulating hormone based on the augmentation with human chorionic gonadotrophin. *Endocrinology* **53**, 604–616.

Stein, K. (1929). Early embryonic differentiation of the chick hypophysis as shown in chorio-allantoic grafts. *Anat. Rec.* **43**, 221–237.

Stein, K. F. (1933). The location and differentiation of the presumptive ectoderm of the forebrain and hypophysis as shown by chorioallantoic grafts. *Physiol. Zool.* **6**, 205–235.

Stetson, M. H., and Erickson, J. E. (1970). A daily periodicity in pituitary gonadotropin in White-crowned Sparrows. *Z. Vergl. Physiol.* **68**, 263–267.

Stockell Hartree, A., and Cunningham, F. J. (1969). Purification of chicken pituitary follicle-stimulating hormone and luteinizing hormone. *J. Endocrinol.* **43**, 609–616.

Tanaka, K., and Yoshioka, S. (1967). Luteinizing hormone activity of the hen's pituitary during the egg laying cycle. *Gen. Comp. Endocrinol.* **9**, 374–379.

Tanaka, K., Mather, F. B., Wilson, W. O., and McFarland, L. Z. (1965). Effect of photo-periods on early growth of gonads and on potency of gonadotropins of the anterior pituitary in *Coturnix. Poultry Sci.* **44**, 662–665.

Tanaka, K., Wilson, W. O., and McFarland, L. Z. (1966a). Testicular response in Jap-anese Quail as a bioassay of pituitary gonadotropins. *Amer. J. Vet., Res.* **27**, 1067–1069.

Tanaka, K., Wilson, W. O., Mather, F. B., and McFarland, L. Z. (1966b). Diurnal varia-tion in gonadotropic potency of the adenohypophysis of Japanese Quail (*Coturnix coturnix japonica*). *Gen. Comp. Endocrinol.* **6**, 1–4.

Thiery, J. P. (1967). Mise en évidence des polysaccharides sur coupes fines en micro-scopie électronique. *J. Microsc. (Paris)* **6**, 987–1018.

Thommes, R. C., and Russo, R. P. (1959). Vasculogenesis in the adenohypophysis of the developing chick embryo. *Growth* **23**, 205–219.

Tixier-Vidal, A. (1954). Etude histophysiologique de l'hypophyse antérieure de l'em-bryon de Poulet. *Arch. Anat. Microsc. Morphol. Exp.* **43**, 163–186.

Tixier-Vidal, A. (1956). Etude chronologique *in vivo* et *in vitro* des corrélations hypo-physe-thyroïde chez l'embryon de Poulet. *Arch. Anat. Microsc. Morphol. Exp.* **45**, 236–253.

Tixier-Vidal, A. (1958). Etude histophysiologique des relations hypophyse-thyroïde chez l'embryon de Poulet. *Arch. Anat. Microsc. Morphol. Exp.* **47**, 235–340.

Tixier-Vidal, A. (1963). Histophysiologie de l'adénohypophyse des Oiseaux. *In* "Cytologie de l'adénohypophyse" (J. Benoit et C. Da Lage, eds.), pp. 255–274. CNRS, Paris.

Tixier-Vidal, A. (1965). Caractères ultrastructuraux des types cellulaires de l'adénohypophyse du Canard mâle. *Arch. Anat. Microsc. Morphol. Exp.* **54**, 719–780.

Tixier-Vidal, A. (1968). Influence de la testostérone sur la cytologie et l'ultrastructure de l'adénohypophyse du Canard mâle. *Arch. Anat., Histol. Embryol.* **51**, 709–717.

Tixier-Vidal, A. (1970). Cytologie hypophysaire et relations photosexuelles chez les Oiseaux. *Colloq. Int. Cen. Nat. Rech. Sci.* **172**, 211–232.

Tixier-Vidal, A., and Assenmacher, I. (1961). Etude comparée de l'activité thyroïdienne chez le Canard ♂ normal, castré ou maintenu à l'obscurité permanente. I and II. *C. R. Soc. Biol.* **155**, 215–220; 286–290.

Tixier-Vidal, A., and Assenmacher, I. (1963). Action de la métopirone sur la préhypophyse du Canard mâle: Essai d'identification des cellules corticotropes. *C. R. Soc. Biol.* **157**, 1350–1354.

Tixier-Vidal, A., and Assenmacher, I. (1966). Etude cytologique de la préhypophyse du pigeon pendant la couvaison et la lactation. *Z. Zellforsch. Mikrosk. Anat.* **61**, 489–519.

Tixier-Vidal, A., and Benoit, J. (1962). Influence de la castration sur la cytologie préhypophysaire du Canard male. *Arch. Anat. Microsc.* **51**, 265–286.

Tixier-Vidal, A., and Gourdji, D. (1965). Evolution cytologique ultrastructurale de l'hypophyse du Canard en culture organotypique. Elaboration autonome de prolactine par les explants. *C. R. Acad. Sci.* **261**, 805–808.

Tixier-Vidal, A., and Picart, R. (1968). Coloration spécifique des types cellulaires de l'antéhypophyse chez deux espèces d'Oiseaux après inclusion au glycol-méthacrylate et coloration par l'acide phosphotungstique. *J. Microsc. (Paris)* **2**, 59a.

Tixier-Vidal, A., and Picart, R. (1970). Localisation ultrastructurale des glycoprotéines, des phosphatases acides et des structures osmiophiles dans la zone Golgienne des cellules glycoprotidiques de l'adénohypophyse. *C. R. Acad. Sci.* **271**, 767–769.

Tixier-Vidal, A., and Picart, R. (1971). Electron microscopic localization of glycoproteins in pituitary cells of duck and quail. *J. Histochem. Cytochem.* **19**, 775–797.

Tixier-Vidal, A., Herlant, M., and Benoit, J. (1962). La préhypophyse du Canard Pékin mâle au cours du cycle annuel. *Arch. Biol.* **73**, 319–367.

Tixier-Vidal, A., Benoit, J., and Assenmacher, I. (1966). Modifications cytologiques et ultrastructurales de l'antéhypophyse du Canard mâle en fonction de l'âge et de l'exposition prolongée à la lumière ou à l'obscurité permanente. *Arch. Anat. Microsc. Morhol. Exp.* **55**, 539–559.

Tixier-Vidal, A., Follett, B. K., and Farner, D. S. (1968). The anterior pituitary of the Japanese Quail, *Coturnix coturnix japonica.* The cytological effects of photoperiodic stimulation. *Z. Zellforsch. Mikrosk. Anat.* **92**, 610–635.

Tixier-Vidal, A., Picart, R., and Gourdji, D. (1969). Détection de glycoprotéines au niveau des cellules de l'adénohypophyse du Canard et de la Caille par la technique de Thiéry. Valeur signalétique. *J. Microsc. (Paris)* **8**, 88a.

Tixier-Vidal, A., Chandola, A., and Franquelin, F. (1972). "Cellules de thyroïdectomie" et "cellules de castration" chez la Caille japonaise *Coturnix coturnix japonica. Z. Zellforsch. Mikrosk. Anat.* **125**, 506–531.

Tougard, C. (1971). Recherches sur l'origine cytologique de l'hormone mélanophorotrope chez les Oiseaux. *Z. Zellforsch. Mikrosk. Anat.* **116**, 375–390.

Tougard, C., Tixier-Vidal, A., and Picart, R. (1968). Fixation sélective de la sérotonine tritiée au niveau de certaines granulations hypophysaires chez la Caille Japonaise. Etude autoradiographique au microscope électronique. *C. R. Acad. Sci.* **267,** 1405–1408.

van Oordt, P. G. W. J. (1964). Nomenclature of the hormone-producing cells in the adenohypophysis. A report of the International Committee for nomenclature of the adenohypophysis. *Gen. Comp. Endocrinol.* **5,** 131–134.

van Tienhoven, A. (1961). Endocrinology of reproduction in birds. *In* "Sex and Internal Secretions" (W. C. Young, ed.), 3rd ed., Vol. 2, pp. 1088–1169. Williams & Wilkins, Baltimore, Maryland.

van Tienhoven, A., Nalbandov, A. V., and Morton, H. W. (1954). Effect of dibenamine on progesterone-induced and "spontaneous" ovulation in the hen. *Endocrinology* **54,** 605–611.

Vitums, A., Mikami, S., Oksche, A., and Farner, D. S. (1964). Vascularization of the hypothalamo-hypophysial-complex in the White-crowned Sparrow. *Zonotrichia leucophrys gambelii. Z. Zellforsch. Mikrosk. Anat.* **64,** 541–569.

Wilson, M. E. (1949). Certain aspects of the developmental and definitive cytology of the chick pituitary gland. *Anat. Rec.* **103,** 521.

Wilson, M. E. (1952). The embryological and cytological basis of regional patterns in the definite epithelial hypophysis of the chick. *Amer. J. Anat.* **91,** 1–50.

Wingstrand, K. G. (1951). "The Structure and Development of the Avian Pituitary." Gleerup, Lund.

Witschi, E. (1955). Vertebrate gonadotrophins. *Mem. Soc. Endocrinol.* **4,** 149–163.

Yanai, R., and Nagasawa, H. (1969). Quantitative analysis of prolactin by disc electrophoresis and its relation to biological activity. *Proc. Soc. Exp. Biol. Med.* **131,** 167–171.

Yasuda, M. (1953). Cytological studies of the anterior pituitary in the broody fowl. *Proc. Jap. Acad.* **29,** 586–594.

Chapter 3

THE PERIPHERAL ENDOCRINE GLANDS

Ivan Assenmacher

The endocrine glands are essential components of the complex neuroendocrine apparatus that, by communication via nerve impulses, synaptic transmitters, and hormones, provides the basis for internal regulation and adjustment to the changing environment. The artificial designation "peripheral endocrine glands," is used here for those endocrines that are neither a part of the brain, such as the hypothalamus, its derivative the neurohypophysis, and the pineal organ, nor closely associated with it, i.e., the pars distalis and the pars tuberalis of the hypophysis.

The avian peripheral endocrine glands (see below) are, in general, typical of those of the higher vertebrate scheme, although they are by no means without specialized functions and adaptations that are typically avian. The high calcitonin content of avian ultimobranchial glands and its high specific activity seem related to the very intense calcium metabolism that appears characteristic of avian species, possibly linked to their high calcium requirement during the reproductive period. On the other hand, the very low binding capacity of avian plasma proteins to the circulating thyroid hormones, and the resulting short biological half-life of these hormones in birds, has been related to the high rate of heat production and to the high body temperature, another characteristic of avian physiology, that preadapts them better than mammals, for instance, to live in areas of intense heat. Other adaptations may be observed in special groups. The adrenal size, probably associated with a higher hormonal activity, of marine bird species, as compared with freshwater or terrestrial species, has been claimed to reflect the need of marine birds to stimulate chronically the extrarenal (nasal) salt excretory pathway. Other avian endocrinological pecularities (e.g., the overwhelming lipolytic role of the pancreatic hormone glucagon versus the adrenomedullary catecholamines) are of unknown evolutionary or adaptative significance.

Certain "peripheral endocrine glands" have been, for obvious reasons, treated in other chapters; endocrine testis and endocrine ovary in Chapter 1 of this volume; in part, endocrine pancreas in Chapter 8 of Volume II; and the hormones of the digestive tract in Chapter 6 of Volume II.

I.   Adrenal Glands

A.   MORPHOLOGY

The adrenals are a pair of oval, pear-shaped, or triangular glands, yellow or orange in color, that lie just anterior to the postcaval vein

TABLE I. PERIPHERAL ENDOCRINE GLANDS OF BIRDS

Endocrine gland	Hormones produced	Principal functions	Control of secretion
Adrenal cortex	Corticosterone	Liver glucogenesis, gluconeogenesis; lipogenesis; degradation of proteins; salt gland stimulation	Pituitary (and median eminence?) ACTH under hypothalamic control (CRF)
	8-Hydroxycorticosterone	Unknown	Unknown
	Aldosterone	Sodium retention	Probably angiotensin II
Adrenal medulla	Norepinephrine Epinephrine	Moderate lipolysis; Moderate Hyperglycemia; Rise in systolic and diastolic blood pressure	Cholinergic innervation
Thyroid gland	Thyroxine (T4)	Heat production; Hyperglycemia, liver glycogenolysis; Growth promotion	Pituitary TSH, under hypothalamic control
	Triiodothyronine (T3)	Development of skin epithelium and of feathers; molt (?); Migratory behavior (?)	
Endocrine pancreas	Insulin	Hypoglycemia; glycogenesis; Moderate hypolipemia	Blood sugar level
	Glucagon	Hyperglycemia; glycogenolysis; Main lipolytic hormone	Blood sugar level; Plasma free fatty acids (FFA)
Parathyroid gland	Parathormone (PTH)	Hypercalcemia	Blood calcium level
Ultimobranchial bodies	Calcitonin	Hypocalcemia	Blood calcium level
Thymus	Thymic "hormones" (?)	Lymphocytosis (immunologically competent lymphocytes); (Growth promotion and Gametogenesis)?	Unknown
Bursa of Fabricius	Bursal "hormones" (?)	Maturation of lymphoid cells producing humoral antibodies	Unknown
Small intestine	Digestion controlling hormones	Humoral control of gastrointestinal secretory and motor processes	Content of gastrointestinal tract (HCl, peptones, fats, etc.)
Gonads	Glucagon	Unknown	Unknown
	Sex steroid hormones	See Chapter 1	

and to the cephalic lobe of the kidney. The avian adrenal gland, like that of most higher vertebrates, is formed of two components, the cortex (interrenal tissue) and the medulla (chromaffin tissue), which are of different origins and distinct functions.

The cortical portion is the first to be differentiated. The primordium buds off from the coelomic epithelium, ventrally and medially to the mesonephros. Soon after their formation, the prospective cortical cells leave the epithelium and move dorsally to form paired masses of scattered cell groups in the mesenchyma on each side of the aorta. During their further development, the cortical cells become arranged in cords, and blood cells and vessels appear between the cell cords. Medullary (chromaffin) cells appear several days later, arising from the primordium of the sympathetic nervous system. These cells migrate singly, ventrally between the aorta and groups of cortical masses, collect, and tend to arrange themselves in cords between the cortical tissue, before separating into groups of variable size around blood vessels (Romanoff, 1960).

In the adult animal, the ratio of cortical to chromaffin tissue is about 2 to 1. In most species, the cortical tissue is arranged in strands, which are surrounded by a highly vascular connective tissue. In longitudinal section, each strand appears as a double layer of columnar epithelial cells with the thin long axes perpendicular to the surface of the cell strand. Unlike the mammalian cortical cells, the avian interrenal cells have a marked polarity in the distribution of cytoplasmic organelles (Kjaerheim, 1968). The majority of lipid droplets, which are typical of cortical cells, and of the smooth-surfaced reticulum, occur in the basal cytoplasm adjacent to the connective tissue, whereas the Golgi area, dense bodies, and specialized attachment structures occur apically in the nuclear region.

Stimulation of cortical tissue, e.g., by ACTH injections, induces marked depletion of the lipid droplets, as classically observed with the light microscope. The lipid depletion also is evident at the electron-microscope level, together with (1) increased density of mitochondria; (2) increase of smooth-surface reticulum (SER), which is diffusely dispersed in the cytoplasm separating the mitochondria from the lipid droplets; (3) increase of the rough ergastoplasmic reticulum, which is absent in nonstimulated cells; (4) marked increase in size of the Golgi apparatus, associated with dense bodies. Correlatively, the nucleus and nucleolus increase in size, and numerous vacuoles appear at the surface of the nucleolus. Finally, numerous coated vesicles and invaginations appear, along with a fibrillar substance in intercellular spaces, as signs of uptake of material (proteins) into the cell. From a cytofunctional point of view, it is generally accepted that the close relationship among lipid droplets, SER, and mitochondria points to the basal cytoplasm as the main site of steroidogenesis, while other functions, such as protein synthesis, required by an increased need in enzymes involved in steroidogenesis, increased production of primary lysosomes, and increased degradation of lipids, may be attributed to the apical structures.

Another peculiarity of the arrangement of avian cortical tissue is a much greater homogeneity in cell population. Early investigations did not recognize any zonation in the avian adrenal cortex. More recently, several authors have described two different zones, a thin subcapsular zone and a thicker inner zone, e.g., in the pigeon (Miller, 1949; Sinha and Ghosh, 1961), the Brown Pelican (*Pelecanus occidentalis*) (Knouff and Hartman, 1951), and the duck (Tixier-Vidal and Assenmacher, 1963). The two zones were easily demonstrated at the ultrastructural level in the fowl (Kondics and Kjaerheim, 1966). From a histochemical standpoint, a thorough exploration of a number of species seems to indicate that the glands of certain birds are fully zonated (e.g., the pigeon), while others are completely homogeneous (e.g., Little Cormorant, *Phalacrocorax niger*) (Ghosh, 1962). Similar variations have been observed in respect of the histological organization of the interrenal tissue in 22 species of 17 different orders (Bhattacharyya *et al.*, 1972). On the other hand, the subcapsullar layer seems particularly reactive to experimental saltwater balance alterations (Kondics, 1963; Sinha and Ghosh, 1964), and species that consume water in saline areas and on the seashore have been found to have less active peripheral cortical cells than do freshwater-adapted species (Péczely, 1964).

In contrast with the cortical cells, the medullary chromaffin cells have no definite pattern of arrangement. They may be intimately intermingled with cortical tissue (e.g., in Passeriformes) or more or less condensed in medullary islets, intermixed with interrenal tissue (e.g., in Galliformes).

When studied with the electron microscope, two medullary cell types have been described in the domestic fowl (Kano, 1959; Fujita and Machino, 1962) and in the Gentoo Penguin (*Pygoscelis papua*) (Cuello, 1970), which differ in the size and shape of electron-dense granules. One cell type contains granules of very irregular size (800–5000 Å) and shape, while the other type is characterized by regular, spherical granules, that are also smaller than the others.

The former are considered to correspond to norepinephrine cells, the latter cell type being epinephrine cells (Cuello, 1970). On the other hand, the cells with irregular granules have been claimed to form the greater part of the medulla in the fowl (Kano, 1959), while an opposite ratio was observed in the Gentoo Penguin (Cuello, 1970). Both types of granules consist of an electron-dense, central or eccentric osmiophilic deposit, which appears as a mass of numerous fine grains surrounded by a light halo, and finally by a limiting membrane. These granules, which seem to contain catecholamines, lie close to the Golgi apparatus, and it has been suggested that the outer membrane of the granules may originate from Golgi vesicles. Sometimes the limiting membrane of the granules contacts the cell membrane, and their content seems to be excreted through the opening of this contact zone, into the intracellular or perisinusoidal spaces. Between the medullary-cell columns, there are bundles of unmyelinated nerve fibers invested by Schwann cells. At the contact zone between the nerve endings and medullary cells, a typical vesicular component, consisting of synaptic vesicles, is observed. Strong stimulation of medullary cells (e.g., by insulin) leads to a remarkable reduction in number, size, and electron density of the granules, with many smooth-surfaced vacuoles containing only diffuse fine microgranules appearing in most cells (Fujita et al., 1959).

B.   Cortical Tissue

1. Hormones and Metabolism

Corticosterone, as the main corticosteroid secreted by the avian adrenal, was first demonstrated in the efferent adrenal blood of castrated chickens (Phillips and Chester-Jones, 1957), and of intact fowl, pheasants, and turkeys (Urist and Deutsch, 1960; Brown, 1960), and thereafter in a number of other species. Plasma corticosterone can be routinely measured by fluorometry (Silber et al., 1958), which is currently applied to mammalian species. However, in most avian species, except for ducks (Bouillé et al., 1969), a preliminary chromatographic purification of the plasma extract is necessary in order to obtain accurate assays. Aldosterone, another important avian corticoid, has also been shown to occur in adrenal efferent blood (Phillips and Chester-Jones, 1957). The latter authors, and others (Urist and Deutsch, 1960) also have identified small amounts of cortisol in the adrenal and peripheral blood of pullets, while Nagra et al. (1960) were unable to detect 17-oxycorticosteroids in the adrenal effluent blood from chicken, pheasant, or turkey. Corticosterone and aldosterone were also measured by the double isotopic dilution method in adrenal extracts from ducks, pigeons, and Japanese Quail (Daniel and Assenmacher, 1969a; see Table IV). No significant interspecific variations could be detected, either in the concentration of both hormones or in the corticosterone to aldosterone ratio, which is approximately 8:1. Using the same accurate method, Daniel (1970) has shown that cortisol is present in the duck adrenal at a very low level (about 1.3 ng per gram of adrenal tissue).

A deeper insight into the metabolic sequences of the biogenesis of corticosteroids by the avian adrenal has been obtained from *in vitro* studies, which have been performed in a number of bird species (De Roos, 1961; Sandor *et al.*, 1963; Donaldson *et al.*, 1965; Lamoureux, 1966; Macchi, 1967; Frankel *et al.*, 1967a; Whitehouse and Vinson, 1967; Sandor, 1969). First, it has been confirmed that the avian adrenal synthesizes almost exclusively 17-deoxycorticosteroids, i.e., corticosterone, 18-hydroxycorticosterone, and aldosterone. Further investigations, using labeled presumed precursors, have shown that the metabolic pathways of steroidogenesis seem to conform to the classic route of other vertebrates: acetate → cholesterol → pregnenolone → progesterone → 11-deoxycorticosterone (DOC) → corticosterone → 18-hydroxycorticosterone (18 OH-B) and aldosterone. However, several pecularities of the avian biogenesis of corticosteroids *in vitro* have recently been demonstrated with duck adrenal slices or intracellular fractions (Sandor, 1969) (Fig. 1):

1. While exogenous [^{14}C]progesterone, as well as [^{3}H]pregnenolone, are easily converted to the three main corticosteroids, significant accumulation of labeled progesterone never occurs if [^{3}H]-pregnenolone is used. A metabolic sequence originating from pregnenolone, but without an oxidation step to progesterone, thus seems to exist, at least in the duck adrenal.

2. It has also been proved that, while both progesterone and 11-deoxycorticosterone are easily transformed to corticosterone, the hydroxylation at C-11 of progesterone might precede that of C-21, i.e., 11-deoxycorticosterone is not a necessary intermediate between progesterone and corticosterone. Duck adrenals convert either pregnenolone or progesterone to 11β-hydroxyprogesterone, which in turn leads to corticosterone without any production of 11-deoxycorticosterone. Corticosterone is further transformed to 18-OH-B and aldosterone, both hormones being also obtained from incubations with either progesterone or 11β-hydroxyprogesterone.

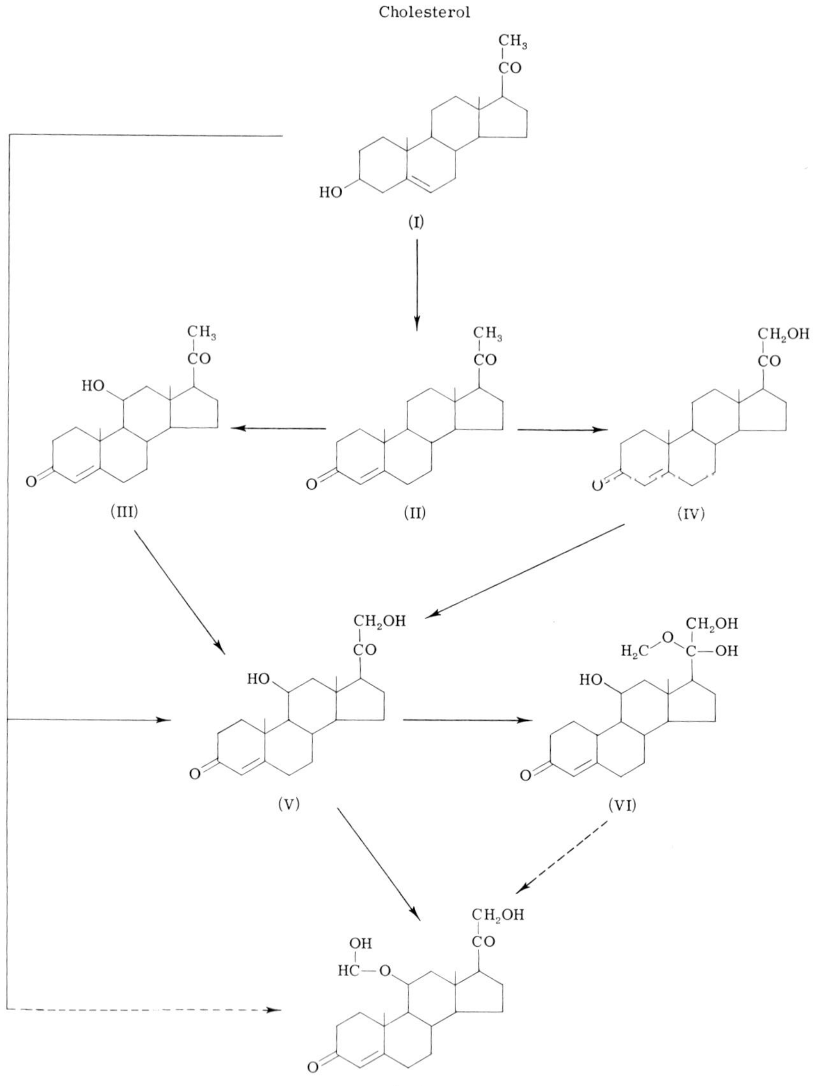

FIG. 1. Proposed biosynthetic route of corticosteroid synthesis in the adrenal gland of the domestic duck. (I) pregnenolone; (II) progesterone; (III) 11-hydroxyprogesterone; (IV) 11-deoxycorticosterone; (V) corticosterone; (VI) 18-hydroxycorticosterone; (VII) aldosterone. (From Sandor, 1969. Courtesy of Excerpta Medica Foundation. Modified after Sandor and Lanthier, 1970.)

3. The synthesis of aldosterone from corticosterone is always accompanied by the simultaneous production of 18-hydroxycorticosterone. However, the question whether the latter is to be considered an obligatory precursor of aldosterone is still open since it has never been demonstrated to be converted by duck adrenal slices to aldosterone.

As in mammals, the disappearance rate of corticosterone from blood is rapid. In the duck, studies on the disappearance rate from plasma of [^{3}H]costicosterone indicate a biological half-life of about 11 minutes. The corresponding calculated metabolic clearance rate, which provides a better insight into the peripheral metabolism of the hormone, is 104 liters per day or 26.7 ml min^{-1} kg^{-1} (Daniel et al., 1970) (Fig. 2). With the assumption that during the short time of the experiment

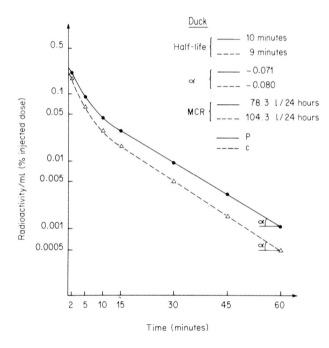

FIG. 2. Biological half-life of [^{3}H]corticosterone in the duck, and therefrom the calculated metabolic clearance rate (MCR). The radioactivities (percent of the injected dose) were measured, either on neutral lipid extracts of plasma samples (P), or on plasma samples purified by paper chromatography (C) on Bush's B5 system. (From Daniel and Assenmacher, 1970.)

the animals were in steady state with respect to corticosteroid metabolism, the secretion rate of the hormone is given by

secretion rate = metabolic clearance rate × plasma corticosterone

i.e., in the former example, $26.7 \times 0.115 = 3.1$ μg min^{-1} kg^{-1}. In similar experiments on pigeons Chan *et al.* (1971) found a metabolic clearance rate of corticosterone of 13.8 ml min^{-1} kg^{-1}. In view of the high metabolic clearance rate of the hormone, the low amount of the hormone available within the gland cannot maintain a normal plasma-corticosterone level for more than some minutes, a feature that is in agreement with the mammalian pattern. Further investigations in the duck have shown that exogenous corticosterone (tritiated) is very rapidly metabolized, and several labeled metabolites appear within the blood stream, in increasing number and concentration with time, e.g., 30 minutes after a single intravenous injection, only 52% of the plasma radioactivity still belongs to corticosterone. Among the newly formed metabolites, 11-dehydrocorticosterone could be identified (Daniel *et al.*, 1970). When measurements of disappearance rate of intravenously injected labeled corticosterone are used as an index to the peripheral metabolism of the hormone, chromatographic purification of each plasma sample is required.

As has first been shown in mammals, plasma corticosteroids are strongly bound by a specific binding protein, corticosteroid-binding globulin (CBG) (Daughaday, 1957), or transcortin (Sandberg *et al.*, 1957). Independently of transcortin, which exhibits high affinity but low capacity for corticosteroids, and which has the major role in binding of corticosteroids, albumin constitutes a second corticosteroid-binding system in the plasma, characterized by a low affinity together with a high capacity. On the other hand, it has been shown that transcortin-bound corticosteroids are biologically inert (Sandberg and Slaunwhite, 1962), while the free, nonprotein-bound forms are the immediately active forms (Doe *et al.*, 1960). Relevant information on the presence of transcortin activity in avian plasma has been provided by Seal and Doe (1963, 1965, 1966) and by Steeno and De Moor, 1966).

1. From a systematic study performed by the latter authors on 28 mammalian species (7 orders) and 16 avian species (11 orders), it appears that, whereas the avian plasma corticosteroid (corticosterone) level is generally lower than that of mammals, the corticosteroid-binding capacity, expressed in micrograms bound per 100 ml plasma ranges between comparable extreme values in mammals (white-faced

chimpanzee, *Pan schweinfurthii*, 30.5; red kangaroo, *Megaleia rufa*, 1.9) and in birds (Mallard, *Anas platyrhynchos*, 35.5; African Wood-Stork, *Ibis ibis*, 6.5). Similar conclusions result from a comparison in domestic animals, between 55 species of mammals and 18 species of birds (Seal and Doe, 1965). As the plasma protein content is generally lower in birds than in mammals, the corticosterone-binding capacity, when expressed in micrograms per 100 gm protein is generally higher in the latter (extreme values for mammals — chimpanzee, 433; red kangaroo, 19; for birds — Mallard, 859; African Woodstork, 196) (Steeno and De Moor, 1966).

2. Whereas in most mammals the relationship of plasma corti-costeroid level to corticosterone-binding capacity lies near or above 1, in almost all avian species studied this ratio is less than 1 (e.g., pigeons, 0.58; ducks, 0.30), indicating that the protein-binding capacity of transcortin exceeds markedly the endogeneous corticosteroid level (Steeno and De Moor, 1966).

3. Sex differences are apparent in the domestic fowl, in which roosters are found to bind 4 μg cortisol per 100 ml plasma, compared with 8 μg% in nonlaying hens and 12 μg% in laying hens (Seal and Doe, 1966). Estrogen would then appear to have stimulating action on transcortin synthesis in birds similar to that in mammals. On the other hand, no sex differences have been found in the species studied by Steeno and De Moor (1966).

4. Intraspecies differences in corticosterone-binding capacity have also been found among ducks (e.g., 35.5 μg per 100 ml in Mal-lards; 33.9 in Indian runners; and 20.5 in other domestic ducks (Steeno and De Moor, 1966).

2. Functions of Adrenocortical Hormones

In birds, as in other vertebrates, adrenal function is vital. Adrenal-ectomized ducks and chickens usually die within 6–60 hours, unless given replacement therapy (Brown *et al.*, 1958b). Although less in-tensive investigations have been performed on the various actions of adrenocortical hormones in birds than in mammals, there is good evi-dence that the avian corticosteroids have the same major metabolic roles (e.g., on carbohydrate and electrolyte metabolism) as in mammals.

a. Carbohydrate Metabolism. Riddle (1937) was the first to show clearly the hyperglycemic effect of adrenal extracts on hypophysec-tomized, thyroidectomized, or partially adrenalectomized pigeons. Later, several authors have demonstrated in the domestic fowl that corticosterone, and also hydrocortisone (but not cortisone), induce marked hyperglycemia, together with enhanced accumulation of glycogen in the liver (Golden and Long, 1942; Stamler *et al.*, 1954; Brown *et al.*, 1958a; Greenman and Zarrow, 1961; Snedecor *et al.*, 1963). That increased liver glycogen is due to stimulation of hepatic glycogenesis rather than to impaired utilization of blood sugar is proved by the fact that corticosterone administration does not affect the *in vivo* oxidation of [^{14}C]glucose to CO_2 (Nagra and Meyer, 1963). On the other hand, the modifications of carbohydrate metabolism are closely correlated with increased nitrogen output, indicating a stimulation of protein catabolism (gluconeogenesis) (Brown *et al.*, 1958a). Thus, except for the relative ineffectiveness of cortisone in birds, the glucocorticoids behave much as in mammals.

Finally, it must be stated that, despite the still questionable participation of the peculiar avian sacral "glycogen body" in carbohydrate metabolism, corticosterone is one of the few hormones that has been found to be active in glycogen enrichment of this organ in chickens (Snedecor *et al.*, 1963).

b. Lipid Metabolism. Corticosterone and hydrocortisone (but not cortisone) induce lipogenesis in the fowl, as indicated by elevation of plasma lipids, increased carcass (mainly subcutaneous) and visceral fat deposition, and generally enhanced liver fat (Stamler *et al.*, 1954; Dulin, 1956; Baum and Meyer, 1960; Nagra and Meyer, 1963). From histochemical studies it appears that when increased deposition of hepatic lipid occurs, the fat accumulates around the blood vessels, without any evidence of degenerative changes in liver cells (Dulin, 1956). Corticosterone-induced lipogenesis does not result from changes in rate of fatty-acid metabolism, as no difference was evident in either the amount of *in vivo* oxidation of [^{14}C] palmitic acid, or in the rate of related $^{14}CO_2$ exhalation. On the other hand, injected [^{14}C]-glucose was directed mainly into lipid metabolism (Nagra and Meyer, 1963). As the same lipogenetic action of corticosterone treatment occurs in hypophysectomized chickens, this effect cannot be related to inhibition of some hypophysial lipolytic factor (ACTH, prolactin) (Nagra, 1965). *In vitro* studies on 3–4-week-old ducklings have also shown that, up to very high concentrations (10 μg per milliliter of medium), corticosterone has no direct effect on adipose tissue (Desbals, 1972).

c. Protein Metabolism—Growth. Despite a significant hyperphagia, corticosterone- or hydrocortisone-treated chickens exhibit marked inhibition of growth (Dulin, 1956; Baum and Meyer, 1960; Nagra and Meyer, 1963). The stimulating action of corticosteroids on nitrogen output (Brown *et al.*, 1958a), has already been mentioned, and Nagra and Meyer (1963) found in growing chickens that after corticosterone treatment there was a reduction in carcass protein and a decreased rate of transformation of injected [^{14}C]glucose into proteins.

d. Electrolyte Metabolism. 1. Kidney. As in mammals, the avian kidney is one of the major target organs of the corticosteroids, aldosterone appearing in birds also to be very potent in causing sodium retention. According to Brown *et al.* (1958a), in chickens with ureters surgically exteriorized, both cortexone acetate and cortisone increased urine flow, but deoxycorticosterone acetate decreased markedly the excretion of sodium and of potassium. In salt-loaded ducks, aldosterone lowers markedly the volume and sodium concentration of urine. Adrenalectomy in the duck causes increased sodium loss in the urine, while large doses of aldosterone or corticosterone reduce sodium excretion (Phillips *et al.*, 1961).

2. Salt glands. Recent investigations have demonstrated that adrenocortical hormones are also involved in the regulation of extrarenal electrolyte excretion in birds with active nasal glands. The nasal (supraorbital) gland of several saltwater species is known to be able to secrete a fluid, mainly a sodium chloride solution, that is hypertonic to the plasma. Due to its concentrating power, which is higher than that of the kidney, the nasal salt gland permits the animal to remain in water balance, even for prolonged periods of high salt diet (Schmidt-Nielsen *et al.*, 1958). Except for the Falconiformes (Cade and Greenwald, 1966), no similar function has been described for the nasal glands of terrestrial birds, although the microscopic anatomy of the gland of the latter is very similar to that of aquatic birds (McLelland *et al.*, 1968). The size of nasal gland, the length of the secretory tubules, as well as the salt-concentrating capability of the glands from a number of marine species were found to be closely correlated with the ecology of the species (Staaland, 1967). Several investigations on ducks and gulls have placed emphasis on the controlling role of the adrenal cortex on nasal-gland secretion. Acute salt load is followed by a biphasic reaction, with an initial diuresis followed by increased output of nasal secretion. The extrarenal response is increased by administration of adrenocorticosteroids or ACTH (Holmes *et al.*, 1961). On the other hand, hypertonic saline loading increases synthesis and release of corticosterone, which compensates for a correlative enlargement of the extracellular fluid volume (Donaldson and Holmes, 1965; Holmes and Phillips, 1965; Macchi *et al.*, 1967). In turn, secretion by the nasal gland is significantly decreased by adeno-hypophysectomy (Wright *et al.*, 1966; Holmes *et al.*, 1972) and is suppressed by adrenalectomy (Phillips *et al.*, 1961). However, treatment of adenohypophysectomized ducks with ACTH restored the plasma corticosterone concentration, together with the extrarenal excretion of water and electrolytes, to normal (Holmes *et al.*, 1972). Finally, corticosterone has been found more effective in salt-gland stimulation than cortisol or cortexone (Phillips and Bellamy, 1963), whereas neither aldosterone nor neurohypophyseal hormones seem necessary for the normal function of the nasal glands (Phillips and Bellamy, 1962). The major glucocorticoid component of avian adreno-cortical secretion seems thus to display a preeminent part of the controlling mechanism of extrarenal electrolyte excretion in birds. More recently, prolactin has also been shown to be capable of increasing nasal salt-gland secretion in ducks (Peaker *et al.*, 1970).

e. Reproduction. In view of the marked metabolic effects of corticosteroids, it is not surprising that adrenalectomy or corticosteroid administration have been reported to interfere with reproduction. Adrenalectomy has been claimed to induce pronounced atrophy of the testes (Herrick and Finerty, 1941; Hewitt, 1947) and to inhibit the development of the right gonad in ovariectomized chickens (Taber *et al.*, 1956). On the other hand, administration of cortisone acetate was followed by testicular atrophy (Selye and Friedman, 1941), and corticosterone injections by depression of egg laying in chickens (Greenman and Zarrow, 1961); whereas others have reported an androgenic effect of deoxycorticosterone acetate on comb size (Hooker and Collins, 1940; Boas, 1958) or a stimulating effect on testes (Boas, 1958). Induced states of hyper- and hypoadrenocorticalism elicited pathomorphic changes in reproductive systems of the pigeon (Bhattacharyya and Ghosh, 1970a).

3. Regulation of Adrenal Cortical Function

a. Hypothalamic–Hypophyseal Control. 1. Pituitary control. Although a number of experiments with mammalian ACTH or avian pituitary extracts have suggested the occurrence of a classic pituitary–adrenocorticotropic control scheme in birds as in other vertebrates, this control may differ in some respects from that of mammals, as the level of adrenal-cortical function of hypophysectomized birds differs from that in hypophysectomized mammals. Whereas in the latter, hypophysectomy leads to a drastic atrophy of the adrenals, and an associated fall of 70–80% in plasma corticosterone, Tables II and III and Fig. 3 reveal that, in hypophysectomized birds, adrenal atrophy

TABLE II
ADRENAL WEIGHT FOLLOWING HYPOPHYSECTOMY

| Species | Observations | |
	Normal	Atrophy
Duck	—	Benoit and Assenmacher, 1953; Assenmacher, 1958a; Assenmacher, and Baylé, 1964; Boissin, 1967; Baylé et al., 1971
Fowl	Baum and Meyer, 1956; Brown et al., 1958a; Newcomer, 1959; Nagra et al., 1963	Nalbandov and Card, 1943; Ma and Nalbandov, 1963; Resko et al., 1964; Frankel et al., 1967a, 1967b
Brown Pelican (Pelecanus occidentalis)	—	Knouff and Hartman, 1951
Ring-necked Pheasant (Phasianus colchicus)	—	Nagra et al., 1963
Pigeon	Bates et al., 1962; Miller, 1967; Baylé et al., 1971	Schooley et al., 1941; Miller and Riddle, 1942
Japanese Quail	Baylé et al., 1971	—

TABLE III
PLASMA CORTICOSTERONE CONTENT FOLLOWING HYPOPHYSECTOMY

| Species | Observations | |
	Normal	Decreased (% of controls)
Duck	—	(50%) Boissin, 1967; (36–43%) Baylé et al., 1971; (10%) Holmes et al., 1972
Fowl	—	(64%) Resko et al., 1964; (60%) Nagra et al., 1960; (48%) Frankel and Nalbandov, 1966; (30–37%) Frankel et al., 1967a,b
Ring-necked Pheasant (Phasianus colchicus)	—	(60%) Nagra et al., 1963
Pigeon	—	(54%) Baylé et al., 1971
Japanese Quail	—	(46%) Baylé et al., 1971
Turkey	Brown, 1961	—

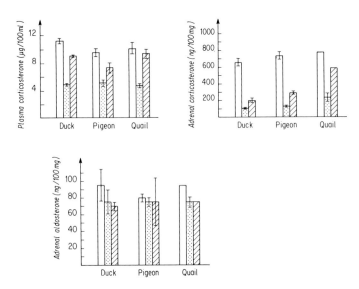

FIG. 3. Plasma and adrenal corticosterone and aldosterone in control (clear), hypophysectomized (dotted), and pituitary autografted (hatched) ducks, pigeons, and quail. Mean ± SD for 9–18 individuals. Where no SD are indicated, two pools of 5 animals each are measured. (From Daniel and Assenmacher, 1969a; Baylé *et al.*, 1971.)

is variable and, moreover, that plasma-corticosterone levels remain at about 40–60% of controls. The adrenal corticosterone content appears more affected, whereas the aldosterone level remains unaltered (Daniel and Assenmacher, 1969a) (Fig. 3). One may speculate that the high residual plasma-corticosterone level in hypophysectomized birds, as compared with that of hypophysectomized mammals, may be due to a lowered metabolic clearance rate. In fact, hypophysectomized ducks exhibit an 83% increment of the half-life of corticosterone, and a correlative 42% reduction of metabolic clearance rate (Baylé *et al.*, 1971). However in hypophysectomized rats, the half-life of corticosterone also exhibits a 50% increase (Mialhe-Voloss *et al.*, 1965), so that the obvious discrepancy in plasma corticosterone in either birds or mammals may more likely be due to higher residual corticosterone secretion rather than to some quantitative difference in the peripheral metabolism of the hormone.

Two theories have been proposed to account for these differences: (1) the occurrence of an autonomous adrenal cortex secretory activity, independent of pituitary ACTH (Newcomer, 1959; Nagra *et al.*, 1963; Boissin *et al.*, 1966), and (2) the possible action of an extrahypophyseal corticotropic substance, either ACTH or an ACTH analogue (Miller, 1961; Resko et al., 1964; Frankel et al., 1967a,b; Péczely and Zboray, 1967; Péczely, 1969). The latter hypothesis receives support from the fact that in both the domestic fowl (Salem et al., 1970a,b) and pigeon (Péczely and Zboray, 1967; Péczely, 1969; Péczely et al., 1970) an ACTH-like activity has been demonstrated in the median eminence of the hypothalamus. Péczely et al. (1970) have shown that this ACTH-like activity in the median eminence of the pigeon, 1 month after hypophysectomy, is still as high as in control animals. Furthermore, hypophysectomized pigeons respond to formalin stress by morphological activation of cortical tissue (Miller, 1967), while short-term surgical stress increased plasma corticosteroid levels in hypophysectomized chickens (Frankel et al., 1967a). On the other hand, Frankel et al. (1967a) have recorded in the hypophysectomized fowl a drop in plasma corticosteroid to nearly zero after injection of dexamethasone. It seems possible, therefore, that in birds, a production of ACTH or some related substance may persist after hypophysectomy and may be responsible to some extent for the high residual plasma corticosteroid values observed and for adrenal stimulation in the absence of the pituitary.

2. *Hypothalamic control.* In the hypothalamic control of ACTH release, there is a further discrepancy between birds and mammals. In mammals, numerous investigations have shown that pituitary autografts are ineffective in preventing a drastic fall in plasma corticosteroids to hypophysectomy levels. In birds, on the other hand, although pituitary autografting is often accompanied by morphological atrophy (Assenmacher, 1958a; Assenmacher and Baylé, 1964; Resko et al., 1964). The basal plasma corticosteroid level, in long-term autografted ducks, pigeons, and Japanese Quail, lies far above the hypophysectomy level (Baylé et al., 1967b, 1971). Similarly, the adrenal corticosterone level is significantly raised (Fig. 3), and the half-life and metabolic clearance rate for corticosterone is restored to normal values (Baylé et al., 1971). Moreover, the ACTH content of pituitary transplants, in autografted pigeons, was found to equal that of the pituitary *in situ* (Péczely et al., 1970). This is in contrast to the observations of Resko et al. (1964), who reported that, in the fowl, autografts have no effect on plasma corticosteroid levels in the adrenal effluent blood. This may reflect a species peculiarity. However, the post-operative time lapse seems very important, as in ducks the plasma corticosteroid level decreases equally within the first 3 weeks after

either autograft or hypophysectomy, but 2 months later the plasma corticosterone level is elevated to the typical 80% of controls in autografted animals (Baylé *et al.*, 1971). Furthermore, the reestablishment of plasma corticosteroid levels after autograft is concomitant with the recovery phase of the cytofunctional characteristics of the grafted tissue in this species (Tixier-Vidal *et al.*, 1973). Thus, the abovementioned apparent inability of pituitary grafts to promote stimulation of adrenal cortical function in chickens may be due to an incomplete recovery of the grafted tissue, since measurements were not recorded beyond 40 days postoperatively. In fact, corticotropin-releasing factor (CRF) similar to that formed in mammals, has been reported to occur in median eminence extracts of pigeons (Péczely and Zboray, 1967; Péczely, 1969; Péczely *et al.*, 1970), hens (Salem *et al.*, 1970a,b), and ducks (Stainer and Holmes, 1969). In the pigeon, hypophysectomy leads to the disappearance of the CRF activity in the median eminence extracts, suggesting a rapid depletion of CRF into the general circulation, while in autografted pigeons, CRF activity in the median eminence is retained (Péczely *et al.*, 1970). The possible high release of CRF into the general blood stream by hypophysectomized or autografted animals may account for the maintenance of some ACTH release by the ectopic pituitary. As a matter of fact the destruction of the mediolateral posterior area of the hypothalamus in pituitary-autografted pigeons prevents the recovery of the plasma corticosterone concentrations to the near normal levels that are progressively attained in the simply autografted controls (Bouillé and Baylé, 1973). On the other hand, it has been emphasized that destruction of the median eminence in addition to autografting in the duck was without effect on the subnormal plasma corticosteroid level, which is usual in animals with pituitary autografts (Baylé *et al.*, 1971). The site of origin of the CRF is still unknown. However, Frankel *et al.* (1967b) have noticed marked depression of corticosterone secretion in the fowl after lesions within the ventral hypothalamus, whereas similar results were obtained in the pigeon after destruction of a more dorsal area of the hypothalamus (Bouillé and Baylé, 1973). Finally it must be emphasized that the neuroendocrine control of the adrenocortical function probably involves central catecholaminergic and serotoninergic neurons, since their pharmacological suppression, which can be traced within the median eminence (Calas, 1972), coincides with a 40 to 60% fall in the mean level of the plasma corticosterone (Boissin and Assenmacher, 1971a).

b. Rhythmicity of Adrenal Cortical Function.   1. Diurnal rhythm.
The occurrence of diurnal fluctuations in adrenal-cortical function, as
measured by plasma or urinary corticosteroids, have been clearly
demonstrated in mammals. In general, mammalian species, such as
monkey and also man, that are diurnally active has more active
adrenals in early morning than in late evening hours, the reverse
occurring in species such as rats or mice that are predominantly nocturnal animals. Extensive investigations have been made in this respect in a typical diurnally active bird, the Japanese Quail, with
simultaneous measurements of plasma and adrenal-corticosterone
levels, together with actographic and body-temperature records.
Quail reared in isothermic rooms but under natural light conditions
(June), exhibit a marked daily rhythm for both adrenal-cortical parameters studied very similar to that of diurnal mammals, with a steep
increase during the second half of the night, a peak toward the end of
the night hours, and a progressive lowering during day hours (Fig.
4) (Boissin and Assenmacher, 1968). The same adrenal-cortical cycle

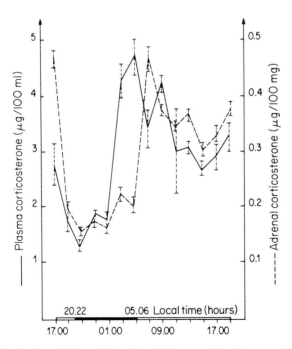

FIG. 4. Daily rhythm of adrenal function in Japanese Quail exposed to natural light
conditions (June) in Montpellier (France). (From Boissin and Assenmacher, 1968.)

has been observed in ducks (Assenmacher and Boissin, 1972). On the other hand, if quail are submitted to constant environmental conditions (constant light at 3 lux) an "endogenous" rhythm of plasma corticosterone and locomotor activity takes place, with a "circadian" periodicity of 22.50 hours (Boissin, 1973). The cyclic evolution of the plasma and adrenal corticosterone has also been studied in quail submitted to various photoperiodic schedules, e.g., 12L:12D (130 lux), 12L:12D (3–5 lux); 6L:18D (130 lux, starting either at 7:00 AM or at 7:00 PM); 18L:6D (130 lux) (Boissin and Assenmacher, 1970). The adrenal rhythms were always strictly synchronized by the photoperiod, the daily increment of corticosterone occurring always during the scotophase, followed by a progressive decrease during the photophase. The adrenal rhythm also remains in constant phase relationship with the light–dark-entrained rhythms of locomotor activity and of body temperature. Furthermore, the corticosterone rhythms, as well as the locomotor-activity rhythm, of quail remain in phase with the environmental light–darkness cycles, even when the birds are submitted to more sophisticated ahemeral photoperiods, such as 6L:7D (3 lux), 9L:26D (3 lux), or 26L:9D (3 lux), which range definitely below (13 hours), or beyond (35 hours), the classic circadian span (20–28 hours) (Boissin and Assenmacher, 1971b; Assenmacher and Boissin, 1972). That the diurnal rhythm of adrenocortical function is dependent on its complex neuroendocrine control machinery is shown by (a) the occurrence of a diurnal rhythm in the pituitary ACTH content, which precedes by a few hours, the phase of the plasma corticosterone rhythm, and (b) the suppression of a 24 hour rhythm of the plasma corticosterone by either autografting the pituitary onto the kidney, or blocking pharmacologically the catecholaminergic or serotoninergic neurons in the central nervous system (Boissin, 1973; Boissin and Assenmacher, 1971a).

2. *Annual (seasonal) rhythm.* In past years, a number of investigators have claimed the occurrence of seasonal variations of morphologic parameters (e.g., weight and histological aspect) of the bird adrenal. In most species, histological activation of the adrenal cortical tissue has been noticed during the breeding season, e.g., in the duck (Höhn, 1947; Höhn et al., 1965; Phillips and van Tienhoven, 1960), the hen (Riddle, 1923; Legait and Legait, 1959), the Brown Pelican (Knouff and Hartman, 1951), the European Starling (*Sturnus vulgaris*) (Burger, 1938), the European Blackbird (*Turdus merula*) (Fromme-Bouman, 1962), and the House Sparrow (*Passer domesticus*) (Bhattacharyya and Ghosh, 1965; Moens and Coessens, 1970). However,

in the duck a second period of adrenal histological activation has been observed during fall and winter (Höhn *et al.*, 1965). On the other hand, Lorenzen and Farner (1964) observed in the White-crowned Sparrow (*Zonotrichia leucophrys*) that the adrenal-cortical tissue showed a maximum activity during the quiescent phase of the testicular cycle and vice versa.

Only a few data based on biochemical measurements of adrenal cortical function deal with that problem. Resko *et al.* (1964) claimed that in the chicken the corticosterone level in adrenal venous blood was higher in winter (December–January: 24 μg per 100 ml plasma), than in summer (May–June: 6 μg per 100 ml). Machi *et al.* (1967), working on ducks, observed an increase of peripheral plasma corticosterone levels from February–March (21.05±2.53 μg per 100 ml) to April–May (33.57±1.86 μg per 100 ml). A systematic study on monthly fluctuations of peripheral plasma corticosterone levels has been performed on ducks reared outdoors under field conditions of the Mediterranean region and fed *ad libitum*. The plasma corticosterone revealed marked annual variations, with a minimum in March (11.83±1.61 μg per 100 ml) and a maximum in October–November (27.30±1.56 μg per 100 ml). During the summer months (May–September), the plasma corticosteroid level was at a plateau at about 17 μg per 100 ml (Soulé and Assenmacher, 1966).

The marked discrepancy among the cited results obviously may reflect species variations, as well as differences in climatic, nutritional, or other environmental factors to which the birds were submitted. Finally, the different techniques used make difficult a valid comparison of the available data. However, three types of correlations may be considered: (1) a positive correlation between the adrenal cortical cycle and the photoperiodically induced breeding season, (2) a negative correlation between the same parameters, (3) a positive correlation between cortical stimulation and low environmental temperatures. The problems raised herein will be discussed in the following sections.

c. Gonadoadrenal Interrelationships. Although in a number of bird species the adrenal glands enlarge annually during the reproductive season, some species — e.g., the male White-crowned Sparrow (Lorenzen and Farner, 1964) and the male duck (Soulé and Assenmacher, 1966) — show an inverse phase relation between the two endocrine cycles, which may point to a possible antagonism, at least between testes and adrenals. For instance, in ducks reared under field conditions, the extreme annual plasma corticosterone levels were

27.30 μg per 100 ml in November, as compared with 11.8 μg per 100 ml in March (Soulé and Assenmacher, 1966), while the extreme, plasma testosterone levels, measured by gas chromatography, were 4.6 ng per 10 ml in November as compared with 16.1 ng per 10 ml in March (Jallageas and Assenmacher, 1970). Such an antagonism, in fact, has been shown experimentally.

Most of the scant data available on gonadoadrenal relationships in birds are related to the male, and the hitherto known effects of testosterone on avian adrenocortical function seem very close to those described in mammals. In cockerels (Breneman, 1941; Kar, 1947; Nagra *et al.*, 1965) and in White-crowned Sparrows (Stetson and Erickson, 1971) castration leads to adrenal hypertrophy, and testosterone injections were found to prevent this effect according to Breneman (1941) and Kar (1947), but not according to Nagra *et al.* (1965). In fowl, castration also elevates the corticosterone level in adrenal venous blood, while testosterone injections lower the adrenal effluent corticosterone to control values (Nagra *et al.*, 1965). On the other hand, testosterone-injected gonadectomized fowls exhibited the same biological half-life of corticosterone as untreated castrates, when measured by the unlabeled corticosterone-loading method of Kitay (1961) (Nagra *et al.*, 1965). As in several investigations on mammals, testosterone administration in intact androgen-secreting fowls had no effect on adrenal weight (Schomberg *et al.*, 1964; Nagra *et al.*, 1965) nor on the amount of corticosterone in adrenal effluent blood (Nagra *et al.*, 1965), but the same investigators showed that the biosynthesis of corticosterone by adrenal tissue *in vitro* was inhibited by the male hormone.

Table IV brings further evidence for a strong androgen inhibition on corticosterone secretion without any impairment of aldosterone secretion in the duck (Boissin *et al.*, 1968; Daniel and Assenmacher, 1969b). It may also be recalled here that castration in quail and ducks submitted to either long or to short days elicits significant activation of the ACTH cells within the anterior pituitary (Tixier-Vidal, 1970). On the other hand, male pigeons seem to behave in a different way than fowls and ducks, since castration as well as testosterone injections have been found to elicit adrenal atrophy, together with cytological and histochemical pictures of adrenal regression (Bhattacharyya, 1968).

Estrogens have been less extensively investigated thus far. However, adrenal hypertrophy, which is classic in mammals, also occurs in estrogen-treated fowls (Breneman, 1942), and pigeons (Miller and Riddle, 1939; Bhattacharyya, 1968), together with histochemical

TABLE IV

INHIBITION OF ADRENAL CORTICAL FUNCTION BY ANDROGEN IN THE DUCK[a]

No.	Treatment	Plasma corticosterone (μg/100 ml)	No.	Treatment	Adrenal[b] Corticosterone (ng/100 mg)	Aldosterone (ng/100 mg)
9	Intact; indoors; +20°C; November; natural light	17.8±0.9	8	Intact; indoors; +20°C; December; natural light	810±160	90±20
9	Intact; indoors; +20°C; November; photoperiod LL	11.5±1	7	Intact; indoors; +20°C; December; testosterone 10 mg/day	400±40	115±5
9	Intact; indoors; +20°C; November; testosterone 1 mg/day	14.4±0.9	9	Castrated; indoors; +20°C; December	1560±100	115±15
12	Intact; outdoors; March	11.8±1.6	8	Castrated; indoors; +20°C; December; testosterone 10 mg/day	870±60	120±30
3	Castrated; outdoors; March	17.6±0.6				
7	Castrated; outdoors; March; testosterone 1 mg/day	8.8±0.7				

[a] From Boissin et al. (1969).
[b] From Daniel and Assenmacher (1969b).

pictures of cortical stimulation (Bhattacharyya, 1968). According to the latter, no significant adrenal effect was noticed in spayed pigeons. Zarrow and Baldini (1952) failed to induce adrenal hypertrophy in estrogen-treated Bobwhite Quail (*Colinus virginianus*). Progesterone, on the other hand, has been shown to induce increased phosphatase activity in the adrenal cortex of pigeons of both sexes (Sarkar and Ghosh, 1959).

   d. Environmental (External) Factors. 1. Photoperiod. Despite the many observations of the adrenal-cortex activity as a function of the seasonal (see above) or experimental (Busheikin, 1951) increase of day length, no decisive conclusion can be drawn as to whether or not a direct causal relationship exists, i.e., whether photoperiod acts as a specific source of information in the control of the adreno-corticotropic axis as for other functions (e.g., the gonadal function). Enhanced locomotor activity and nutritional functions, which are simultaneously activated throughout the seasonal long-day period could conceivably act as light-induced secondary stimulators for adrenal-cortical function. On the other hand, for those species, in which an inverse correlation exists between increasing light–dark ratios and adrenal-cortical activity, it has been seen in Section I,B,3,c that there are arguments to consider these correlations as a result of gonadoadrenocortical interactions.

   The diurnal periodicity in adrenal cortical function probably provides a better model than do seasonal rhythms for the study of photoperiod on the function of the adrenal cortex. Whereas all of the above cited light–dark schedules entrained the plasma and adrenal corticosterone rhythms in the quail, the testes of the birds were either developed—e.g., in the natural summer environment, and in the schedules: 12L:12D (130 lux), 18L:6D (130 lux); 6L:7D (3 lux); 26L:9D (3 lux)—or regressed—6L:18D (130 lux); 12L:12D (3 lux); 9L:26D (3 lux) (Boissin and Assenmacher, 1968, 1970, 1971). Because of the limiting action of either day length or intensity on the gonadotropic axis, the two neuroendocrine mechanisms were dissociated in these experiments, and the high degree and plasticity of light–dark dependence of the adrenal-cortical function appears here to be clearly independent of gonadal function. However, here again it is not possible to assess the direct light sensitivity of the adrenocorticotropic machinery since the effects of different light–dark ratios may really be derived from other direct photoperiodic regulations, such as in the control of locomotor activity and food intake, which have marked metabolic impacts, and which are entrainable by light–dark cycles.

2. *Temperature.* Höhn's description (Höhn *et al.*, 1965) of a second peak of morphological stimulation of the duck's adrenal during the cold season points to an adrenal cortical adaptation to *chronic cold environment.* Knouff and Hartman (1951) also have reported on a histochemical activation of cortical tissue in Brown Pelicans exposed to cold. The annual adrenal rhythm in the duck (Soulé and Assenmacher, 1966), which has been analyzed above, does not show any definite correlation between environmental temperature and function of the adrenal cortex. However, the comparison between the plasma-corticosterone level of the animals studied in this experiment and held outdoors in November (26.40 μg per 100 ml) and ducks from the same flock that had been kept at the same period for 2 weeks indoors at +20°C (18.8 μg per 100 ml) reveals a possible stimulating action of chronic exposure to cold. On the other hand, Boissin (1967) has studied the adrenal repercussion of *acute exposure to cold* in the duck. When measured at 4, 12, and 24 hours after cold exposure (from +25°C to +4°C), plasma corticosterone was significantly increased respectively by 25, 40, and 21%, showing a rapid but transient response of the adrenal cortex. Hypophysectomized ducks and specimens bearing pituitary autografts were submitted to the same treatment. The hypophysectomized ducks, which had plasma corticosteroid levels at 47% of intact controls, failed to exhibit adrenal reaction to cold, whereas the autografted birds had a positive reaction, though shorter than the intact controls. The 25°C autografted ducks had plasma corticosterone levels at 76% of controls. After a 4-hour exposure to cold the plasma corticosteroids increased by 28%, after 12 hours by 48%. However, 24 hours later the plasma corticosteroid values had returned to normal levels for birds with pituitary autografts.

3. *Feeding.* Although there has been much interest in the gonadal effects of underfeeding, few studies have dealt with possible adrenal involvement. Nevertheless, several species, when fed a restricted diet, show some evidence of adrenal-cortical stimulation. Underfed pigeons (Vincent and Hollenberg, 1920) and chickens (Breneman, 1942; Conner, 1959) showed adrenal hypertrophy, while Hartman *et al.* (1954) noted a higher rate of mitoses associated with lipid depletion in the adrenal cortex of undernourished Brown Pelicans. In the turkey, starvation leads to a significant increase in plasma corticosterone (Brown, 1960). Enhanced plasma corticosterone levels also were observed in male ducks starved for 17 days, and the absence of a significant concomitant alteration of the biological half-life of

corticosterone in those animals points to a definite stimulation of the adrenal-cortical function (Table V) (Malaval and Assenmacher, 1973).

TABLE V
ADRENOCORTICAL EFFECTS OF STARVATION IN DUCKS KEPT
INDOORS IN DECEMBER[a]

Treatment	No.	Plasma corticosterone (μg/100 ml) (mean ± SD)	No.	Biological half-life of [³H] corticosterone (minutes) (mean ± SD)
Controls	17	12.28 ± 0.42	5	10.69 ± 0.25
Starved 17 days	12	19.75 ± 1.32	4	12.00 ± 0.52

[a] From Malaval and Assenmacher (1973).

C. MEDULLA

1. Hormones

As in other vertebrates, the medullary cells secrete catecholamines. The avian medulla has first been considered as secreting mainly norepinephrine, as the predominance of that hormone has been evidenced by histochemical methods in the pigeon (Wright and Chester-Jones, 1955; Ray and Ghosh, 1961) and in the Gentoo Penguin (Cuello, 1970), and by chromatographic procedures in the fowl (70–80% of the total catecholamine content of the gland; Shepherd and West, 1951). On the other hand, adrenaline-containing granules and noradrenaline-carrying granules have been isolated from homogenates of chicken chromaffin tissue by centrifugation in a sucrose-density gradient (Schümann, 1957). Systematic investigations using histochemical assays have shown that there are marked species differences in the norepinephrine–epinephrine ratio (Ghosh, 1962; 1973); (1) medullas of Pelecaniformes, Ciconiiformes, and Galliformes are essentially norepinephrine producers; (2) Passeriformes are mainly epinephrine secretors; and (3) Columbiformes and Cuculiformes have both hormones present in their adrenals, with a predominance of norepinephrine in the former and of epinephrine in the latter. According to Ghosh, this could reflect an evolutionary trend, the hormonal methylation being mainly absent in more primitive avian orders. On the other hand, the higher levels of epinephrine in comparison with norepinephrine in peripheral blood of the chicken, turkey, and pigeon have been interpreted as a reflection of a higher release of epinephrine by the adrenals of those species, an inverse ratio existing in the duck blood (Sturkie *et al.*, 1970). However, the neuronal source of both catecholamines might be partly involved in the plasma levels of the hormones, and the question hence needs further clarification. On the other hand, the concentration of norepinephrine has been found to be significantly higher in plasma of female than male chickens, but there was no sex difference in concentrations of epinephrine (Sturkie and Lin, 1968).

Regarding the control of epinephrine biosynthesis, a cytochemically assessed rise of epinephrine and a concomitant depletion of norepinephrine in cortisone-treated pigeon adrenal medullas indicates the possibility of conversion of the latter to the former by a methylation mechanism similar to that which exists in mammals (Ghosh, 1973).

2. Role of the Adrenal Medulla

Although some effects of epinephrine and norepinephrine in birds have been reported, little is known about the precise physiological role of the avian adrenal medulla, as other (nervous) sources actively participate to the production of catecholamines.

a. Heat Production. Norepinephrine may have a thermogenic function in mammals with brown adipose tissue, whereas in species without brown fat it does not. The hitherto studied birds behave like the latter mammals. In the young chick, neither noradrenaline nor adrenaline arrest the fall of body temperature induced by cold and anesthesia (Freeman, 1970a). Indeed, noradrenaline lowers the rectal temperature of anesthetized chicks exposed to cold (Allen and Marley, 1967; Freeman, 1970a), and the same hypothermic effect has also been shown in quail 6–9 weeks old (Freeman, 1970b). The physiological significance of the 60% release of the norepinephrine stored in the adrenal medulla of pigeons exposed to acute cold (Lahiri and Banerji, 1969) therefore remains obscure.

b. Lipid Metabolism. In the neonatal fowl noradrenaline (300 μg/kg) induces a rise in plasma free fatty acids. This response increases from hatching to 2 weeks of age. By 8 weeks, the response is much reduced (Freeman, 1969b). In adult fowls (Carlson *et al.*, 1964; Heald, 1966; Langslow and Hales, 1969), in ducklings (Desbals, 1972), and in adult ducks and turkeys (Grande, 1969), neither epinephrine nor norepinephrine, when used at physiological doses, either *in vivo* or *in vitro*, bring about release of free fatty acids from adipose tissue, nor are the catecholamines able to cause any significant change of liver triglycerides in the duck (Grande and Prigge, 1970). However, epinephrine stimulates lipolysis in pigeon adipose tissue *in vitro* (Goodrige and Ball, 1965) and norepinephrine induces a steep increase in plasma free fatty acids in 6–9-week-old quail (Freeman, 1970b). In geese, both catecholamines, whether injected or infused, cause an elevation of plasma free fatty acids, although the response is considerably smaller than that produced by glucagon (Grande, 1969).

As norepinephrine is not a thermogenic factor, but induces hypothermia in birds exposed to cold, it seems unlikely that it is directly responsible for the increase in plasma free fatty acids that results from cold exposure (Freeman, 1967).

c. *Carbohydrate Metabolism.* The hyperglycemic effect of epinephrine has been known for years in the duck (Fleming, 1919), in the chicken (Golden and Long, 1942; Langslow *et al.*, 1970), in the pigeon (Ghosh, 1973), and in the goose (Grande, 1969). In the duck, infusion of norepinephrine causes also a significant hyperglycemia, while the hormone produces only minimal changes in geese and turkeys (Grande, 1969). In pigeons, norepinephrine is practically as potent as epinephrine with regard to hyperglycemia (Ghosh, 1973). In the neonatal fowl, norepinephrine (300 μg/kg) increases plasma glucose, but no hyperglycemia developed in 1–2-week-old chicks. A significant hyperglycemia again occurs in 4–8-week-old chickens (Freeman, 1969b). On the other hand, in 6–9-week-old Japanese Quail, there is an unexpected effect of norepinephrine—a significant fall in plasma glucose within 30 minutes after epinephrine injection has been observed (Freeman, 1970b). Finally, epinephrine in large doses also has been found to enhance significantly the glycogen content of the sacral glycogen body in chickens, whereas the liver glycogen was decreased (Snedecor *et al.*, 1963).

d. *Blood Pressure.* Perfusion of anesthetized chickens with norepinephrine and even more so with epinephrine, leads to significant rise in the systolic and diastolic blood pressure (Akers and Peiss, 1963; Carlson *et al.*, 1964; Dewhurst and Marley, 1965). From histochemical studies Ghosh (1973) concluded that both catecholamines are equally effective as pressor hormones in the House Crow (*Corvus splendens*) and the fowl. Heart rate is unaltered by norepinephrine or even decreased with epinephrine unless the vagi are cut; in the latter case both catecholamines increase heart rate slightly (Akers and Peiss, 1963).

II. Thyroid Gland

A. MORPHOLOGY

The avian thyroid gland is paired, and its two dark red, ovoid lobes are located low in the neck, internal to the jugular vein, and external to the carotid (at its junction with the subclavian artery) and to the trachea. Structurally the avian thyroid gland is composed, as in other vertebrates, of roughly spherical follicles, each consisting of a colloid-containing lumen surrounded by a single layer of cuboidal epithelial cells. An extensive vascular supply is provided to the follicles.

As in all vertebrates, the avian thyroid gland arises as a midventral outpocketing from the floor of the pharynx, which forms the longitudinal mesobranchial groove. Later on, the mesobranchial groove disappears anteriorly, leaving as its only remnant an epithelial thickening, which is the first visible primordium of the thyroid gland. The original primordium soon becomes bilobed and gradually migrates back as paired organs toward its adult location. Progressively invading mesenchyme and blood vessels crowd the epithelial mass of each lobe into solid cylindrical cords. Midway in the incubation period the cords begin to break up in small cell groups that acquire central lumina as a result of secretion and coalescence of colloid droplets originating in the follicular cells. From then on, the thyroid follicles are fully differentiated, and their functional state already is evidenced by a more or less strong radioiodine (^{131}I) uptake (Hansborough and Khan, 1951; Blanquet et al., 1952). During the latter half of incubation, the main feature consists in follicular growth in size and number (see Romanoff, 1960).

Normal and TSH-stimulated thyroid glands of the fowl have been investigated by Fujita (1963) with the electron microscope. Two types of intracellular droplets, respectively of high and of low electron densities, were identified. Both are enclosed by a limiting membrane composed of two dense and a less dense layer, similar to that of the other cytoplasmic membrane systems. The dense granules appear to be formed in the Golgi field, while the less dense ones appear to derive from the endoplasmic reticulum. The latter droplets are considered to contain colloid, as they increase markedly in number and size, together with endoplasmic sacs, after TSH stimulation, and several droplets seem to be secreted into the follicular lumen through an opening occurring at a junction between the apical plasma membrane and the limiting membrane of the droplet. Some cellular processes containing numerous low-density droplets and rough-surfaced endoplasmic sacs are also observed after TSH administration. On the other hand, after TSH treatment, a distinct endoplasmic reticulum and large droplets of low density may be seen close to the basal plasma membrane, suggesting a direct secretion of the content of droplets and endoplasmic sacs into the pericapillary or interfollicular space from the basal part of the follicular cell. Finally, a central flagellum with a basal corpuscule is observed in some follicular cells of young and adult fowls. Its function has not been clarified.

B. HORMONES AND METABOLISM

There is an extensive literature in recent years on thyroid function in birds, especially since the development of radioisotopic methods (Taurog et al., 1950). From these data, it appears that the hormones of avian thyroid and the sequences of their synthesis, metabolism, and catabolism are fundamentally similar to those of mammals despite some peculiar features, which generally concern quantitative rather than qualitative aspects of metabolism. It is well established that the biosynthesis of thyroid hormones occurs within the polypeptide chain of the specific protein thyroglobulin. As in mammals, the major thyroid iodoprotein present in avian thyroid follicles has a sedimentation coefficient close to 19 S. In the duck and chick, the 19 S thyroglobulin concentration among soluble thyroid iodoproteins is about 94% (it is 92% in man, 93% in rats, and 94% in dogs). Other minor thyroid iodoproteins have also been identified in birds as in mammals — e.g., a 27 S thyroglobulin, in a concentration of 5% in ducks and 6% in chickens (8% in man, 7% in rats, and 6% in dogs), together with traces of 12 S thyroglobulin (a compound that has not been detected in man or dog and is present in traces in the rat (Roche et al., 1968).

Using radioisotopes, most investigators have identified the two classical thyroid hormones within thyroid hydrolysates, i.e., triiodothyronine and thyroxine, the latter being largely predominant — e.g., in the domestic fowl (Shellabarger and Pitt-Rivers, 1958; Vlijm, 1958; Mellen and Wentworth, 1959; Kobayashi and Gorbman, 1960; Spronk, 1961; Laguë and van Tienhoven, 1969), in the Japanese Quail (Astier, 1973b), in the duck (Tixier-Vidal and Assenmacher, 1958, 1965; Astier et al., 1966; Astier, 1973b; Rosenberg et al., 1967), in the pigeon (Lachiver and Poivilliers de la Querière, 1959; Baylé et al., 1966), in the Golden Bishop (*Euplectes afer*), and in the White-throated Sparrow (*Zonotrichia albicollis*) (Kobayashi et al., 1960). Similarly the two hormones were shown to occur in the plasma of the fowl (Vlijm, 1958; Mellen and Wentworth, 1959; Wentworth and Mellen, 1961), the Japanese Quail (Astier, 1973b), the turkey and the duck (Wentworth and Mellen, 1961; Astier, 1973b). However only thyroxine has been identified by others in the thyroid gland of the domestic fowl (Rosenberg et al., 1963, 1964). Whether the discrepancies noted in domestic fowl may originate from technical differences, from iodine content in the diet (Leloup and Lachiver, 1955), or from variations in the strains studied needs further clarification.

From the numerous reports on avian thyroid function, and from several systematic comparative studies performed in birds and mammals an avian pattern of thyroid function may be outlined.

1. At the thyroid-gland level, the bird is characterized, comparatively with the mammal by (1) a higher total thyroid ^{127}I content (Rosenberg et al., 1963, 1967; Astier, 1973b); associated with (2) a higher thyroid T4 ^{127}I concentration, since the distribution of ^{127}I among iodinated thyroid compounds as well as the equilibrium of thyroid ^{131}I with ^{127}I in chickens is similar to that in rats fed a similar iodine diet (Rosenberg et al., 1963, 1964); (3) an elevated net rate of trapping of ^{127}I, as calculated from specific activities of serum iodide and thyroid ^{131}I uptake at early times after injection of radioiodide (Rosenberg et al., 1964); and (4) a prominent iodine storage, leading to a prolonged thyroid retention of ^{131}I and to an almost flat thyroid ^{131}I output curve, both definite characteristics of birds [Wahlberg, 1955; Vlijm, 1958, in the domestic fowl; Tixier-Vidal and Assenmacher, 1958, in the duck; Poivilliers de la Querière and Lachiver, 1957, in the pigeon; Baylé et al., 1967a, in the Japanese Quail; Morgan and Mraz, 1970, in the Bobwhite Quail; Fink, 1957, in the White-crowned Sparrow and in the Brown Towhee (Pipilo fuscus); Kobayashi and Tanabe, 1959, in the canary; Kobayashi et al., 1960, in the White-throated Sparrow and in the Golden Bishop; Chandola, 1972, in the Nutmeg Mannikin (Lonchura punctulata)].

2. At the peripheral-plasma level, avian thyroid metabolism exhibits further marked peculiarities: (1) In the absence of a thyroxine-binding globulin (TBG) in birds, the only available binding proteins lie within the prealbumin (TBPA) and albumin (TBA) zone (Tata and Shellabarger, 1959; Farer et al., 1962; Refetoff et al., 1970). Chicken and pigeon sera bind both T4-^{125}I and T3-^{125}I to TBA, but among all vertebrates hitherto studied, the pigeon is the only species that binds T3 to prealbumin (Refetoff et al., 1970). On the other hand, although

Tata and Shellabarger (1959) had claimed that both T3 and T4 were equally bound to plasma albumin in chickens, others have pointed to a greater affinity to T4 than T3 for the plasma protein binding sites (Dubowitz *et al.*, 1962; Heninger and Newcomer, 1964). These peculiarities of serum-binding proteins for thyroid hormones may account for some further characteristics, such as (2) lowered amounts for either total serum iodine, or PB ^{127}I (Rosenberg *et al.*, 1963, 1964; Refetoff *et al.*, 1970; Astier *et al.*, 1972, 1973b). (3) While avian levels of total thyroxine range between classic values found in other warm-blooded vertebrates, the free plasma thyroxine, which might be considered a more accurate index for thyroid status than total T4, is relatively high in birds in comparison with most mammals (Refetoff *et al.*, 1970). (4) As a consequence of the foregoing, the peripheral metabolism of both thyroid hormones is remarkably rapid in birds, as evidenced by their short biological half-lives (Shellabarger and Tata, 1961; Heninger and Newcomer, 1964; McFarland *et al.*, 1966; Hendrich and Turner, 1967; Singh *et al.*, 1967; Assenmacher *et al.*, 1968; Astier *et al.*, 1972, 1973b) (Fig. 5). Since the clearance rate of labeled iodide from the plasma appears equally slow in birds and in rats with perchlorate-blocked thyroids (Astier, 1973b), the fast peripheral metabolism of the thyroid hormones in birds (almost four times as rapid as in rats) may account for the marked difference in the half-life of thyroxine-*I, if measured either from total plasma samples, or from only the protein-bound *I fraction of the plasma, that is usual in birds (see Table VI) but not in rats (Astier, 1973b). (5) On the other hand, emphasis has also been laid recently on the fact that, unlike in rats, the separation by Sephadex gel chromatography of the ^{125}I-labeled compounds in sera of several bird species (fowl, duck, pigeon, Japanese Quail) points to a large amount of nonhormonal iodoproteins within the avian PBI. This parameter is therefore a rather inaccurate index of thyroid hormone concentrations in the blood of birds (Astier, 1973a, 1973b).

Very little is known as yet about the role of storage and discharge of thyroid hormones from extrathyroid tissues in the regulation of the effective level of circulating thyroid hormones in birds. On the other hand, it has been well established from studies performed on the domestic fowl with either artificial anus (Tanabe *et al.*, 1965) or canulated bile ducts and ureters (Hutchins and Newcomer, 1966) that, as in mammals, the main route of excretion for both thyroid hormones is via the bile and feces, the urine containing almost exclusively iodine in inorganic form. Furthermore, chromatographic and electrophoretic analyses of bile samples from chickens injected with T4-^{131}I or T3-^{131}I have revealed that the assortment of ^{131}I-labeled metabolites iden-

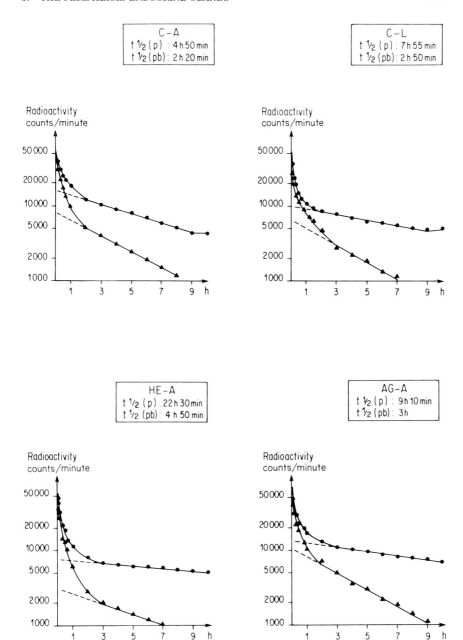

FIG. 5. Biological half-life (t ½) of [^{131}I]thyroxine in the duck. C = controls; HE = hypophysectomy; AG = pituitary autograft; A = iodine-adequate diet; L = low-iodine diet. (From Astier, 1973.)

TABLE VI

THYROID AND PLASMA ^{127}I CONTENT

Species	Diet (I⁻) (μg/gm)	(μg/100 gm body weight)	^{127}I Thyroid content			Reference
			(μg)	(μg/mg thyroid)		
Pigeon	0.87		1.61			Poivilliers de la Querière and Lachiver, 1957
Duck	adequate		822.5	4.7 ±0.3		Rosenberg et al., 1967
	low		934.5	4.13±0.24		Astier, 1973b
			233.7	1.73±0.22		Astier, 1973b
Domestic fowl (6 week)	0.14		89	0.92		Rosenberg et al., 1963
adult	2.14		310	3.61		Rosenberg et al., 1963
	low	1.74				Singh et al., 1968b
		7.90				Singh et al., 1968b
Japanese Quail	adequate		6.32	1.14±0.18		Astier, 1973b

Species	Diet (I⁻) (μg/gm)		Thyroid [^{127}I]thyroxine content			Reference
			(μg)	(μg/mg)		
Domestic fowl	0.14		18.7	0.59±0.03		Rosenberg et al., 1963
	2.14		30.3	—		Rosenberg et al., 1963
Duck	adequate		132.6	0.59±0.03		Astier, 1973b
	low		41.8	0.30±0.03		Astier, 1973b
Japanese Quail	low		5.09	0.22±0.03		Astier, 1973b

		Protein-bound ^{127}I (μg/100 ml plasma)	
Domestic fowl		1.13–1.22	Mellen and Hardy, 1957
	0.14	0.51±0.02	Rosenberg et al., 1963
	2.14	0.87±0.02	Rosenberg et al., 1963
		2.0	Refetoff et al., 1970
	low	1.0135±0.0483	Singh et al., 1967
	adequate	1.1226±0.0367	Singh et al., 1967
Colinus virginianus	adequate	1.7572±0.0637	Singh et al., 1967
Japanese Quail	adequate	1.2603±0.1320	Singh et al., 1967
	low	1.36±0.11	Astier, 1973b
Pigeon		1.4–1.8	Refetoff et al., 1970
Duck		1.49	Taurog and Chaikoff, 1946
		1.17–1.49	Mellen and Hardy, 1957
	low	1.05±0.05	Astier, 1972; 1973b
	adequate	1.90±0.06	Astier, 1972; 1973b

	Plasma [^{127}I]thyroxine control	
Domestic fowl	total (μg/100 ml) 1.4–1.6	Refetoff et al., 1970
	free (ng/100 ml) 5.5	Refetoff et al., 1970
Pigeon	total (μg/100 ml) 2.4–3.3	Refetoff et al., 1970
	free (ng/100 ml) 6.0–6.9	Refetoff et al., 1970

tified in the bile, and hence the peripheral metabolism of both hormones, is similar to that known in mammals. Metabolism was found to occur by conjugation (mainly glucurono binding, and to a lesser extent, sulfate binding), deamination, and deiodination (with formation of diiodothyronine).

Similarly, extensive studies have shown that goitrogenic drugs have the same effect in birds as in mammals, both at the thyroid and at the peripheral level. Thiouracil, for instance, has marked extrathyroid effects in chickens as in rats, in reducing the peripheral utilization of thyroxine, while neither methimazole nor potassium chlorate have such effects (Tanabe et al., 1965).

TABLE VII
THYROID SECRETION RATE (TSR)

Species	Method[a]	TSR (L-thyroxine μg/100 gm body weight/day)	Reference
Domestic fowl	GP	2.32	Singh et al., 1968a
	TS	2.00	Singh et al., 1968a
	DO	1.20	Singh et al., 1968a
	TD	2.03	Singh et al., 1968a
	GP	1.39	Mellen, 1964
	TS	1.91–2.27	Mellen and Wentworth, 1960
	GP	1.5	Tanabe et al., 1965
	TS	1.5	Tanabe et al., 1965
	DO(I-adequate diet)	1.30±0.15	Singh and Reinecke, 1968
	DO(I-low diet)	0.46±0.05	Singh and Reinecke, 1968
	GP	1.04–1.18	Kleinpeter and Mixner, 1947
	GP	1.30–1.40	Odell, 1952
Colinus virginianus	TD	2.49±0.49	Singh et al., 1967
Japanese Quail	TD	2.78±0.34	Singh et al., 1967
Duck	GP	1.90	Hoffmann, 1950
	GP	1.28–1.70	Biellier and Turner, 1950
	TD	2.05±0.17	Astier et al., 1972
Domestic turkey	GP	1.88	Mellen, 1964
	GP	1.31–1.49	Smyth and Fox, 1951
	GP	0.76–1.33	Biellier and Turner, 1955
Pigeon	TS	1.94±0.07	Grosvenor and Turner, 1960

[a] GP = goiter prevention method (thyroxine necessary to yield thyroid weight of goitrogen-treated birds equal to controls) (Mixner et al., 1944). TS = thyroxine substitution method (thyroxine necessary to block the release of ^{131}I) (Pipes et al., 1958). DO = direct output method (calculated from daily ^{131}I thyroid output rate after ^{131}I injection) (Singh et al., 1968a). TD = thyroxine degradation method (calculated from plasma half-life of injected [^{131}I]thyroxine) (Singh et al., 1968a).

TABLE VIII. BIOLOGICAL HALF-LIFE OF THYROXINE.[°I]

Species	Radioactivity	Time of measurements (hours after °I-thyroxine injection)	Half-life	Reference
Domestic fowl	Whole body	0–24 hours	9.4 hours	Shellabarger and Tata, 1961
		24–54 hours	24.0 hours	
	Plasma	1–4 minutes	3±0.6 minutes	Heninger and Newcomer, 1964
		5–20 minutes	28.9±7.5 minutes	Heninger and Newcomer, 1964
		20–50 minutes	31.1±4 minutes	Heninger and Newcomer, 1964
		1–24 hours	8.3±1.6 hours	Heninger and Newcomer, 1964
		24–96 hours	16.5±2 hours	Heninger and Newcomer, 1964
	Total blood	3–40 hours (cold)	7.6±0.50 hours	Hendrich and Turner, 1967
		3–40 hours (heat)	11.4±0.49 hours	Hendrich and Turner, 1967
	Plasma	3–12 hours (iodine-adequate diet)	3.44 hours	Singh et al., 1968b
		3–12 hours (iodine-deficient diet)	3.86 hours	Singh et al., 1968b
Duck	Plasma	2–18 hours (iodine-adequate diet)	4.00 hours	Assenmacher et al., 1968
		2–18 hours (iodine-deficient diet)	8.47 hours	Astier et al., 1969
	Protein-bound ^{125}I	2–12 hours (iodine-adequate diet)	2.68 hours	Astier, 1973b
		2–12 hours (iodine-deficient diet)	3.29 hours	Astier, 1973b
Colinus virginianus Japanese Quail	Plasma	3–12 hours	4.60 hours	Singh et al., 1967
	Plasma	3–12 hours	5.40 hours	Singh et al., 1967
	Total blood	17–65 hours	18.4 hours	McFarland et al., 1966

Finally, the total thyroid activity, as measured by daily thyroid-hormone secretion rate ranges within limits close to those found in mammals. Tables VI, VII, and VIII provide some numerical data that have been reported on the main parameters of thyroid function in birds. Unfortunately, data published until recently lack information on the iodine content of the diet of the animals studied.

Now very thorough investigations on the domestic fowl by Rosenberg *et al.* (1963, 1964) afford evidence of the definite importance of the alimentary iodine. The comparison of chickens fed an iodine-deficient diet (0.14 μg I$^-$ per gram) with others receiving an iodine-supplemented diet (+2 μg I$^-$ per gram) lead to the following conclusions: (1) The deficient chickens had consistently lowered (one-third) amounts of total ^{127}I in their thyroids. Despite an increased fraction of [^{127}I]thyroxine in the thyroid glands of the deficient chickens, the total intraglandular amount of the hormone was low in these birds. (2) The deficient chickens exhibited enhanced thyroid ^{131}I uptake (52% at 24 hours compared with 12% for the supplemented group). (3) With the deficient diet, ^{131}I incorporated among iodinated compounds of the thyroid gland reached an apparent equilibrium distribution with the preexisting stored ^{127}I compounds at 48 hours, while equilibration required approximately 4 days with the supplemented diet. (4) Within the plasma, the deficient diet lead to a marked fall in the total ^{127}I content (20% of the supplemented diet), as well as in the protein bound ^{127}I (60% of the supplemented diet). Similar results have been described in the duck (Astier, 1973b).

Using an open two-compartment model devised by Wollman and Reed (1959), an attempt has been made to characterize the transport of radioiodide between blood and thyroid gland in chicken, in relation to iodide levels of the diet (Newcomer, 1967). This method demonstrated that the thyroid/serum radioiodide ratio after a single ^{131}I injection and the thyroid clearance constant for radioiodide varied inversely with the concentration of alimentary iodide.

All these results are of primary importance, as they clearly reveal that no valid assumption on possible intra- and interspecific differences in thyroid metabolism among the various bird species hitherto studied may be assessed without a precise statement on the dietary iodine intake of the birds.

C. ROLE OF THYROID HORMONES

In mammals, triiodothyronine has been shown to possess 2–6 times the potency of an equimolar quantity of thyroxine as measured by a number of physiological tests. In birds, controversial results have been obtained. According to Gilliland and Strudwick (1953), triiodothyronine has more potency than thyroxine in blocking TSH release, whereas thyroxine was found more potent than triiodothyronine in preventing goiter in thiouracil-treated birds (Newcomer, 1957; Mellen and Wentworth, 1959) and in promoting oxygen uptake by myocardium in ducks (Newcomer and Barnett, 1960). On the other hand, both hormones were claimed to be equipotent in preventing goiter (Shellabarger, 1955), in stimulating heart rate (Newcomer, 1957), and in counteracting the effects of propylthiouracil or radiothyroidectomy on body and comb growth rates and on liver glycogen (Raheja and Snedecor, 1970). A possible explanation for the different relative potencies of T4 and T3 could lie, as in mammals, in different binding affinities of the hormones for their carrier plasma proteins. But, as was mentioned previously, there is also some discrepancy among the available data concerning this point, so that further clarification is still needed.

1. Growth

As in other vertebrates, thyroid hormones are essential for regulating harmonious growth. Thyroidectomy in ducks (Woitkewitch, 1938) and in domestic fowl (Blivaiss and Domm, 1942; Blivaiss, 1947) caused marked retardation in growth, leading to the classic hypothyroid shortleg dwarfism, associated with abnormal fat deposition and sometimes obesity (Blivaiss, 1947). Similar results are produced by radiothyroidectomy (Winchester and Davis, 1952), while triiodothyronine or thyroxine restore the body weight of the operated chicks (Raheja and Snedecor, 1970). Goitrogen feeding yields less constant results (Astwood et al., 1944; Glazener and Jule, 1946; Hebert and Brunson, 1957; Wilson and MacLaury, 1961). However, the growth rate of 3-week-old chickens was significantly retarded by administration of tapazole, and thyroxine (2–3 μg per 100 gm/day) counteracted completely the effect of the goitrogen (Singh et al., 1968b). Similar results have been obtained by Raheja and Snedecor (1970) using propylthiouracil and triiodothyronine or thyroxine (0.3 μg per 100 gm) as replacement therapy.

On the other hand, thyroid-hormone administration to intact birds results only in moderate acceleration, if any, of growth (Parker, 1943; Boone et al., 1950; Hebert and Brunson, 1957). Singh et al., (1968b) noted improved growth rate in 3-week-old chickens injected with thyroxine (1–4 μg per 100 gm/day). However, higher doses (6 μg per 100 gm/day) accelerated catabolic processes and depressed the growth rate of the birds.

2. Heat Production

In adult birds, the main role of the thyroid seems to be the maintenance of a generally normal metabolic rate. Although thyroidectomy is
not lethal, the removal of thyroid hormones leads to a marked depression of heat production in the chicken (Winchester, 1939; Mellen and
Wentworth, 1962), the pigeon (Marvin and Smith, 1943), and the
goose (Lee and Lee, 1937). Similar depression of the metabolic rate
is obtained by chemical (goitrogen) hypothyroidism (Sulman and
Perek, 1947; McCartney and Shaffner, 1950; Mellen, 1958), while
administration of thyroprotein enhances heat production in the pigeon
(Sierens and Noyons, 1926), and in the fowl (McCartney and Shaffner,
1950; Mellen, 1958). Singh *et al.* (1968b) have observed that single
injections of thyroxine and triiodothyronine or of a combination of the
two produced only a small and transient (2-hour) rise in metabolic rate
of chickens. The metabolic rate was depressed 24 hours after injection.

In the neonatal chick, both thyroxine and triiodothyronine, when injected intraperitoneally (300 $\mu g/kg$) were thermogenic. Rectal temperature increased significantly within 30 minutes when the chicks
were maintained in a thermally neutral environment. Both hormones
were equally efficacious in delaying the fall in rectal temperature
when the chicks were exposed to cold (Freeman, 1970a). The stimulating effect of thyroid hormones on tissue metabolism has been well
demonstrated in an older experiment by Haarmann (1936). Oxygen
consumptions by heart, liver, and breast-muscle slices from chickens
and pigeons were elevated up to 30% when incubated with minimal
concentrations (10^{-11}–10^{-18}) of thyroxine.

3. Carbohydrate Metabolism

Riddle and Opdyke (1947) found that in pigeons thyroxine reduces
liver-glycogen level and produces mild hyperglycemia. Inversely,
thyroidectomy causes hypoglycemia in the pigeon (Riddle and Opdyke, 1947) and in the duck (Ensor *et al.*, 1970). Radiothyroidectomy
or feeding propylthiouracil to chicks induces hepatic hypertrophy and
enhanced liver-glycogen levels (Snedecor and King, 1964; Snedecor,
1968), while injections of triiodothyronine or thyroxine prevent all
of these effects (Snedecor and Camyre, 1966; Raheja and Snedecor,
1970).

4. Skin and Feathers; Molting

The thyroid gland exerts marked influence on the skin and its derivatives. The morphogenetic effect of thyroxine on the epidermis is

already apparent in the chick embryo. At 14 days of incubation, the skin epithelium of untreated embryos is restricted to three undifferentiated cell layers, whereas embryos treated with 20 μg thyroxine showed seven cell layers with distinct germinative and corneal layers, a picture similar to that of 16-day-old untreated controls (Bartels, 1944).

In the adult bird, the importance of thyroid hormone on plumage has been amply verified. Thyroidectomy- or goitrogen-induced hypothyroidism affects feather morphology markedly; the feathers become elongated, thinner, and lose their regular, rounded shape, acquiring an almost completely irregularly fringed contour, due to impaired development of the barbules. In some species, e.g., the domestic fowl, most melanin pigments disappear (Greenwood and Blyth, 1929; Parkes and Selye, 1937; Woitkewitch, 1938), whereas in others, thyroid glands have no marked effects on the feather pigments [pigeon: Höhn, 1961; Nutmeg Mannikin (*Lonchura punctulata*): Thapliyal and Pandha, 1967a)]. These alterations are reversed by injections of thyroxine (Juhn and Barnes, 1931; Emmens and Parkes, 1940; Singh *et al.*, 1968b). Moreover, Kraetzig (1937), Svetsarov and Streich (1940), and Thapliyal *et al.* (1968) have shown a direct stimulating action of thyroxine on the feather follicle, where numerous mitoses are observable from the first day of treatment.

These classic observations have lead to the assignment to the thyroid gland of a major role in the onset of molt. From the considerable literature published on this topic (see reviews by Assenmacher, 1958b; Jöchle, 1962; Tanabe and Katsuragi, 1962) (see, also, Chapter 3, Volume II) several conclusions may be drawn: (1) There is general agreement that administration of either thyroid extract, as was first demonstrated in the hen by Zawadowsky (1922), or thyroxine (Akiyoshi, 1932) in the pigeon, brings on artificial molts, whatever the season may be. (2) On the other hand, surgical removal of the thyroid delays or even prevents the natural molt, providing that thyroidectomy occurs long before the normal onset of molt (Crew, 1927, in the fowl; Svetsarov and Streich, 1940, in the duck). The same occurs after chemical suppression of the thyroid gland (Glazener and Jull, 1946). (3) A number of investigators have produced evidence of cytofunctional activation of the thyroid before the normal onset of molt (Riddle, 1925, in the pigeon; Zawadowsky, 1926, in the fowl; Höhn, 1949, in the duck). Using the goitrogen prevention method, Reineke and Turner (1945a) demonstrated a maximum in secretion of thyroxine during the molting period in the hen. On the other hand, whereas

Tanabe *et al.* (1957) were unable to observe any increase in thyroid ^{131}I uptake in molting hens, Astier *et al.* (1970) have shown that, in the duck, a seasonal peak in thyroid ^{131}I uptake and in the ^{131}I conversion ratio (protein bound ^{131}I) occur within the month prior to the onset of the annual molt.

The thyroid gland seems thus definitely involved in molting by its stimulating effect on the development of young feathers, albeit that other endocrine glands, such as the gonads, seem to contribute to the complex endocrine mechanism of feather shedding and replacement, possibly by exerting a protecting effect on the old feather (Vaugien, 1955; Tanabe *et al.*, 1957). Whether one or another of these glands plays the major role in the onset of molt or eventually some balance between hormones (e.g., thyroxine and sex steroids, or TSH and gonadotropins), needs further clarification. However, it can be stated here, that experimentally forced molt that always occurs in ducks some weeks after pituitary autograft, i.e., at a time when the circulating androgens fall to near zero values while thyroid hormones remain much less affected, can be consistently prevented either by moderate (0.4 mg/day) testosterone treatment (Assenmacher and Baylé, 1968), or by feeding propylthiouracil (Baylé, 1972) (see Volume II, Chapter 3).

5. Reproduction

For several decades a number of converging investigations have demonstrated pronounced beneficial effects of thyroid hormones on gonadal function in domestic birds. However, Woitkewitsch's (1940) observations on inverse thyroid–gonad relationships in the European Starling, which were followed by very intensive investigations on several species of subtropical Indian finches by Thapliyal's group (see below), have led us to distinguish at least two main categories of birds in regard to their thyroid–gonad relationships: (1) birds that require thyroid hormones for a normal gonadal development and (2) birds that may achieve gonadal development without the thyroid glands.

The domestic species obviously belong to the first group. In male domestic birds, injections of moderate doses of thyroid hormones promote growth of the testes in young chickens (Wheeler *et al.*, 1948; Kumaran and Turner, 1949), in adult domestic fowls (Greenwood and Blyth, 1942), and in drakes (Jaap, 1933; Aron and Benoit, 1934), whereas thyroidectomy inhibits testicular development in young cockerels (Woitkewitch, 1940) and ducklings (Benoit and Aron, 1934) blocks the phototesticular response in adult ducks (Benoit, 1936) and causes gonadal regression in cockerels (Benoit and Aron, 1934; Blivaiss and Domm, 1942; Caridroit, 1943; Payne, 1944). Although their effects are less constant, goitrogens may also lead to depression of spermatogenesis, comb size, fertility in cocks (Shaffner and Andrews, 1948; Kumaran and Turner, 1949), and to inhibition of the phototesticular response in adult ducks (Assenmacher and Tixier-Vidal, 1962b).

Several wild species exhibit similar thyroid–gonad relationships. Thyroid hormones induce testicular growth "off-season" in House Sparrows (*Passer domesticus*) (Miller, 1935; Vaugien, 1954), while in the Baya Weaver (*Ploceus philippinus*) the gonadal cycle remains suppressed indefinitely after thyroidectomy (Thapliyal and Garg, 1969).

In female domestic fowl, thyroidectomy also induces delayed gonadal maturation (Blivaiss, 1947) and decreases egg production (Winchester, 1939; Blivaiss, 1947), and thiouracil treatment reduces ovarian weight in ducks (Berg and Bearse, 1951) and egg production and fertility in hens. However, thyroid extracts exert marked inhibition on the ovary of pigeons that are photostimulated or treated with pregnant mare serum (Clavert, 1953) and on light-stimulated Japanese White-eyes (*Zosterops japonica*) (Kobayashi, 1954b), but the inhibition of ovarian growth may result from a depression of the concentration of yolk precursors for the ovary (Clavert, 1953; van Tienhoven, 1961).

The assumption that in domestic species the thyroid exerts exclusively a promoting role on gonadal development was recently challenged when Lehman (1970) claimed that administration of a very low dosage of thyroxine to 2-week-old cockerels had no effect on testis weight and ^{32}P uptake, but lowered significantly the pituitary gonadotropic potency. On the other hand, administration of thyroxine, either during the vernal progressive phase or during autumn in ducks that were stimulated by artificial long days, had no marked effect on testis weight, but caused a severe decrease in plasma testosterone, associated with an augmented metabolic clearance rate of the hormone (Jallageas and Assenmacher, 1973). Since similar alterations of testosterone production and metabolism occur normally in early summer (Jallageas and Assenmacher, 1973), when the thyroid gland undergoes an annual peak of activity in ducks (see below, Astier *et al.*, 1970), the seasonal activation of the thyroid could appear at least as part of the complex mechanism whereby gonadal regression comes about in this domestic but seasonally cycling species.

The European Starling (Woitkewitch, 1940; Wieselthier and van

Tienhoven, 1972) and several species of subtropical Indian finches belong to the second type of thyroid–gonad relationships, in which gonadal growth takes place even in the absence of thyroid hormones. Moreover, in such birds thyroid hormones play an essentially inhibiting role on gonadal (gonadotropic?) function. Regarding the effects of thyroidectomy, these birds fall into two subgroups. In the male Chestnut Mannikin (*Lonchura malacca*) and the Red Avadavat (*Amandava amandava*) (Thapliyal and Pandha, 1967b) removal of the thyroid gland before the onset of gonadal development leads to a marked extension (over 5 months) of the active phase of the gonad, resulting in a delayed and shortened regression phase. In the second subgroup of birds, including the females of the Baya Weaver (Thapliyal and Bageshwar, 1970), the Chestnut Mannikin (Chandola, 1972), and both male and female Nutmeg Mannikin (Pandha and Thapliyal, 1964; Thapliyal and Pandha, 1965, 1967a; Thapliyal, 1969; Chandola, 1972), the gonadal cycle of thyroidectomized birds is fixed at a maximum level for an indefinite period (more than 3 years), while regression can be precipitated any time of the year by thyroxine administration (Thapliyal *et al.*, 1968). Since thyroidectomized starlings were not followed over year-long periods, they cannot be aligned with either of the subgroups.

Several patterns of interactions of the thyroid gland on the reproductive cycle thus seem to exist, and a deeper insight into these relationships obviously will depend on more accurate measurements of blood concentration of thyroid hormones throughout the reproductive cycle of the different species.

6. Migratory Behavior

The problem of the relation between thyroid activity and migration has been raised repeatedly; Häcker (1926), Küchler (1935), and Merkel (1938) showed that migrating birds have very active thyroid glands, while resting individuals have inactive glands. Merkel (1938) also was the first to simulate migratory behavior (*Zugunruhe*) with injections of thyroxine or TSH.

D. REGULATION OF THYROID FUNCTION

1. Hypothalamic–Hypohyseal Control[*]

a. Pituitary Control. Adenohypophysectomy in birds elicits marked atrophy of the thyroid gland (domestic fowl: Hill and Parkes,

[*]See also Chapters 2 and 4.

1934; Rosenberg *et al.*, 1963; Ma and Nalbandov, 1963; duck: Assenmacher, 1958a; Tixier-Vidal and Assenmacher, 1962; Assenmacher and Baylé, 1964; Rosenberg *et al.*, 1967; Japanese Quail: Baylé and Assenmacher, 1967b; Baylé, 1968; pigeon: Baylé, 1968). Several functional investigations have confirmed the considerable impairment of the thyroid function.

1. Hypophysectomized chickens on an iodine-supplemented diet ($+2$ μg/gm), 6 weeks postoperative, had approximately one-third as much thyroid ^{127}I as corresponding intact ones, but the same amount as either intact or hypophysectomized chickens on a low-iodine diet (0.14 μg/gm). On the other hand, in chickens on the low-iodine diet, hypophysectomy produced a 65% decrease of thyroid [^{127}I]thyroxine, with a twofold increase of [^{127}I]diiodotyrosine. No such effects were observed in iodine-supplemented hypophysectomized chickens, the controls of which having lowered thyroid iodine content, as was stated above (Rosenberg *et al.*, 1963). In hypophysectomized ducks, the thyroid content in ^{127}I was also found decreased, but the effect was less striking than in chickens (Rosenberg *et al.*, 1967).

2. Hypophysectomy markedly impairs uptake of ^{131}I by the thyroid gland. In growing chickens, this parameter of thyroid function, was reduced 6–8 weeks postoperative to 15% of controls, 19 hours after injection of ^{131}I (Ma and Nalbandov, 1963), and to 20–25% of controls, irrespective of the amount of iodine in the diet, 24 hours after the tracer injection (Rosenberg *et al.*, 1963). In 2–3-week hypophysectomized ducks, a number of investigations indicated an ^{131}I uptake of 3% of controls, at 1 hour, which did not rise above 10% of controls 72 hours after injection (Tixier-Vidal and Assenmacher, 1963; Rosenberg *et al.*, 1967; Baylé *et al.*, 1967a; Baylé, 1968). In Japanese Quail, 2–3 weeks after the operation, the corresponding values were 7% of controls at 12 hours and 22% at 72 hours (Baylé *et al.*, 1967a; Baylé, 1968). Under similar conditions, pigeons 12–25 weeks postoperative had an ^{131}I uptake of 15% of controls at 24 hours, and of 66% at 72 hours after the tracer injection (Baylé *et al.*, 1966, 1967a; Baylé, 1968). When measured *in vivo* in hypophysectomized pigeons, the thyroid radioactivity curve was also markedly depressed, the maximum of radioactivity being only one-third of the control maximum, and this peak was attained 19 hours after injection of ^{131}I in hypophysectomized birds compared with 9 hour controls (Fig. 6). It is interesting to note that the depression of the thyroid ^{131}I uptake after hypophysectomy is observable in the chick embryo (Maraud *et al.*, 1957).

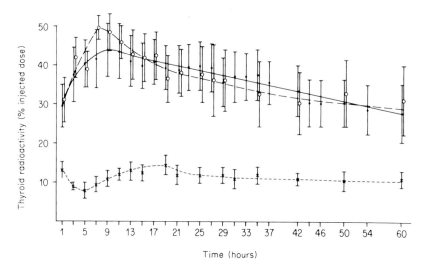

FIG. 6. *In vivo* measured radioactivity of thyroid glands in ^{131}I-injected pigeons, after hypophysectomy (×--×) or pituitary autografting (O--O) (percentage of injected dose) and in controls (●——●). (From Baylé and Assenmacher, 1968; Baylé, 1968.)

3. Radiochromatographic studies, on the other hand, have shown that in the hypophysectomized chicken, regardless of the level of iodine in the diet, 24 hours after injection of radioiodine, the extent of labeling of the thyroid pool of thyroxine amounted to one-tenth that of intact birds, while the percentage of diiodotyrosine was increased (Rosenberg *et al.*, 1963). Under similar conditions, the percentage of ^{131}I in thyroxine was 15–20% in ducks (Baylé *et al.*, 1967a; Baylé, 1968; Rosenberg *et al.*, 1967), 10% in pigeons, but 64% in the Japanese Quail (Baylé *et al.*, 1967a; Baylé, 1968).

4. At the peripheral level, the plasma protein-bound [127]I was lowered to 21% of controls in hypophysectomized chickens on low iodine intake, and to 35% of controls in iodine-supplemented birds (Rosenberg *et al.*, 1963). Furthermore the ^{131}I conversion ratio curve (protein-bound ^{131}I) also showed severe depression. In hypophysectomized ducks, the protein-bound ^{131}I was 20% of controls 24 hours after injections of ^{131}I, reaching 50% of controls at 72 hours (Tixier-Vidal and Assenmacher, 1962; Rosenberg *et al.*, 1967; Baylé *et al.*, 1967a). For hypophysectomized pigeons and Japanese Quail, the corresponding values were 15% of controls at 24 hours, and, respectively, 55% and 85% of controls at 72 hours (Baylé *et al.*, 1967a); Baylé, 1968).

5. Finally, measurements of the biological half-life of $[^{125}I]$L-thyroxine in hypophysectomized ducks have shown a marked depression of peripheral metabolism of the hormone; the half-life was 5.05 hours compared with 2.10 hours in controls (Astier and Baylé, 1970). The delayed appearance of protein-bound ^{131}I within the plasma may thus reflect the reduced output of thyroid hormone, as well as the reduction of peripheral metabolism.

All these findings, which point to the preeminently controlling role of the pituitary in thyroid function in birds, are reminiscent of similar conclusions for mammals. No significant difference can be detected between these classes of vertebrates in this respect.

b. Hypothalamic Control. Although the hypothalamus has an important role in the control of the thyrotropic function of the anterior pituitary, the intimate vascular relationship between them appears less important than in mammals. Thus, whereas ectopic pituitary autografting improves rather moderately thyroid function of hypophysectomized mammals, recovery is much more extensive in birds. (1) Although thyroid weights may still be reduced in autografted pigeons and Japanese Quail (Baylé, 1968), atrophic glands are found in neither ducks or chickens with sectioned hypophyseal portal veins (Benoit and Assenmacher, 1953; Shirley and Nalbandov, 1956), pituitary autografted ducks (Assenmacher, 1958a), nor chickens (Ma and Nalbandov, 1963). (2) The thyroid glands of ducks with autografts or severed portal vessels exhibit iodine accumulation, as indicated by high ^{127}I content (Rosenberg *et al.*, 1967). And despite the fact that thyroid-^{131}I uptake is scarcely affected in the autografted chicken (Ma and Nalbandov, 1963), in ducks with autografts or severed hypophyseal portal veins (Assenmacher and Tixier-Vidal, 1962a; Rosenberg *et al.*, 1967; Baylé, 1968), and in pigeons with pituitary autografts (Baylé *et al.*, 1967a; Baylé, 1968), a more detailed analysis of thyroid function in these three species has revealed a decreased rate of iodine metabolism. (3) The relative thyroid content in thyroxine-^{131}I is always significantly depressed 24 hours after ^{131}I injection. Subnormal values may also be observed after 48 hours. (4) Correlatively, the ^{131}I conversion ratio (protein-bound ^{131}I) was generally depressed during the first 24 hours after the tracer injection. (5) The half-life of $[^{125}I]$thyroxine in ducks with pituitary autografts also showed an appreciable restoration of peripheral metabolic rate of the hormones; the half-life was 3.13 hours, compared with 2.10 hours in controls and 5.05 hours in hypophysectomized birds (Astier and Baylé, 1970).

In summary, the basal pituitary thyrotropic function of birds has a fair degree of autonomy with respect to its hypothalamic control. However, the absence of the normal hypothalamohypophyseal connection does lead to reduction of metabolism in the thyroid gland, even when the animal is at the basal state. The role of the hypothalamopituitary axis in the regulation of thyrotropic function by environmental factors will be discussed in Section IV. But it may be mentioned here that ducks with disconnected hypophyseal veins fail to exhibit goitrogenic reaction to propylthiouracil treatment (Assenmacher, 1958a).

2. Rhythmicity of Thyroid Function

A number of morphological investigations on the avian thyroid glands have led to the assumption of the occurrence of seasonal variations of thyroid function. Several observers have claimed that histological stimulation of the thyroid gland coincides with the onset of the cold season, e.g., Riddle et al. (1932) in the pigeon; Burger (1938) in the European Starling; Höhn (1949, 1950) in the Mallard; Davis and Davis (1954) in the House Sparrow; Oakeson and Lilley (1960) in Zonotrichia leucophrys nuttalli; Wilson and Farner (1960), Farner (1965) in Zonotrichia leucophrys gambelii; Legendre and Rakotondrainy (1963) in the Red Fody (Foudia madagascariensis).

On the other hand, Bigalke (1956) and Fromme-Bouman (1962) have shown, in the European Blackbird, a morphological activation in thyroid in May, whereas the Nutmeg Mannikin exhibits an annual thyroid cycle with morphological indications of lower activity in August–September (Thapliyal, 1965). Now a systematic month-to-month study of the annual cycle of thyroid-^{131}I uptake and of plasma protein-bound [131]I 24 hours after injection of radioiodine has been carried out in male ducks reared under field conditions within the mild climatic conditions of the Mediterranean region. An annual peak of both parameters occurred in June, together with an increased relative thyroid content of [^{131}I]thyroxine 24 hours after injection of [131]I (Astier et al., 1970) and with a decreased thyroid [127]I content (Rosenberg et al., 1967), indicating a marked thyroid activation at that period. In a similar study on the Nutmeg Mannikin, Chandola (1972) noted a depressed thyroid ^{131}I uptake in August–September (breeding season) as compared with other months.

These observations indicate that the avian thyroid gland has seasonal functional fluctuations. However, they do not provide decisive explanations of the possible factors involved herein, such as the ambient temperature, the photoperiod, and the sexual cycle, all of which will be discussed in the following sections.

Finally, it must be stated that up to now no diurnal or circadian rhythms in thyroid function have been identified in birds. The only statement on short periodic functions is that of a 12 hour rhythmicity in thyroid ^{131}I uptake recently demonstrated in the duck (Dainat *et al.*, 1969).

3. Gonadothyroid Relationships

The annual cycle in thyroid activity has been related to the reproductive cycle. There is morphological evidence of depression of thyroid function during the reproduction period in *Zonotrichia leucophrys nuttalli* (Oakeson and Lilley, 1960), *Foudia madagascariensis* (Legendre and Rakotondrainy, 1963), *Lonchura punctulata* (Chandola and Thapliyal, 1968), and in most finches studied by Thapliyal (1969). In other species, such as the European Starling (Burger, 1938), the Mallard (Höhn, 1950), and the European Blackbird (Fromme-Bouman, 1962), the thyroid gland goes through a stage of stimulated activity as the gonads pass into their seasonal regression. The same pattern has been demonstrated with radioisotopic techniques in the duck (Astier *et al.*, 1970) and in *Lonchura punctulata* (Chandola, 1972). Such inverse phase relationships between the seasonal cycle of the thyroid gland and the photoperiodically induced sexual cycle could be relevant to gonadothyroid relationships and/or to photoperiodic regulation of thyroid function.

Although gonadothyroid relationships undoubtedly exist, at least in the male, the precise mechanism is still controversial. In the male, Kobayashi (1954a) observed stimulation of the thyroid parenchyma in testosterone-treated pigeons, and Stetson and Erickson (1971) found no effect on thyroid weight in castrated White-crowned Sparrows (*Zonotrichia leucophrys gambelii*) subjected to either long or short days. On the other hand, capons exhibit greater thyroid weight than intact chickens (Breneman, 1954), but Oddel (1952) found no modification of the thyroxine secretion rate in castrated chickens. Chandola (1972) measured an increased thyroid ^{131}I uptake in *Lonchura punctulata* receiving testosterone, whereas castration had the opposite effects. In the duck, castration has been found to increase the circulating TSH level (Benoit and Aron, 1931), together with the pituitary thyrotropic cells (Tixier-Vidal and Benoit, 1962), to increase thyroid weight, thyroid-^{131}I uptake, and protein-bound [131]I (Tixier-Vidal and Assenmacher, 1961a,b), while testosterone treatment had the opposite effects on both thyroid radioiodine tests, in both castrated and intact birds (Jallageas and Assenmacher, 1972). A similar depressive effect of testosterone was observed on pituitary thyrotropic cells of the Japanese Quail (Tixier-Vidal *et al.*, 1967). However, at the peripheral level, testosterone accelerates the metabolism of thyroxine, decreasing its biological half-life in the duck (Jallageas and Assenmacher, 1972).

The seasonal decrease in testosterone secretion could thus be at least partially responsible of the annual increment in thyroid activity in ducks. However, since, vice versa, hyperthyroidism also inhibits testosterone secretion (*vide supra*), it appears difficult, at the present time, to assess which of the endocrine alterations, hyperthyroidism or hypoandrogenism, may be considered as the primary inducer of this peculiar seasonal endocrine balance.

Less attention has been given to gonadothyroid relationships in females. Estrogens have been claimed to have no effect on thyroid weight in chickens (Breneman, 1942) but to depress the thyroid follicles in pigeons (Kobayashi, 1954a) and to lower the thyroid ^{131}I uptake in *Lonchura punctulata* (Chandola, 1972). Progesterone induces marked inhibition of thyroid-^{131}I uptake in hens (Tanabe *et al.*, 1957) and in Bengalese Finches (*Lonchura striata*) (Kobayashi and Tanabe, 1959).

4. Environmental Factors

a. Light and Day Length. Despite the occurrence of definite and probably inhibitory repercussions of photoperiodically stimulated gonadal activity on the thyroid gland of most species light has been claimed to act somehow as a stimulator of thyroid function. This statement is mainly based on histological observations of thyroid glands of European Starlings (Burger, 1938) and of ducks (Radnot and Orban, 1956; Hollwich and Tilgner, 1963) and of the hypophyseal thyrotropic cells of ducks (Tixier-Vidal and Assenmacher, 1965) and of Japanese Quail (Tixier-Vidal *et al.*, 1968). Light has also been shown to enhance thyroxine-secretion rate in the chicken (Kleinpeter and Mixner, 1947). However, radioisotopic investigations have led to more ubiquitous conclusions.

1. An experimental long-day regimen depressed markedly thyroid ^{131}I uptake in intact drakes (Tixier-Vidal and Assenmacher, 1959) and in male (Baylé and Assenmacher, 1967a) and female (Follett and Riley, 1967) Japanese Quail, in which photostimulated sexual steroids

could be interacting, as well as in castrated ducks (Jallageas and Assenmacher, 1972).

2. The same investigators also observed a mild depression of the conversion ratio (protein-bound ^{131}I) in intact or castrated ducks and in male Japanese Quail. However, Follett and Riley (1967) concluded from an increased slope of the protein-bound ^{131}I within the first hours after the injection of tracer into female quail that the hormonal output from the thyroid was stimulated somewhat by the light regimen.

3. Long days exert a slight inhibition on peripheral thyroxine catabolism evaluated by the half-life of thyroxine in castrated ducks, and inhibit the above-mentioned decrease in thyroxine half-life due to exogenous testosterone in intact controls (Jallageas and Assenmacher, 1972).

Although no decisive conclusions result from these observations, the radio-iodine tests bring more arguments in favor of an inhibition of thyroid activity than of stimulation by light, at least in birds with photoregulated gonadal cycles. Of special interest are, therefore, the recent investigations with similar methods on Nutmeg Mannikins, which are not dependent on photoperiod for the regulation of the reproductive cycle (Chandola, 1972). In this species long-day photoperiods, that did not lead to gonadal recrudescence, induced a significant stimulation of the thyroid ^{131}I uptake and of the plasma protein-bound ^{131}I, indicating a definite activation of thyroid metabolism.

In addition to the problem of the eventually stimulating or inhibiting action of the photoperiod on thyroid function and metabolism, another unanswered question concerns the pathway whereby light may act on the thyroid gland. It is tempting to admit *a priori* that a route similar to that of the photoneuroendocrine mechanism controlling the gonadal cycles of many avian species may also convey photostimuli to the thyroid gland. Alternatively some neural or metabolic factors linked to locomotor activity, which is definitely controlled by the light–dark cycle (see Boissin, 1973), could be involved in this mechanism. The maintenance of the depressive effect of long days on ^{131}I uptake and protein-bound ^{131}I in hypophyseal autografted quail could actually point to the latter hypothesis (Baylé and Assenmacher, 1967a).

b. Temperature. Several previously cited investigations on seasonal fluctuations of thyroid function have reported an annual maximum of thyroid activity during the cold season. In less systematic studies, comparing winter to summer trials in domestic fowl, it has been stated that thyroid glands were larger in winter than in summer (Cruickshank, 1929; Podhradsky, 1935; Galpin, 1938). The thyroid iodine content was equally depressed in summer (Cruickshank, 1929), while the thyroxine secretion rate declined about 15% from March to May (Turner, 1948), the total reduction from winter to summer reaching 58% (Stahl and Turner, 1961). Thyroid ^{131}I uptake in ducks also was found higher during the coldest months (January–February) than under the milder March temperatures (Tixier-Vidal and Assenmacher, 1959). Similarly, Kobayashi et al. (1960) noted in the White-throated Sparrow a higher ^{131}I uptake associated with increased thyroid [^{131}I]throxine levels in February than in May. On the other hand, the biological half-life of [^{131}I]thyroxine, as measured from the decrease of the total plasma radioactivity, was found to be lower in winter in domestic fowls (7.6±0.50 hours) than in summer birds (11.4+0.50 hours) (Hendrich and Turner, 1967).

The stimulating effect of low environmental temperature on thyroid function thus demonstrated has been studied experimentally. Cockerels kept for 3 weeks at 7.5°C had higher acinar-cell levels and higher thyroxine secretion rates (goiter-prevention method) than others kept at 23.5° or 31.5°C (Hoffmann and Shaffner, 1950). Similarly, pullets reared at 13°C had a greater thyroxine-secretion rate (radioiodine method) (3.47 μg/100 gm) than at 29.5°C (2.59 μg/100 gm) (Mueller and Anezcua, 1959), and ducks exposed to 7.2°C exhibited an enhanced rate of thyroxine secretion (thyroxine-degradation method) (2.05±0.17 μg/100 gm per day), compared with birds, at 25°C (1.57± 0.05 μg/100 gm per day) (Astier et al., 1972). Indirect evaluations of TSH secretion rate also indicated higher values in lower as compared to higher environmental temperatures (Hendrich and Turner, 1965). Acute cold exposure (4.4°C for summer temperature-adapted chickens) caused increased rates of uptake and release of thyroid ^{131}I after at least 24 hours. Both parameters reached a maximum after 72 hours and returned to normal in a longer term (1 month) of exposure to cold (Hendrich and Turner, 1963). The time required for thyroid stimulation by cold exposure thus appears to be much longer in birds than in mammals (Harris, 1959).

The effect of acute cold exposure (24 hours from 25°C to 4°C) has also been investigated in adult ducks by measuring the equilibration rate of radioactive [^{131}I]iodoamino acids (monoiodotyrosine, diiodotyrosine, thyroxine) with the respective [127]I components in thyroid extracts after ^{131}I injection (Astier, 1973b). Equilibrium was reached after 48 hours in the cold-exposed individuals, as well as in the controls. However, the slope of the specific radioactivity curve of the total iodine content of the thyroid gland, of mono- and diiodotyrosine and thyroxine was steeper in the cold-exposed animals, the corresponding maxima were higher, and the maxima were attained earlier (14 hours compared with 24 hours) in the cold-exposed ducks than in the controls. On the other hand, adult ducks exposed for 30 days to 5°C exhibited decreased thyroid [127]I]thyroxine (0.39±0.02 μg per milligram thyroid compared with 0.59±0.03), decreased protein-bound [127]I (1.71±0.07 μg per 100 ml plasma compared with 1.90± 0.06), and decreased half-life of ^{125}I]thyroxine (2 hours 42 minutes compared with 3 hours 23 minutes), compared to the controls kept at 23°C (Astier, 1973b). Similarly, McFarland et al. (1966) measured a decrease in half-life of the radiothyroxine (18.4 hours) in cooled (21°C) Japanese Quail, as compared to birds kept at warmer temperatures (32°C for 30.4 hours), while Hendrich and Turner (1967) claimed that reduction of a mean winter environmental temperature at 12.8°C or a warm summer temperature (32.3°C) to a constant 4.4°C had no significant effect on the half-life of ^{131}I]thyroxine in the domestic fowl.

The opposite effect of change in temperature, namely, from cold to warmer environments, has received less attention. However, the return of long-term cold-adapted chickens (2 years in 4.4°C) to a warm environment (12.8–23.9°C) resulted in a rapid reduction of thyroxine secretion rate (from 0.91 to 0.53 μg per 100 gm) (Hahn et al., 1966).

The effects of high environmental temperatures on thyroid function have also been studied in the domestic fowl. Chickens maintained at 31.5°C showed reduced thyroid weights, thyroid-^{131}I radioactivities, and thyroid-secretion rates, together with reduced oxygen consumption, than controls kept at 19°C (Huston, 1960; Huston and Carmon, 1962; Huston et al., 1962). Similarly, cockerels grown for 3 weeks at 24°C had a thyroid-secretion rate (goiter-prevention method) of 3.65 μg/bird/day, compared with 1.60 μg/day at 35°C, and 0.7 μg/day at 40.5°C, the last animals being in hyperthermia (Heninger et al., 1960).

c. Feeding. In contrast to mammals, in which emphasis has been placed on the depressive effects of malnutrition on thyroid function, little attention has been given to the thyroid repercussions of caloric malnutrition in birds. Underfeeding has been claimed to have no influence on relative thyroid weight in the chicken (Breneman, 1942; Mellen et al., 1954), although the metabolic rate is depressed (Mellen et al., 1954). On the other hand, starvation for 5 days was found to inhibit completely thyroid release of ^{131}I in the domestic fowl (Tanabe *et al.*, 1957), and food restriction inducing a 23% body weight loss also reduced the thyroid secretion rate to half normal values in the pigeon (Premachandra and Turner, 1960). However, in order to avoid interference with thyroid function by caloric malnutrition and iodide deprivation due to restricted food intake, more systematic investigations were performed in ducks previously adapted to an iodine-deficient diet. After 17 days of severe malnutrition (food intake reduced to 50% or 25% of controls or starvation) there was no evidence of marked impairment of ^{131}I uptake by the thyroid gland and of ^{131}I distribution among the intrathyroid amino acids (Assenmacher *et al.*, 1968). However more accurate methods of investigation such as the measurement of the "specific activity" [ratio of the labeled (^{131}I) to the stable ([127]I) compounds] of the intrathyroid amino acids have yielded evidence of a depressed metabolism of the thyroid gland in undernourished ducks. Starvation delayed the peak values of the specific activity of the iodotyrosines from 6 hours to 24 hours, and that of the iodothyronines from 8 hours to 48 hours (Astier, 1973b). In the plasma the ^{131}I conversion ratio (protein-bound ^{131}I) was markedly depressed (Assenmacher *et al.*, 1968) in proportion to the extent of inanition. There was also a decrease in the plasma [^{131}I]-thyroxine level 18 hours after ^{131}I injection (Astier *et al.*, 1969).

III.   Endocrine Pancreas

A.   MORPHOLOGY

The avian pancreas is an elongated gland that lies primarily between the two limbs of the duodenal loop. It consists of two main lobes, the dorsal and the ventral lobes, which may be independent or fused, and a third, smaller splenic lobe, that extends, from the dorsal lobe toward the spleen. There are usually three secretory ducts that lead to the ascending limb of the duodenal loop.

The pancreas originates from three primordial evaginations of the gut, one dorsal rudiment, and two ventral anlagen, the dorsal evagination usually appearing first. The distal portion of the three rudiments proliferate masses of glandular tissue that later fuse to form the definitive pancreas. Histologically, the primordium of the pancreas first appears as a compact mass of tissue that differentiates later into cellular cords that gradually acquire lumina. The acinic cells, as well as the endocrine islets, differentiate from the trabeculae, or from the walls of the tubules throughout the entire organ, beginning at an early stage (8 days in the chick).

1. Light Microscopy

In adults, the insular endocrine tissue generally is more abundant in the dorsal than in the ventral lobe; moreover, it is the main component of the splenic lobe. As in most vertebrates, the avian islet of Langerhans appears in two dimensions as a roughly circular aggregate of cells with a diameter ranging from 50 to 200 μ. The endocrine cells are accompanied by a thin layer of connective tissue that extends into the islet, dividing it into lobules. The majority of islets cells contain varied numbers of small granules that stain (e.g., with aldehyde-fuchsin and counterstain), and according to the cells, either blue (β, or B cells), or red to pink (α, α_2, or A cells), or sometimes orange to gray (α_1 or D cells). A few cells contain cytoplasm free of either granules (C cells) (Epple, 1968; Falkmer and Patent, 1970). The arrangement of the two main cell types exhibits in birds a very peculiar pattern, since two types can be distinguished: the dark islands, containing mainly α cells, and the light islands, containing predominantly β cells (Clara, 1924; Epple, 1968).

2. Electron Microscopy

Björkmann and Hellman (1964), Sato et al. (1966), and others have shown that α and β cells are clearly identifiable in all species. They are already differentiated by the day of hatching (Peter, 1970). In general, the β cells appear spherical to ellipsoidal and contain fewer aggregations of secretory granules than the α cells. An inverse relationship generally exists between number of secretory granules and number of other cytoplasmic organelles (unattached ribosomes, mitochondria, Golgi apparatus, and endoplasmic reticulum). In turn, the α-cell cytoplasm is characterized by a predominance of secretory granules, and a smaller proportion of other organelles. In both cell types, the granules are formed in the Golgi zone. β Granules are less electron dense, less homogeneous, and slightly larger than the α granules in all species studied. In the domestic fowl, most granules are present as bar-shaped crystalloids, whereas those of the turkey appear as numerous overlapping rods. Most β granules lie within a membrane-enclosed space. As in other vertebrates, the avian α granules appear spherical. The lining membrane also is more closely apposed to the granule surface than in β granules. Many β granules and a few α granules are surrounded by fine filaments. Dense multivacuolated cytoplasmic bodies 0.5–0.8 μ in diameter are present in both β and α cells.

B. HORMONES

As in other vertebrates, the avian pancreas produces at least two hormones — insulin and glucagon.

1. Insulin

Emphasis has been placed on the low yield of insulin extractable from the avian pancreas in comparison with mammalian glands (1–2 mg insulin per 100 gm of pancreas in chicken, as compared with 10–15 mg/100 gm in the steer) (Kimmel et al., 1968). Nevertheless, crystalline insulin has been recently isolated from chick (Mirsky et al., 1963; Smith, 1966; Falkmer and Wilson, 1967; Kimmel et al., 1968), and turkey (Weitzel et al., 1969) pancreas. Its amino acid sequence is close to that of pig and dog insulin, except for the amino acids in position 8, 9, and 10 in chain A, and 1, 2, and 27 in chain B (Smith, 1966). This difference in structure between chick and mammalian insulin may conceivably account, at least partially, for the much higher effectiveness of the avian hormone in promoting hypoglycemia in intact chickens, when compared with equivalent amounts of mammalian hormone (Fig. 7). In fact, 0.02 units/kg produced a de-

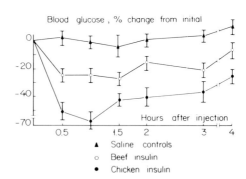

FIG. 7. Effect of intravenous injection of chicken and beef insulin (0.1 units) on the blood-glucose levels of the chick. Initial blood-glucose levels ranged from 196 to 230 mg/100 ml blood. (From Hazelwood and Barksdale, 1970, courtesy of Pergamon Press.)

gree of hypoglycemia equal to that observed in chickens injected with a beef-insulin dose that is tenfold greater (Hazelwood et al., 1968), and vice versa, mammalian (beef) insulin causes a stronger hypoglycemia in intact rats than does chicken insulin (Fig. 8) (Hazelwood et al.,

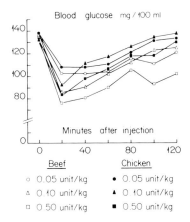

FIG. 8. Effects of intravenous injection of beef and chicken insulin on the blood-glucose levels of normal nonfasted rats. (From Hazelwood *et al.*, 1971, courtesy of Pergamon Press.)

1971). On the other hand, *in vitro* assay systems appear to respond equally well to all insulins (Hazelwood *et al.*, 1968). More recently, some light has been thrown on the possible cause of such interspecific resistance toward heterologous insulins (Hazelwood *et al.*, 1971). In this experiment, preincubation of beef insulin with chicken plasma led to severe inhibition of glucose uptake by the isolated rat diaphragm, while preincubation of chicken insulin with rat plasma similarly reduced glucose uptake potential *in vitro* (Figs. 9 and 10). These findings suggest that, in addition to differences in the amino acid sequence of insulins, and hence interspecific receptor-site differences, the occurrence of inhibitory plasma factor(s) may play a role in determining species specificity in the function of insulin. The

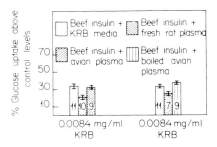

FIG. 9. Effects of avian and rat plasma incubated with beef insulin on glucose uptake by rat diaphragm. Insulin dose, 0.2 unit/ml KRB employed. (From Hazelwood *et al.*, 1971, courtesy of Pergamon Press.)

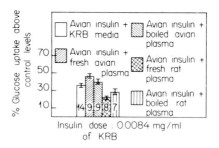

FIG. 10. Effects of avian insulin incubated with avian or rat plasma on glucose uptake by rat diaphragm. Insulin dose equivalent to 0.2 unit/ml KRB. (From Hazelwood *et al.*, 1971, courtesy of Pergamon Press.)

normal plasma-insulin level (I.R.I., immunoreactive insulin) is about 18 μ unit/ml in the duck (Mialhe, 1971) (Table IX).

TABLE IX

BLOOD SUGAR CONTROL BY INSULIN AND GLUCAGON IN THE DUCK[a]

Treatment	Glucose (mg/100ml plasma)	Glucagon (ng/ml plasma)	Insulin (μU/ml plasma)
Intact fasting	208±4	1.15±0.09	18±2
Pancreatectomized (3–22 hours)	96±5	0.20±0.03 (gut glucagon)	6±0.6
Eviscerated (4–7 hours)	235±4.5	1.14±0.13 (pancreas glucagon)	28±3
Glucose load 1.75 gm/kg	596±40	0.75±0.07	47±5

[a] From Mialhe (1971).

2. *Glucagon*

Glucagon arises, as in mammals, from two main sources, the pancreas and the gut (Samols *et al.*, 1969b). The hormone has been extracted from the pancreas of the pigeon (Al Gauhari, 1960), the domestic duck, two carnivorous species, the Barn Owl (*Tyto alba*), and the Common Kestrel (*Falco tinnunculus*) by Vuylsteke and de Duve (1953); almost ten times as much glucagon was found as in the pancreas of mammals. From recent biochemical studies on duck glucagon (Krug and Mialhe, 1970) it appears that two forms of glucagon are found in the pancreas, one with a molecular weight (3000) equivalent to crystalline mammalian glucagon, and the other with approximately double that molecular weight, which seems similar to that secreted in very small amounts by the dog pancreas. Intestinal (jejunum or ileum) extracts contain large amounts of the large-molecule glucagon, but none of the smaller type, which contributes to the total glucagon level in the blood (1 ng per ml plasma in the fasting duck, e.g., more than four times the level found in fasting humans, Mialhe, 1971). In serum, the only large glucagon (6000) has been observed in either normal, totally pancreatectomized, or eviscerated ducks (Krug et al., 1971). Despite the fact that glucagon from both origins share some biological properties, the role of gut glucagon seems to be of restricted importance, since its plasma concentration is rather low (Table IX).

C. ROLE OF PANCREATIC HORMONES[*]

1. Carbohydrate Metabolism

For years, the endocrine pancreas has been considered to be of limited importance in the regulation of carbohydrate metabolism in birds, since a number of investigators have claimed that pancreatectomy in the chicken, pigeon, duck, and goose induced either no diabetes at all, or at most, a mild diabetes lasting 1 week without any further alteration of carbohydrate metabolism (Minkowski, 1893; Kausch, 1896; Ivy et al., 1926; Mirsky et al., 1941). This classic assumption has been questioned by the more recent and extensive investigations of Mialhe (1958, 1969, 1971) in the duck, Sitbon (1967), and Karrman and Mialhe (1968) in the goose, and Mikami and Kazuyuki (1962) in the chicken. The data from these experiments leave no doubt concerning the preeminent role of the two pancreatic hormones in avian carbohydrate metabolism.

Whereas less than total pancreatectomy induces a transient (duck) (Mialhe, 1958, 1969, 1971) or severe (goose) (Karrman and Mialhe, 1968) diabetes (hyperglycemia, glucosuria, polydipsia, and polyphagia), total pancreatectomy leads to an opposite situation. In both species, the operation is followed by severe hypoglycemia (and convulsions), associated with a markedly impaired glucose tolerance (Fig. 11). Hypoglycemia persists if the birds are held in a fasting state. Food intake or glucose injections cause marked oscillations of blood sugar, from hypo- to hyperglycemia, due to decreased glucose

[*]See also Volume II, Chapter 8.

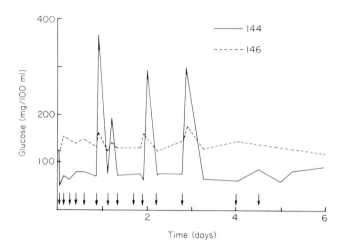

FIG. 11. Hypoglycemia and impaired glucose tolerance after total pancreatectomy in the duck. Intravenous glucose injections (1.75 gm/kg), indicated by arrows, given during the first two postoperative days when hypoglycemic convulsions occur or are imminent in the pancreatectomized fasting animal (no. 144). When measured 1 hour after the injections (4 peaks of the curve), glucose tolerance is impaired. Compare with the curve obtained for a normal duck (No. 146) receiving identical injections. (From Mialhe, 1969, courtesy of Excerpta Medica Foundation.)

tolerance. Survival may also be prolonged by administration of hyperglycemic hormones such as glucagon, adrenaline, or ACTH. In the fasting pancreatectomized duck, the blood glucose level can be maintained in normal range by repeated injections of glucagon and insulin in a w/w ratio of 1.5:3. Glucose tolerance is still impaired, but can be restored to normal by increasing the insulin dose.

Mikami and Kazuyuki (1962) have made observations on the domestic fowl that similarly emphasized the role of glucagon. They showed that extirpation of the two lobes of the pancreas that contain the α cells induced severe hypoglycemia, which could be restored by glucagon injections, whereas partial pancreatectomy involving other lobes of the gland elicited a mild and transient hyperglycemia.

The hypoglycemic role of insulin and the hyperglycemic effect of glucagon have now been amply verified not only in domestic birds, such as domestic fowl (Grande, 1968; Hazelwood et al., 1968; Langslow et al., 1970) and ducks and geese (Grande and Prigge, 1970), but also in finches (Goodridge, 1964). The glycogenic effect of insulin together with the glycogenolytic effect of glucagon have been demonstrated in pancreatectomized ducks and in intact fowls (Mialhe, 1958; Hazelwood *et al.*, 1968). It can, therefore, be concluded that:

1. In fasting pancreatectomized birds, hypoglycemia is due to a lack of glucagon, and decreased glucose tolerance is due to a lack of insulin (Fig. 12). The discrepancy between recent and earlier findings

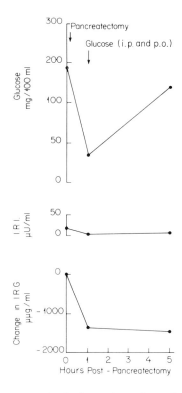

FIG. 12. Total pancreatectomy in duck causing a large decrease in plasma glucose, immunoreactive glucagon (I.R.G.) and immunoreactive insulin (I.R.I.). There was no increase in I.R.G. after alimentary glucose. (From Samols *et al.*, 1969b, courtesy of Excerpta Medica Foundation.)

on pancreatectomized birds presumably is due to incomplete extirpation of the gland by the earlier investigators. However, further investigation will be needed to throw light on the behavior of pancreatectomized carnivorous species, such as owls, which have been claimed to become definitively diabetic after pancreatectomy (Minkowski, 1893; Nelson *et al.*, 1942).

2. In the normal bird, as in mammals, glucagon seems to play an important role in maintaining a normal blood sugar level in the fasting state, whereas insulin prevents hyperglycemia after feeding.

3. The ratio of the plasma concentrations of glucagon and insulin (G/I) seems to be of major importance for the regulation of the blood-sugar level.

On the other hand, *in vivo* and *in vitro* studies have shown that the glycogen body, which is a small mass lying dorsal to the sacral spinal cord within the vertebral column and contains 60–80% glycogen of unknown role, fails to react to avian as well as to mammalian insulin (Hazelwood and Barksdale, 1970).

2. *Lipid Metabolism*

Recent investigators have emphasized the prominent lipolytic action of glucagon in birds. Glucagon has a marked stimulatory effect on the concentration of free fatty acids in the plasma *in vivo* in the domestic fowl (Heald *et al.*, 1965; Grande, 1968; Langslow *et al.*, 1970), the duck (Grande, 1968; Desbals *et al.*, 1970a,b) and the goose and the turkey (Grande, 1968), as well as on the release of free fatty acids *in vitro* from adipose tissue in *Passer domesticus* and *Zonotrichia leucophrys gambelii* (Goodridge, 1964), pigeons (Goodridge and Ball, 1965), domestic fowl (Langslow and Hales, 1969), and ducks (Desbals *et al.*, 1970b).

According to Grande and Prigge (1970), glucagon infusion (2 hours) in geese and ducks produced elevations of plasma free fatty acids and plasma triglycerides together with significant increase of the liver triglyceride content, while epinephrine or norepinephrine infused at the same rate failed to elicit any modification. In ducks, Desbals *et al.* (1970a,b) have shown that administration of glucagon to either intact fasting, to totally pancreatectomized ducks, or even to hypophysectomized birds, brings about an early and important increase of free fatty acids in the plasma. On the other hand, total pancreatectomy induces a rapid decline (25%) in lipemia, together with a decrease (50%) of free fatty acids in the plasma. According to Lepkovsky *et al.* (1967) pancreatectomized chickens would behave in a different manner, but here again, the maintenance of normal blood levels of free fatty acids raises the question of completeness of the operation; further experiments are needed before conclusions on species differences may be drawn. In fact, glucagon appears to be the major adipokinetic factor in birds. By contrast, catecholamines, which elevate plasma free fatty acids and hepatic triglycerides in mammals, have no such effects in the avian species hitherto studied.

On the other hand, insulin, when tested *in vitro* in passerine birds, has no detectable effects on labeled glucose uptake by abdominal or furcular fat pads, nor upon fatty acid or glyceride glycerol synthesis from labeled glucose by the adipose tissue (Goodridge, 1964). Similarly, fatty-acid synthesis in adipose tissue of the pigeon studied *in vitro* and *in vivo* was not appreciably influenced by insulin (Goodridge and Ball, 1966, 1967). Moreover, insulin was found to have no antagonistic (antilipolytic) effect on glucagon, either *in vitro* or *in vivo* in the duck (Desbals *et al.*, 1970a,b) and in the chick (Langslow and Hales, 1969), nor did it counteract epinephrine-induced lipolysis in pigeon adipose tissue (Goodridge and Ball, 1965). However, in isolated chick adipose tissue, insulin stimulates the incorporation of glucose-U-^{14}C into glyceride glycerol (O'Hea and Leveille, 1968).

In vivo insulin injections into intact fasting ducks, domestic fowl, (Heald *et al.*, 1965; Langslow *et al.*, 1970), and ducks (Desbals *et al.*, 1969, 1970) reduce lipemia but induce a delayed (30 minute) increase in free fatty acids in plasma. This might be attributed to an indirect release of glucagon, as the latter effect does not occur in pancreatectomized ducks (Desbals *et al.*, 1970).

D. REGULATION OF PANCREATIC HORMONE SECRETION

1. Glucose

As in mammals, blood-sugar level serves in birds as a major regulator of secretion of pancreatic hormones. Acute or chronic glucose injections, or intrajejunal infusions of glucose, induce a sharp increase in immunoreactive insulin in the goose (Mialhe, 1969), in the duck (Samols *et al.*, 1969b) and in the domestic fowl (Langslow *et al.*, 1970) (Fig. 13) (Table IX). In correlation, the concentration of immunoreactive glucagon in the plasma decreases following intravenous injection or intrajejunal infusion of glucose. The feedback mechanisms glucose–glucagon and glucose–insulin have been shown to apply to physiological variations of blood-glucose level (Sitbon and Mialhe, 1972).

2. Glucagon–Insulin Interactions

Measurements of immunoreactive insulin after injections of glucagon in ducks have provided direct evidence for a glucagon-induced insulin secretion (Fig. 14). This insulinogenic effect of glucagon is even higher than that of glucose infusions, inducing similar increases in blood glucose (Mialhe, 1969). On the other hand, Samols *et al.* (1969b) have demonstrated a significant increase in concentration of

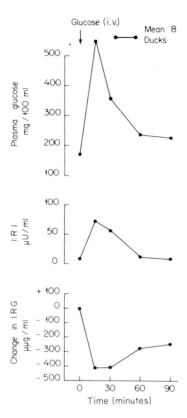

FIG. 13. Effect of intravenous glucose (1.75 gm/kg body weight) in ducks. (From Samols *et al.*, 1969b, courtesy of Excerpta Medica Foundation.)

immunoreactive glucagon after intravenous injection of insulin in the duck. However they could not observe a specific insulinogenic effect of intestinal glucagon in response to alimentary glucose.

3. *Free Fatty Acids*

It has been stated that enhanced levels of free fatty acids may be a potent stimulator of plasma glucagon in the duck (Samols *et al.*, 1969b), a fact that enlarges the concept of metabolic feedback regulation of glucagon secretion.

4. *Permissive Role of Other Endocrines*

As in mammals, pituitary or pituitary-controlled hormones have been shown in the duck to interfere with regulation by pancreatic hormones (Mialhe, 1958, 1969).

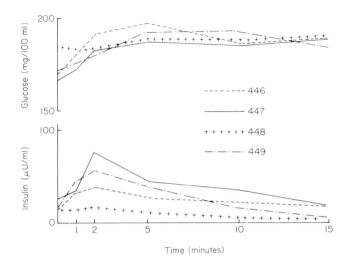

FIG. 14. Insulinogenic effect of glucagon in ducks. Intravenous injection of 1.25 μg/kg of glucagon increased the level of immunoreactive insulin in 3 out of 4 animals. With similar rises of serum glucose induced by intravenous glucose infusion, there is no change in serum level of immunoreactive insulin. (From Mialhe, 1969, courtesy of Excerpta Medica Foundation.)

1. Hypophysectomy improves the glucose tolerance of totally pancreatectomized birds, while it enhances the hypoglycemic response to total pancreatectomy.

2. Hypophysectomized ducks are hypersensitive to insulin, i.e., they show a supernormal glucose tolerance with an insulin-induced hypoglycemia following the hyperglycemic peak. This hypersensitivity can be corrected by administration of either growth hormone or corticosterone.

3. Hypophysectomy markedly impairs the hyperglycemic effect of glucagon. The normal response is restored if the hypophysectomized animals are treated with growth hormone or corticosterone.

4. The anterior pituitary, or at least growth hormone, and corticosterone, are also necessary for a normal insulinogenic effect of glucagon.

5. Gastrointestinal Hormones

In mammals, this peculiar group of hormones has a marked stimulatory effect on glucagon and insulin secretion (Unger *et al.*, 1967). Pertinent data also are available for the duck, in which pancreozymin injections induce a definite and prolonged increase in glucagon secretion, whereas insulin secretion shows only a very transient stimulation (Mialhe, 1970).

6. *Drugs*

The administration of hypoglycemic "antidiabetic" sulphonylureas (e.g., tolbutamide) to ducks produces (1) a rapid decrease in plasma glucagon level and (2) a sharp increase in plasma insulin. As either effect may occur before any hypoglycemic response, and as in some instances, the suppression of pancreatic glucagon is associated with only negligible change in insulin levels, the drugs seem to have at least partly direct and independent effects on both major islet components (Samols *et al.*, 1969a). On the other hand, diazoxide, a potent hyperglycemic sulfamide in mammals, also induces hyperglycemia associated with a transient lowering of the plasma insulin in the domestic fowl (Langslow *et al.*, 1970).

IV.   Parathyroid Glands

A.   MORPHOLOGY

The parathyroids arise from the dorsal wings of the third and fourth branchial pouches that thicken very early into a solid mass of cells, the primordium of a parathyroid gland. Two glands are thus found on each side close to the posterior pole of the thyroid. The number and location of the four parathyroids remain unaltered in most species, although partial fusion of two homolateral glands may occur in some of them (domestic fowl). In addition to this general pattern, chickens often possess functional accessory parathyroid tissue, which is embedded within the ultimobranchial glands. This peculiarity may account for the occasional failure of parathyroidectomy to induce tetany as reported by several authors for this species (Hurst and Newcomer, 1969).

Light microscope studies (Benoit, 1950) have revealed that the main difference between mammalian and avian parathyroid glands lies in the absence, in the latter, of oxyphilic cells between the chief cells. The chief cells are arranged in elongated or branching cords of polygonal to columnar cells, separated by a thin stroma of connective tissue, with numerous sinusoid capillaries. When active, the nucleus of the chief cells may be hyperchromatic and the cytoplasm very granular, and hence rather dark with few vacuoles. When inactive, the nucleus and cytoplasm may stain very lightly, and the poorly granulated cytoplasm contains numerous vacuoles. It is interesting to note that generally chief cells belonging to the same area of the gland exhibit the same functional characteristics, while different levels of activity may be observed within distant areas of the gland. Hyperplasia and hypertrophy of the parathyroids occur if the bird is deprived of dietary calcium or vitamin D, or of ultraviolet light. Pronounced hypertrophy of the gland also parallels maximum reproductive activity. On the other hand, fasting, aging, and hypophysectomy cause structural regression.

Electron microscopic study of the parathyroid gland of laying hens, i.e., of a presumably actively secreting gland, has recently been performed (Nevalainen, 1969). The ultrastructural organization of the parathyroid cells corresponding to the mammalian chief cells appear to be, in principle, very similar to that of other vertebrates hitherto studied. The parenchymal cells show a tortuous plasma membrane, and the highly organized cytoplasm contains well-developed organelles and rod-shaped mitochondria, and vesicles and cisternae of the smooth endoplasmic reticulum are dispersed throughout the cytoplasm. Aggregations of free ribosomes and a well-developed rough endoplasmic reticulum, which consists of parallel cisternae with attached ribosomes, are also present. The Golgi apparatus is prominent and consists of dilated concentric cisternae and vesicles of variable size. Numerous small prosecretory membrane-bound granules 500 Å in diameter are observed in the Golgi area as well as in the cytoplasm outside the Golgi complex. A few large granules, 1000–4000 Å in diameter, which correspond to the mature secretory granules described in other vertebrates, are also observed in the cytoplasm, but no fusion of prosecretory granules to form larger droplets was seen. Finally, there are also in the cytoplasm coated vesicles that sometimes appear to be fused with the plasma membrane. The hormone is believed to be synthesized in the granular ergastoplasm and transferred to the Golgi apparatus, where it is packed into small prosecretory granules, and finally discharged into the intercellular space. The formation of the coalesced larger storage granules does not seem to be necessary.

B. HORMONE

The parathyroid hormone, or parathormone, appears to be a straight-chain polypeptide. In mammals, parathormone has been claimed to contain 73 amino-acids residues, with a molecular weight of approximatively 8500, and a biological activity of 2500 U.S.P. units/mg (Potts and Auerbach, 1965), but little is known of the precise structure of the avian hormone.

C. ROLE

As in mammals, the vital importance of the parathyroid in birds is evidenced by the fact that the total removal of parathyroid tissue leads to death in about 24 hours (Doyon and Jouty, 1904; Benoit et al., 1941). Clinically, death is due to tetany, and biochemically, the main syndrome lies in hypocalcemia, with blood calcium concentrations reaching 5 mg% in about 20 hours instead of 10 mg% in the normal bird, and is associated with increased phosphoremia.

On the other hand, despite possible species differences in the biochemical structure of avian and mammalian parathormone, injections of mammalian parathormone have been shown to induce hypercalcemia in the duck (Benoit, 1950) and in the fowl (see Table X).

TABLE X

HYPERCALCEMIC ROLE OF PARATHYROID EXTRACTS (PE)[a]

Subject and treatment	Total calcium (mg%)	Ultrafiltered calcium (mg%)
Roosters	10.1 ± 0.2	6.0 ± 1
+ 500 IU PE	19.5 ± 3	10.5 ± 3
Nonlaying hens	13.4 ± 2	6.3 ± 1
+ 500 IU PE	19.5 ± 4	9.8 ± 1
Laying hens	29.8 ± 11	11.5 ± 2
+ 500 IU PE	47.7 ± 9	12.0 ± 2

[a]After Urist (1967).

Calcium homeostasis is the major function of the parathyroid gland. Calcium ions are involved in many functions, such as regulation of neuromuscular irritability, muscular contraction, modifying the permeability of capillary membranes to water, clotting of blood, calcification of bone and of eggshells, and as a cofactor of various enzymes. The parathyroid glands control the plasma concentration of calcium ions, which in turn is in equilibrium with the protein-bound blood calcium. Plasma calcium is adjusted principally by extraction of the calcium from the skeleton, control of the renal excretion of calcium phosphate, and increase of the rate of absorption of calcium by the gut. Emphasis has been placed on the synergistic action of parathormone and vitamin D, which also increases the rate of absorption of calcium from the gut, together with the rate of resorption of bone calcium. In fact, bone is the primary target organ of parathyroid hormone. Parathyroid hormone-induced hypercalcemia may depend on the resorption of metaphyseal and endosteal bone tissue, the skeleton acting as the most extensive calcium reservoir of the organism. Moreover, the parathormone has been found to induce differentiation of a significant number of active osteoclasts (Benoit and Clavert, 1947), which contribute to resorption of bone tissue.

Calcium homeostasis and parathyroid function exhibit marked activation in female birds during the reproduction cycle (Table X). Eggshell production requires an adequate calcium supply to the shell glands of the uterus. How the movement from the plasma through the shell gland is regulated is not yet clear, although there is some evidence that the shell gland acts as a calcium pump involving a protein-binding transfer mechanism similar to that of the gut. By this mechanism, calcium is removed from the blood during the main period of eggshell formation (16 hours in the fowl) at a mean rate of approximately 125 mg/hour for an average shell containing 2 gm of calcium. It seems unlikely that a rate of absorption of such magnitude from the digestive tract can be maintained throughout the full period of rapid shell calcification, and the evidence suggests that some degree of mobilization of calcium occurs for at least part of the time of shell formation. It has actually been shown that after intravenous injections of parathormone into laying hens, there were significant increases in the excretion of three main components of bone, namely, (1) calcium, derived from the hydroxyapatite crystals, (2) hydroxyproline from the fibrillar collagenous matrix, and (3) uronic acids from the chondroitin sulfates in the amorphous fraction of the matrix. Moreover, the ratios of calcium–hydroxyproline and calcium–uronic acids in the urine increase up to 1 hour. Thereafter, there was a reversal, and the urine became enriched with the organic end products of bone catabolism. After about 2 hours, a new reversal occurred, with signs of a return to the basal state (Candlish, 1970). Parathormone seems thus to cause active resorption of both the mineral and organic phases of bone, but not necessarily simultaneously. The mobilization of calcium from the skeleton of laying hens by parathormone is thus extremely rapid, and this mechanism is obviously capable of providing calcium for eggshell formation within a short time after release of the hormone (Candlish and Taylor, 1970).

On the other hand, there has evolved in female birds a complex, specific mechanism that facilitates provision of the enhanced calcium

requirement of the animal during the period of egg laying. It consists
in definite alterations in calcium metabolism elicited by a high output
of estrogen by the growing follicle. The female hormone induces a
drastic hypercalcemia. Experimentally, blood calcium levels as high
as 145 mg% have been obtained in the duck by estrogen injections
(Benoit, 1950). The total calcium concentration in the plasma of the
laying hen is thus regulated by estrogen as well as by parathormone,
but the effects of the two hormones are additive rather than mutually
potentiating. Parathormone maintains a suitable level of ionic calcium in the plasma, while estrogen increases total plasma calcium,
without increasing ionic calcium, by the formation of a calcium–
phosphoprotein complex (Urist et al., 1960). This seems due to the
large numbers of calcium-binding sites available in the estrogen-
dependent yolk proteins as compared with the common plasma proteins (25-fold greater than albumin) (Urist, 1967). The marked
estrogen-induced hypercalcemia appears thus as a reflection of an
enhanced transport of protein–calcium complex from the liver to
the ovary.

Simultaneously, a very specific modification occurs in the skeleton,
with extended deposition of a secondary system of highly calcified
spongy bone within the marrow cavities, the so-called medullary bone
(Pfeiffer and Gardner, 1938, in the pigeon; Landauer et al., 1939, in
the fowl; Benoit et al., 1941, in the duck). This medullary bone serves
as an advanced, highly labile reserve supply of calcium throughout
the laying period. The precise type of interaction of parathormone
and estrogen in the control of the metabolism of medullary bone is
controversial. Benoit (1950) claimed that in the duck parathyroidectomy inhibits estrogen-induced hypercalcemia, as well as the minerali-
zation of the spongy medullary bone, and brings estrogen-induced
hypercalcemia, once obtained, to a dramatic, and finally fatal, decline.
However, it has been shown later, in the fowl, that this does not
necessarily occur, provided that the birds are massively supplemented
with calcium or with vitamin D. Some light may again be thrown on
the question if one assumes that estrogen-induced synthesis of yolk
proteins requires an adequate concentration of serum calcium ions
and that there is an equilibrium between calcium and protein in solution in the plasma, the fine regulation of it being a main function of
the parathyroid hormone.

The nature of the control of the resorption of medullary bone, and
hence the calcium output from the reservoir, is also still unclear. According to Urist (1967), one striking fact is that parathyroid hormone,

which is so active in rapid resorption of metaphyseal and endosteal bone, seems to be without effect on resorption of estrogen-induced intramedullary bone. The fluctuations in rate of deposition or resorption of calcium salts from medullary bone could reflect mainly the corresponding fluctuations of the level of plasma estrogens. On the other hand, as stated above, to others (Candlish, 1970; Candlish and Taylor, 1970), the parathyroid plays a major role in the control of skeletal resorption during the egg-laying cycle.

D. REGULATION

The chief regulatory role of parathormone synthesis and release undoubtedly is exerted by the feedback action of the plasma calcium level. Thus, high calcium diets (3%) fed to chickens caused, together with a significant hypercalcemia, a marked reduction in the dry, defatted weight of the parathyroid glands, and a correlative decrease in the amino acid uptake by the glands, while low calcium diets (0.3%) led to opposite modifications (Mueller *et al.*, 1970).

Hypophysectomy has also been claimed to lead to moderate depression of the parathyroid glands (Benoit, 1950); hypophysectomy-induced atrophy of the glands is corrected by pituitary autografting (Assenmacher and Baylé, 1964). The nature of these relationships, which are probably indirect, is unknown.

V. Ultimobranchial Bodies

A. MORPHOLOGY

The paired ultimobranchial bodies arise from a caudal pharyngeal complex corresponding to the fifth and sixth branchial pouches. Both terminal branchial pouches are essentially evaginations of the posterior wall of the fourth visceral pouch, and in contrast with the first four pouches, they remain small and rudimentary. These primordia develop early into ductless glands in the vicinity of the parathyroid and the thyroid glands. Whereas in adult mammals the ultimobranchial bodies become completely embedded within the thyroid gland, forming therein the parafollicular C cells, both glands remain distinct in most avian species, except in the pigeon, in which a partial incorporation of ultimobranchial tissue within the typical thyroid gland has been claimed to occur (Matthews *et al.*, 1968). The typical position of the avian ultimobranchial glands (e.g., domestic fowl, turkey) lies bilaterally near the origin of the subclavian and common carotid arteries, and in line with the two parathyroid glands and the thyroid gland, which are located more anteriorly along the carotid artery. In the domestic fowl the ultimobranchial bodies are well delineated rounded organs 1–2 mm in diameter. In laying hens, the weight of a single gland is about 17 mg, as compared with 3 mg for a parathyroid gland and 70 mg for one thyroid gland (Copp and Parkes, 1968).

Under the light microscope, the most conspicuous components of the adult ultimobranchial bodies are cystic structures of varying shape and size. Between the cavities, there are cords and sheets of parenchymal cells, mainly composed of light cells in association with a second, scarcer cell type characterized by dense cytoplasm. Numerous blood vessels are present among the cell aggregates. With the electron microscope, the cells bordering the cavities appear columnar to cuboidal, with short microvilli at their luminal pole. Some cells contain large amounts of light granules that give them the aspect of a mucoid cell. More typical are the parenchymal cells of the aggregates that lie between the cavities. Here again, two cell types are present— the dominant light cells that are ovoid to polygonal in shape and contain a pale cytoplasm, and scattered dark cells that are more irregular in shape and have a dense cytoplasm, dilated ergastoplasmic cavities, and numerous and large mitochondria. The light cells contain numerous electron-dense granules which fall into two size groups of about 2000 Å in diameter for the smaller one and about 3000 Å for the larger, and which are lined by a smooth limiting membrane. The ultimobranchial bodies of the chick contain light cells in various states of functional activity. Active cells are characterized by a well-developed rough endoplasmic reticulum, a prominent Golgi complex, and round- to rod-shaped mitochondria throughout the cytoplasm (Stoeckel and Porte, 1967, 1969; Malmquist et al., 1968; Chan et al., 1969). The fine structure of the light cells strongly resembles that of the light, parafollicular C cells of the mammalian thyroid gland, the granules of which have been shown to have hypocalcemic activity (Cooper and Tashjian, 1966).

The embryonic development of the ultimobranchial bodies of the chick begins with the organization of the compact glandular primordium from the wall of the branchial pouches. From the eleventh day of incubation, ultrastructural indications of secretory activity become evident, with 1100–1600 Å granules appearing within the Golgi complex. The secretory activity increases thereafter with the development of large ergastoplasmic cysternae and increased numbers of granules with diameters of about 2000 Å. Others become as large as 3000 Å before hatching, and even 4500 Å in the newly hatched chick. From the thirteenth day on, some dark cells differentiate; they contain mainly the larger type of granules. After hatching, the principal changes are related (1) to the size of the granules, which decreases to that of the above-mentioned adult types; (2) to the relative number of dark cells, which increases; and (3) to the formation of follicular and cystic structures, which are characteristic of the adult gland and seem to represent degenerative formations (Stoeckel and Porte, 1969).

B. HORMONE

The distinctive anatomical site of the avian ultimobranchial bodies in relation to the thyroid gland has been used extensively in recent years to elucidate the precise site of origin of the hypocalcemic principle that had been previously extracted in mammals from the parathyroid gland (Copp, 1962) or from the thyroid gland (Hirsch and Munson, 1963). In fact, this hypocalcemic hormone, calcitonin, is the specific hormone of the ultimobranchial gland. Table XI clearly indicates the ultimobranchial origin of calcitonin in those species in which the glands are definitively separate from the thyroid.

TABLE XI

CALCITONIN CONTENT OF ULTIMOBRANCHIAL BODIES
OR THYROID

Species	Ultimobranchial bodies (MCR mU/mg)	Thyroid (MCR mU/mg)
Swine[a]	—	10.0
Pigeon[a]	3.5	3.4
Domestic fowl[a]	37.8	0.9
Domestic fowl[b]	30–120	—
Domestic fowl[c]	125–180	—
Turkey[b]	100	—

[a] After Matthews et al. (1968).
[b] After Copp and Parkes (1968).
[c] After Witterman et al. (1969).

Chemically, calcitonin appears to be a polypeptide. In mammals, the hormone is estimated to contain 32–35 amino-acid residues (32 in swine calcitonin), with a molecular weight of approximately 4000 (Potts et al., 1968). According to O'Dor et al. (1969), avian calcitonins have molecular weights very similar to the mammalian hormones, ranging between 4300 (domestic fowl) and 4600 (turkey), whereas mammalian calcitonins lie between 3600 (swine) and 4500 (bovine).

C. ROLE

Although very little information is available on the physiological significance of calcitonin in birds, its hypocalcemic effect has been clearly demonstrated. Injections of 40 mU/100 gm of calcitonin into 2-month-old cockerels caused a fall of 1.7–2.1 mg/100 ml in plasma calcium in 1 hour (Copp, 1968). Moreover, injections of extracts of 800 μg of ultimobranchial tissue in rats produced a fall in the serum calcium of 1.17±0.17 mEq/liter associated with a definite hypophosphatemia (−2.95±0.52 mg%) (Sørensen and Nielsen, 1969; Witterman et al., 1969). On the other hand, chronic high calcium diet in chickens, with moderate hypercalcemia (increasing from 10.39± 0.19 to 13.89±0.39 mg%) led to marked hypertrophy and hyperplasia of the dominant light cells of the ultimobranchial glands. These cells also showed extensive degranulation and ultrastructural modifications that were indicative of increased metabolic and synthesizing activity (Chan et al., 1969). Similarly, high calcium diets (3%) that induced hypercalcemia were associated with a significant increase in the dry, lipid-free weight of the chick ultimobranchial gland, and a marked stimulation of amino-acid uptake by the glandular tissue, whereas low calcium diets (0.3%) had the opposite effects (Mueller et al., 1970).

The high calcitonin content of the avian ultimobranchial glands (Table XI), together with their high activity (15,000 MCR units/gm extracts for chickens against 250 for hogs, 110 for rats, and 5 for man) (O'Dor et al., 1969), the hypertrophy of the glands in laying hens, and the reactivity of their main parenchymal cell type to chronic hypercalcemia, suggest that the ultimobranchial bodies may be involved in regulation of calcium metabolism. Very intense calcium metabolism appears, in fact, as a characteristic of avian species, as it has been shown that birds are able to remove calcium from the blood five times more rapidly than mammals (Simkiss, 1961).

However, calcitonin seems to lack any major role in the growth of the skeleton. After removal of the ultimobranchial glands in 1-day-old male chicks, Brown et al. (1969, 1970) were unable to detect any significant difference, at age 1–14 months, compared with sham-operated animals, neither in body weight nor in serum calcium and alkaline phosphatase. Only serum phosphorus was decreased. Furthermore, removal of the ultimobranchial bodies did not alter the amounts of calcium, magnesium, phosphorus, hydroxyproline, and hexosamine in bone, nor did it affect the X-ray density of the tibia or the thickness of cortical bone at 3 months of age. However, the decrease in serum calcium to hypocalcemic levels following parathormone-induced hypercalcemia was prevented. Despite the lack of involvement in skeletal growth and in the bone turnover of the growing chicken, calcitonin seems thus essential in the prevention of hypercalcemia, and may also have a significant role in serum phosphorus homeostasis.

On the other hand, removal of the ultimobranchial bodies from laying hens led to a slight difference in egg size, a trend toward reduced eggshell thickness, a significant decrease in plasma calcium, and trends toward lower plasma phosphorus and lower alkaline-phosphatase activity, which suggest lower rates of bone-calcium turnover in these birds (Speers *et al.*, 1970).

Further investigations are needed to understand the mechanism of action of calcitonin. In mammals (rats), the primary effect of calcitonin, like parathormone, appears to be on bone, where it inhibits bone resorption with correlative calcium and phosphorus release and increases the ratio of osteoblasts to osteoclasts.

D. REGULATION

As for parathormone, the main regulation of calcitonin secretion by the ultimobranchial glands depends on the serum-calcium level. This was nicely demonstrated by an experiment with geese in which Bates *et al.* (1969) measured the rate of calcitonin secretion from a perfused ultimobranchial–parathyroid complex. Perfusion of the complex by hypercalcemic blood caused a net increase in the rate of secretion of the hormone, whereas the removal of the ultimobranchial gland from the tissue perfused with hypercalcemic blood reduced the calcitonin output to zero.

The possible involvement of a neural regulatory mechanism has been recently raised by Hodges and Gould (1969) who have placed emphasis on the rich vagal innervation of the ultimobranchial gland of the domestic fowl. Electron-microscopic studies have demonstrated direct connections between vagal fibers and the densely granulated epithelioid cells of the gland. Moreover, electrical stimulation of the vagus was followed by a very significant decrease of total plasma calcium, an effect that was prevented by atropinization, but could also be initiated by injections of parasympathomimetic drugs.

VI. Thymus

The inclusion of the thymus among the endocrine gland may appear questionable to modern biologists in light of the recently demonstrated role of the thymus in immunobiology. However, the recent isolation of very potent fractions extracted from mammalian thymus, some of which behave like hormones, might lead the way to a reconciliation of the older and newer concepts of the thymus.

A. MORPHOLOGY

In birds, the thymus is an elongated lobulated gland that lies on each side along the course of the jugular vein and the vagus nerve, from the third cervical segment downward to the thoracic cavity. There are usually seven thymic lobes on each side.

The embryological origin of the thymus is similar to that of the parathyroid gland in that the third and the fourth branchial pouches are the principal contributors to it. The ventral portion of the third pouch fuses with the dorsal portion of the fourth pouch to form a mass of epithelial cells on each side, the thymus primordium. This primordium elongates then to form an epithelial cord, extending progressively, cranially and caudally, along the jugular vein. The elongated thymus is syncytial and forms a reticular framework as anastomoses of fine protoplasmic threads develop among the stellate epithelial cells. The epithelial cells produce lobules of tissue that progressively give the thymus its typical aspect of irregular broken chains of lobes extending on both sides from the heart to the thoracic region. The thymocytes, i.e., the thymic lymphocytes, appear only secondarily within the syncytial mass of epithelial tissue (Romanoff, 1960). The precise origin of the thymocytes still remains an open question. Despite conspicuous evidence pointing to a possible transformation of "undifferentiated" epithelial cells from the primordial thymus into lymphocytes (Ackerman and Knouff, 1964; Auerbach, 1964), there are also strong experimental arguments, from thymus grafts, parabionts, and radiation chimeras, that favor the alternate possibility that lymphoid precursor cells, or some type of lymphocytes, invade the thymus epithelial rudiment very early and behave there as typical thymocytes. From the second half of embryonic development on, the thymus lobules have acquired the distinct cortical (dense) and medullary (loose) zones that are characteristic of the adult thymus, the latter zone already bearing rudiments of degenerating Hassal's corpuscules.

As in other vertebrates, the avian thymus increases in size until sexual maturity and then undergoes a marked regression. However, while in mammals the postpuberal regression of the thymus is irreversible, Höhn (1956) has shown that the thymus of a number of birds recovers an enlarged size and a juvenile histological aspect for some weeks, following at least the first sexual cycle. As in mammals, several hormones influence the size and aspect of the bird thymus, although the observed effects may be different. Thus, thyroxine elicits thymic enlargement, while cortisone, corticosterone, and deoxycorticosterone cause atrophy; estrone and testosterone are without effect (Höhn, 1961).

B. ROLE

The postnatal avian thymus has a well-known role in the formation of lymphocytes, and in several immunological processes. This also applies to mammals. However, there is now a good deal of evidence that the avian thymus shares with the bursa of Fabricius the immunological function restricted to the thymus in mammals (Chang *et al.*, 1955; Warner and Szenberg, 1964b; Cain *et al.*, 1968). Thus, neonatal thymectomy in chickens has no consistent effects on immunoglobulin and antibody production, but it diminishes markedly the population of small lymphocytes, which are immunologically competent cells in the blood and in the spleen. By contrast, surgical removal of the bursa of Fabricius at hatching, or inhibition of its development *in ovo* (e.g., by intraallantoic injections of testosterone) is associated with considerable impairment of the capacity to produce immunoglobulin and antibody responses. These findings have suggested the existence of two immunological systems: one, a thymus-dependent system, composed of the thymus and the circulating small lymphocytes that act as recognition agents and produce cell-mediated immune reactions (cellular immunity), such as in delayed hypersensitivity and transplantation immune reactions (coping with homografts); the other, a bursa-dependent system, composed of the bursa of Fabricius, that is responsible for the development of different classes of lymphoid cells that produce immunoglobulins and humoral antibodies (humoral immunity). Both organs direct the differentiation of lymphoid precursor or stem cells into immunologically competent cells capable of reacting to antigens in an appropriate manner within the secondary lymphoid organs (lymph nodules, spleen, etc.).

For many years, oil extracts of the thymus had been claimed to produce lymphocytosis in pigeons (Bomskov and Sladović, 1940). More recently, several mammalian thymic protein fractions with similar actions have been purified; included are the lymphocytosis-stimulating hormone (Metcalf, 1958; Hand *et al.*, 1967), the competence-inducing hormone (Miller, 1965) and thymosin (White *et al.*, 1969), a highly purified thymic protein, which acts, when injected, like the two former substances.

A number of other thymic "hormones" have been recently postulated, e.g., an erythrocyte-inhibiting factor (Field *et al.*, 1968); an insulin-like factor (Pansky *et al.*, 1965); calcitonin (Galante *et al.*, 1968), which could be secreted by cells of ultimobranchial origin enclosed within the thymus; thymin (Goldstein, 1968), a myolytic substance blocking neuromuscular transmission; promine (Szent-Gyorgyi *et al.*, 1962), a growth-stimulating factor; and a sterilizing factor (Szent-Gyorgyi *et al.*, 1962), which inhibits gametogenesis.

Summarizing, it may be concluded that the main roles of the avian thymus (i.e., lymphocytopoiesis and immunological processes directed against homografts) could conceivably be associated with the production of one or more thymic hormones, like the mammalian thymosin. Further investigation will be needed to ascertain a specific role of the avian thymus in growth promotion and gametogenesis, in accordance with the older observations of Parhon and Cahane (1937, 1939).

VII. Bursa of Fabricius

The bursa of Fabricius is a peculiar avian paracloacal derivative, which has a lymphoepithelial structure closely related to that of the thymus. As stated above, it has been definitely established that the bursa confers immunological competence to birds (Chang *et al.*, 1955; Warner and Szenberg, 1964a). More precisely, the bursa is responsible for the maturation of the lymphoid cells, which produce humoral rather than cellular antibodies (Szenberg and Warner, 1962; Cooper *et al.*, 1966). There is some evidence to suggest that this prominent effect of the bursa on the development of the bursa-dependent immunological system is mediated through a humoral agent secreted by the bursa (Glick, 1960b; St. Pierre and Ackerman, 1965; Jankovic *et al.*, 1967; May *et al.*, 1967).

Extensive research has stressed several relationships between the bursa and other endocrines. It is thus generally admitted that a negative correlation exists between growth of the bursa and development of gonads (Jolly, 1913; Riddle, 1928; Glick, 1955, 1960a) and of the adrenals (Glick, 1960a). The bursa also has been shown to regress in the presence of exogenous androgens in the Ring-necked Pheasant (Kirkpatrick and Andrews, 1944) and in the chick (Glick, 1957). Similarly, administration of adrenal-cortical hormone, as well as stress, result in involution of the bursa of the domestic fowl (Selye, 1943; Garren and Shaffner, 1956; Glick, 1959), and the pigeon (Bhattacharyya and Ghosh, 1970). On the other hand, Riddle and Tange (1928) were unable to detect any change in rate of body growth or onset of sexual maturity in bursectomized pigeons. This was confirmed in the chick by Woodward (1931). However, the data of Taibel (1941) and Glick (1955) reveal moderately larger gonads in bursectomized fowl, and bursectomized chicks aged 3–13 weeks had higher body weights than control birds (Freeman, 1969a). Neonatal bursectomy also was associated in 5-week-old chickens with enlarged thyroidal follicles and depressed thyroid-^{131}I uptake (Pintea and Pethes, 1967) and with a 50% lowered thyroxine-secretion rate (Pethes and Fodor, 1970).

A peculiar relationship between the bursa and the adrenal gland has been suggested by some recent investigations. For years it has been found that the ascorbic-acid depletion response of the adrenal following ACTH treatment, which is classic in mammals, did not occur in intact immature fowls (Jailer and Boas, 1950; Elton and Zarrow, 1955; Newcomer, 1959). More recently, Freeman (1969a, 1970c) showed that in immature chickens submitted to the stress of handling, adrenal ascorbic-acid depletion occurred very early (10 minutes after the stimulus), whereas ascorbic-acid repletion was completed 1 hour after the initial stimulus despite continued stimulation (handling) at 5 minute intervals. However, ACTH injection superimposed on the handling-stress induced no ascorbic-acid depletion. Indeed, significant increase in ascorbic-acid content of the gland was found; this was interpreted as a result of larger ascorbic-acid uptake than utilization by the gland. On the other hand, the ascorbic-acid depletion response to ACTH was demonstrated, both in adult birds in which the bursa had naturally regressed, or in immature bursectomized chicken (Perek and Eckstein, 1959; Perek and Eilat, 1960; Freeman, 1970c), and pigeons (Bhattacharyya et al., 1967; Ghosh and Bhattacharyya, 1967; Bhattacharyya and Ghosh, 1970). The hyperglycemic response to exogenous ACTH also appears depressed, or at least delayed, in bursectomized chicks, as compared to intact controls (Freeman, 1969a). The mechanism of the interrelationships between the bursa and the adrenal glands is still speculative. Freeman (1969a) raised the hypothesis of a loss of adrenal reactivity to ACTH in bursectomized chicks, which could result from the deprivation of some bursal hormone.

Recently, three fractions have been identified chromatographically from saline bursal homogenates of chicks. One of them is assumed to be a steroid derivative (Kemény et al., 1968).

REFERENCES

Ackerman, G. A., and Knouff, R. A. (1964). Lymphocyte formation in the thymus of the embryonic chick. Anat. Rec. 148, 253–254.

Akers, T. K., and Peiss, C. N. (1963). Comparative study of effect of epinephrine and norepinephrine on cardiovascular system of turtle, alligator, chicken and opossum. Proc. Soc. Exp. Biol. Med. 112, 396–399.

Akiyoshi, Y. (1932). Experimentelle Forschung der Veränderung des Gefieders an Tauben bei Darreichung von Schilddrüsenhormon und ihrer verwandten Verbindungen. Nagasaki Igakkai Zasshi 10, 491–498.

Al-Gauhari, A. E. I. (1960). Die Widerstandsfähigkeit der Tauben gegen Insulin im Zusammenhang mit den α-Zellen der Langerhansschen Inseln. Z. Vergl. Physiol. 44, 41–59.

Allen, D. J., and Marley, E. (1967). Effect of sympathomimetic and allied amines on temperature and oxygen consumption in chickens. Brit. J. Pharmacol. 31, 290–312.

Aron, M., and Benoit, J. (1934). Sur le conditionnement hormonique du développement testiculaire chez les Oiseaux: Rôle de la thyroïde. C. R. Soc. Biol. 116, 218–220.

Assenmacher, I. (1958a). Recherches sur le contrôle hypothalamique de la fonction gonadotrope préhypophysaire chez le Canard. Arch. Anat. Microsc. Morphol. Exp. 47, 447–452.

Assenmacher, I. (1958b). La mue des Oiseaux et son déterminisme endocrinien. Alauda 26, 241–289.

Assenmacher, I., and Baylé, J. D. (1964). Répercussions endocriniennes de la greffe hypophysaire ectopique chez le Canard mâle. C. R. Acad. Sci. 259, 3848–3850.

Assenmacher, I., and Baylé, J. D. (1968). Balance endocrinienne et mue du plumage chez le Canard Pékin mâle adulte. Arch. Anat., Histol. Embryol. 51, 67–73.

Assenmacher, I., and Boissin, J. (1972). Circadian endocrine and related rhythms in birds. Gen. Comp. Endocrinol. Suppl. 3, 489–498.

Assenmacher, I., and Tixier-Vidal, A. (1962a). A comparison between the effects of sectioning the pituitary portal vessels with those of hypophysectomy on thyroid function in the male duck. Gen. Comp. Endocrinol. 2, Abstr. 38.

Assenmacher, I., and Tixier-Vidal, A. (1962b). Le réflexe photosexuel après thyroïdectomie chimique chez le Canard mâle. C. R. Soc. Biol. 156, 18–21.

Assenmacher, I., Astier, H., and Jougla, N. (1968). Répercussions thyroïdiennes de la sous-alimentation chez le Canard domestique. J. Physiol. (Paris) 60, 342–343.

Astier, H. (1973a). Présence d'une fraction importante de protéines iodées "non hormonales" dans le P.B.I. des Oiseaux. C. R. Acad. Sci. 276, 793–796.

Astier, H. (1973b). Etude comparée de la fonction thyroïdienne chez le Canard et chez le Rat. D.Sc. Thesis. University of Montpellier, France.

Astier, H., and Baylé, J. D. (1970). Epuration plasmatique de la ^{125}I-L-Thyroxine après hypophysectomie et autogreffe hypophysaire chez le Canard. *J. Physiol. (Paris)* **62**, 237.

Astier, H., Jougla, N., and Assenmacher, I. (1969). Effets de la sous-alimentation sur le taux plasmatique et sur la demi-vie de la thyroxine chez le Canard. *C. R. Soc. Biol.* **163**, 1886.

Astier, H., Jougla, N., and Assenmacher, I. (1972). Variation du "taux de sécrétion" (T.S.) de la thyroxine chez le Canard mâle soumis à des modifications du niveau alimentaire et de la température ambiante. *C. R. Acad. Sci.* **275**, 2531–2534.

Astier, H., Halberg, F., and Assenmacher, I. (1970). Rythmes circanniens de l'activité thyroïdienne chez le Canard Pékin. *J. Physiol. (Paris)* **62**, 219–230.

Astier, H., Rosenberg, L. L., Lissitzky, S., Simon, C., Assenmacher, I., and Tixier-Vidal, A. (1966). Recherches sur la séparation chromatographique des iodothyronines intrathyroïdiennes chez le Canard Pékin. *Ann. Endocrinol.* **27**, 571–574.

Astwood, E. B., Bissell, A., and Hughes, A. M. (1944). Inhibition of the endocrine function of the chick thyroid. *Fed. Proc., Fed. Amer. Soc. Exp. Biol.* **3**, 2.

Auerbach, R. (1964). Experimental analysis of mouse thymus and spleen morphogenesis. *In* "The Thymus in Immunobiology" (R. A. Good and A. E. Gabrielsen, eds.), pp. 95–113. Harper (Hoeber), New York.

Bartels, E. A. (1944). Morphogenetische Wirkungen des Schilddrüsenhormons auf das Integument von Vogelembryonen. *Wilhelm Roux' Arch. Entwicklungsmech. Organismen* **142**, 763–816.

Bates, R. F. L., Bruce, J., and Care, A. D. (1969). Measurement of calcitonin secretion rate in the goose. *J. Endocrinol.* **45**, xiv–xv.

Bates, R. W., Miller, R. A., and Garrisson, M. M. (1962). Evidence in hypophysectomized pigeon of a synergism among prolactin, growth hormone, thyroxine and prednisone upon weight of the body, digestive tract, kidney and fat stores. *Endocrinology* **71**, 345–360.

Baum, G. J., and Meyer, R. K. (1956). The influence of diethylstilbestrol on lipid in intact and hypophysectomized cockerels. *Endocrinology* **58**, 338–346.

Baum, G. J., and Meyer, R. K. (1960). Effect of adrenal steroids and diethylstilbestrol on growth and fat content of cockerels. *Amer. J. Physiol.* **198**, 1263–1266.

Baylé, J. D. (1968). Importance relative des différents niveaux de commande des régulations neuroendocriniennes chez les Oiseaux. D.Sc. Thesis, University of Montpellier (France) No. C.N.R.S. A 02191.

Baylé, J. D., and Assenmacher, I. (1967a). Le contrôle hypothalamo-hypophysaire de la fonction thyroïdienne chez les Oiseaux. *Gen. Comp. Endocrinol.* **9**, 433.

Baylé, J. D., and Assenmacher, I. (1967b). Contrôle hypothalamo-hypophysaire du fonctionnement thyroïdien chez la Caille. *C. R. Acad. Sci.* **264**, 125–128.

Baylé, J. D., Astier, H., and Assenmacher, I. (1966). Activité thyroïdienne du Pigeon après hypophysectomie ou autogreffe hypophysaire. *J. Physiol. (Paris)* **58**, 459.

Baylé, J. D., Astier, H., and Assenmacher, I. (1967a). Métabolisme thyroïdien du radioiode chez l'oiseau après hypophysectomie ou déconnexion hypothalamo-hypophysaire. *J. Physiol. (Paris)* **59**, 210.

Baylé, J. D., Boissin, J., and Assenmacher, I. (1967b). Le taux plasmatique de la corticostérone libre chez l'Oiseau après hypophysectomie ou déconnexion hypothalamo-hypophysaire. *C. R. Acad. Sci., Ser. D* **265**, 1524–1526.

Baylé, J. D., Boissin, J., Daniel, J. Y., and Assenmacher, I. (1971). Hypothalamic-hypophysial control of adrenal cortical function in birds. *Neuroendocrinology* **7**, 308–321.

Benoit, J. (1936). Rôle de la thyroïde dans la gonadostimulation par la lumière artificielle chez le Canard domestique. *C. R. Soc. Biol.* **123**, 243.

Benoit, J. (1950). Les glandes endocrines. *In* "Traité de Zoologie" (P.-P. Grassé, ed.), Vol. 15, pp. 290–334. Masson, Paris.

Benoit, J., and Aron, M. (1931). Influence de la castration sur le taux d'hormone pré-hypophysaire excito-sécrétrice de la thyroïde présent dans le milieu intérieur chez le coq et le canard. Notion d'un cycle saisonnier de l'activité pré-hypophysaire chez les Oiseaux. *C. R. Soc. Biol.* **108**, 786.

Benoit, J., and Aron, M. (1934). Sur le conditionnement hormonique du développement testiculaire chez les Oiseaux. Rôle de la thyroïde. *C. R. Soc. Biol.* **116**, 215.

Benoit, J., and Assenmacher, I. (1953). Rapport entre la stimulation sexuelle pré-hypophysaire et la neurosécrétion chez l'Oiseau. *Arch. Anat. Microsc. Morphol. Exp.* **42**, 334–386.

Benoit, J., and Clavert, J. (1947). Anatomie, histologie et histophysiologie des glandes parathyroïdes du Canard domestique. *Acta Anat.* **4**, 49–53.

Benoit, J., Messerschmitt, J., and Grangaud, R. (1941). Action ostéogénétique du ben-zoate d'oestradiol chez le Canard domestique. Etude morphologique. *C. Rend. Soc. Biol.* **135**, 1593.

Benoit, J., Stricker, P., and Fabiani, G. (1941). Technique et résultats de la parathy-roïdectomie chez le Canard domestique. *C. R. Soc. Biol.* **135**, 1600.

Berg, L. R., and Bearse, G. E. (1951). Effect of iodinated casein and thiouracil on the performance of laying hens. *Poultry Sci.* **30**, 21.

Bhattacharyya, T. K. (1968). Studies on cortico-gonadal relationship in the pigeon. *Zool. Anz.* **180**, 154–162.

Bhattacharyya, T. K., and Ghosh, A. (1965). Seasonal histophysiologic study of the inter-renal of the House Sparrow. *Acta Biol. Hung.* **16**, 69–77.

Bhattacharyya, T. K., and Ghosh, A. (1970a). Histomorphic changes following chronic adrenocortical activation and inhibition in the pigeon. *J. Morph.* **130**, 257–270.

Bhattacharyya, T. K., and Ghosh, A. (1970b). Influence of surgical and steroidal bur-sectomy on the behavior of adrenal ascorbic acid during stress in juvenile pigeons. *Gen. Comp. Endocrinol.* **15**, 420–424.

Bhattacharyya, T. K., Sinha, D., and Ghosh, A. (1972). A comparative histological sur-vey of the avian adrenocortical homologue. *Arch. Histol. Jap.* **34**, 419–432.

Bhattacharyya, T. K., Sarkar, A., Ghosh, A., and Ganguli, A. (1967). A comparative study on avian adrenocortical response to exogenous and endogenous corticotrophin. *J. Exp. Zool.* **165**, 301–308.

Biellier, H. V., and Turner, C. W. (1950). The thyroxine secretion rate of growing white Pekin ducks. *Poultry Sci.* **29**, 248–257.

Biellier, H. V., and Turner, C. W. (1955). The thyroxine secretion rate of growing turkey poults. *Poultry Sci.* **34**, 1158–1162.

Bigalke, R. (1956). Über die zyklischen Veränderungen der Schilddrüse und des Kör-pergewichtes bei einigen Singvögeln im Jahresablauf. Inaugural Dissertation, University of Frankfurt am Main.

Björkmann, N., and Hellman, B. (1964). Ultrastructure of the islets of Langerhans in the duck. *Acta Anat.* **56**, 348–367.

Blanquet, P., Stoll, R., and Capol, L. (1952). Sur l'activité des thyroïdes de l'embryon de poulet étudiée dans l'oeuf "marqué" par l'administration de radio-iode à la poule. *C. R. Soc. Biol.* **146**, 1103.

Blivaiss, B. B. (1947). Development of secondary sexual characters in the thyroidecto-mized Brown Leghorn hen. *J. Exp. Zool.* **104**, 267.

Blivaiss, B. B., and Domm, L. V. (1942). Relation of thyroid gland to plumage pattern and gonad function in the Brown Leghorn male. *Anat. Rec.* **84,** 529.

Boas, N. F. (1958). The effects of desoxycorticosterone acetate on testes size and function in the cockerels. *Endocrinology* **63,** 323.

Boissin, J. (1967). Le contrôle hypothalamo-hypophysaire de la fonction cortico-surrénalienne chez le Canard. *J. Physiol. (Paris)* **59,** 423–444.

Boissin, J. (1973). Photorégulation des rythmes circadiens de la fonction corticosur-rénalienne et de l'activité générale chez la Caille. D.Sc. Thesis. University of Montpellier, France, No. A.O. 7972.

Boissin, J., and Assenmacher, I. (1968). Rythmes circadiens des taux sanguin et sur-rénalien de la corticostérone chez la Caille. *C. R. Acad. Sci.* **267,** 2193.

Boissin, J., and Assenmacher, I. (1970). Circadian rhythms in adrenal cortical activity in the quail. *J. Interdisc. Cycle Res.* **1,** 251–265.

Boissin, J., and Assenmacher, I. (1971a). Implication des mécanismes aminergiques centraux dans le déterminisme du rythme circadien de la corticostéronémie. *C. R. Acad. Sci.* **273,** 1744–1747.

Boissin, J., and Assenmacher, I. (1971b). Entrainment of the adrenal cortical rhythm and of the locomotor activity rhythm by anhemeral photoperiod in the quail. *J. Inter-disciplinary Cycle Res. (Amsterdam)* **2,** 437–443.

Boissin, J., Baylé, J. D., and Assenmacher, I. (1966). Le fonctionnement corticosur-rénalien du Canard mâle après pré-hypophysectomie ou autogreffe hypophysaire ectopique. *C. R. Acad. Sci.* **263,** 1127–1129.

Boissin, J., Nouguier-Soulé, J., and Assenmacher, I. (1969). Circannual and circadian rhythms of adrenal cortical functions in birds. *Indian J. Zootomy* **10,** 187–196.

Bomskov, C., and Sladović, L. (1940). Das Thymushormon und seine biologische Aus-wertung. *Pfluegers Arch. Gesamte Physiol. Menschen Tiere* **243,** 611–622.

Boone, M. A., Davidson, J. A., and Reineke, E. P. (1950). Thyroid studies in fast and slow feathering Rhode Island Red chicks. *Poultry Sci.* **29,** 195–203.

Bouillé, C., and Baylé, J. D. (1973). Experimental studies on the adrenocorticotropic area in the pigeon hypothalamus. *Neuroendocrinology* **11,** 73–91.

Bouillé, C., Boissin, J., Daniel, J. Y., and Assenmacher, I. (1969). Fluorometric assay of corticosterone after purification by bidimensional thin layer chromatography in birds. *Steroids* **14,** 7–20.

Breneman, W. R. (1941). Growth of the endocrine glands and viscera in the chick. *Endocrinology* **28,** 946–954.

Breneman, W. R. (1942). Action of prolactin and estrone on weights of reproductive organs and viscera of the cockerel. *Endocrinology* **30,** 609–615.

Breneman, W. R. (1954). The growth of thyroids and adrenals in the chick. *Endo-crinology* **55,** 54–64.

Brown, D. M., Perey, D. Y. E., Dent, P. B., and Good, R. A. (1969). Effect of chronic calcitonin deficiency on the skeleton of chicken. *Proc. Soc. Exp. Biol. Med.* **130,** 1001–1004.

Brown, D. M., Perey, D. Y. E., and Jowsey, J. (1970). Effects of ultimobranchialectomy on bone composition and mineral metabolism in the chicken. *Endocrinology* **87,** 1282–1291.

Brown, K. I. (1960). Response of turkey adrenals to ACTH and stress measured by plasma corticosterone. *Anat. Rec.* **137,** 344.

Brown, K. I. (1961). Validity of using plasma corticosterone as a measure of stress in the turkey. *Proc. Soc. Exp. Biol. Med.* **107,** 538–542.

Brown, K. I., Brown, D. J., and Meyer, R. K. (1958a). Effect of surgical trauma, ACTH and adrenal cortical hormones on electrolytes, water balance and gluconeogenesis in male chickens. *Amer. J. Physiol.* **192**, 43.

Brown, K. I., Meyer, R. K., and Brown, D. J. (1958b). A study of adrenalectomized male chickens with and without adrenal hormone treatment. *Poultry Sci.* **37**, 680.

Burger, J. W. (1938). Cyclic changes in the thyroid and adrenal cortex of the male starling *Sturnus vulgaris* and their relation to the sexual cycle. *Amer. Natur.* **72**, 562–570.

Busheikin, J. C. (1951). Unpublished M.Sc. Thesis, Department of Zoology, University of Alberta (cited in Höhn *et al.*, 1965).

Cade, T. J., and Greenwald, L. (1966). Nasal gland secretion in falconiform birds. *Condor* **68**, 338–350.

Cain, W. A., Looper, M. D., and Good, R. A. (1968). Cellular immune competence of spleen, bursa and thymus cells. *Nature (London)* **217**, 87–89.

Calas, A. (1972). Identification des diverses catégories axonales dans la neurohypophyse des Oiseaux. *Gen. Comp. Endocrinol.* **18**, 580.

Candlish, J. K. (1970). The urinary excretion of calcium, hydroxyproline and uronic acid in the laying fowl after administration of parathyroid extracts. *Comp. Biochem. Physiol.* **32**, 703–707.

Candlish, J. K., and Taylor, T. G. (1970). The response-time to the parathyroid hormone in the laying fowl. *J. Endocrinol.* **48**, 143–144.

Caridroit, F. (1943). Effets de la thyroïdectomie complète sur le crête et le plumage du coq Leghorn doré. *C. R. Soc. Biol.* **137**, 163.

Carlson, L. A., Liljedahl, S., Verdy, M., and Wirsén, C. (1964). Unresponsiveness to the lipid mobilizing action of catecholamines in vivo and in vitro in the domestic fowl. *Meta. Clin. Exp.* **13**, 227–231.

Chan, A. S., Cipera, J. D., and Bélanger, L. F. (1969). The ultimo-branchial gland of the chick and its response to a high calcium diet. *Rev. Can. Biol.* **28**, 19–31.

Chan, M. Y., Bradley, E. L., and Holmes, W. N. (1972). The effects of hypophysectomy on the metabolism of adrenal steroids in the pigeon (*Columba livia*). *J. Endocrinol.* **52**, 435–450.

Chandola, A. (1972). Thyroid in reproduction. Reproductive physiology of *Lonchura punctulata* in relation to iodin metabolism and hypothyroidism. Ph.D. Thesis, Banaras Hindu University (India), No. 97747.

Chandola, A., and Thapliyal, J. P. (1968). Further studies on the endocrine regulation of the body weight of Spotted Munia, *Uroloncha punctulata. Gen Comp. Endocrinol.* **11**, 272–277.

Chang, T. S., Glick, B., and Winter, A. R. (1955). The significance of the bursa of Fabricius of chickens in antibody production. *Poultry Sci.* **34**, 1187.

Clara, M. (1924). Studie zur Kenntnis der Langerhansschen Inseln. *Z. Mikrosk.-Anat. Forsch.* **1**, 513–562.

Clavert, J. (1953). A propos de la vitellogenèse de la phase de grand accroissement des follicules chez la Pigeonne. *C. R. Ass. Anat.* **74**, 397–401.

Conner, M. H. (1959). Effect of various hormone preparations and nutritional stresses in chicks. *Poultry Sci.* **38**, 1340–1343.

Cooper, C. W., and Tashjian, A. H., Jr. (1966). Subcellular localization of thyrocalcitonin. *Endocrinology* **79**, 819–822.

Cooper, M. D., Peterson, R. D., South, M. A., and Good, R. A. (1966). The function of the thymus system and the bursa system in the chicken. *J. Exp. Med.* **123**, 75–102.

Copp, D. H. (1962). Calcitonin: A second hormone from the parathyroid and its function in regulating blood calcium. *In* "Yearbook of Endocrinology" (G. S. Gordan, ed.), pp. 10–18. Yearbook Pub., Chicago, Illinois.

Copp, D. H. (1968). Parathyroid hormone, calcitonin and calcium homeostasis. *In* "Parathyroid Hormone and Thyrocalcitonin (Calcitonin)" (R. V. Talmage and L. F. Bélanger, eds.), pp. 25–39. Excerpta Med. Found., Amsterdam.

Copp, D. H., and Parkes, C. O. (1968). Extraction of calcitonin from ultimobranchial tissue. *In* "Parathyroid Hormone and Thyrocalcitonin (Calcitonin)" (R. V. Talmage and L. F. Bélanger, eds.), pp. 74–82. Excerpta Med. Found., Amsterdam.

Crew, F. A. E. (1927). Die Wirkungen der Schilddrüsenektomie am hennengefiederten Hahn. *Arch. Geflügelkunde* **1**, 234–239.

Cruickshank, E. M. (1929). Observations on the iodine content of the thyroid and ovary of the fowl during growth, laying and molting periods. *Biochem. J.* **23**, 1044–1049.

Cuello, A. C. (1970). Occurrence of adrenaline and noradrenaline cells in the adrenal gland of the Gentoo Penguin (*Pygoscelis papua*). *Experientia* **26**, 416–418.

Dainat, J., Nouguier-Soulé, J., and Assenmacher, I. (1969). Mise en évidence d'un rythme ultradien de la fixation thyroïdienne du radioiode chez le Canard. *C. R. Soc. Biol.* **163**, 684.

Daniel, J. Y. (1970). Dosage des corticostéroïdes dans la surrénale du canard Pékin mâle. *Ann. Endocrinol.* **31**, 209–216.

Daniel, J. Y., and Assenmacher, I. (1969a). Dosage de la corticostérone et de l'aldostérone dans la surrénale du Canard après hypophysectomie ou autogreffe. *C. R. Acad. Sci.* **269**, 1308–1311.

Daniel, J. Y., and Assenmacher, I. (1969b). Interrelations testiculo-surrénaliennes chez le Canard. *Gen. Comp. Endocrinol.* **13**, 499–500.

Daniel, J. Y., Boissin, J., and Assenmacher, I. (1970). Transformation rapide de la ³H-corticostérone en métabolites tritiés dans le sang circulant, chez le Lapin et le Canard. *C. R. Acad. Sci.* **271**, 111–114.

Daughaday, W. H. (1957). Corticosteroid binding by a plasma alpha globulin. *J. Clin. Invest.* **36**, 881–882.

Davis, J., and Davis, B. S. (1954). The annual gonad and thyroid cycle of the English Sparrow in Southern California. *Condor* **56**, 328.

De Roos, R. (1961). The corticoids of the avian adrenal gland. *Gen. Comp. Endocrinol.* **1**, 494–512.

Desbals, P. (1972). Effets de la pancréatectomie et de l'hypophysectomie sur la circulation des lipides chez le Canard. D.Sc. Thesis. University of Toulouse, France.

Desbals, P., Desbals, B., and Miahle, P. (1970a). Pancreatic regulation of lipolysis in the duck. *Diabetologia* **6**, 65.

Desbals, P., Mialhe, P., and Desbals, B. (1970b). Lipolytic "in vitro" effect of glucagon on adipose tissue of ducks. *Diabetologia* **6**, 625.

Dewhurst, W. G., and Marley, E. (1965). Action of sympathomimetic and allied amines on the central nervous system of the chicken. *Brit. J. Pharmacol. Chemother.* **25**, 705–727.

Doe, R. P., Zinneman, H. H., Flink, E. B., and Ulstrom, R. A. (1960). Significance of concentration of non-protein-bound plasma cortisol in normal subjects, Cushing's syndrome, pregnancy, and during estrogen therapy. *J. Clin. Endocrinol. Metab.* **20**, 1484–1492.

Donaldson, E. M., and Holmes, W. N. (1965). Corticosteroidogenesis in the fresh-water and saline-maintained duck (*Anas platyrhynchos*). *J. Endocrinol.* **32**, 329–336.

Donaldson, E. M., Holmes, W. N., and Stachenko, J. (1965). In vitro corticosteroidogenesis by duck (*Anas platyrhynchos*) adrenal. *Gen. Comp. Endocrinol.* **5**, 542.

Doyon, M., and Jouty, A. (1904). Ablation des parathyroïdes chez l'Oiseau. *C. R. Soc. Biol.* **66**, 11.

Dubowitz, L. M. S., Myant, N. B., and Osorio, C. (1962). A comparison of the binding of thyroid hormone by rat, chicken, and human serum. *J. Physiol. (London)* **162**, 358–366.

Dulin, W. E. (1956). Effects of corticosterone, cortisone and hydrocortisone on fat metabolism in the chick. *Proc. Soc. Exp. Biol. Med.* **92**, 253–255.

Egge, A. S., and Chiasson, R. B. (1963). Endocrine effects of diencephalic lesions in the White Leghorn hens. *Gen. Comp. Endocrinol.* **3**, 346–361.

Elton, R. L., and Zarrow, M. L. (1955). Analysis of adrenal ascorbic acid and cholesterol in different species. *Anat. Rec.* **122**, 473–474.

Emmens, C. W., and Parkes, A. S. (1940). The endocrine system and plumage types. II. The effects of thyroxine injections to normal, caponized and thyroidectomized caponized birds. *J. Genet.* **39**, 485.

Ensor, D. M., Thomas, D. M., and Phillips, J. G. (1970). The possible role of the thyroid in extrarenal secretion following a hypertonic saline load in the Duck (*Anas platyrhynchos*). *J. Endocrinol.* **46**, x.

Epple, A. (1968). Comparative studies on the pancreatic islets. *Endocrinol. Japon.* **15**, 107–122.

Falkmer, S., and Wilson, S. (1967). Comparative aspects of the immunology and biology of insulin. *Diabetologia* **3**, 519–528.

Farer, L. S., Robbins, J., Blumberg, B. S., and Rall, J. E. (1962). Thyroxine-serum protein complexes in various animals. *Endocrinology* **70**, 686–696.

Farner, D. S. (1965). Annual endocrine cycles in temperate-zone birds with special attention to the White-crowned Sparrow *Zonotrichia leucophrys gambelii*. *Proc. Int. Congr. Endocrinol., 2nd, 1964* pp. 114–118.

Field, E. O., Caughi, M. N., Blackett, N. M., and Smithers, D. W. (1968). Marrow-suppressing factors in the blood in pure red-cell aplasia, thymoma and Hodgkin's disease. *Brit. J. Haematol.* **15**, 101–110.

Fink, B. A. (1957). Radioiodine, a method for measuring thyroid activity. *Auk* **74**, 487–493.

Fleming, G. B. (1919). Carbohydrate metabolism in ducks. *J. Physiol. (London)* **53**, 236–246.

Follett, B. K., and Riley, J. (1967). Effect of the length of the daily photoperiod on thyroid activity in the female Japanese Quail (*Coturnix coturnix japonica*). *J. Endocrinol.* **39**, 615–616.

Frankel, A. I., and Nalbandov, A. V. (1966). Adrenal function in adenohypophysectomized and intact cockerels. *Excerpta Med. Internat. Congr. Ser.* No. 111, 106.

Frankel, A. I., Graber, J. W., and Nalbandov, A. V. (1967a). Adrenal function in cockerels. *Endocrinology* **80**, 1013–1019.

Frankel, A. I., Graber, J. W., and Nalbandov, A. V. (1967b). The effect of hypothalamic lesions on adrenal function in intact and adenohypophysectomized cockerels. *Gen. Comp. Endocrinol.* **8**, 387–396.

Freeman, B. M. (1967). Some effects of cold on the metabolism of the fowl during the perinatal period. *Comp. Biochem. Physiol.* **20**, 179–193.

Freeman, B. M. (1969a). The Bursa of Fabricius and adrenal cortical activity in *Gallus domesticus*. *Comp. Biochem. Physiol.* **29**, 639–646.

Freeman, B. M. (1969b). Effect of noradrenaline on the plasma free fatty acid and glucose levels in *Gallus domesticus*. *Comp. Biochem. Physiol.* **30**, 993–996.

Freeman, B. M. (1970a). Thermoregulatory mechanisms of the neonate fowl. *Comp. Biochem. Physiol.* **33**, 219–230.

Freeman, B. M. (1970b). Some aspects of thermoregulation in the adult Japanese Quail (*Coturnix coturnix japonica*). *Comp. Biochem. Physiol.* **34**, 871–881.

Freeman, B. M. (1970c). The effects of adrenocorticotrophic hormone on adrenal weight and adrenal ascorbic acid in the normal and bursectomized fowl. *Comp. Biochem. Physiol.* **32**, 755–761.

Fromme-Bouman, H. (1962). Jahresperiodische Untersuchungen an der Amsel (*Turdus merula*). *Vogelwarte* **21**, 188–198.

Fujita, H. (1963). Electron microscope studies on the thyroid gland of domestic fowl with special reference to the mode of secretion and the occurrence of a central flagellum in the follicle cell. *Z. Zellforsch. Mikrosk. Anat.* **60**, 615–632.

Fujita, H., and Machino, M. (1962). Electron microscopic observations on the secretory granules of the adrenal medulla of domestic fowl. *Arch. Histol. Jap.* **23**, 67–77.

Fujita, H., Kano, M., Kunishima, I., and Kido, J. (1959). Electron microscopic observations on the adrenal medulla of the chick after injection of insulin. *Arch. Histol. Jap.* **18**, 411–419.

Galante, L., Gudmunsson, T. V., Matthews, E. W., Tse, A., Williams, E. D., Woodhouse, N. J. Y., and McIntyre, I. (1968). Thymic and parathyroid origin of calcitonin in man. *Lancet* **2**, 537–538.

Galpin, N. (1938). Factors affecting the hatching weight of Brown Leghorn chickens. *Proc. Roy. Soc. Edinburgh* **58**, 98–113.

Garren, H. W., and Shaffner, C. S. (1956). How the period of exposure to different stress stimuli affects the endocrine and lymphatic gland weights of young chickens. *Poultry Sci.* **35**, 2066–2073.

Ghosh, A. (1962). A comparative study of histochemistry of the avian adrenals. *Gen. Comp. Endocrinol., Suppl.* **1**, 75–80.

Ghosh, A. (1973). Histophysiology of the avian adrenal medulla. *Proc. 60th Indian Science Congress*, Part II, pp. 1–24.

Ghosh, A., and Battacharyya, T. K. (1967). Comparative study on cortical ascorbic acid response in birds to exo- and endogenous corticotrophin. *Proc. III. Asia Oceania Cong. Endocrinol.* pp. 315–316.

Gilliland, I. C., and Strudwick, J. I. (1953). Comparison of the effects of equimolar solutions of L-thyroxine and L-triiodothyronine in blocking TSH secretion in the chick's pituitary. *Mem. Soc. Endocrinol.* **1**, 14.

Glazener, E. W., and Jull, M. A. (1946). Effects of thiouracil, desiccated thyroid, and stilboestrol derivatives on various glands, body weight and dressing appearance in the chicken. *Poultry Sci.* **25**, 236.

Glick, B. (1955). Growth and function of the bursa of Fabricius in the domestic fowl. Ph.D. Dissertation, Ohio State University, Columbus.

Glick, B. (1957). Experimental modification of the growth of the bursa of Fabricius. *Poultry Sci.* **36**, 18–23.

Glick, B. (1959). The experimental production of the stress picture with cortisone and the effect of penicillin in young chickens. *Ohio J. Sci.* **59**, 81–87.

Glick, B. (1960a). Growth of the bursa of Fabricius and its relationship to the adrenal in the White Pekin duck, White Leghorn outbred and inbred New Hampshire. *Poultry Sci.* **39**, 130–132.

Glick, B. (1960b). Extracts of bursa of Fabricius- lymphoepithelial gland of the chicken stimulate the production of antibodies in bursectomized chickens. *Poultry Sci.* **39**, 1097–1101.

Golden, W. R. C., and Long, C. N. H. (1942). The influence of certain hormones on the carbohydrate levels of the chick. *Endocrinology* **30**, 674–686.

Goldstein, G. (1968). The thymus and neuromuscular function. A substance in thymus which causes myositis and myasthenic neuromuscular block in Guinea pigs. *Lancet* **2**, 119–122.

Goodridge, A. G. (1964). The effect of insulin, glucagon and prolactin on lipid synthesis and related metabolic activity on migratory and non-migratory finches. *Comp. Biochem. Physiol.* **13**, 1–26.

Goodridge, A. G., and Ball, E. C. (1965). Studies on the metabolism of adipose tissue. XVIII. In vitro effects of insulin, epinephrine and glucagon on lipolysis and glycolysis in pigeon adipose tissue. *Comp. Biochem. Physiol.* **16**, 367–381.

Goodridge, A. G., and Ball, E. G. (1966). Lipogenesis in the pigeon: In vitro studies. *Amer. J. Physiol.* **211**, 803–808.

Goodridge, A. G., and Ball, E. G. (1967). Lipogenesis in the pigeon: In vivo studies. *Amer. J. Physiol.* **213**, 245–249.

Grande, F. (1968). Effect of glucagon on plasma free fatty acids and blood sugar in birds. *Proc. Soc. Exp. Biol. Med.* **128**, 532–536.

Grande, F. (1969). Effects of catecholamines on plasma free fatty acids and blood sugar in birds. *Proc. Soc. Exp. Biol. Med.* **131**, 740–744.

Grande, F., and Prigge, W. F. (1970). Glucagon infusion, plasma FFA and triglycerides, blood sugar and liver lipids in birds. *Amer. J. Physiol.* **218**, 1406–1411.

Greenman, D. L., and Zarrow, M. X. (1961). Steroids and carbohydrate metabolism in the domestic bird. *Proc. Soc. Exp. Biol. Med.* **106**, 459–462.

Greenwood, A. W., and Blyth, J. S. S. (1929). An experimental analysis of the plumage of the brown leghorn fowl. *Proc. Roy. Soc. Edinburgh* **49**, 313.

Greenwood, A. W., and Blyth, J. S. S. (1942). Some effects of thyroid and gonadotropic preparations in the fowl. *Quart. J. Exp. Physiol.* **31**, 175–186.

Grosvenor, C. E., and Turner, C. W. (1960). Measurement of thyroid secretion rate of individual pigeons. *Amer. J. Physiol.* **198**, 1–3.

Haarmann, W. (1936). Über den Einfluss von Thyroxin auf den Sauerstoffverbrauch überlebender Gewebe. *Arch. Exp. Pathol. Pharmak.* **180**, 167–182.

Häcker, V. (1926). Über jahreszeitliche Veränderungen und klimatisch bedingte Verschiedenheit der Vogelschilddrüse. *Schweiz. Med. Wochenschr.* **56**, 337–341.

Hahn, D. W., Ishibashi, T., and Turner, C. W. (1966). Alteration of thyroid hormone secretion rate in fowls changed from a cold to a warm environment. *Poultry Sci.* **45**, 31–33.

Hand, T., Caster, P., and Luckey, T. D. (1967). Isolation of a thymus hormone LSH. *Biochem. Biophys. Res. Commun.* **26**, 18–23.

Hansborough, L. A., and Khan, M. (1951). The initial function of the chick thyroid gland with the use of radioiodine ^{131}I. *J. Exp. Zool.* **116**, 447–452.

Harris, G. W. (1959). Neuroendocrine control of TSH regulation. *In* "Comparative Endocrinology" (A. Gorbman, ed.), pp. 202–222. Chapman & Hall, London.

Hartman, F. A., Knouff, R. A., and Howard, G. A. (1954). Response of the pelican adrenal to various stimuli. *Anat. Rec.* **120**, 469–493.

Hazelwood, R. L., and Barksdale, B. K. (1970). Failure of chicken insulin to alter polysaccharide levels of the avian glycogen body. *Comp. Biochem. Physiol.* **36**, 823–827.

Hazelwood, R. L., Kimmel, J. R., and Pollock, H. G. (1968). Biological characterization of chicken insulin in rats and domestic fowl. *Endocrinology* **83**, 1331–1336.

Hazelwood, R. L., Kimmel, J. R., and Pollock, H. G. (1971). Influence of chicken and rat plasma on in vitro activity of chicken and beef insulin. *Comp. Biochem. Physiol.* **38B**, 111–122.

Heald, P. J. (1966). The effects of glucagon and insulin on the plasma glucose and unesterified fatty acids of the domestic fowl. *In* "Physiology of the Domestic Fowl" (C. Horton-Smith and E. C. Amoroso, eds.), pp. 113–124. Oliver & Boyd, Edinburgh.

Heald, P. J., McLachlan, P. M., and Rookledge, K. A. (1965). The effects of insulin, glucagon and adrenocorticotrophic hormone on the plasma glucose and free fatty acids of the domestic fowl. *J. Endocrinol.* **33**, 83–95.

Hebert, B. A., and Brunson, C. C. (1957). The effects of diethylstilboestrol, testosterone, thiouracil and thyroprotein on the chemical composition of broiler carcass. *Poultry Sci.* **36**, 898.

Hendrich, C. E., and Turner, C. W. (1963). Time relations in the alteration of thyroid gland function in fowl. *Poultry Sci.* **42**, 1190–1195.

Hendrich, C. E., and Turner, C. W. (1965). Estimation of thyroid-stimulating hormone (TSH) secretion rates of New Hampshire fowls. *Proc. Soc. Exp. Biol. Med.* **117**, 218–222.

Hendrich, C. E., and Turner, C. W. (1967). A comparison of the effect of environmental temperature changes and 4.4°C cold on the biological half-life ($t_{1/2}$) of thyroxine ^{131}I in fowls. *Poultry Sci.* **46**, 3–5.

Heninger, R. W., and Newcomer, W. S. (1964). Plasma proteins binding, half-life, uptake of thyroxine and triiodothyronine in chickens. *Proc. Soc. Exp. Biol. Med.* **114**, 624–628.

Heninger, R. W., Newcomer, W. S., and Thayer, R. H. (1960). The effect of elevated ambient temperatures on the thyroxine secretion rate of chickens. *Poultry Sci.* **39**, 1332–1337.

Herrick, E. H., and Finerty, J. C. (1941). The effect of adrenalectomy on the anterior pituitary of fowl. *Endocrinology* **27**, 279.

Hewitt, W. F. (1947). The essential role of the adrenal cortex in the hypertrophy of the ovotestis following ovariectomy in the hen. *Anat. Rec.* **98**, 159.

Hill, E. T., and Parkes, A. S. (1934). Hypophysectomy of birds. IV. Plumage changes in hypophysectomized fowls. *Proc. Roy. Soc., Ser. B* **117**, 202–209.

Hirsch, P. F., and Munson, P. L. (1963). Hypocalcemic effect of thyroid extracts in rats. *Pharmacologist* **5**, 272.

Hodges, R. D., and Gould, R. P. (1969). Partial nervous control of the avian ultimobranchial body. *Experientia* **25**, 1317–1319.

Hoffmann, E. (1950). Thyroxine secretion rate and growth in the White Pekin duck. *Poultry Sci.* **29**, 109–114.

Hoffmann, E., and Shaffner, C. S. (1950). Thyroid weight and function as influenced by environmental temperature. *Poultry Sci.* **29**, 365.

Höhn, E. O. (1947). Sexual behavior and seasonal changes in the gonads and adrenals of the mallard. *Proc. Zool. Soc. London* **117**, 182–304.

Höhn, E. O. (1949). Seasonal changes in the thyroid gland and effects of thyroidectomy in the Mallard, in relation to molt. *Amer. J. Physiol.* **158**, 337–344.

Höhn, E. O. (1950). Physiology of the thyroid gland in birds: a review. *Ibis* **92**, 464–473.

Höhn, E. O. (1956). Seasonal recrudescence of the thymus in sexually mature birds. *Can. J. Biochem. Physiol.* **34**, 90–101.

Höhn, E. O. (1961). Endocrine glands, thymus and pineal body. *In* "Biology and Comparative Physiology of Birds" (A. J. Marshall, ed.), Vol. 2, pp. 87–114. Academic Press, New York.

Höhn, E. O., Sarkar, A. K., and Dzubin, A. (1965). Adrenal weight in wild Mallard and domestic ducks and seasonal adrenal weight change in the Mallard. *Can. J. Zool.* **43**, 475–487.

Hollwich, F., and Tilgner, S. (1963). Über die gonadotrope und thyreotrope Wirkung der Bestrahlung des Auges mit monochromatischem Licht. *Endokrinologie* **44**, 167–188.

Holmes, W. N., and Phillips, J. G. (1965). Adrenocortical hormones and electrolyte metabolism in birds. *Excerpta Med. Found. Int. Congr. Ser.* No. 83, pp. 158–161.

Holmes, W. N., Phillips, J. G., and Butler, D. G. (1961). The effect of adrenocortical steroids on the renal and extrarenal responses of the domestic duck (*Anas platyrhynchos*) after hypertonic saline loading. *Endocrinology* **69**, 483–495.

Holmes, W. N., Bradley, E. L., Helton, E. D., and Chan, M. Y. (1972). The distribution and metabolism of corticosterone in birds. *Gen. Comp. Endocrinol.* Suppl. **3**, 266–278.

Hooker, C. W., and Collins, V. J. (1940). Androgenic action of DOCA. *Endocrinology* **26**, 269.

Hurst, J. G., and Newcomer, W. S. (1969). Functional accessory parathyroid tissue in ultimobranchial bodies in chickens. *Proc. Soc. Exp. Biol. Med.* **132**, 555–557.

Huston, T. M. (1960). The effects of high environmental temperature upon blood constituents and thyroid activity of domestic fowl. *Poultry Sci.* **39**, 1260.

Huston, T. M., and Carmon, J. L. (1962). The influence of high environmental temperature on thyroid size of domestic fowl. *Poultry Sci.* **41**, 175–179.

Huston, T. M., Edwards, H. M., Jr., and Williams, J. J. (1962). The effects of high environmental temperature on thyroid secretion rate of domestic fowl. *Poultry Sci.* **41**, 640–645.

Hutchins, M. O., and Newcomer, W. S. (1966). Metabolism and excretion of thyroxine and triiodothyronine in chickens. *Gen. Comp. Endocrinol.* **6**, 239–248.

Ivy, A. C., Tatum, A. C., and Jung, F. T. (1926). Studies in avian diabetes and glycosuria. *Am. J. Physiol.* **78**, 666.

Jaap, R. G. (1933). Testis enlargement and thyroid administration in Ducks. *Poultry Sci.* **12**, 233–241.

Jailer, J. W., and Boas, N. F. (1950). The inability of epinephrine or adrenocorticotrophic hormone to deplete the ascorbic acid content of the chick adrenal. *Endocrinology* **46**, 314–318.

Jallageas, M., and Assenmacher, I. (1970). Testostéronémie du Canard photostimulé ou soumis à des injections répétées de testostérone. *C. R. Soc. Biol.* **242**, 164–167.

Jallageas, M., and Assenmacher, I. (1972). Effets de la photopériode et du taux d'androgène circulant sur la fonction thyroïdienne du Canard. *Gen. Comp. Endocrinol.* **19**, 331–340.

Jallageas, M., and Assenmacher, I. (1973). Thyroid gonadal interactions in the male domestic duck in relationship with the sexual cycle. *Gen. Comp. Endocrinol.* In press.

Jankovic, B. D., and Leskowitz, S. (1965). Restoration of antibody producing capacity in bursectomized chickens by bursal grafts in millipore chambers. *Proc. Soc. Exp. Biol. Med.* **118**, 1164–1166.

Jankovic, B. D., Isakovic, K., and Hŏrvat, J. (1967). Antibody production in bursectomized chickens treated with lipid and protein fractions from bursa, thymus and liver. *Experientia* **23**, 1062.

Jöchle, W. (1962). Experimentelle Untersuchungen zur neuroendokrinen Steuerung der Mauser beim Haushuhn. *Proc. Symp. Deu. Ges. Endokrinol., 8th, 1962* pp. 416–421.

Jolly, J. (1913). L'involution physiologique de la bourse de Fabricius et ses relations avec l'apparition de la maturité sexuelle. *C. R. Soc. Biol.* **75**, 638–648.

Juhn, M., and Barnes, B. O. (1931). The feather germ as indicator for thyroid preparations. *Amer. J. Physiol.* **98**, 463–466.

Kano, M. (1959). Electron microscopic study of the adrenal medulla of domestic fowl. *Arch. Histol. Jap.* **18**, 25–56.

Kar, A. B. (1947). The adrenal cortex testicular relations in the fowl: The effect of castration and replacement therapy on the adrenal cortex. *Anat. Rec.* **99**, 177.

Karrman, H., and Mialhe, P. (1968). Pancréatectomie subtotale et diabète permanent chez l'oie. *Diabetologia* **4**, 394.

Kausch, W. (1896). Über den Diabetes mellitus der Vögel nach Pankreasextirpation. *Naunyn-Schmiedebergs Arch. Exp. Pathol. Pharmakol.* **37**, 274.

Kemény, V., Pethes, G., and Kozma, M. (1968). Chromatographically separable fractions in the extract of bursa of Fabricius in the chicken. *Comp. Biochem. Physiol.* **26**, 757–759.

Kimmel, J. R., Pollock, H. G., and Hazelwood, R. L. (1968). Isolation and characterization of chicken insulin. *Endocrinology* **83**, 1323–1330.

Kirkpatrick, C. M., and Andrews, F. N. (1944). The influence of sex hormones on the bursa of Fabricius in the Ring-necked Pheasant. *Endocrinology* **134**, 340–345.

Kitay, J. I. (1961). Sex differences in adrenal cortical secretion in the rat. *Endocrinology* **68**, 818–824.

Kjaerheim, A. (1968). Studies of adrenocortical ultrastructure. 5-Effects of metopirone on interrenal cells of the domestic fowl. *J. Microsc. (Paris)* **7**, 739–754.

Kleinpeter, M. E., and Mixner, J. P. (1947). The effect of the quantity and quality of light on the thyroid activity of the baby chick. *Poultry Sci.* **26**, 494–498.

Knouff, R. A., and Hartman, F. A. (1951). A microscopic study of the adrenal of the Brown Pelican. *Anat. Rec.* **109**, 161–178.

Kobayashi, H. (1954a). Studies on molting in the Pigeon. VIII. Effects of sex steroids on molting and thyroid gland. *Annot. Zool. Jap.* **27**, 22–26.

Kobayashi, H. (1954b). Inhibition by sex steroids and thyroid substance of light-induced gonadal development in the passerine bird, *Zosterops palpebrosa japonica*. *Endocrinol. Jap.* **1**, 51–55.

Kobayashi, H., and Gorbman, A. (1960). Radioiodine utilization in the chick. *Endocrinology* **66**, 795–804.

Kobayashi, H., and Tanabi, Y. (1959). ^{131}I uptake by the thyroid and effect of progesterone on it in passerine birds. *Tori (Tokyo)* **15**, 55–60.

Kobayashi, H., Gorbman, A., and Wolfson, A. (1960). Thyroidal utilization of radioiodine in the White-throated Sparrow and weaver finch. *Endocrinology* **67**, 153–161.

Kondics, L. (1963). Über die Wirkung des Kochsalzes auf die interrenalen Zellen der Nebenniere bei Haustauben. *Ann. Univ. Sci. Budapest. Rolando Eotvos Nominatae, Sect. Biol.* **6**, 101–107.

Kondics, L., and Kjaerheim, A. (1966). The zonation of interrenal cells in fowls. (An electron microscopical study.). *Z. Zellforsch. Mikrosk. Anat.* **70**, 81–90.

Kraetzig, H. (1937). Histologische Untersuchungen zur Frage der Struktur und Farbveränderungen an Federn nach künstlicher (Thyroxin-)Mauser. *Wilhelm Roux' Arch. Entwicklungsmech. Organismen* **137**, 86–150.

Krug, E., and Mialhe, P. (1971). Pancreatic and intestinal glucagon in the duck. *Horm. Metab.* 3, 24–27.

Krug, E., Bielher, O., and Mialhe, P. (1971). Molecular weight of gut and pancreatic circulating glucagon in the duck. *Horm. Metab.* 3, 258–261.

Küchler, W. (1935). Jahreszyklische Veränderungen im histologischen Bau der Vogelchilddrüse. *J. Ornithol.* 83, 414–461.

Kumaran, J. D. S., and Turner, C. W. (1949). The endocrinology of spermatogenesis in Birds, III. Effects of hypo and hyperthyroidism. *Poultry Sci.* 28, 653–665.

Lachiver, F., and Poivilliers de la Querière, F. (1959). Etude du rapport ^{131}I hématie/^{131}I plasma et du rapport de conversion ^{131}I organique plasmatique/^{131}I plasmatique en vue de leur application à l'étude de la fonction thyroïdienne chez les Colombidés. *Z. Vergl. Physiol.* 42, 6–16.

Laguë, P. C., and van Tienhoven, A. (1969). Comparison between chicken and guinea pig thyroid with respect to the incorporation of ^{131}I in vitro. *Gen. Comp. Endocrinol.* 12, 305–312.

Lahiri, P., and Banerji, H. (1969). *Ind. J. Physiol. Allied Sci.*, 23, 100–106 [cited after Ghosh, A. (1973)].

Lamoureux, J. (1966). La biosynthèse des stéroïdes surrénaliens chez les vertébrés inférieurs aux mammifères. Ph.D. Thesis, Faculty of Medicine, University of Montreal, Montreal.

Landauer, W., Pfeiffer, C. A., Gardner, W. V., and Man, E. B. (1939). Hypercalcification, calcemia and lipemia in chickens following administration of estrogen. *Proc. Soc. Exp. Biol. Med.* 41, 80–82.

Langslow, D. R., and Hales, C. N. (1969). Lipolysis in avian adipose tissue in vitro. *J. Endocrinol.* 43, 285–294.

Langslow, D. R., Butler, E. J., Hales, C. N., and Pearson, A. W. (1970). The response of plasma insulin, glucose and non-esterified fatty acids to various hormones, nutrients and drugs in the domestic fowl. *J. Endocrinol.* 46, 243–260.

Lee, M., and Lee, R. C. (1937). Effect of thyroidectomy and thyroid feeding in geese on the basal metabolism at different temperatures. *Endocrinology* 21, 790.

Lehman, G. C. (1970). The effects of hypo- and hyperthyroidism on the testes and anterior pituitary gland in cockerels. *Gen. Comp. Endocrinol.* 14, 567–577.

Legait, H., and Legait, E. (1959). Variations d'activité du système hypothalamoneurohypophysaire et modifications surrénaliennes chez la Poule au cours du cycle annuel. *C. R. Soc. Biol.* 153, 668–670.

Legendre, R., and Rakotondrainy, A. (1963). Variation de l'épaisseur de l'épithélium thyroïdien en rapport avec la coloration nuptiale chez le mâle du plocéidé malgache *Foudia madagascariensis C. R. Acad. Sci.* 256, 1019.

Leloup, J., and Lachiver, F. (1955). Influence de la teneur en iode du régime sur la biosynthèse des hormones thyroïdiennes. *C. R. Acad. Sci.* 241, 509.

Lepkovsky, S., Dimick, M. K., Furuta, F., Snapir, N., Park, R., Narita, N., and Komatsu, K. (1967). Response of blood glucose and plasma free fatty acids to fasting and to injection of insulin and testosterone in chickens. *Endocrinology* 81, 1001–1006.

Lorenzen, L., and Farner, D. S. (1964). An annual cycle in the interrenal tissue of the White-crowned sparrow, *Zonotrichia leucophrys gambelii. Gen. Comp. Endocrinol.* 4, 253–263.

Ma, R. C. S., and Nalbandov, A. V. (1963). Hormonal activity of the autotransplanted adenohypophysis. *Advan. Neuroendocrinol., Proc. Symp., 1961* pp. 306–311.

McCartney, M. G., and Shaffner, C. S. (1950). The influence of altered metabolism upon fertility and hatchability in the female fowl. *Poultry Sci.* **29,** 67–77.

Macchi, I. A. (1967). Regulation of reptilian and avian adrenocortical secretion: Effect of ACTH on the in vitro conversion of progesterone to corticoids. *Proc. Int. Congr. Horm. Steroids, 2nd, 1966* pp. 1094–1103.

Macchi, I. A., Phillips, J. G., and Brown, P. (1967). Relationship between the concentration of corticosteroids in avian plasma and nasal gland function. *J. Endocrinol.* **38,** 319–329.

McFarland, L. Z., Yousef, M. R., and Wilson, W. O. (1966). The influence of ambient temperature and hypothalamic lesions on the disappearance rate of thyroxine ^{131}I in the Japanese Quail. *Life Sci.* **5,** 309–315.

McLelland, J., Moorhouse, P. D. S., and Pickering, E. C. (1968). An anatomical and histochemical study of the nasal gland of *Gallus domesticus. Acta Anat.* **71,** 122–133.

Malaval, F., and Assenmacher, I. (1973). Unpublished data.

Malmqvist, E., Ericson, L. E., Almqvist, S., and Ekholm, R. (1968). Granulated cells, uptake of amine precursors, and calcium lowering activity in the ultimobranchial body of the domestic fowl. *J. Ultrastruct. Res.* **23,** 457–461.

Maraud, R., Stoll, R., and Blanquet, P. (1957). Sur le rôle de l'hypophyse dans la concentration du radioiode par la thyroïde de l'embryon de Poulet. *C. R. Soc. Biol.* **151,** 572–574.

Marvin, H. N., and Smith, G. C. (1943). Technique for thyroidectomy in the pigeon and the early effect of thyroid removal on heat production. *Endocrinology* **32,** 87.

Matthews, E. W., Moseley, J. M., Breed, T. U., Godmundsson, T. V., Byfield, P. G. H., Galante, L., Tse, A., and MacIntyre, I. (1968). Ultimobranchial and thyroid calcitonin. *In* "Parathyroid Hormone and Thyrocalcitonin (Calcitonin)" (R. V. Talmage and L. F. Bélanger, eds.), pp. 68–73. Excerpta Med. Found., Amsterdam.

May, J. D., Hill, C. H., and Garren, H. W. (1967). Partial restoration of precipitin production in bursectomized chicks. *Fed. Proc., Fed. Amer. Soc. Exp. Biol.* **26,** 769.

Mellen, W. J. (1958). Duration of effect of thyroxine and thiouracil in young chickens. *Poultry Sci.* **37,** 672–679.

Mellen, W. J. (1964). Thyroxine secretion rate in chicks and poults. *Poultry Sci.* **43,** 776–777.

Mellen, W. J., and Hardy, L. B., Jr. (1957). Blood protein-bound iodine in the fowl. *Endocrinology* **60,** 547–551.

Mellen, W. J., and Wentworth, B. C. (1959). Thyroxine vs triiodoithyronine in the fowl. *Poultry Sci.* **38,** 228–230.

Mellen, W. J., and Wentworth, B. C. (1960). Comparison of methods for estimating thyroid secretion rate in chickens. *Poultry Sci.* **39,** 678–686.

Mellen, W. J., and Wentworth, B. C. (1962). Observations on radiothyroidectomized chickens. *Poultry Sci.* **41,** 134–141.

Mellen, W. J., Hill, F. W., and Dukes, H. H. (1954). Studies of the energy requirements of chickens. 2. Effect of dietary energy level on the basal metabolism of growing chickens. *Poultry Sci.* **33,** 791.

Merkel, F. W. (1938). Zur Physiologie der Zugunruhe bei Vögeln. *Ber. Ver. Schles. Ornithol.* **23,** 1–72.

Metcalf, D. (1958). The thymic lymphocytosis-stimulating factor. *Ann. N. Y. Acad. Sci.* **73,** 113–119.

Mialhe, P. (1958). Glucagon, insuline et régulation endocrine de la glycémie chez le Canard. *Acta Endocrinol. (Copenhagen)* Suppl. **36,** 1–154.

Mialhe, P. (1969). Some aspects of the regulation of carbohydrate metabolism in birds. *In* "Progress in Endocrinology" (C. Gual, ed.), Int. Congr. Ser. No. 184, pp. 158–164. Excerpta Med. Found., Amsterdam.

Mialhe, P. (1971). Does an understanding of experimental diabetes advance the knowledge of spontaneous diabetes? *In* "Progress in Endocrinology" (C. Gual, ed.), Int. Congr. Ser. No. 231, pp. 843–853. Excerpta Med. Found., Amsterdam.

Mialhe-Voloss, C., Koch, B., and Kamoun, A. (1965). Effet de la greffe surrénalienne, de la surrénalectomie et de l'hypophysectomie sur la demi-vie de la corticostérone chez le rat femelle. *J. Physiol. (Paris)* **57,** 661.

Mikami, F., and Kazuyuki, O. (1962). Glucagon deficiency induced by extirpation of alpha islets of the fowl pancreas. *Endocrinology* **71,** 464–473.

Miller, D. S. (1935). Effects of thyroxine on plumage of the English sparrow, *Passer domesticus* (Linnaeus). *J. Exp. Zool.* **71,** 293–309.

Miller, J. F. A. P. (1965). Influence of the thymus on the development of the immune system. *Ser. Haematol.* **8,** 41.

Miller, R. A. (1961). Hypertrophic adrenals and their response to stress after lesions in the median eminence of totally hypophysectomized pigeons. *Acta Endocrinol. (Copenhagen)* **37,** 565–576.

Miller, R. A. (1967). Regional responses of interrenal tissue and of chromaffin tissue to hypophysectomy and stress in pigeons. *Acta Endocrinol. (Copenhagen)* **55,** 108–118.

Miller, R. A., and Riddle, O. (1939). Stimulations of adrenal cortex of pigeons by anterior pituitary hormones and by their secondary products. *Proc. Soc. Exp. Biol. Med.* **41,** 518–522.

Miller, R. A., and Riddle, O. (1942). Rest activity and repair in cortical cells of the pigeon adrenal. *Anat. Rec.* **75,** 103.

Minkowski, O. (1893). Untersuchungen über den Diabetes mellitus nach Pankreas-extirpation. *Naunyn-Schmiedebergs Arch. Exp. Pathol. Pharmakol.* **31,** 85.

Mirsky, I. A., Jinks, R., and Perisutti, G. (1963). The isolation and crystallization of human insulin. *J. Clin. Invest.* **42,** 1869–1872.

Mirsky, I. A., Nelson, N., Grayman, I., and Korenberg, M. (1941). Studies on normal and depancreatized domestic ducks. *Amer. J. Physiol.* **135,** 223.

Mixner, J. P., Reineke, E. P., and Turner, C. W. (1944). Effects of thiouracil and thio-urea on the thyroid gland of the chick. *Endocrinology* **34,** 168–174.

Moens, L., and Coessens, R. (1970). Seasonal variations in the adrenal cortex cells of the House Sparrow, *Passer domesticus* (L), with special reference to a possible zonation. *Gen. Comp. Endocrinol.* **15,** 95–100.

Morgan, E. L., and Mraz, F. R. (1970). Effects of dietary iodine on the thyroidal uptake and elimination of [131]Iodine in Coturnix and Bobwhite Quail. *Poultry Sci.* **49,** 161–164.

Mueller, G. L., Anast, C. S., and Breitenbach, R. P. (1970). Dietary calcium and ulti-mobranchial body and parathyroid gland in the chicken. *Amer. J. Physiol.* **218,** 1718–1722.

Mueller, W. J., and Amezcua, A. A. (1959). The relationship between certain thyroid characteristics of pullets and their egg production, body weight and environment. *Poultry Sci.* **38,** 620–624.

Nagra, C. L. (1965). Effect of corticosterone on lipids in hypophysectomized male chickens. *Endocrinology* **77,** 221–222.

Nagra, C. L., and Meyer, R. K. (1963). Influence of corticosterone on the metabolism of palmitate and glucose in cockerels. *Gen. Comp. Endocrinol.* **3**, 131–138.

Nagra, C. L., Baum, G. J., and Meyer, R. K. (1960). Corticosterone levels in adrenal effluent blood of some gallinaceous birds. *Proc. Soc. Exp. Biol. Med.* **105**, 68–70.

Nagra, C. L., Birnie, J. G., and Meyer, R. K. (1963). Suppression of the output of corticosterone in the pheasant by methopyranone (metopirone). *Endocrinology* **73**, 835–837.

Nagra, C. L., Sauers, A. K., and Wittmaier, H. W. (1965). Effect of testosterone, progestagens and metopirone on adrenal activity in cockerels. *Gen. Comp. Endocrinol.* **5**, 69–73.

Nalbandov, A. V., and Card, L. E. (1943). Effects of hypophysectomy of growing chicks. *J. Exp. Zool.* **94**, 387–413.

Nelson, N., Elgart, S., and Mirsky, I. A. (1942). Pancreatic diabetes in the owl. *Endocrinology* **31**, 119.

Nevalainen, T. (1969). Fine structure of the parathyroid gland of the laying hen (*Gallus domesticus*). *Gen. Comp. Endocrinol.* **12**, 561–567.

Newcomer, W. S. (1957). Relative potencies of thyroxine and triiodothyronine based on various criteria in thiouracil-treated chickens. *Amer. J. Physiol.* **190**, 413–418.

Newcomer, W. S. (1959). Effect of hypophysectomy on some functional aspects of the adrenal gland of the chicken. *Endocrinology* **65**, 133–135.

Newcomer, W. S. (1967). Transport of radioiodine between blood and thyroid in chickens. *Amer. J. Physiol.* **212**, 1391–1396.

Newcomer, W. S., and Barrett, P. A. (1960). Effects of various analogues of thyroxine on oxygen uptake of cardiac muscle from chicks. *Endocrinology* **66**, 409–415.

Oakeson, B. B., and Lilley, B. R. (1960). Annual cycle of thyroid histology in two races of White-crowned Sparrow. *Anat. Rec.* **136**, 41–50.

Odell, T. T., Jr. (1952). Secretion rate of thyroid hormone in White Leghorn castrates. *Endocrinology* **51**, 265–266.

O'Dor, R. K., Parkes, C. O., and Copp, D. H. (1969). Biological activities and molecular weights of ultimobranchial and thyroid calcitonin. *Comp. Biochem. Physiol.* **29**, 295–300.

O'Hea, E. K., and Leveille, G. A. (1968). Lipogenesis in isolated adipose tissue of the domestic chick (*Gallus domesticus*). *Comp. Biochem. Physiol.* **26**, 111–120.

Pandha, S. K., and Thapliyal, J. P. (1964). Effect of thyroidectomy upon the testes of Indian Spotted Munia, *Uroloncha punctulata*. *Naturwissenschaften* **8**, 51.

Pansky, B., House, E. L., and Cone, L. A. (1965). An insulin-like thymic factor. *Diabetes* **14**, 325.

Parhon, C. J., and Cahane, M. (1937). L'influence de la thymectomie et de cette dernière associée au traitement thyroïdien sur les gonades des oiseaux. *Bull. Sect. Endocrinol. Soc. Roum. Neurol. (Bucharest)* **3**, 263–266.

Parhon, C. J., and Cahane, M. (1939). Nouvelles recherches sur la fonction génitale des oiseaux éthymisés. *Bull. Sect. Endocrinol. Soc. Roum. Neurol. (Bucharest)* **5**, 36–37.

Parker, J. E. (1943). Influence of thyroactive iodocasein on growth of chicks. *Proc. Soc. Exp. Biol. Med.* **52**, 234–236.

Parkes, A. S., and Selye, H. (1937). The endocrine system and plumage types. I: some effects of hypothyroidism. *J. Genet.* **34**, 298–306.

Payne, F. (1944). Anterior pituitary-thyroid relationships in the fowl. *Anat. Rec.* **88**, 337–350.

Peaker, M., Phillips, J. G., and Wright, A. (1970). The effects of prolactin on the secretory activity of the nasal salt-gland of the domestic duck. *J. Endocrinol.* **47**, 123–127.

Péczely, P. (1964). The adaptation to salt water conditions of the adrenal structure in various bird species. *Acta Biol. Ac. Sci. Hungar. (Budapest)* **15**, 171–179.

Péczely, P. (1969). Effect of median eminence of the Pigeons (*Columba livia domestica* L), on the regulation of adenohypophysial corticotropin secretion. *Acta Physiol. Ac. Sci. Hungar.* **35**, 47–57.

Péczely, P., Baylé, J. D., Boissin, J., and Assenmacher, I. (1970). Activité corticotrope et "CRF" dans l'éminence médiane, et activité corticotrope de greffes hypophysaires chez le Pigeon. *C. R. Acad. Sci.* **270**, 3264–3267.

Péczely, P., and Zboray, G. (1967). Die Rolle der Eminentia mediana in der Zentralregulation der Nebenniere des Vogels. *Gen Comp. Endocrinol.* **9**, 517–518.

Perek, M., and Eckstein, B. (1959). The adrenal ascorbic acid content of molting hens, and the effect of ACTH on the adrenal ascorbic acid content of laying hens. *Poultry Sci.* **38**, 996–999.

Perek, M., and Eilat, A. (1960). The bursa of Fabricius and adrenal ascorbic acid depletion following ACTH injections in chicks. *J. Endocrinol.* **20**, 251–255.

Peter, S. (1970). Die Feinstruktur des Inselorgans im Pankreas des Huhnes in den ersten Lebenstagen und Wochen. *Z. Mikrosk.-Anat. Forsch.* **81**, 387–404.

Pethes, G., and Fodor, A. (1970). Alterations in thyroid secretion rate after removal of the bursa of Fabricius in the fowl. *Acta Physiol.* **37**, 368.

Pfeiffer, C. A., and Gardner, W. U. (1938). Skeletal changes and blood serum calcium level in pigeons receiving estrogens. *Endocrinology* **23**, 485–491.

Phillips, J. G., and Bellamy, D. (1962). Aspects of the hormonal control of nasal gland secretion in birds. *J. Endocrinol.* **24**, vi–vii.

Phillips, J. G., and Bellamy, D. (1963). Adrenocortical hormones. *In* "Comparative Endocrinology" (U. S. von Euler and H. Heller, eds.), Vol. 1, pp. 208–257. Academic Press, New York.

Phillips, J. G., and Chester-Jones, I. (1957). The identity of adrenocortical secretion in lower vertebrates. *J. Endocrinol.* **16**, iii.

Phillips, J. G., and van Tienhoven, A. (1960). Cited in Höhn *et al.* (1965).

Phillips, J. G., Holmes, W. N., and Butler, D. G. (1961). The effect of total and subtotal adrenalectomy on the renal and extrarenal response of the domestic duck (*Anas platyrhynchos*) in saline loading. *Endocrinology* **69**, 958–959.

Pintea, V., and Pethes, G. (1967). The effect of bursectomy on thyroidal ^{131}I uptake in the chicken. *Acta Univ. Agr., Brno, Fac. Vet.* **37**, 449–452.

Pipes, G. W., Premachandra, B. N., and Turner, C. W. (1958). Measurement of the thyroid hormone secretion rate of individual fowls. *Poultry Sci.* **37**, 36–41.

Podhradsky, J. (1935). Die Veränderungen der inkretorischen Drüsen und einiger innerer Organe bei der Legeleistung. *Z. Zücht. Reihe B* **33**, 76–103.

Poivilliers de la Querière, F., and Lachiver, F. (1957). Mesure in vivo de la radioactivité thyroïdienne après injection d'iode radioactif ^{131}I chez les colombidés. *Z. Vergl. Physiol.* **40**, 479–491.

Potts, J. T., Jr., and Auerbach, G. D. (1965). The chemistry of parathyroid hormone. *In* "The Parathyroid Glands: Ultrastructure, Secretion and Function" (P. J. Gaillard, R. V. Talmage, and A. M. Budy, eds.), pp. 53–67. Univ. of Chicago Press, Chicago, Illinois.

Potts, J. T., Jr., Brewer, H. B., Jr., Reisfeld, R. A., Hirsch, P. F., Schlueter, R., and Munson, P. L. (1968). Isolation and chemical properties of porcine thyrocalcitonin. *In* "Parathyroid Hormone and Thyrocalcitonin (Calcitonin)" (R. V. Talmage and L. F. Bélanger, eds.), pp. 54–67. Excerpta Med. Found., Amsterdam.

Premachandra, B. N., and Turner, C. W. (1960). Thyrotropic hormone secretion rate (TSH) in the fowl. *Poultry Sci.* **39**, 1286.

Radnot, M., and Orban, T. (1956). Die Wirkung des Lichtes auf die Funktion der Schilddrüse. *Acta Morphol. Ac. Sci. Hungar.* **6**, 375–379.

Raheja, K. L., and Snedecor, J. G. (1970). Comparison of subnormal multiple doses of L-thyroxine and L-triiodothyronine in propylthiouracil-fed and radiothyroidectomized chicks (*Gallus domesticus*). *Comp. Biochem. Physiol.* **37**, 555–563.

Ray, I., and Ghosh, A. (1961). The chemical cytology of the adrenal medulla in the domestic pigeon. *Acta Histochem.* **11**, 68–77.

Refetoff, S., Robin, N. I., and Fang, V. S. (1970). Parameters of thyroid function in serum of 16 selected vertebrate species: Study of PBI, serum T4, Free T4 and the pattern of T4 and T3 binding to serum proteins. *Endocrinology* **86**, 793–805.

Reineke, E. P., and Turner, C. W. (1945a). Seasonal rhythms in the thyroid hormone secretion of the chick. *Poultry Sci.* **24**, 499.

Reineke, E. P., and Turner, C. W. (1945b). The relative thyroidal potency of L- and D,L-thyroxine. *Endocrinology* **36**, 200.

Resko, J. A., Norton, H. W., and Nalbandov, A. V. (1964). Endocrine control of the adrenal in chickens. *Endocrinology* **75**, 192–200.

Riddle, O. (1923). Studies on the physiology of reproduction in birds. XIV. Suprarenal hypertrophy coincident with ovulation. *Amer. J. Physiol.* **66**, 322–339.

Riddle, O. (1925). Studies on thyroid. XX. Reciprocal size changes of gonads and thyroid in relation to season and ovulation rate in pigeon. *Amer. J. Physiol.* **73**, 5–16.

Riddle, O. (1928). Growth of the gonads and bursa Fabricii in doves and pigeons with data for body growth and age at maturity. *Amer. J. Physiol.* **86**, 248–365.

Riddle, O. (1937). On carbohydrate metabolism in pigeons. *Cold Spring Harbor Symp. Quant. Biol.* **5**, 362–374.

Riddle, O., and Opdyke, D. F. (1947). The action of pituitary and other hormones on the carbohydrate and fat metabolism of young pigeons. *Carnegie Inst. Wash. Publ.* **569**, 49.

Riddle, O., and Tange, M. (1928). On the extirpation of the bursa Fabricii in young doves. *Amer. J. Physiol.* **86**, 266–273.

Riddle, O., Smith, G. S., and Benedict, F. G. (1932). Seasonal endocrine and temperature factors which determine percentage metabolism change per degree of temperature change. *Amer. J. Physiol.* **101**, 88.

Riddle, O., Rauch, V. M., and Smith, G. C. (1945). Action of estrogen on plasma calcium and endosteal bone formation in parathyroidectomized pigeons. *Endocrinology* **36**, 41.

Roche, J., Salvatore, G., Sena, L., Aloj, S., and Covelli, J. (1968). Thyroid iodoproteins in vertebrates: Ultracentrifugal pattern and iodination rate. *Comp. Biochem. Physiol.* **27**, 67–82.

Romanoff, A. L. (1960). "The Avian Embryo: Structural and Functional Development." Macmillan, New York.

Rosenberg, L. L., Dimick, M. K., and La Roche, G. (1963). Thyroid function in chickens and rats: Effect of iodine content of the diet and hypophysectomy on iodine metabolism in White Leghorn cockerels and Long-Evans rats. *Endocrinology* **72**, 749–758.

Rosenberg, L. L., Goldman, M., La Roche, G., and Dimick, M. K. (1964). Thyroid function in rats and chickens. Equilibration of injected iodide with existing thyroidal iodine in Long-Evans rats and White Leghorn chickens. *Endocrinology* **74**, 212.

Rosenberg, L. L., Astier, H., La Roche, G., Baylé, J. D., Tixier-Vidal, A., and Assenmacher, I. (1967). The thyroid function of the drake after hypophysectomy or hypothalamic pituitary disconnection. *Neuroendocrinology* **2**, 113–125.

St. Pierre, R. L., and Ackerman, G. A. (1965). Bursa of Fabricius in chickens: Possible hormonal factor. *Science* **147**, 1307–1308.

Salem, M. H. R., Norton, H. W., and Nalbandov, A. V. (1970a). A study of ACTH and CRF in chicken. *Gen. Comp. Endocrinol.* **14**, 270–280.

Salem, M. H. R., Norton, H. W., and Nalbandov, A. V. (1970b). The rôle of vasotocin and of CRF in ACTH release in the chicken. *Gen. Comp. Endocrinl.* **14**, 281–289.

Samols, E., Tyler, J. M., and Mialhe, P. (1969a). Suppression of pancreatic glucagon release by the hypoglycaemic sulphonylureas. *Lancet* **2**, 174.

Samols, E., Tyler, J. M., Marks, V., and Mialhe, P. (1969b). The physiological role of glucagon in different species. *In* "Progress in Endocrinology" (C. Gual, ed.), Int. Congr. Ser. No. 184, pp. 206–219. Excerpta Med. Found., Amsterdam.

Sandberg, A. A., and Slaunwhite, W. R., Jr. (1962). Biological potential of transcortin. *J. Clin. Invest.* **41**, 1396–1397.

Sandberg, A. A., Slaunwhite, W. R., Jr., and Antoniades, H. N. (1957). The binding of steroids and steroid conjugates to human plasma proteins. *Recent Progr. Horm. Res.* **13**, 209–267.

Sandor, T. (1969). A detailed study of the biogenesis of corticosteroids in the avian adrenal by in vitro techniques. *Gen. Comp. Endocrinol., Suppl.* **2**, 284.

Sandor, T., and Lanthier, A. (1970). Studies on the sequential hydroxylation of progesterone to corticosteroids by domestic duck (*Anas platyrhynchos*) adrenal gland preparation *in vitro*. *Endocrinology* **86**, 552–559.

Sandor, T., Lamoureux, J., and Lanthier, A. (1963). Adrenocortical function in birds: In vitro biosynthesis of radioactive corticosteroids from pregnenolone-7-³H and progesterone 4-¹⁴C by adrenal glands of the domestic duck (*Anas platyrhynchos*) and the chicken (*Gallus domesticus*). *Endocrinology* **73**, 629–636.

Sarkar, A. K., and Ghosh, A. (1959). Non-sex specific action of progesterone on the adrenocortical alkaline phosphatase concentration in pigeons. *J. Exp. Med. Sci.* **3**, 19–20.

Sato, T., Herman, L., and Fitzgerald, P. J. (1966). The comparative ultrastructure of the pancreatic islets of Langerhans. *Gen. Comp. Endocrinol.* **7**, 132–157.

Schmidt-Nielsen, K., Jørgensen, C. B., and Osaki, H. (1958). Extrarenal salt excretion in birds. *Amer. J. Physiol.* **193**, 101–107.

Schomberg, D. W., Strob, M., and Andrews, F. N. (1964). Effects of 6α-methyl-17α-autoxyprogesterone, 17α-éthynyl-19 nortestosterone, progesterone and testosterone propionate on the adrenals, gonads and bursa of Fabricius of the chicken. *Gen. Comp. Endocrinol.* **4**, 54–60.

Schooley, J. P., Riddle, O., and Bates, R. W. (1941). Replacement therapy in hypophysectomized juvenile pigeons. *Amer. J. Anat.* **69**, 123.

Schümann, H. J. (1957). The distribution of adrenaline and noradrenaline in chromaffin granules from the chicken. *J. Physiol. (London)* **137**, 318–326.

Seal, U. S., and Doe, R. P. (1963). Corticosteroid-binding globulin: Species distribution and small scale purification. *Endocrinology* **73**, 371–376.

Seal, U. S., and Doe, R. P. (1965). Vertebrate distribution of corticosteroid-binding globulin and some endocrine effects on concentrations. *Steroids* **5**, 827.

Seal, U. S., and Doe, R. P. (1966). Corticosteroid-binding globulin: Biochemistry, physiology, and phylogeny. *In* "Steroid Dynamics" (G. Pincus, J. F. Tait, and T. Nakeo, eds.), pp. 63–90. Academic Press, New York.

Selye, H. (1943). Morphological changes in the fowl following chronic overdosages with various steroids. *J. Morphol.* **73**, 401–421.

Selye, H., and Friedman, S. (1941). The action of various steroid hormones on the testis. *Endocrinology* **28**, 229.

Shaffner, C. S., and Andrews, F. N. (1948). The influence of thiouracil on semen quality in the fowl. *Poultry Sci.* **27**, 91.

Shellabarger, C. J. (1955). A comparison of triiodothyronine and thyroxine in the chick goiter-prevention tests. *Poultry Sci.* **34**, 1437–1440.

Shellabarger, C. J., and Pitt-Rivers, R. H. (1958). Presence of triiodothyronine in fowls. *Nature (London)* **181**, 546.

Shellabarger, C. J., and Tata, J. R. (1961). Effect of administration of human serum thyroxine-binding globulin on the disappearance rates of thyroid hormones in the chickens. *Endocrinology* **68**, 1056–1058.

Shepherd, D. M., and West, G. B. (1951). Noradrenaline and the suprarenal medulla. *Brit. J. Pharmacol. Chem. Ther.* **6**, 665–674.

Shirley, H. V., Jr., and Nalbandov, A. V. (1956). Effects of transecting hypophyseal stalks in laying hens. *Endocrinology* **58**, 694.

Sierens, A., and Noyons, A. K. (1926). L'influence de la thyroxine sur le métabolisme du pigeon. *C. R. Soc. Biol.* **94**, 789–792.

Silber, R. H., Bush, R. D., and Oslapas, R. (1958). Practical procedure for estimation of corticosterone and hydrocorticosterone. *Clin. Chem.* **4**, 278–285.

Simkiss, K. (1961). Calcium metabolism and avian reproduction. *Biol. Rev. Cambridge Phil. Soc.* **36**, 321.

Singh, A., and Reineke, E. P. (1968). Estimation of thyroid secretion rate in the chicken by direct output method, with related observations on thyroidal iodine, body weight in the chicken and quail. *Gen. Comp. Endocrinol.* **10**, 296–303.

Singh, A., Reineke, E. P., and Ringer, R. K. (1967). Thyroxine and triiodothyronine turnover in the chicken and the Bob-white and the Japanese Quail. *Gen. Comp. Endocrinol.* **9**, 353–361.

Singh, A., Reineke, E. P., and Ringer, R. K. (1968a). Comparison of thyroid secretion rate in chickens as determined by (1) goiter prevention, (2) thyroid hormone substitution, (3) direct output, (4) thyroxine degradation methods. *Poultry Sci.* **47**, 205–211.

Singh, A., Reineke, E. P., and Ringer, R. K. (1968b). Influence of thyroid states of the chick on growth and metabolism with observations on several parameters of thyroid function. *Poultry Sci.* **47**, 212–219.

Sinha, D., and Ghosh, A. (1961). Some aspects of adrenocortical cytochemistry in the domestic fowl. *Endokrinologie* **40**, 270–280.

Sinha, D., and Ghosh, A. (1964). Cytochemical study of the suprarenal cortex of the pigeon under altered electrolyte balance. *Acta Histochem.* **17**, 222–229.

Sitbon, G. (1967). La pancréatectomie totale chez l'Oie. *Diabetologia* **3**, 427.

Sitbon, G., and Mialhe, P. (1972). Les mécanismes du feed-back glucose-glucagon et glucose-insuline chez l'Oie. *Gen. Comp. Endocrinol.* **18**, 624.

Smith, L. F. (1966). Species variation in the amino acid sequence of insulin. *Amer. J. Med.* **40**, 662–666.

Smyth, J. R., Jr., and Fox, T. W. (1951). The thyroxine secretion rate of turkey poults. *Poultry Sci.* **30**, 607–614.

Snedecor, J. G. (1968). Liver hypertrophy, liver glycogen accumulation and organ-weights changes in radio-thyroidectomized and goitrogen-treated chicks. *Gen. Comp. Endocrinol.* **10**, 277–291.

Snedecor, J. G., and Camyre, M. F. (1966). Interaction of thyroid hormone and androgens on body weight, comb and liver in cockerels. *Gen. Comp. Endocrinol.* **6**, 276–287.

Snedecor, J. G., and King, D. B. (1964). Effects of radiothyroidectomy in chicks with emphasis on glycogen body and liver. *Gen. Comp. Endocrinol.* **4**, 144–154.

Snedecor, J. G., King, D. B., and Hendrikson, R. C. (1963). Studies on the chick glycogen body: Effects of hormones and normal glycogen turnover. *Gen. Comp. Endocrinol.* **3**, 176–183.

Sørensen, O. H., and Nielsen, S. P. (1969). The hypocalcemic and hypophosphatemic effect of extracts from chickens ultimobranchial glands. *Acta Endocrinol. (Copenhagen)* **60**, 689–695.

Soulé, J., and Assenmacher, I. (1966). Mise en évidence d'un cycle annuel de la fonction corticosurrénalienne chez le Canard mâle. *C. R. Acad. Sci.* **263**, 983–985.

Speers, G. M., Perey, D. Y. E., and Brown, D. M. (1970). Effects of ultimobranchialectomy in the laying hen. *Endocrinology* **87**, 1292–1297.

Spronk, N. (1961). Thyroid gland fractions of cockerels and radioiodine metabolism. *Arch. Neer. Zool.* **14**, 1–44.

Staaland, H. (1967). Anatomical and physiological adaptations of the nasal glands in Charadriiformes birds. *Comp. Biochem. Physiol.* **23**, 933–944.

Stahl, P., and Turner, C. W. (1961). Seasonal variation in thyroxine secretion rate in two strains of New Hampshire chickens. *Poultry Sci.* **40**, 239–242.

Stainer, I. M., and Holmes, N. N. (1969). Some evidence for the presence of a corticotrophin releasing factor (CRF) in the duck (*Anas platyrhynchos*). *Gen. Comp. Endocrinol.* **12**, 350–359.

Stamler, J., Pick, R., and Katz, L. N. (1954). Effects of cortisone, hydrocortisone and corticosterone on lipemia, glycemia, and atherogenesis in cholesterol-fed chicks. *Circulation* **10**, 237–246.

Steeno, O., and De Moor, P. (1966). The corticosteroid binding capacity of plasma transcortin in mammals and aves. *Bull. Soc. Roy. Zool. Anvers* **38**, 3–24.

Stetson, M. H., and Erickson, J. E. (1971). Endocrine effects of castration in White-crowned Sparrow. *Gen. Comp. Endocrinol.* **17**, 105–114.

Stoeckel, M. E., and Porte, A. (1967). Sur l'ultrastructure des corps ultimobranchiaux du poussin. *C. R. Acad. Sci.* **265**, 2051–2053.

Stoeckel, M. E., and Porte, A. (1969). Etude ultrastructurale des corps ultimobranchiaux du poulet. I. Aspect normal et développement embryonnaire. *Z. Zellforsh. Mikrosk. Anat.* **94**, 495–512.

Sturkie, P. D., and Lin, Y. C. (1968). Sex difference in blood norepinephrine of chickens. *Comp. Biochem. Physiol.* **24**, 1073–1075.

Sturkie, P. D., Poorvin, D., and Ossorio, N. (1970). Levels of epinephrine and norepinephrine in blood and tissues of duck, pigeon, turkey and chicken. *Proc. Soc. Exp. Biol. Med.* **135**, 267–274.

Sulman, F., and Perek, M. (1947). Influence of thiouracil on the basal metabolism rate and on molting in hens. *Endocrinology* **41**, 514.

Svetsarov, E., and Streich, G. (1940). Hormonal mechanism of the moult in birds. *Dokl. Acad. Nauk SSSR* **27**, 393.

Szenberg, A., and Warner, N. L. (1962). Dissociation of immunological responsiveness in fowls with a hormonally arrested development of lymphoid tissues. *Nature (London)* **194**, 146–147.

Szent-Györgyi, A., Hegyeli, A., and McLaughlin, I. A. (1962). Constituents of the thymus gland and their relation to growth, fertility, muscle and cancer. *Proc. Nat. Acad. Sci. U.S.* **48**, 1439–1442.

Taber, E., Salley, K. W., and Knight, J. S. (1956). The effects of hypoadrenalism and chronic inanition on the development of the rudimentary gonad in sinistrally ovariectomized fowl. *Anat. Rec.* **126**, 177.

Taibel, A. M. (1941). Il timo nei polli bursectomizatti. *Riv. Biol.* **31**, 1429–1444.

Tanabe, Y., and Katsuragi, T. (1962). Thyroxine secretion rate of molting and laying hens and general discussion on the hormonal induction of molting in hens. *Bull. Nat. Inst. Agr. Sci., Ser. G* **21**, 49–59.

Tanabe, Y., Himeno, K., and Nozaki, H. (1957). Thyroid and ovarian function in relation to molting in the hen. *Endocrinology* **61**, 661–666.

Tanabe, Y., Komiyama, T., Kobota, D., and Tamaki, Y. (1965). Comparison of the effects of thiouracil, propylthiouracil and methimazole on ^{131}I metabolism by the chick thyroid, and measurements of thyroxine secretion rate. *Gen. Comp. Endocrinol.* **5**, 60–68.

Tata, J. R., and Shellabarger, C. J. (1959). An explanation for the difference between the response of mammals and birds to thyroxine and triiodothyronine. *Biochem. J.* **72**, 608–613.

Taurog, A., and Chaikoff, J. L. (1946). On the determination of plasma iodine. *J. Biol. Chem.* **163**, 313–320.

Taurog, A., Tong, W., and Chaikoff, J. L. (1950). The monoiodotyrosine content of the thyroid gland. *J. Biol. Chem.* **184**, 83.

Thapliyal, J. P. (1965). Body weight cycle of the Spotted Munia *Uroloncha punctulata*. *Symp. Comp. Endocrinol.* Nat. Inst. Sci. India, Jaîpur, cited in Chandola, A. (1972).

Thapliyal, J. P. (1969). Thyroid in avian reproduction. *Gen. Comp. Endocrinol. Suppl.* **2**, 111.

Thapliyal, J. P., and Bageshwar, K. (1970). Light responses of thyroidectomized common weaver birds. *Condor* **72**, 190.

Thapliyal, J. P., and Garg, R. K. (1967). Thyroidectomy in the juveniles of the Chestnut-bellied Munia (*Munia atricapilla*). *Endokrinologie* **52**, 75–79.

Thapliyal, J. P., and Garg, R. K. (1969). Relation of thyroid to gonadal and body weight cycles in male weaver bird, *Ploceus philippinus*. *Arch. Anat. Histol. Embryol.* **516**, 689.

Thapliyal, J. P., and Pandha, S. K. (1965). Thyroid-gonad relationship in Spotted Munia, *Uroloncha punctulata*. *J. Exp. Zool.* **158**, 253–261.

Thapliyal, J. P., and Pandha, S. K. (1967a). The thyroid and the hypophysial gonadal axis in the female Spotted Munia, *Uroloncha punctulata. Gen. Comp. Endocrinol.* **8**, 84–93.

Thapliyal, J. P., and Pandha, S. K. (1967b). Thyroidectomy and gonadal recrudescence in Lal Munia, *Estrilda amandava. Endocrinology* **81**, 915–918.

Thapliyal, J. P., Garg, R. K., and Pandha, S. K. (1968). Effect of thyroxine on the gonad and body weight of Spotted Munia, *Uroloncha punctulata. J. Exp. Zool.* **169**, 279–286.

Tixier-Vidal, A. (1970). Cytologie hypophysaire et relations photosexuelles chez les Oiseaux. *Coll. Int. Cent. Nat. Rech. Sci. Montpellier, 1967, No. 172*, 211–232.

Tixier-Vidal, A., and Assenmacher, I. (1958). Données préliminaires sur le fonctionnement thyroïdien du Canard Pékin mâle étudié à l'aide du radioiode [131]I. *C. R. Acad. Sci.* **247**, 2035–2038.

Tixier-Vidal, A., and Assenmacher, I. (1959). Etude des synthèses iodées thyroïdiennes chez le Canard Pékin mâle à température constante. Premiers résultats sur l'influence de la lumière et de l'obscurité permanente. *C. R. Soc. Biol.* **153**, 721–726.

Tixier-Vidal, A., and Assenmacher, I. (1961a). Etude comparée de l'activité thyroïdienne chez le Canard ♂ normal, castré, ou maintenu à l'obscurité permanente. I. Période de l'activité sexuelle saisonnière. *C. R. Soc. Biol.* **155**, 215–220.

Tixier-Vidal, A., and Assenmacher, I. (1961b). Etude comparée de l'activité thyroïdienne chez le canard ♂ normal, castré ou maintenu à l'obscurité permanente. II. Période de repos sexuel saisonnier. *C. R. Soc. Biol.* **155**, 286–290.

Tixier-Vidal, A., and Assenmacher, I. (1962). The effect of anterior hypophysectomy on thyroid metabolism of radioactive iodine (^{131}I) in male ducks. *Gen. Comp. Endocrinol.* **2**, 574–585.

Tixier-Vidal, A., and Assenmacher, I. (1963). Action de la métopirone sur la préhypophyse du canard mâle. Essai d'identification des cellules corticotropes. *C. R. Soc. Biol.* **157**, 1350.

Tixier-Vidal, A., and Assenmacher, I. (1965). Some aspects of the pituitary-thyroid relationship in birds. *Excerpta Med. Found. Int. Congr. Ser. No. 83*, pp. 172–182.

Tixier-Vidal, A., and Benoit, J. (1962). Influence de la castration sur la cytologie préhypophysaire du Canard mâle. *Arch. Anat. Micros. Morphol. Exp.* **51**, 265–286.

Tixier-Vidal, A., Follett, B. K., and Farner, D. S. (1967). Identification cytologique et fonctionnelle des types cellulaires de l'adénohypophyse chez le Caille mâle, *"Coturnix coturnix japonica"* soumise à différentes conditions expérimentales. *C. R. Acad. Sci.* **264**, 1739–1742.

Tixier-Vidal, A., Follett, B. K., and Farner, D. S. (1968). The anterior pituitary of the Japanese Quail, *Coturnix coturnix japonica.* The cytological effects of photoperiodic stimulation. *Z. Zellforsch. Mikrosk. Anat.* **92**, 610–635.

Tixier-Vidal, A., Baylé, J. D., and Assenmacher, I. (1972). Etude chronologique de l'évolution structurale de l'hypophyse après autogreffe ectopique chez le Canard. Unpublished data.

Turner, C. W. (1948). Effects of age and season on the thyroxine secretion rate of White Leghorn hens. *Poultry Sci.* **27**, 146–154.

Unger, R. H., Ketterer, H., Dupré, J., and Eisentraut, A. M. (1967). The effects of secretin, pancreozymin and gastrin on insulin and glucagon secretion in anesthetized dogs. *J. Clin. Invest.* **46**, 630.

Urist, M. R. (1967). Avian parathyroid physiology including special comments on calcitonin. *Amer. Zool.* **7**, 883–895.

Urist, M. R., and Deutsch, N. M. (1960). Influence of ACTH upon avian species and osteoporosis. *Proc. Soc. Exp. Biol. Med.* **104**, 35–39.

Urist, M. R., Deutsch, N. M., Pomerantz, G., and McLean, F. C. (1960). Interrelations between actions of parathyroid hormone and estrogens on bone and blood in avian species. *Amer. J. Physiol.* **199**, 851–855.

van Tienhoven, A. (1961). Endocrinology of reproduction in birds. *In* "Sex and Internal Secretions" (W. C. Young, ed.), 3rd ed., Vol. 2, pp. 1088–1169. Williams and Wilkins, Baltimore, Maryland.

Vaugien, L. (1954). Influence de l'obscuration temporaire sur la durée de la phase réfractaire du cycle sexuel du Moineau domestique. *Bull. Biol. Fr. Belg.* **88**, 294–309.

Vaugien, L. (1955). Sur les réactions testiculaires du jeune Moineau domestique illuminé à diverses époques de la mauvaise saison. *Bull. Biol. Fr. Belg.* **89**, 218–244.

Vincent, S., and Hollenberg, M. S. (1920). The effects of inanition upon the adrenal bodies. *Endocrinology* **4**, 408–410.

Vlijm, L. (1958). On the production of hormones in the thyroid gland of birds (cockerels). *Arch. Neer. Zool.* **12**, 467–531.

Vuylsteke, C. A., and de Duve, C. (1953). Le contenu en glucagon du pancréas aviaire. *Arch. Int. Physiol.* **61**, 273–274.

Wahlberg, P. (1955). The effect of thyrotropic hormone on thyroid function. *Acta Endocrinol. (Copenhagen)* Suppl. **23**, 1–79.

Warner, N. L., and Szenberg, A. (1964a). The immunological function of the bursa of Fabricius in the chicken. *Annu. Rev. Microbiol.* **18**, 235–268.

Warner, N. L., and Szenberg, A. (1964b). Immunologic studies on hormonally bursectomized and surgically thymectomized chickens: Dissociation of immunologic responsiveness. *In* "Thymus in Immunobiology" (R. A. Good and A. E. Gabrielsen, eds.), pp. 395–413. Harper, New York.

Weitzel, G., Oertel, W., Rager, K., and Kemmler, W. (1969). Insulin vom Truthuhn (*Meleagris gallopavo*). *Hoppe-Seyler's Z. Physiol. Chem.* **350**, 57–62.

Wentworth, B. C., and Mellen, W. J. (1961). Circulating thyroid hormones in domestic birds. *Poultry Sci.* **40**, 1275–1276.

Wheeler, R. S., Hoffman, N. E., and Graham, C. L. (1948). The value of thyroprotein in starting, growing and laying rations. I. Growth, feathering and food consumption in Rhode Island Red broilers. *Poultry Sci.* **27**, 103–111.

White, A., Goldstein, A. L., Banerjee, S., and Asanuma, Y. (1969). Preparation, purification and properties of thymosin, a biologically active thymic hormone. *In* "Progress in Endocrinology" (C. Gual, ed.), Int. Congr. Ser. No. 184, pp. 1228–1235. Excerpta Med. Found., Amsterdam.

Whitehouse, B. J., and Vinson, G. P. (1967). Pathways of corticosteroid biosynthesis in duck adrenal glands. *Gen. Comp. Endocrinol.* **9**, 161–170.

Wieselthier, A. S., and van Tienhoven, A. (1972). The effect of thyroidectomy on testicular size and on the photorefractory period in the starling (*Sturnus vulgaris.* L.). *J. Exp. Zool.* **179**, 331–338.

Wilson, A. C., and Farner, D. S. (1960). The annual cycle of thyroid activity in White-crowned Sparrows of eastern Washington. *Condor* **62**, 414.

Wilson, H. R., and MacLaury, D. W. (1961). The effects of tapazole on growth of hybrid cockerels. *Poultry Sci.* **40**, 890.

Winchester, C. F. (1939). Influence of thyroid on egg production. *Endocrinology* **24**, 697.

Winchester, C. F., and Davis, G. K. (1952). Influence of thyroxine on growth of chicks. *Poultry Sci.* **31**, 31–34.

Witterman, E. R., Cherian, G., and Radde, I. C. (1969). Calcitonin content of ultimobranchial body tissue in chicks, pullets, and laying hens. *Can. J. Physiol. Pharmacol.* **47**, 175–180.

Woitkewitsch, A. A. (1938). The influence of thyroidectomy on the development of ducklings. *Dokl. Akad. Nauk SSSR* **21**, 202–205.

Woitkewitsch, A. A. (1940). Dependence of seasonal periodicity in gonadal changes on the thyroid gland in *Sturnus vulgaris* L. *Dokl. Akad. Nauk SSSR* **27**, 741–745.

Wollman, S. H., and Reed, F. E. (1959). Transport of radioiodide between thyroid gland and blood in mice and rats. *Am. J. Physiol.* **196**, 113–120.

Woodward, M. (1931). Studies in bursectomized and thymectomized chickens. Ph.D. Thesis, Kansas State College, Lawrence.

Wright, A., and Chester-Jones, I. (1955). Chromaffin tissue in the lizard adrenal gland. *Nature (London)* **175,** 1001–1002.

Wright, A., Phillips, J. G., and Huang, D. P. (1966). The effect of adenohypophysectomy on the extrarenal and renal excretion of the saline-loaded duck (*Anas platyrynchos*). *J. Endocrinol.* **36,** 249–256.

Zarrow, M. X., and Baldini, J. T. (1952). Failure of adrenocorticotrophin and various stimuli to deplete ascorbic acid content of adrenal gland of the quail. *Endocrinology* **50,** 555–561.

Zawadowsky, B. M. (1926). Eine neue Gruppe der morphogenetischen Funktionen der Schilddrüse. *Arch. Entwicklungsmech. Organismen* **107,** 329–354.

Zawadowsky, M. M. (1922). "Das Geschlecht und die Entwicklung der Geschlechts-Merkmale." Imp. Government, Moscow.

Chapter 4

NEUROENDOCRINOLOGY IN BIRDS

Hideshi Kobayashi and Masaru Wada

I. Introduction

External and internal environmental changes may stimulate the receptors of animals. Some of the information generated at the receptors in the form of nerve impulses may be transferred ultimately to neurosecretory cells in the hypothalamus. The neurosecretory cells transform the afferent neural signals into neurohormonal information through neurohypophyseal hormones and adenohypophyseal hormone-releasing factors. The former are released into the systemic circulation via the neural lobe and act mainly on the kidney, blood vessels, and uterus. This neurosecretory system can be demonstrated by means of Gomori's aldehyde fuchsin or chrome alum hematoxylin staining methods (Fig. 1). The sites of production of the neurosecretory material are well known. The releasing factors are accumulated in the median eminence and pass into the adenohypophysis through the portal vessels. The neurosecretory system producing the releasing factors is not, or at least largely not, stained with Gomori's staining methods. Therefore, the location of the nuclei that produce the releasing factors and the pathways involved have not yet been determined. In addition to these two neurosecretory systems, the ependymal cells of the median eminence seem to be involved in the transport of information from the third ventricle to the adenohypophysis (see Kobayashi et al., 1970). The main motive of recent research in neuroendocrinology has been to discover the series of events in the processes mentioned above and to clarify the mechanisms whereby the processes are integrated.

Investigations in this field are most advanced in mammals, and much less is known in birds. In order to understand neuroendocrine mechanisms in general, it is necessary to know basic patterns common to animals of all vertebrate classes and, further, to recognize characteristic features that are specific to animals of a certain class. In birds, several characteristic features have already been demonstrated: (1) nerve cells in the hypothalamus are generally diffusely distributed; (2) the median eminence is morphologically distinct; (3) the adenohypophysis is completely separated from the brain; (4) portal vessels generally lie external to the median eminence and the adenohypophysis; and (5) gonadal growth is readily regulated by photoperiodic manipulation in some species. Except for (1), these features are advantageous for studying neuroendocrine mechanisms in birds. Some investigators have considered these advantages and engaged in avian neuroendocrinology, although the number is much smaller than those that investigate mammalian species. Excellent reviews of the avian studies have been published by Farner and Oksche (1962) and Farner et al. (1967). Therefore, papers published before 1962 are not cited in this chapter unless they are specifically necessary. Further, the pineal and subcommissural organs are not mentioned here, since they are described in other chapters.

II. Anatomy of the Avian Neurosecretory System

A. LIGHT MICROSCOPY OF THE NEUROSECRETORY SYSTEM

1. Neurosecretory Cells and Their Axonal Pathways

a. Gomori-Positive Neurosecretory System. The hypothalamic neurosecretory nucleus of fishes and amphibians consists of one cell group, the preoptic nucleus (Fig. 1). In fishes, the neurosecretory cells send their axons mostly to the posterior part of the neurohypophysis, which is equivalent to the pars nervosa of higher vertebrates, but a few axons terminate in the anterior part, which corresponds to the median eminence. In amphibians, the neurosecretory

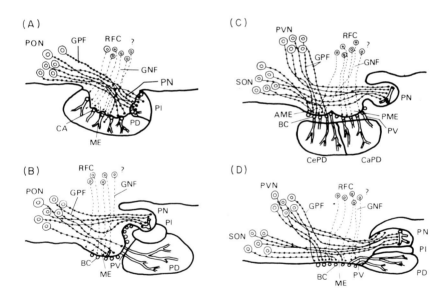

FIG. 1. Diagram of sagittal section of the median eminence–hypophyseal region, (A) teleosts; (B) anurans; (C) reptiles and birds; (D) mammals. AME = anterior median eminence; BC = capillaries of primary plexus; CA = capillaries draining into pars distalis; CaPD = caudal pars distalis; CePD = cephalic pars distalis; GNF = Gomori-negative fiber; GPF = Gomori-positive fiber; ME = median eminence; PD = pars distalis; PI = pars intermedia; PME = posterior median eminence; PN = pars nervosa; PON = preoptic nucleus; PV = portal vessel; PVN = paraventricular nucleus; RFC = cells producing releasing or inhibiting factors (the site of these cells in the hypothalamus is not known); SON = supraoptic nucleus.

cells send their axons to the median eminence and the pars nervosa (Fig. 1). In fishes and amphibians, these axons form the preopticohypophyseal tract. In reptiles and birds, the hypothalamic neurosecretory nuclei consist of the supraoptic nucleus and the paraventricular nucleus. The neurosecretory neurons of both nuclei send axons to the median eminence and the pars nervosa. The axonal tract, which is clearly seen in the median eminence, leading to the pars nervosa is designated as the supraopticohypophyseal tract (Fig. 1) (Wingstrand, 1951; Oksche *et al.*, 1959). It should be noted here that the median eminence of reptiles and birds is so well developed that it is excellent material for morphological investigations (Figs. 1 and 3).

Several investigators have demonstrated that both the supraoptic and paraventricular nuclei contain several clusters of neurosecretory cells. In the Budgerigar *(Melopsittacus undulatus)*, for instance, three neurosecretory cell groups are recognized in the supraoptic nucleus (Kobayashi *et al.*, 1961). Matsui (1966a) divided the neurosecretory cells into ten groups in the Tree Sparrow *(Passer montanus saturatus)*. In *Zosterops japonica*, Arai (1963) and Uemura and Kobayashi (1963) divided the neurosecretory neurons into several groups. The number of neurosecretory cells of each group in *Zosterops* has been estimated by Arai (1963). Rossbach (1966) identified nine groups of neurosecretory cells in the hypothalamus of the European Blackbird *(Turdus merula)*: nucleus entopeduncularis anterior, nucleus entopeduncularis medialis, nucleus entopeduncularis posterior, nucleus entopeduncularis ventralis, nucleus lateralis externus hypothalami, nucleus magnocellularis interstitialis dorsalis, nucleus paraventricularis dorsalis, nucleus paraventricularis ventralis, and nucleus supraopticus medialis lateralis. Thus it is typical for the neurosecretory cells to occur in several groups or clusters. However, functional differences among these groups of neurosecretory cells are not known.

Almost all of the cells of those groups send axons into the pars nervosa and/or the median eminence. Then, a question arises as to which group or groups send axons to the median eminence and which to the pars nervosa. There have been attempts to clarify this problem. As will be mentioned later, in *Passer montanus* the neurosecretory material in the pars nervosa, but not in the median eminence, decreases in amount when they are deprived of drinking water (Matsui, 1964), and the material in the median eminence, but not in the pars nervosa, decreases when they are exposed to long days (Matsui, 1966a). By examining the changes in the neurosecretory cell perikarya in the hypothalamus under two different experimental conditions,

Matsui (1966a) has demonstrated that neurosecretory cells of both supraoptic and paraventricular nuclei send their axons to both regions, but in different numerical ratios.

In the Rhode Island fowl, Legait (1959) reported the presence of two extrahypothalamohypophyseal pathways: the hypothalamoseptal tract, which ends in a subseptal organ (subfornical organ), and the hypothalamohabenular tract, which connects the hypothalamic and epithalamic regions of the brain. The functional significance of these pathways is not known.

b. Gomori-Negative Neurosecretory System. It has recently been demonstrated in mammals that the brain produces hypophysiotropic neurohormones that induce or inhibit hormone release from the adenohypophysis (Fig. 1) (see Meites, 1970). However, specific sites in the brain for the production of these releasing and inhibiting factors are not known. Judging from pituitary cytology, the avian adenohypophysis seems to be similar to that of the mammals and apparently produces six hormones (Tixier-Vidal, 1965; Matsuo *et al.*, 1969; Mikami *et al.*, 1969; Tixier-Vidal *et al.*, 1968). Therefore, at least six hypophysiotropic neurohormones must be produced in the brain. Actually, adrenocorticotropic hormone-, gonadotropin-, prolactin-, and growth hormone-releasing factors have been demonstrated in birds (see Section VII,B). However, the nuclei that produce these neurohormones have not yet been determined. Detailed physiology of the releasing factors in birds will be mentioned later (Section VII).

2. Median Eminence

A definition of the median eminence has recently been presented by the senior author (see Kobayashi *et al.*, 1970). According to him, the median eminence is exteriorly the portion covered by the capillaries of the primary plexus of the hypophyseal portal veins, and interiorly the basal portion of the hypothalamus occupied by the processes of the secretory ependymal cells (Fig. 2; Section VI). In this portion there are some perikarya of the infundibular nucleus. However, these neurons are excluded from the definition of the median eminence in order to generalize its definition. For instance, the anterior portion of the teleost neurohypophysis, the equivalent of the median eminence of higher vertebrates, does not contain any neurons (see Kobayashi *et al.*, 1970). Usually, the median eminence begins anteriorly immediately caudal to the optic chiasm and ends at the beginning of the infundibular stem, on the surface of which the capillaries of the primary plexus no longer occur. The avian median emi-

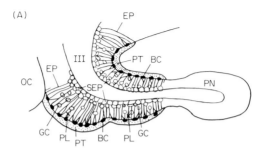

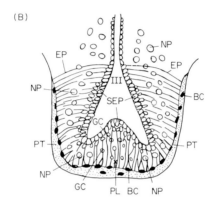

FIG. 2. Diagram of the median eminence of the pigeon, (A) sagittal section, (B) cross section. BC = capillaries of primary plexus; GC = glial cells; EP = ependymal processes; NP = perikaryon of neuron; OC = optic chiasma; PL = palisade layer; PN = pars nervosa; PT = pars tuberalis; SEP = secretory ependymal cells; III = third ventricle.

nence is clearly divided into anterior and posterior divisions. The anterior division contains abundant aldehyde fuchsin-positive neurosecretory material, whereas the posterior division rarely contains such material (Fig. 3).

In the median eminence, the following layers can be recognized: (1) ependymal layer, (2) hypendymal layer, (3) fiber layer, (4) reticular layer, and (5) palisade layer (Fig. 4). On the basal surface of the palisade layer lie the capillaries of the primary plexus. Further, the pars tuberalis covers the network of the primary plexus. The ependymal layer is a single layer of ependymal cells. The ependymal cells extend their processes to the basal surface of the hypothalamus and also to the capillaries of the primary plexus (Fig. 2). Below the ependymal layer are located the hypendymal cells forming one- or two-cell layers.

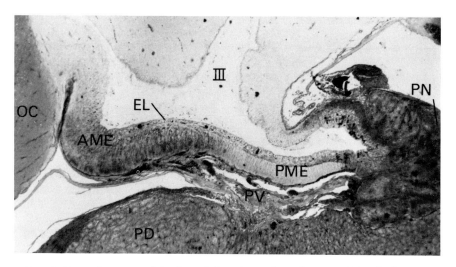

FIG. 3. Sagittal section of the hypothalamohypophyseal region of the duck. AME = anterior median eminence; EL = ependymal layer; OC = optic chiasma; PD = pars distalis; PME = posterior median eminence; PN = pars nervosa; PV = portal vessels; III = third ventricle. AF staining; × 50.

The degree of development of this layer varies with the species. Their processes proceed to the outer surface of the median eminence. The fiber layer contains numerous unmyelinated fibers. Among them, the supraopticohypophyseal tract proceeds toward the pars nervosa. From this tract, many neurosecretory axons proceed down toward the surface of the median eminence. The processes of the ependymal and hypendymal cells run downward through this layer. In the fiber layer there are elongated glial cells with processes that extend in various directions. There is a reticular layer below the fiber layer; here, the neurosecretory axons, which run down from the supraopticohypophyseal tract, Gomori-negative axons, and the ependymal and hypendymal processes are intermingled. However, in the posterior median eminence, the reticular layer is not prominent, partly due to the paucity of neurosecretory axons. There are many round glial cells in this layer. The outermost layer is the palisade layer in which the ependymal and hypendymal processes and Gomori-positive and -negative nerve fibers are arranged in a radially palisade fashion. There are many glial cells, and the neurosecretory material is accumulated around them, just as in the neural lobe. The components of the median eminence are shown in Table I.

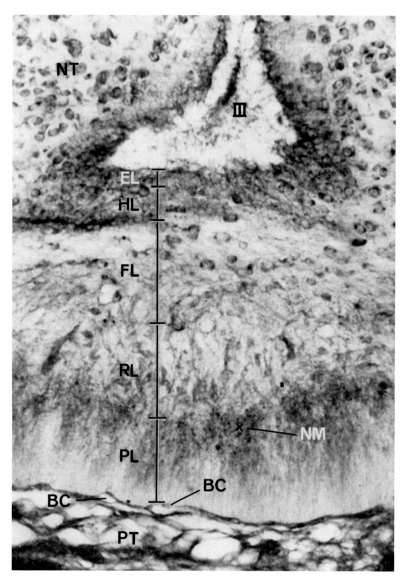

FIG. 4. Anterior median eminence of the Japanese Quail. BC = blood capillary; EL = ependymal layer; FL = fiber layer; HL = hypendymal layer; NM = neurosecretory material; NT = perikarya of nucleus tuberis; PL = palisade layer; PT = pars tuberalis; RL = reticular layer; III = third ventricle. AF and toluidine blue O staining; × 300.

TABLE I

STRUCTURAL COMPONENTS OF THE MEDIAN EMINENCE OF BIRDS

Zone	Layer	Components
Internal zone	Ependymal layer	Ependymal cells
	Hypendymal layer	Hypendymal cells
		Glial cells
	Fiber layer	Ependymal processes
		Supraopticohypophysial tract
		Glial cells
		Ependymal, hypendymal, and glial processes
External zone	Reticular layer	Supraopticohypophyseal tract
		Tuberohypophyseal tract or infundibulohypophyseal tract
		Glial cells
		Ependymal, hypendymal, and glial processes
	Palisade layer	Same as above. After leaving the reticular layer, all the fibers and processes proceed in palisade fashion to the basal surface of the median eminence in this layer

3. The Neural Lobe

Wingstrand (1951) grouped the avian neural lobe into four categories according to the thickness of the diverticular wall and the size of the lumen. The primitive neural lobe has many thin-walled diverticula and a large lumen (*Gallus, Phasianus*). The most complex is that with thick-walled diverticula and with a very restricted lumen (Anseriformes, *Larus*). The thin diverticular wall of the neural lobe consists of an ependymal layer facing the lumen, a fiber layer, and a "glandular zone" (external layer) that contains aldehyde fuchsin (AF)-positive neurosecretory axon endings. The connective tissue and the vessels are on the surface of the external layer. The glial cells (pituicytes) are distributed in the fiber and external layers. However, the neural lobe with the thick-walled diverticula contains more neural and glial elements and is very compact in appearance. In all cases, the neural lobe contains numerous terminals of neurosecretory axons and, therefore, has a strong affinity for aldehyde fuchsin or chrome alum hematoxylin (Figs. 3 and 5).

The neurosecretory material is dense around the glial cells, especially around those cells that are near blood vessels (Kobayashi *et al.*, 1961). This suggests some functional relationship between the glial

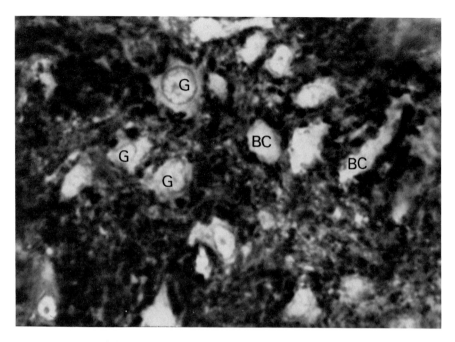

FIG. 5. A part of the pars nervosa of the Budgerigar. The neurosecretory material is dense around the capillaries and the glial cells. BC=blood capillary; G=glial cells (pituicyte). AF staining; × 1700.

cells and the neurosecretory axons. Immediately adjacent to the blood vessels there is a clear zone in which neurosecretory material is almost absent. Electron microscopy has revealed that this zone is composed of the terminals of the pituicyte and ependymal processes as mentioned later. In other words, the ependymal and pituicyte processes usually intervene between the neurosecretory axon endings and the capillaries of the primary plexus. These findings raise a question as to whether or not the ependymal and glial (pituicyte) processes are involved in the release of neurosecretory substance from the neurosecretory endings into the capillaries.

It should be noted here that destruction of the neurohemal organs of the stainable hypothalamohypophyseal neurosecretory system in the neural lobe and/or palisade layer of the anterior median eminence of the White-crowned Sparrow (*Zonotrichia leucophrys gambelii*) results in the formation of large aggregations of stainable neurosecretory material surrounding the hypertrophied capillaries. In the Japanese Quail, this appears not to occur and there is no indication of repair (Stetson, 1969b). It seems that the degree of regeneration of the neurosecretory system depends upon the species involved.

4. Fiber Connections in the Neurosecretory System

Wingstrand (1951) described four efferent hypothalamic tracts terminating in the neurohypophysis of the pigeon: (1) tractus (tr.) hypophyseus anterior, (2) tr. supraopticohypophyseus, (3) tr. tubero-hypophyseus, and (4) tr. hypophyseus posterior. The tr. supraoptico-hypophyseus is formed by neurosecretory fibers that originate from the supraoptic and paraventricular nuclei. As already mentioned, some fibers of this tract proceed to the anterior median eminence, but most fibers terminate in the neural lobe. The tr. tuberohypophyseus is formed by the fibers of the infundibular nucleus. The tr. hypophys-eus anterior seems to include all hypophyseal fibers behind the chiasma which do not belong to the tr. supraopticohypophyseus or the tr. tuberohypophyseus (Wingstrand, 1951). However, it is possible that the tr. hypophyseus anterior contains a part of a bundle in the supraopticohypophyseal tract, which is derived from the most rostral neurosecretory cells (Farner et al., 1967). The tr. hypophyseus pos-terior proceeds to the pars nervosa. This tract is distinct in Anser (Wingstrand, 1951). However, the origin of the fibers of this tract is not known.

Through the use of fluorescence microscopy, monoaminergic fibers terminating in the median eminence have recently been demonstrated (Björklund et al., 1968; Sharp and Follett, 1968, 1970; Warren, 1968; Oehmke et al., 1969; Oksche et al., 1970). The median eminence receives the fluorescent fibers from the nucleus tuberis (NT) and possibly from the nucleus hypothalamicus posterior medialis (NHPM). These fibers form part of the tuberohypophyseal tract (Fig. 6). The two nuclei (NT and NHPM) are connected by two tracts; one tract runs in the stratum cellulare internum close to the wall of the third ven-tricle and the other curves round through the lateral regions of the hypothalamus (Sharp and Follett, 1970).

Monoaminergic fibers associated with the Gomori-positive axons proceed down behind the posterior border of the optic chiasma toward the median eminence in the Japanese Quail (Sharp and Fol-lett, 1968, 1970), the White-crowned Sparrow (Warren, 1968), and the House Sparrow (Passer domesticus) (Oehmke et al., 1969). These fibers must form in part the tractus hypophyseus anterior of Wing-strand. This fiber tract can be traced to the supraoptic nucleus (Oehmke, 1969; Sharp and Follett, 1968, 1970). Oehmke (1969) ob-

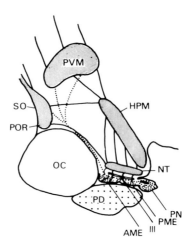

FIG. 6. A diagram of a sagittal section of the hypothalamus of the Japanese Quail showing the regions containing monoamines and the fiber connections between them. The Gomori-positive neurosecretory tracts are shown by dotted lines and monoamine fiber tracts are shown by solid lines. HPM = nucleus hypothalamicus posterior medialis; NT = nucleus tuberis; POR = preoptic recess; PVM = nucleus paraventricularis magnocellularis; SO = nucleus supraopticus. See legend of Fig. 3 for other abbreviations. (Modified from Sharp and Follett, 1970.)

served green-yellow fluorescence (indicating the presence of monoamines) in the infundibular and ventromedial nuclei of the Greenfinch *(Carduelis chloris)* and the Mallard *(Anas platyrhynchos)*. He divided the infundibular nucleus into three regions (Oehmke, 1968). The basal layer showed stronger fluorescence than the middle and dorsal layers. Fluorescence was also observed in the subependymal zone of the median eminence of both species, as well as other birds (Björklund *et al.*, 1968; Sharp and Follett, 1968; Warren, 1968; Oehmke *et al.*, 1969). The fluorescence in the palisade layer shows considerable variation in the intensity depending on the species. The fluorescent fibers in the palisade layer are probably derived from the monoaminergic fibers in the subependymal layer, which originate from the nucleus tuberis and possibly from the nucleus hypothalamicus posterior medialis (Sharp and Follett, 1970). In the reticular layer, a few fluorescent fibers run in a rostrocaudal direction; these can be traced to the neural lobe in the domestic fowl (Enemar and Ljunggren, 1968) and the Japanese Quail (Sharp and Follett, 1970).

    The neurosecretory cells of the supraoptic and paraventricular nuclei are embedded in monoaminergic nerve fibers (Sharp and Follett, 1968; Warren, 1968; Oehmke *et al.*, 1969). Monoamine-oxidase activity is also strong around the neurosecretory cells of the Japanese Quail (Urano, 1968). These monoaminergic fibers have been found to form synaptic contacts with the neurosecretory cells (Oehmke *et al.*, 1969; Priedkalns and Oksche, 1969). The origin of these fibers is still unknown, but it has been suggested that the fibers may be derived from those in the forebrain bundle as was observed in mammals (Fuxe and Ljunggren, 1965). In the pigeon and the Japanese Quail, the fluorescent tract in the forebrain bundle continues anteriorly to the nucleus basalis and the forebrain (Sharp and Follett, 1970).

As is shown in Fig. 6, there are monoamingergic fiber connections between the following regions, as mentioned above: nucleus (n.) tuberis, n. hypothalamicus posterior medialis, n. paraventricularis, n. supraopticus, preoptic recess, n. basalis, and median eminence. These different monoaminergic fiber connections in the avian hypothalamus may be a part of the integrative mechanism in the hypothalamic control of the adenohypophysis (Sharp and Follett, 1970).

Cholinergic innervation in the median eminence and neural lobe remains obscure, but acetylcholinesterase has been shown in the neurosecretory cells, cells of the n. infundibularis, the palisade layer of the median eminence, and the pars nervosa (see Kobayashi *et al.*, 1970).

B. ELECTRON MICROSCOPY OF THE NEUROSECRETORY SYSTEM

Since the nuclei that produce the hypophysiotropic neurohormones have not yet been identified, this description will be confined mostly to the Gomori-positive neurosecretory system.

1. Neurosecretory Cells and Their Fibers

The neurosecretory cells of the supraoptic nucleus of the House Sparrow contain 2000–2500 Å granules (Oehmke *et al.*, 1969). The granules are formed in the Golgi apparatus. The neural lobe contains 2000–2500 Å granules. Therefore, 2000–2500 Å granules must be carriers of the neurohypophyseal hormones. Priedkalns and Oksche (1969) found synaptic terminals of the monoamine fibers at the cell perikarya of the supraoptic nucleus and the infundibular nucleus of *Passer domesticus*.

In the neuropile and the perikarya of the nucleus infundibularis, in the tuberoinfundibular tract and in the reticular and palisade layers of the median eminence, 500–1000 Å granules are found in *Passer domesticus* (Oehmke *et al.*, 1969). Since these granules disappeared after reserpine treatment, they seem to contain monoamines.

2. Anterior Median Eminence

There have been several investigations of the fine structure of the avian median eminence (Kobayashi *et al.*, 1961; Oota and Kobayashi, 1962; Nishioka *et al.*, 1964; Bern and Nishioka, 1965; Bern *et al.*, 1966; Matsui, 1966b; Matsui and Kobayashi, 1968; Oehmke *et al.*, 1969; Calas and Assenmacher, 1970; Péczely and Calas, 1970). Some ependymal cells (Fig. 7) have microvilli and bulbous protrusions on the apical surfaces in the White-crowned Sparrow, duck, and pigeon (Matsui, 1966b; Matsui and Kobayashi, 1968; Calas and Assenmacher, 1970; Oota, 1970). The physiological meaning of this phenomenon will be mentioned later (see Section VI). In the subependymal layer, the hypendymal (subependymal) cells show a structure similar to the ependymal cells except for the apical protrusions. Among the processes of the ependymal cells in this layer there are sometimes fibers containing granules of 1000 Å. The presence of this type of granule coincides with the presence of monoamine fluorescence in the subependymal portion (Calas and Assenmacher, 1970). In the fiber layer (Fig. 8), there are many unmyelinated axons, each containing small (1000 Å), intermediate (1200–1500 Å), or large (1500–2500 Å) electron-dense granules. In addition, in this layer, small axons without granules are numerous. The large granules and intermediate granules occur in the neural lobe (see Kobayashi *et al.*, 1970). Therefore, these granules are doubtless carriers of the neurohypophyseal hormones. The small granules are probably carriers of monoamines (see Kobayashi *et al.*, 1970). However, there is the possibility that these three kinds of granules contain releasing factors.

In the upper portion of the palisade layer, there are fibers or processes that are continuations from the reticular layer. They proceed of the ependymal, hypendymal, and glial cells are interwoven in this layer.

In the upper portion of the palisade layer, there are fibers or processes that are continuations from the reticular layer. They proceed perpendicularly to the ventral surface of the median eminence. The axons with or without the three kinds of granules in this layer frequently contain synaptic vesicle-like structures with diameters of 300–500 Å. The axons containing exclusively these vesicles may be cholinergic fibers.

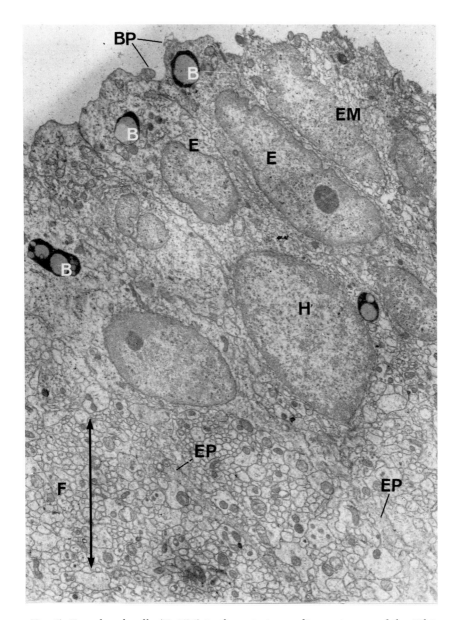

Fig. 7. Ependymal cells (E, EM) in the anterior median eminence of the White-crowned Sparrow. Some cells (B) contain black droplets with clear center. BP = bulbous protrusion; EM = cell showing possible merocrine secretion; EP = ependymal process; F = fiber layer; H = hypendymal cells. × 6400.

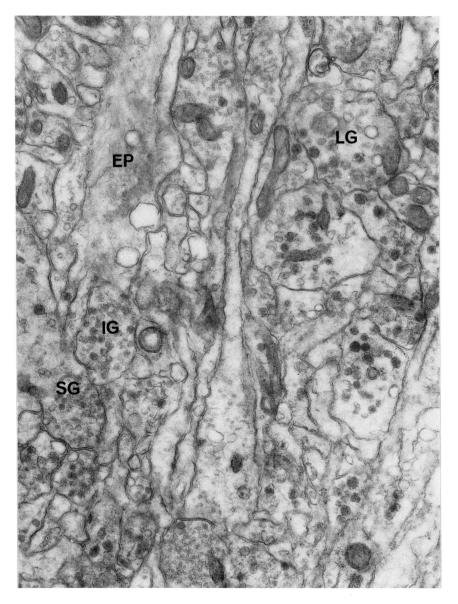

FIG. 8. Upper portion of the palisade layer of the anterior median eminence of the White-crowned Sparrow. EP = ependymal process; IG = intermediate granule; LG = large granule; SG = small granule. × 24,000.

The peripheral portion of the palisade layer of the anterior median eminence consists of numerous unmyelinated fibers with or without the granules, nerve endings, and processes of glial and ependymal cells (Fig. 9). Most of the nerve fibers contain one of three kinds of granules and synaptic vesicle-like structures. However, the fibers containing large granules are far fewer than in the upper portion. In the White-crowned Sparrow, *Passer domesticus, Agelaius tricolor,* and *Zonotrichia atricapilla,* the feet of the ependymal processes are frequently interposed between the nerve endings and the perivascular space of the primary plexus (Bern and Nishioka, 1965). However, in the Budgerigar (Kobayashi *et al.,* 1961) and the duck (Calas and Assenmacher, 1970), both the nerve endings and the ependymal processes are in contact with the perivascular space. The biological significance of this difference is obscure.

The perivascular space of the capillaries covering the surface of the median eminence is very thick (about 1 μ), a condition generally true of other endocrine organs. The space may be a reservoir for neurohormones (Hartmann, 1958). Sometimes it protrudes into the parenchyma of the median eminence, thus increasing the contact area between the space and the parenchyma. There are parenchymal and endothelial basement membranes in the space, and among them are fibroblasts and many collagen fibrils. The endothelial cells of the capillaries have many pinocytotic vesicles on the cell membrane facing the perivascular space. This suggests that neurohormones that are released into the perivascular space from the axons are absorbed by the endothelial cells and then released into the capillary lumen.

3. Posterior Median Eminence

The fine structure of the posterior median eminence is basically the same as that of the anterior median eminence. The principal differences are that (1) in the former there are few fibers that contain the large (Matsui, 1966b; Calas and Assenmacher, 1970) and intermediate granules in the reticular and the palisade layers; (2) synaptoid contacts between the axons containing granules of 1000 Å and the ependymal processes are frequent in the hypendymal and palisade layers in the posterior median eminence but seldom in the anterior median eminence (Matsui, 1966b); (3) in the posterior median eminence, the axon endings are more frequently in contact with the perivascular space than in the anterior median eminence; (4) the peri-

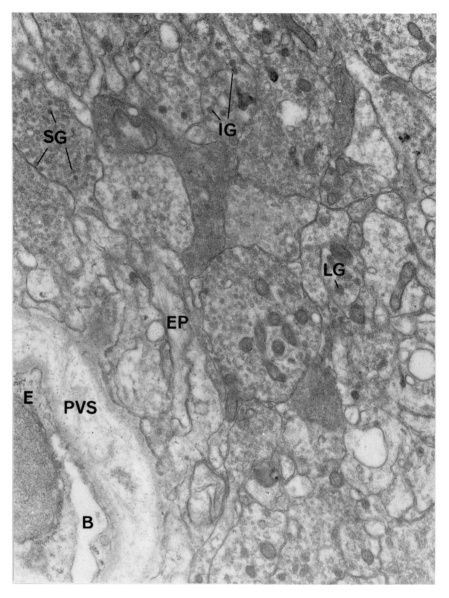

FIG. 9. Neurohemal region of the anterior median eminence of the White-crowned Sparrow. B = blood capillary; E = endothelial cell; EP = ependymal process; IG = intermediate granule; LG = large granule; PVS = perivascular space; SG = small granule. × 24,000.

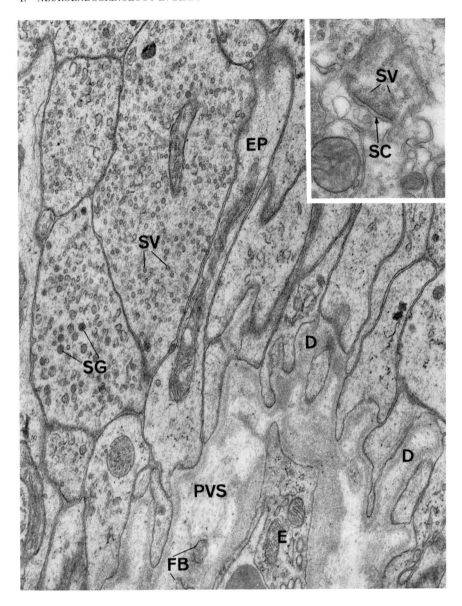

FIG. 10. Neurohemal region of the posterior median eminence of the White-crowned Sparrow. The neurohemal region of the posterior median eminence contains only small granules (SG) and small vesicles (SV). Note many digitations (D) of the perivascular space (PVS). EP = ependymal process; FB = fibroblast; SC = synaptoid contact; SG = small granule; SV = synaptic vesicle. × 24,000.

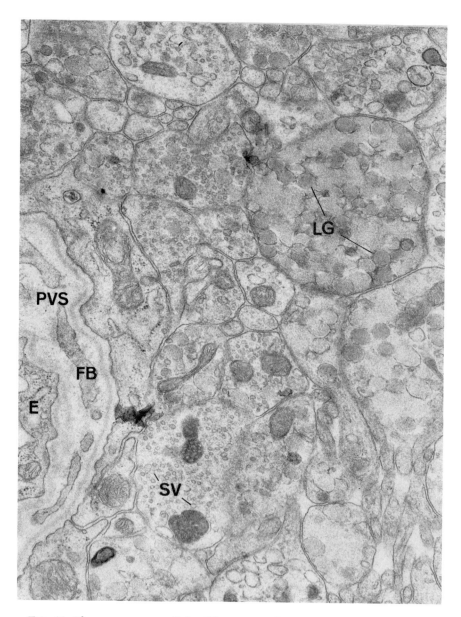

FIG. 11. The pars nervosa of the White-crowned Sparrow. E = endothelial cell; FB = fibroblast; LG = large granule; PVS = perivascular space; SV = synaptic vesicle. × 32,000.

vascular space of the posterior median eminence protrudes more digitations in complex form into the parenchyma than in the anterior median eminence (Fig. 10) (Matsui, 1966b; Calas and Assenmacher, 1970).

4. Neural Lobe

Duncan (1956) first examined the avian neural lobe with an electron microscope. In the neural lobe of the chicken he observed many neurosecretory endings containing many granules with diameters of 1000–2000 Å and pituicytes and their processes. In 1959, Legait and Legait observed granules of 1000–1800 Å in the pituicytes of the fowl neural lobe. Both investigators pointed out the intimate contact of the neurosecretory fibers with the pituicytes and suggested the involvement of the pituicyte in the release of the neurohypophyseal hormones from the endings. Since these investigations, electron microscopy has been done on the neural lobe of several species of birds —Budgerigar (Kobayashi et al., 1961; Oota and Kobayashi, 1962), White-crowned Sparrow (Nishioka et al., 1964; Bern et al., 1966), and pigeon (Péczely and Calas, 1970). All the investigators showed the presence of two kinds of granules therein; one with diameters of about 1200 Å and the other about 1600 Å (Fig. 11). Interposition of the feet of the pituicytes between the nerve endings and the pericapillary space is not always found. It depends on the species. The axons in the neural lobe contain the small vesicles of about 500 Å in addition to the granules. There are some axons containing only the small vesicles of about 500 Å, the granules being absent. These axons may be cholinergic fibers terminating in the neural lobe.

With fluorescent microscopy, it was found that in the reticular layer of the Japanese Quail monoaminergic fibers proceed to the neural lobe (Sharp and Follett, 1970). Therefore, it is expected that the granules of about 1000 Å should be present in the avian neural lobe.

The perivascular space of the neural lobe is thick as in the median eminence. There are many pinocytotic vesicles on the surface of the endothelial cells facing the perivascular space. The vesicles may be involved in the transendothelial transport of the neurohypophyseal hormones.

III. Hypothalamohypophyseal Neurovascular Link

The details of the blood supply and drainage of the hypothalamo-hypophyseal region have been described by Wingstrand (1951) and

Assenmacher (1952, 1953). More recently, Vitums *et al.* (1964) have described in detail the distribution of the blood vessels of the hypothalamohypophyseal system. Their most interesting observation is the distribution of the capillaries of the primary plexus on the surface of the median eminence. There are distinct anterior and posterior capillary plexuses corresponding to the anterior and posterior divisions of the median eminence (Fig. 1). Anterior and posterior groups of portal vessels collect the blood from the anterior and posterior primary capillary plexuses, respectively. The anterior group of portal vessels is mainly distributed to the cephalic lobe of the pars distalis and the posterior group of the portal vessels supplies the caudal lobe of the pars distalis. The development of the portal blood system has been described in the White-crowned Sparrow by Vitums *et al.* (1966). Sharp and Follett (1969a) have also observed that the primary plexus of the median eminence of the Japanese Quail is divisible into anterior and posterior parts and that it shows a point-to-point distribution to the pars distalis. Dominic and Singh (1969) have also observed the same phenomena in fifteen species of birds. Since the cell types of the cephalic lobe of the pars distalis are not always the same as those of the caudal lobe (see Farner *et al.*, 1967; Brasch and Betz, 1971), it is supposed that different adenohypophyseal hormone releasing factors may be delivered to the anterior and posterior median eminence.

Duvernoy *et al.* (1969) studied the organization of the primary plexus of several species of birds. According to them, the primary capillary bed is composed of superficial vessels and deep vessels. The deep vessels include capillary loops and a subependymal network, both of which anastomose with hypothalamic vessels. They argue that the deep vessels connect the primary plexus and the third ventricle. This idea is important because there is the possibility that the ventricular fluid contains adenohypophyseal hormone releasing factors in mammals (Halász *et al.*, 1962; Szenthágothai *et al.*, 1968).

IV. Biologically Active Substances in the Neurohypophysis

As mentioned before, Gomori-positive neurosecretory material is stored in the median eminence and the pars nervosa in birds. The avian neurohypophyseal hormones in the neural lobe are arginine vasotocin and oxytocin (for review, see Farner *et al.*, 1967; Follett, 1970b). Since the function of the median eminence is different from the pars nervosa, the possibility was expected that the Gomori-positive neurosecretory material in the median eminence would contain different neurosecretory hormones from those of the pars nervosa. However, it was found by bioassay methods and chromatography that the median eminence contains the same neurohypophyseal hormones as those of the pars nervosa (Hirano, 1964). This is true for the animals of other vertebrate classes (see Kobayashi et al., 1970).

The neurohypophyseal hormones appear to be present in the large and intermediate granules. It is not known which type of granule contains which neurohypophyseal hormone (arginine vasotocin or oxytocin). Moreover, there is the possibility that these granules also contain releasing factors. Bioassays and chemical analyses of substances in the granules isolated by ultracentrifugation may yield more precise information on the nature of the substances in the granules (Kobayashi and Ishii, 1969; Ishii, 1970). Carrier proteins of the neurohypophyseal hormones have not yet been studied in birds, but they seem to contain cysteine. DL-Cysteine[^{35}S] injected into the third ventricle was taken up by the tissues containing Gomori-positive material (Taguchi et al., 1966).

The small granules of about 1000 Å may be carriers of monoamines, since the small granules in the neural lobe and the median eminence are similar in size and profiles to the granules carrying monoamines in the anterior hypothalamus of the rat (Pellergrino de Iraldi et al., 1963; Aghajanian and Bloom, 1966; Ishii, 1967). The distribution of the small granules is almost the same as that of monoamines and monoamine oxidase in the median eminence.

With respect to the small vesicles in the median eminence and pars nervosa, there is lack of agreement about their nature in the mammalian neurohypophysis. This point is also being discussed with reference to the avian neurohypophysis. Some investigators are inclined to believe that they are carriers of acetylcholine (ACh) (Abrahams et al., 1957; Gerschenfeld et al., 1960; Kobayashi et al., 1961; Koelle and Geesey, 1961; Koelle, 1962; de Robertis, 1962, 1964; de Robertis et al., 1963; Kobayashi, 1965; Oota and Kobayashi, 1966), whereas others argue that they might be derived from the fragmentation of large granules (Holmes and Knowles, 1960; Knowles, 1963; Holmes, 1964; Lederis, 1965; Bern and Knowles, 1966). Herlant (1967) is of the opinion that there are two types of small vesicles — (1) synaptic vesicles carrying ACh and (2) small vesicles derived from the budding of neurosecretory granules. Furthermore, he has suggested that the small vesicles produced by budding contain neurosecretory material and that they are finally expelled into the peri-capillary space. Since acetylcholinesterase is present in the median eminence and neural lobe, ACh should also be present there, possibly being carried by the vesicles.

As will be mentioned below (Section VII,B), the adenohypophysis secretes at least six hormones. Therefore, it is naturally expected that the hypothalamus produces at least six neurohormones (releasing or inhibiting factors) that regulate the adenohypophyseal hormone secretion. However, in the avian median eminence, only three kinds of granules were recognized. Therefore, it is possible that granules with the same diameter carry at least two different releasing factors. For instance, corticotropin releasing factor is present in granules larger than 1500 Å in diameter, which were isolated by ultracentrifugation from the horse median eminence. These granules also possess vasopressor activity (Ishii *et al.*, 1969). Thus, there are neurohypophyseal hormones, monoamines, acetylcholine, and releasing factors in the avian median eminence. The physiological significance of these substances, except for releasing factors, is not known at the present time.

V. Functional Aspects of the Gomori-Positive Neurosecretory System

A. WATER METABOLISM

The presence of an antidiuretic substance in the avian neurohypophysis has been known for a long time (see Follett, 1970b). The neurosecretory system becomes active following dehydration or osmotic stress in birds: the neurosecretory material decreases from the pars nervosa, but not from the median eminence; and the neurosecretory cells increase their nuclear diameters (Fig. 12) (for references, see Farner and Oksche, 1962; Farner *et al.*, 1967). Vasotocin has no effect on the filtration rate but only on the tubular reabsorption of water (Skadhauge, 1964; Dantzler, 1966). It is interesting to note that the Budgerigar (Uemura, 1964b) and Zebra Finch *(Poephila guttata castanotis)* (Oksche *et al.*, 1963), which live in the hot, dry parts of Australia, respond weakly or not at all to water deprivation.

B. GONADAL GROWTH

Studies involving sectioning of the supraopticohypophyseal tract of the duck (Benoit and Assenmacher, 1953a,b; Assenmacher, 1958) and lesions of the neurosecretory nuclei (Assenmacher, 1957a,b,

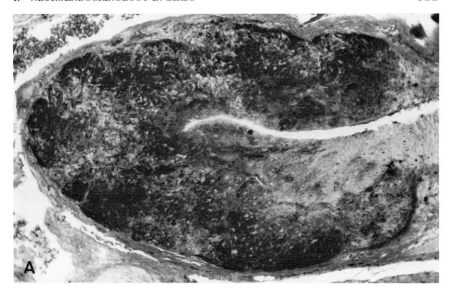

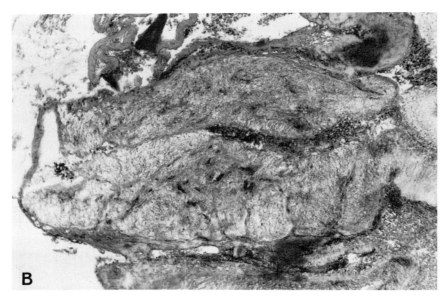

FIG. 12. AF-positive neurosecretory material in the pars nervosa of a control pigeon (A). Note a decrease of the AF-positive material in the pars nervosa of the pigeon subjected to dehydration (B). AF-staining. × 160.

1958) lead to the suggestion that the hypothalamic neurosecretory system, including both the supraoptic and paraventricular nuclei, is involved in the control of gonadotropic activity of the adenohypophysis. These important investigations carried on in Benoit's laboratory are now rather classic; they have stimulated investigations in many laboratories. In the fowl, Legait (1959) observed changes in activity of the neurosecretory cells during periods of sexual inactivity, egg-laying, incubation, and molting. Mikami (1960) reported that in the domestic fowl the neurosecretory cells change their activity after castration. Graber and Nalbandov (1965) found, however, that castration in the fowl has little effect on the neurosecretory system. In the passerine birds, Oksche *et al.* (1959) performed extensive studies on this problem using the White-crowned Sparrow. The neurosecretory system is relatively inactive during the midwinter period of sexual inactivity, and the neurosecretory material is accumulated in the perikarya and in the external layer of the median eminence. During the early summer period of sexual activity, the neurosecretory material decreases in amount in the perikarya and in the median eminence. During the refractory period, the neurosecretory material is stored extensively in both the cells and the median eminence (Oksche *et al.*, 1959). These changes in the neurosecretory system were confirmed under natural photoperiods in the same species (Laws, 1961). Rossbach (1966) has also observed seasonal changes in the hypothalamic neurosecretory system. Konishi and Kato (1967) found a similar rhythm of the neurosecretory activities of the cells and the median eminence in the Japanese Quail. The rhythm could be changed by artificial daily photoperiods. Konishi (1965, 1967) found that neurosecretory material is more abundant in the median eminence of Japanese Quail kept under short days than that of quail held in continuous light. He observed, however, that there was no correlation between the amount of neurosecretory material and the testicular development. Ishii *et al.* (1962) found that there was an increase of neurosecretory material in the median eminence of the duck in which the growth of the gonads was stimulated by long days. Hirano *et al.* (1962) found that the amount of neurosecretory material increased in the median eminence, but not in the pars nervosa, of *Zosterops japonica* subjected to long days. Thus, the results obtained by different investigators concerning the involvement of the neurosecretory system in gonadal growth are contradictory.

As mentioned before, the hypothalamic neurosecretory nuclei are topographically divisible into several cell groups. Uemura and Kobayashi (1963) divided the neurosecretory cells of *Zosterops* into seven groups: the supraoptic nucleus was subdivided into lateral and median groups; the paraventricular nucleus was subdivided into anterior, periventricular, and lateral groups. The sixth group was distributed in the peduncles, and the seventh group was located in the hilar region of the median eminence. When the birds were subjected to long days, the cells of the lateral group of the supraoptic nucleus and those of the anterior and periventricular groups were stimulated, but the other groups were not. Matsui (1966a) has also obtained similar results in *Passer montanus*. In *Zosterops*, estrogen administration activated the neurosecretory cells of the median group of the supraoptic nucleus of the long-day birds and nullified the activating effect of the long days on the neurosecretory cells of the lateral group of the supraoptic nucleus (Uemura and Kobayashi, 1963). Rossbach (1966) observed the parallelism of seasonal activity between the nucleus paraventricularis and the gonadal cycle, and between the nucleus magnocellularis interstitialis dorsalis (neurosecretory) and adrenal cortical tissues. Other neurosecretory nuclei did not show any correlation with the activities of other endocrine glands. Thus, there are functional differences among different neurosecretory cell groups, and some cell groups show activity changes during gonadal growth, whereas others do not. Since releasing factors for the adenohypophyseal hormones have been found (Section VII,B), the changes in neurosecretory activity during gonadal growth may not be directly related to the release of adenohypophyseal hormones, but they may be secondarily induced. In the experiments by Benoit and Assenmacher (1953a,b, 1955) and Assenmacher (1957a,b), there is the possibility that they lesioned the infundibular nucleus and/or sectioned its axons proceeding to the median eminence. It has recently been demonstrated that the infundibular nucleus regulates gonadotropin release from the adenohypophysis (see Section VII,A).

John and George (1967) observed that the neurosecretory cells of the nucleus supraopticus and nucleus paraventricularis contain little neurosecretory material in the postmigratory period, but they are heavily loaded in the premigratory period. A similar tendency was observed in the amount of neurosecretory material in the median eminence and the neural lobe.

C. Functions of Neurohypophyseal Hormones

As mentioned before, both the pars nervosa and the median eminence contain the same neurohypophyseal hormones — arginine vasotocin and oxytocin (see Follett, 1970b). However, the ratio of vasopressor activity to oxytocic activity is higher in the median eminence. It is possible that arginine vasotocin may be released more actively than oxytocin from the median eminence into the portal vessels or that the median eminence receives more axons from the neurosecretory cells producing mainly arginine vasotocin than from those producing oxytocin. Following dehydration in the Japanese Quail, the hormone content of the hypothalamohypophyseal system showed a decrease (Follett and Farner, 1966). Arginine vasotocin showed a greater decrease (82.1%) than did oxytocin (57.9%). Arginine vasotocin in the median eminence decreased by 36.0%. These results showed that vasotocin is an active antidiuretic principle. Oxytocin is diuretic in birds (Sawyer, 1963).

It has been demonstrated that the neurohypophyseal hormones in the median eminence change in amount as the testes grow in the duck and in *Zosterops* (Hirano *et al.*, 1962; Ishii *et al.*, 1962; Kobayashi, 1963; Hirano, 1966). However, Follett and Farner (1966) found a change in one experiment, but could not find it in another experiment in the Japanese Quail. In those experiments, the hormones in the pars nervosa did not show any change in amount, and there were no conspicuous changes in the neurosecretory material. Here again, the relationship between neurosecretory system and gonadal development is not clear in terms of neurohypophyseal hormones. From these experiments, it is obvious that the neurohypophyseal hormones are not the main substance involved in gonadal growth, although vasopressin was once regarded as a substance that releases adenohypophyseal hormones (Martini, 1966). The change in the amount of the hormones in the above experiments may be induced secondarily.

One important finding is that the neurohypophyseal hormones in the pigeon median eminence, but not in the pars nervosa, decrease markedly after the systemic injection of formalin (Hirano, 1966). It is suggested that the hormones in the median eminence might be related to ACTH release rather than to gonadotropin release. Supporting this idea, Kawashima *et al.* (1964) observed a decrease of sudanophilic materials in the adrenal gland with concomitant decrease of the neurosecretory material after water deprivation in the White-crowned Sparrow. Rossbach (1966) observed a correlation between adrenal cortical activity and the nucleus magnocellularis interstitialis dorsalis, which is neurosecretory. However, considering the presence of corticotropin-releasing factor in the hypothalamus, it is likely that the changes in the amount of the neurohypophyseal hormones and neurosecretory material may be secondarily induced by environmental changes (see p. 331).

Tanaka and Nakajo (1962) showed a marked decrease of arginine vasotocin, but a slight decrease of oxytocin, in the fowl neural lobe after oviposition. Sturkie and Lin (1966) found that the plasma vasotocin level rises remarkably at the time of oviposition. Single intravenous injection of arginine vasotocin induces oviposition within 90 seconds (Munsick et al., 1960). Arginine vasotocin is more potent in causing contraction of the fowl oviduct than oxytocin (Berde and Boissonnas, 1966). However, Shirley and Nalbandov (1956) found that neurohypophysectomy does not affect oviposition.

Although the biological significance is obscure, oxytocin induces an increase of glucose and fatty acids in the blood of the domestic fowl (Kook et al., 1964) and of blood sugar in the White-crowned Sparrow (Farner et al., 1967). Oxytocin has a vasodepressor activity in the domestic fowl and other birds (Woolley, 1959); and this phenomenon is used for the bioassay of the peptide (Coon, 1939).

D. ELECTRON MICROSCOPIC OBSERVATIONS

Bern et al. (1966) studied the changes in the ultrastructure of the median eminence of the White-crowned Sparrow subjected to long days and found that there was no correlation between the type, size, or number of vesicles and gonadal growth. However, osmotic stress induced a decrease in the number of elementary granules.

As mentioned earlier, Péczely and Calas (1970) have observed four types of granules (1600–1900 Å, 1200–1400 Å, 1000 Å, and 600–800 Å) and synaptic vesicles in the median eminence of the pigeon. Adenohypophysectomy induced almost complete disappearance of electron-dense granules in the external layer of the anterior median eminence and induced an increase in empty axons and an increase of the granules (1600–1900 Å) in the posterior median eminence. Following the intravenous injection of insulin, the electron-dense granules in the anterior part of the anterior median eminence were largely depleted, but the granules (1200–1400 Å and 1600–1900 Å) were increased in the posterior median eminence. Metopirone caused the granules (1200–1400 Å) to increase in both anterior and posterior divisions of the median eminence and a significant increase in the number of synaptic vesicles in the posterior median eminence. Prednisolone induced a marked increase of the granules of 1200–1400 Å in the anterior median eminence and induced an increase of the granules of 1000 Å in the posterior median eminence. From all these findings Péczely and Calas (1970) concluded that the granules of 1200–1400 Å have an important role in releasing ACTH from the adenohypophysis.

It was difficult to draw other conclusions from the changes in the number of granules of the different types. After hypophysectomy in the pigeon, large granules (1600–1900 Å) increased in number in the posterior median eminence. This phenomenon is in good agreement with our previous observations that AF-positive neurosecretory materials are accumulated in the posterior median eminence after hypophysectomy, where we rarely see AF-positive neurosecretory material. This finding suggests that the AF-positive neurosecretory material may be involved in some obscure function in the posterior median eminence. The changes in the population of different types of granules in the mammalian median eminence are reviewed by Kobayashi *et al.* (1970).

E. ENZYMES IN THE NEUROSECRETORY SYSTEM

Kobayashi and Farner (1960) attempted to assess the activity of the neurosecretory system by measuring the activity of acid phosphatase, since this enzyme reflects the general metabolism of cells. It was found that the activity of acid phosphatase in the median eminence, but not in the pars nervosa, increased during testicular growth in the photosensitive White-crowned Sparrow subjected to long days. The supraoptic region did not show any significant change in enzyme activity following exposure to long days. The enzyme activity of the median eminence of refractory birds subjected to long days showed no increase. These phenomena indicate that the median eminence is involved in the regulation of gonadal growth, and that the median eminence and the pars nervosa function independently. Following dehydration, the acid phosphatase activity increased in the supraoptic region and the pars nervosa, but not in the median eminence region. Again, the independent functions of the eminential and neural lobe components were confirmed (Kawashima *et al.*, 1964). The pigeon responded in a similar way to dehydration (Kobayashi *et al.*, 1962). The proteinase activity was also used for quantitative expression of neurosecretory system. The change in the activity of this enzyme showed a similar tendency to that of acid phosphatase following exposure of the White-crowned Sparrow to long days or to procedures leading to dehydration (Kobayashi *et al.*, 1962; Kawashima *et al.*, 1964).

The White-throated Sparrow (*Zonotrichia albicollis*) showed more activity increase in the supraoptic region, median eminence, and adenohypophysis following exposure to long days than the White-crowned Sparrow (Wolfson and Kobayashi, 1962). Thus, the level of acid phosphatase activity is a useful indicator of the activity of the parts of the hypothalamohypophyseal system. This technique was used for demonstration of a negative gonadal feedback on the hypothalamus. When castrated *Emberiza* and White-crowned Sparrows were subjected to long days, the phosphatase activity of the adenohypophysis increased more than in the intact birds (Uemura, 1964a; Kobayashi and Farner, 1966). Administration of estrogen or androgen decreased the enzyme activity of the adenohypophysis. The increased enzyme activity of the adenohypophysis returned to the initial level earlier in the intact birds than in the castrated ones. These findings show that the gonadal feedback on the hypothalamus is operating, but weakly, during the period of gonadal development, and that the feedback is not a main factor for the termination of photoperiodically induced gonadal development (Uemura, 1964a; Kobayashi and Farner, 1966).

Tanaka and Nakajo (1959) found that the activity of cholinesterase shows different levels at the times just before, during, and just after oviposition in the fowl. They suggested that in the laying hen the enzyme activity might be related to the stimulation of the oviduct by the egg. Kobayashi and Farner (1964) and Follett *et al.* (1966) found that the neurosecretory cells and the palisade layer of the median eminence show strong acetylcholinesterase (AChE) activity and that the pars nervosa shows weak enzyme activity in the White-crowned Sparrow (see Kobayashi *et al.*, 1970). Uemura (1964c, 1965) obtained similar results in *Zosterops japonica*. These findings suggest that the neurosecretory neurons are cholinergic in nature, and that some cholinergic mechanism is functioning in the median eminence. The capillaries of the primary plexus showed no pseudocholinesterase activity, whereas those of the brain showed strong activity. The absence of enzyme activity in the capillaries of the primary plexus may be related to the active transport of the neurohormones through the membrane of the endothelial cells. Russell (1968) and Russell and Farner (1968) measured the enzyme activity in the hypothalamohypophyseal axis of the White-crowned Sparrow. They found that the acetylcholinesterase activity in the adenohypophysis increases during testicular development. There is a consistent daily cycle of AChE activity in the median eminence. Winget *et al.* (1967) measured the activity of acid phosphatase, alkaline phosphatase, and cholinesterase in the diencephalon, pituitary, and plasma of the domestic fowl exposed to light after darkness for 56 days. Illumination increased the activity of all enzymes except hypothalamic alkaline phosphatase and pituitary cholinesterase. They concluded that visible radiation is a *Zeitgeber* for the enzyme systems evaluated.

Matsui and Kobayashi (1965), in the Tree Sparrow (*Passer montanus*), showed the presence of a strong monoamine-oxidase reaction in the subependymal layer and the palisade layer, especially in the tissues near the capillaries of the primary plexus. This indicated that monoaminergic fibers terminate around capillaries. The neurosecretory cells gave a very weak monoamine-oxidase reaction. The distribution of monoamine oxidase (MAO) in the median eminence is different from that of neurosecretory material. Therefore, neurosecretory neurons are probably not monoaminergic. However, there is strong MAO activity around the perikarya, suggesting that the neurosecretory cells may be innervated by monoaminergic fibers (Section II,A,4). Results similar to the above were obtained by Follett *et al.* (1966) in the White-crowned Sparrow, and by Urano (1968) in the Japanese Quail. In both experiments, the infundibular nucleus showed a strong MAO reaction. The distribution of MAO reaction is very similar to that of monoamine fluorescence and also to that of small granules carrying monoamines. Follett (1969) measured quantitatively the activity of MAO of the median eminence and the basal hypothalamus of the Japanese Quail. He found that there is a daily rhythm in the enzyme activity, but he could not find a clear relationship between MAO activity and the release of gonadotropins resulting from photoperiodic stimulation.

VI. Ependymal Function in the Median Eminence

In the domestic fowl, Legait (1959) observed that stainable material is secreted from the ependymal cells of the median eminence. However, this phenomenon was not observed in the White-crowned Sparrow and the Zebra Finch, although it was noticed that the ependymal cells of the median eminence have extended small protrusions into the third ventricle (see Farner and Oksche, 1962). Later, with an electron microscope, the detail of the ependymal protrusions was studied in the median eminence of the White-crowned Sparrow (Matsui and Kobayashi, 1968). The first type is the bleb-like microvillus or bulbous protrusion. It usually contains polysomes, glycogen granules, mitochondria, and small vesicles or vacuoles. Some of these protrusions are detached from the cell body as masses and others have ruptured membranes discharging their inclusions. The second type is the fingerlike microvillus, which rarely includes cytoplasmic organelles.

At the cell surface near the base of these microvilli, small pinocytotic vesicles, seemingly impounding a droplet of cerebrospinal fluid, are often observable. This may be concerned with the absorbing function

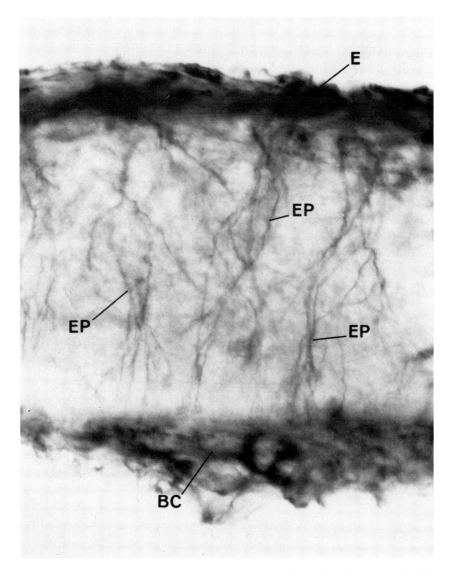

FIG. 13. Ependymal absorption of peroxidase injected into the third ventricle of the Japanese Quail. BC = blood capillary; E = ependymal layer; EP = ependymal process showing peroxidase reaction. × 200.

of the ependymal cells. The third type consists of the surface folds and the marginal folds. The marginal fold protrudes from the margin of the ependymal cell surface and the surface fold protrudes from any point of the apical surface. They recurve their free ends toward the cell surface and entrap a droplet of the ventricular fluid. Matsui (1966b), Oota (1970), and Calas and Assenmacher (1970) have also called attention to protrusions from the ependymal cell surface of the median eminence of the pigeon and duck. Recently, we have demonstrated that the ependymal cells of the median eminence of the Japanese Quail absorb peroxidase injected into the third ventricle (Kobayashi et al., 1972) (Fig. 13). This is the first clear evidence showing the absorption of ventricular fluid by ependymal cells.

As mentioned above, the ependymal processes in the median eminance have synaptic contacts with monoaminergic fibers in the pigeon (Matsui, 1966b). It is probable that the ependymal secretion and absorption may be regulated by the monoaminergic fibers. The origin of the fibers is not known. The ependymal function in the mammalian median eminence is discussed in relation to the adenohypophyseal function in a review by Kobayashi et al. (1970). In the neural lobe of the domestic fowl, Payne (1959) observed that a part of the apical cytoplasm of the ependymal cells is secreted into the lumen.

VII. Functional Aspects of the Gomori-Negative Hypophysiotropic Neurosecretory System

A. Hypothalamic Regions Regulating Adenohypophyseal Function

It is evident that in mammals afferent neural information is transformed into neurohormonal information (releasing factors) in certain Gomori-negative neurosecretory cells, and then the neurohormonal information is transported to the median eminence and conveyed to the adenohypophysis through the portal vessels, resulting in secretion of adenohypophyseal hormones. A similar series of events involved in hypothalamic control of the adenohypophysis seems to occur also in birds. Centers regulating adenohypophyseal functions in birds have not always been as clearly identified as in mammals. However, several attempts have been employed to assess exact sites regulating the release of adenohypophyseal hormones by means of lesions, electric stimulation, and hormone implantation in certain nuclei of the hypothalamus.

1. Gonadotropin

The hypothalamic control of the release of gonadotropin has been demonstrated in many experiments, such as those involving sectioning portal vessels and severing the median eminence (see Farner et al., 1967). The first attempt to introduce lesions into the avian hypothalamus was carried out by Assenmacher (1957a,b, 1958) in the domestic duck. The lesions located in the supraoptic and paraventricular nuclei were always followed by genital atrophy with the same latency as after hypophysectomy (2–3 weeks). From the results and those obtained previously in Benoit's laboratory (Assenmacher, 1952, 1953; Benoit and Assenmacher, 1953a,b, 1955), they concluded that neural information leading to gonadotropin release came from the supraoptic and paraventricular nuclei to the median eminence and was conveyed to the adenohypophysis through the portal vessels. Supporting this idea, Egge and Chiasson (1963) found that electrolytic lesions in the nucleus paraventricularis, median eminence, and the pars tuberalis induced cessation of the egg-laying cycle and reduction in size of the comb and wattles in the domestic fowl. Since the lesions were made in the nucleus paraventricularis, it seems that AF-positive materials are responsible for gonadotropin release. As mentioned before (Section V,B), there have been many histological results supporting this idea in passerine birds. However, Wilson (1967) placed lesions in the hypothalamus of the White-crowned Sparrow and showed that lesions in the anterior median eminence, which is the depot of AF-positive material, gave no effect on photostimulated testicular growth, but lesions in the tuberoin-fundibular region (Fig. 14A,a) completely inhibited photoinduced testicular growth. Stetson (1969a) confirmed by lesioning experiments that the nucleus infundibularis is responsible for photostimulated testicular growth of the White-crowned Sparrow. Lepkovsky and Yasuda (1966) found that the destruction in the ventromedial region induced testicular atrophy in the chicken. Further, Snapir et al. (1969) showed that lesions in the ventromedial region led to regression of the testes and to atrophy of the comb in the chicken. Injections of testosterone propionate restored comb growth but did not prevent testicular regression in the lesioned chickens, suggesting that the lesions inhibited gonadotropin release from the pituitary.

Graber et al. (1967), using more restricted lesions, found that those in the posterior parts of the tuberal nucleus caused testicular regression and inhibited comb growth. In laying hens, Kanematsu et al. (1966) placed lesions in the basal tuber and induced cessation of

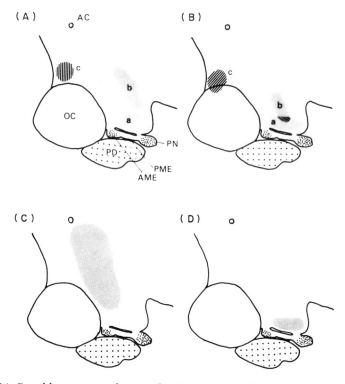

FIG. 14. Possible sites regulating adenohypophyseal function. (A) Gonadotropin regulatory sites revealed by lesions: (a) nucleus tuberis (Sharp and Follett, 1969c, and others) or ventral region of nucleus infundibularis (Wilson, 1967, and others) where gonadotropin release seems to be maintained; (b) nucleus hypothalamicus posterior medialis (Sharp and Follett, 1969c) in which lesioning caused inhibition of photo-induced testicular growth; (c) preoptic area, which seems to be responsible for ovulation in the hen (Ralph, 1959, and others). (B) Androgen sensitive sites: (a) nucleus tuberis or ventral region of nucleus infundibularis (Wilson, 1970), in which testosterone-induced testicular regression has been observed; (b) site sensitive to testosterone in ducks (Gogan, 1968); (c) sensitive site in ducks where testosterone induced partial testicular regression. (C) Possible regulatory site of TSH release (Kanematsu and Mikami, 1970). (D) Possible regulatory site of ACTH release (Frankel et al., 1967a). AC = anterior commissure; AME = anterior median eminence; OC = optic chiasma; PD = pars distalis; PME = posterior median eminence; PN = pars nervosa.

ovulation and atrophy of the adenohypophysis, ovary, oviduct, and comb. Lesions in the preoptic area induced the same results but to a lesser degree. Furthermore, Kanematsu (1968b) showed that destruction in the basal tuber induced severe blockage of the development of the ovary and oviduct. The same results occurred in four of fifteen birds bearing lesions in the preoptic area. There was no case in which premature ovulation occurred. In the male chicken, destruction in the basal tuber induced testicular regression (Kanematsu, 1968b). Sharp and Follett (1969c) placed lesions in the hypothalamus of the Japanese Quail and found that photostimulated testicular growth was completely inhibited when the lesions were located in the nucleus tuberis. Lesions in the anterior median eminence of the Japanese Quail were ineffective in preventing testicular growth as in the White-crowned Sparrow. Testicular development did not occur after destruction of either the nucleus tuberis or the nucleus hypothalamicus posterior medialis (Fig. 14a,b). They found that there are monoaminergic fiber connections between both nuclei (Sharp and Follett, 1968). Perhaps the latter nucleus regulates the former. Stetson (1969c) also reported that electrolytic destruction in the hypothalamus of adult breeding male and female Japanese Quail caused gonadal disfunction. Lesions in the rather posterior infundibular nuclear region seems to inhibit an LH-like hormone secretion in both sexes, and lesions in the anterior infundibular nuclear region seems to inhibit an FSH-like hormone secretion in both. The results suggest that in both male and female Japanese Quail, two separate regions appear to be involved in gonadotropin secretion.

Ralph (1959) found that electrolytic lesions in the ventral preoptic hypothalamus were consistently followed by an immediate cessation and prolonged interruption of ovulation in adult laying hens. Further, Ralph and Fraps (1959) showed that electrolytic destruction in the preoptic area inhibited progesterone-induced ovulation. Electrical stimulation of the anterior median eminence 14 hours before the next ovulation induced delayed ovulation (Opel and Fraps, 1961). Opel (1963) stimulated the preoptic area and found that ovulation was postponed if stimulated 14 hours before the next ovulation. Ralph and Fraps (1960) showed that progesterone injected into the anterior hypothalamus and the preoptic area induced premature ovulation. Thus, the preoptic area (Fig. 14A,c) seems to be involved in the mechanism of ovulation, and progesterone stimulates the preoptic area to induce ovulation. As for the oviposition, if stimulation was applied 4–25 hours before oviposition, premature oviposition was induced (Opel, 1964). He is of the opinion that premature oviposition occurred because of the release of vasotocin in this case. However, the neurohypophyseal hormones do not appear to be essential for oviposition (Shirley and Nalbandov, 1956; Opel, 1965) by neurohypophysectomy.

Androgen implantation was carried out by several investigators to determine the sensitive sites based on the concept of a negative feedback. Kordon and Gogan (1964) found that implants of testosterone in the ventromedial nucleus of the domestic duck (Fig. 14B,b) inhibit testicular growth by photostimulation. Gogan (1968) confirmed this, and further, that in addition to this nucleus, implants in the preoptic area (Fig. 14B,c) partially inhibit testicular growth. In the latter case, spermatogenesis ceased at the stage of spermatocyte I, but the Leydig cells seemed to be active. Photostimulated testicular growth was inhibited in the [American] Tree Sparrow (*Spizella arborea*) by implantation of testosterone in the basal region of the nucleus infundibularis (Fig. 14B,a) (Wilson, 1970). In the Japanese Quail, the situation is the same as in the [American] Tree Sparrow. Implants of testosterone in the nucleus tuberis inhibited photostimulated testicular growth. Implants in other hypothalamic regions including the nucleus hypothalamicus posterior medialis of Sharp and Follett (1969c) did not give clear inhibition of photoinduced testicular development (Wada, 1972). Recently puromycin was implanted into various regions of the hypothalamus of the Japanese Quail. Only when implanted into the nucleus tuberis was puromycin effective in suppression of photoinduced testicular growth (Wada, unpublished). It is known that the avian infundibular nucleus is topographically divisible into several cell groups (Oehmke, 1968; Oksche *et al.*, 1970; Oehmke, 1971). It is not known which cell groups were affected by puromycin.

From the results mentioned above, it could be concluded that the ventromedial region of the hypothalamus, or more precisely the nucleus tuberis, is the regulatory site of gonadotropin secretion from the adenohypophysis (Fig. 14A,a). In the hen, the preoptic area is responsible for gonadotropin release for ovulation (Fig. 14A,c).

2. Prolactin

Since section of the portal vessels induced atrophy of the prolactin cells in the duck (Assenmacher and Tixier-Vidal, 1964; Tixier-Vidal *et al.*, 1966), it is obvious that secretion of prolactin in birds is regulated by the hypothalamic prolactin-releasing factor. No direct evidence for the presence of a regulatory center of prolactin secretion is available.

3. Thyrotropin

Secretion of thyrotropin seems to be regulated by the hypothalamus, since the section of the portal vessels induced a decrease in the thyroid weight in the duck (Assenmacher, 1959; Assenmacher and Tixier-Vidal, 1959, 1963, 1964; Rosenberg *et al.*, 1967) and the pigeon (Baylé *et al.*, 1966). Assenmacher (1958) reported that a certain number of ducks showed atrophy of the thyroid gland after lesioning of the anterior hypothalamus or complete sectioning of the median eminence at the level of just before the anterior median eminence. Egge and Chiasson (1963) found that lesions in the preoptic division of the nucleus supraopticus induced a decrease in thyroid weight. However, there was no apparent morphological difference in the follicular epithelium between the birds with light and heavy thyroid glands, except that the birds with lesions lacked the reabsorption lacunae in the central region of the thyroid glands. McFarland *et al.* (1966) lesioned the anterior hypothalamus including the ventrolateral nuclei and induced a decrease in the thyroxine disappearance rate in the Japanese Quail. Lesions in the basal anterior hypothalamus, which includes the nucleus hypothalamicus posterior medialis, induced atrophy of the thyroid gland in cockerels (Takahara *et al.*, 1967). Another type of experiment has been carried out to assess the hypothalamic regulating site of thyrotropin secretion in the chicken (Kanematsu and Mikami, 1966, 1970; Kanematsu *et al.*, 1967; Mikami *et al.*, 1967; Kanematsu, 1968a). Adenohypophyseal tissue transplanted in the area near the nucleus paraventricularis magnocellularis showed hypertrophy of the basophils after thyroidectomy. Appearance of thyroidectomy cells in the grafted adenohypophysis was inhibited by destruction of the anterior hypothalamus located ventrocaudal to the anterior commissure and dorsocaudal to the optic chiasma. Only four of thirty-nine chickens showed a markedly low thyroidal uptake of ^{131}I. The lesion in these birds was found in the median eminence, except in one bird in which the lesion was located in the nucleus paraventricularis magnocellularis. The protein-bound ^{131}I conversion ratio in the cockerel gave clearer results. Protein-bound ^{131}I was depressed when the region ventrocaudal to the anterior commissure was destroyed. The region includes the nucleus paraventricularis magnocellularis. Destruction of the anterior hypothalamus, which includes the nucleus hypothalamicus posterior medialis, and of the region dorsocaudal to the optic chiasma also depressed the protein-bound ^{131}I conversion ratio. The thyroid gland decreased significantly in weight after destruction of the anterior hypothalamus, including the nucleus hypothalamicus posterior medialis and the nucleus paraventricularis.

Although the evidence obtained so far is not sufficient to draw a conclusion about the exact regulatory site of thyrotropin secretion, the anterior hypothalamus, including the nucleus paraventricularis magnocellularis and the nucleus hypothalamicus posterior medialis (Fig. 14C), seems to be the site responsible for the control of thyrotropin secretion from the adenohypophysis. Thus, the regulating site is not confined to one nucleus.

4. Corticotropin

Section of the hypophyseal portal vessels causes a decrease in adrenal weight and in the ratio of cortical to medullary tissue in sectioned birds. However, the effect of severance of portal vessels is not the same as that of adenohypophysectomy (Assenmacher, 1958). Boissin (1967) sectioned the portal vessels of the duck and showed that adrenal weight and the ratio of cortical to medullary tissue decreased significantly; the decrease was greater in the birds with unregenerated portal vessels than in those in which the vessels were restored. In the sectioned birds, adrenal ascorbic acid did not significantly decrease, and plasma corticosterone concentration of the peripheral blood was not significantly affected. However, corticosterone concentration in the birds with the unregenerated portal vessels had a tendency to be lower than that in the birds with regenerated portal vessels. In the chicken, autotransplantation of the adenohypophysis under the kidney capsule reduced the plasma corticosterone concentration in the adrenal vein to the same level as that of adenohypophysectomized subject (Resko et al., 1964). The peripheral concentration of plasma corticosterone did not differ significantly from that of intact birds after autotransplantation in the duck (Boissin et al., 1966; Boissin, 1967; Baylé et al., 1971), pigeon (Baylé et al., 1967), and Japanese Quail (Baylé et al., 1967). The differences in results among investigations may be due to the difference in species or in experimental period. Resko et al. (1964) killed their chickens 40 days after the implantation and Baylé et al. (1971) killed their ducks 63 days after the implantation.

Miller (1961) found that the adrenals atrophied after hypophysectomy in the pigeon but showed markedly hypertrophic cortical tissue after treatment with formaldehyde or insulin. After large lesions in the pigeon median eminence region including the ventral hypothalamus in addition to total hypophysectomy, the adrenals became hypertrophied rather than atrophied, and the response of these pigeons to formaldehyde was the same as that in hypophysectomized pigeons without lesions. Miller proposed three possible explanations: (1) There may be another source of corticotropin that was not destroyed in this experiment; (2) there may be a stimulus other than corticotropin that evokes adrenal enlargement; (3) the effect of corticotropin may not be direct on the adrenal but mediated via some pathway that can also be aroused by stress and lesion in the median eminence. Egge and Chiasson (1963) placed lesions in the hypothalamus of hens and concluded that no correlation exists between the lesioned area and the corticoid titer in the adrenals. Frankel *et al.* (1967a) concluded from the results obtained with hypothalamic lesions and adenohypophysectomy that the release of corticotropin from the adenohypophysis is controlled by the hypothalamus, and especially by the ventral part of the tuberal nucleus (Fig. 14D). Further, they found that the adrenals are stimulated not only by the adenohypophyseal corticotropin but also by an extrahypophyseal corticotropin-like substance, since lesions in the ventral part of the tuberal nucleus of the hypophysectomized cockerel evoked a higher corticosterone concentration than that in intact lesioned cockerels.

Plasma corticosterone level in the adrenal venous blood of the adenohypophysectomized cockerel decreased immediately after the surgery to about one-third of the level found in intact birds and this level was maintained up to 42 days later (Frankel *et al.*, 1967b). However, the corticosterone level in ducks declined steadily during the first 2 weeks after adenohypophysectomy and then remained at a level that was one-third that of intact birds (Baylé, 1968; Baylé *et al.*, 1971). Experiments show that adenohypophysectomized pigeons can respond to stress (Miller and Riddle, 1942; Miller, 1961). Since the plasma corticosterone concentration remained one-third of that of intact birds after adenohypophysectomy, it is possible that the adrenal gland functions autonomously or that there is extrahypophyseal corticotropin. The latter possibility is more probable than the former, since ACTH or ACTH-like activity has been found in the median eminence or hypothalamus (See Section VII,B). More recently, Bradley and Holmes (1971) have reported that the peripheral concentration of corticosterone falls to only 10% of that of sham-operated ducks 14 days after adenohypophysectomy. Thus, adrenal cortical tissue is greatly dependent on the adenohypophysis. The authors are of the opinion that it is not necessary to postulate either a high degree of adrenal autonomy or an extrahypophyseal corticotropin source to maintain normal adrenocortical function in the duck.

As mentioned above, there are differences of opinion among investigators with respect to the autonomous function of cortical tissue, its dependency on adenohypophysis, and extrahypophyseal corticotropin (for review, see Frankel, 1970). Nevertheless, it is obvious that cortical function is largely regulated by the hypothalamohypophyseal system.

5. Growth Hormone

There seems to have been no experimental attempts to establish the hypothalamic site that controls the release of growth hormone.

B. RELEASING FACTORS AND MONOAMINES

Extensive investigations have been carried out in mammals, and now the presence of the releasing or inhibiting factor for each adenohypophyseal hormones has been demonstrated. The following hypophysiotropic neurohormones are reported: luteinizing hormone-releasing factor (LRF), follicle stimulating hormone-releasing factor (FRF), prolactin-inhibiting and -releasing factors (PIF, PRF), thyrotropin-releasing factor (TRF), growth hormone-releasing and -inhibiting factors (GRF, GIF), and melanocyte stimulating hormone-releasing and -inhibiting factors (MRF, MIF) (for review, see Guillmin, 1964; McCann and Porter, 1969; Geschwind, 1969; Meites, 1970; Schally et al., 1973). Recently, it was found that the active moiety of TRF is composed of only three amino acid residues and that LRF, which also has intrinsic FRF activity, contains ten residues (Schally et al., 1971; Matsuo et al., 1971). These results indicate that the releasing factors are probably small peptides. In birds, although few efforts have been made on this subject, the presence of some releasing factors have been demonstrated.

1. Gonadotropin-Releasing Factors

In birds, it is as yet open to question whether two types of gonadotropin (LH and FSH) are separable or not, although cytologically there appear to be two kinds of gonadotropic cells in the avian adenohypophysis (Tixier-Vidal, 1965; Matsuo et al., 1969; Mikami et al., 1969). Recently, Hartree and Cunningham (1969) effected a partial purification of the chicken pituitary hormones by chromatography and found two fractions, one having FSH activity and the other LH activity. Accordingly, there may be separate releasing factors (LRF and FRF) for LH and FSH.

The first evidence suggesting the presence of LRF is the fact that infusion of hypothalamic extracts into the pituitary of laying hens induces premature ovulation (Opel and Leopore, 1967). Cortical extracts and neurohypophyseal extracts were ineffective. Synthetic oxytocin and arginine vasotocin were also ineffective. On gel filtration (Sephadex G-25) of stalk median eminence, the peak of ovulating hormone-releasing activity was found to overlap with the peak of oxytocin and vasotocin as determined by hen and rat uterine assays. Infusion of 500 μg of lyophylized powder from this fraction induced ovulation in 13 of 21 hens. *In vitro*, hypothalamic extracts induced the release of LH from cock pituitaries (Tanaka *et al.*, 1969). Acetic acid extracts from the hypothalamus of White Leghorn hens (1–2 years of age) were centrifuged and the supernatant was immersed in boiling water for 10 minutes to inactivate the LH that might be present as a contaminant in the extracts. This hypothalamic extract could induce LH release from the pituitaries of White Leghorn cocks incubated in Krebs–Ringer phosphate buffer (pH 7.4) containing 10 mM glucose. Cortical extracts did not induce LH release. Hypothalamic extracts of cockerels induced release of LH from rat pituitaries incubated *in vitro* (Jackson and Nalbandov, 1969a). In this experiment, hydrochloric acid extracts of the hypothalamus of 12-week-old cockerels were centrifuged and the supernatant was used. The extracts induced significant LH release from the rat pituitary incubated in Medium 199. Arginine vasotocin in a concentration equal to that in the extracts (Salem *et al.*, 1970b) had no effect on LH release under the same conditions. In all these experiments, the ovarian ascorbic-acid depleting (OAAD) method was employed for determination of LH. It has been demonstrated that chicken hypothalamic extracts contained another OAAD factor. This effect was not due to the presence of LH, but to arginine vasotocin in the extracts (Jackson and Nalbandov, 1969b). Ishii *et al.* (1970) showed that OAAD activity in the pigeon median eminence is explained by arginine vasotocin therein. The activity was observed in the anterior median eminence where arginine vasotocin was localized. Recently, Jackson (1971a,b) partially purified chicken luteinizing hormone-releasing factor (LRF). Furthermore, he separated chicken LRF from arginine vasotocin and found it is chemically different from mammalian LRF.

Chicken hypothalamic extracts also stimulated FSH release from cock and rat pituitaries incubated *in vitro* (Kamiyoshi *et al.*, 1969). Hypothalamic extracts of turkeys also induced FSH release from the incubated pituitary tissue compared with cerebral cortex extracts (Bixler *et al.*, 1968). FSH was assayed by the Steelman–Pohley method.

Follett (1970a) demonstrated the presence of a gonadotropin-releasing factor in the hypothalamus of the Japanese Quail by an assay method other than OAAD. Hypothalamic extracts of the Japanese

Quail were centrifuged, and the supernatant was added to Medium 199 that contained cockerel pituitaries. After incubation, the gonadotropic activity contained in the incubation media was determined by measuring the uptake of ^{32}P by the testes of 1-day-old chicks after injection of the media. Hypothalamic extracts enhanced gonadotropin release while cortical extracts did not. Extracts of the pars nervosa, synthetic vasopressin, and oxytocin did not evoke gonadotropin release from the incubated cockerel pituitaries. It is interesting to note that gonadotropin-releasing activity in the quail hypothalamus varies in accordance with reproductive states. The birds held under a non-photostimulatory short daily photoperiod did not show detectable gonadotropin-releasing activity, whereas the birds under a long daily photoperiod showed considerable activity. There was no difference between the intact long-day males and castrated long-day males. Testosterone-implanted long-day males possessed high gonadotropin-releasing activity in the hypothalamus, although the testes of these birds were small and the gonadotropin content of the adenohypophysis was low. It is concluded from these experiments that in avian species gonadotropin-releasing factor(s) are present in the hypothalamus.

2. Prolactin-Releasing Factor

The control of prolactin release in birds differs from that of mammals. When the pigeon adenohypophysis was incubated for 6 days *in vitro*, there was no significant increase in the prolactin content of the incubation media (Nicoll and Meites, 1962; Nicoll and Bern, 1965; Nicoll, 1965). Nicoll (1965) examined the effects of acid extracts of cerebral and hypothalamic tissues of the Tricolored Blackbird (*Agelaius tricolor*) on prolactin secretion from *Agelaius* adenohypophysis *in vitro*. The hypothalamic extracts induced prolactin secretion, indicating that the hypothalamus of *Agelaius* contained a stimulating factor for prolactin secretion. Cerebral cortex extracts had no effect on prolactin secretion. Kragt and Meites (1965) have demonstrated that hypothalamic extracts of pigeons stimulate the pigeon pituitary to secrete prolactin into the incubation media. Meites and Nicoll (1966) also showed that hypothalamic extracts from the chicken and quail increased prolactin release from the pigeon pituitary incubated *in vitro*. Gourdji and Tixier-Vidal (1966) demonstrated prolactin-releasing activity in the hypothalamus of the duck. Prolactin-releasing activity has also been demonstrated in the turkey hypothalamus *in vitro* (Chen *et al.*, 1968).

Prolactin-releasing activity in the hypothalamus was much higher in the parent pigeons on the day of hatching than in 4- to 6-week-old pigeons, suggesting that the prolactin-releasing factor is present in different concentrations in the hypothalamus at the different reproductive states (Kragt and Meites, 1965). This phenomenon coincides with the observations on prolactin cells in pigeons that are feeding their young (Tixier-Vidal and Assenmacher, 1966). In the case of the turkey, although total prolactin content in the adenohypophysis of 20-day-old male and female poults was much less than in hens killed at the end of the laying season, their hypothalami appeared to contain as much prolactin releasing activity as the hens (Chen *et al.*, 1968).

Thus, in avian species, prolactin secretion appears to be controlled primarily by a prolactin-releasing factor in the hypothalamus, unlike a prolactin-inhibiting factor in mammals.

3. *Corticotropin-Releasing Factor*

As mentioned above, release of corticotropin in birds has been suggested to be controlled by the hypothalamus. Recently, it has been demonstrated that a corticotropin-releasing factor is present in the avian hypothalamus, as is adrenocorticotropic hormone (ACTH) or ACTH-like activity in the hypothalamus.

Péczely and Zboray (1967) demonstrated *in vitro* a corticotropin-releasing factor in the median eminence of the pigeon. They incubated the pigeon adenohypophysis with pigeon median eminence tissue homogenate. The incubation medium was then added to the other incubation media containing the rat adrenal tissue. Corticosterone secretion from the rat adrenal was enhanced. This suggests that the median eminence homogenate contained CRF and induced release of ACTH from the pigeon adenohypophysis into the incubation medium. In addition to CRF, they found that the pigeon median eminence contains more ACTH or ACTH-like substance than the adenohypophysis (Péczely and Zboray, 1967; Péczely, 1969; Péczely *et al.*, 1970). Péczely *et al.* (1970) have demonstrated in the pigeon that ACTH-like activity in the median eminence is still as high as controls 1 month after adenohypophysectomy.

Stainer and Holmes (1969) demonstrated corticotropin-releasing activity in the hypothalamus of the duck, according to the following method. First, brain tissue of the duck was extracted and lyophylized, suspended in Krebs–Ringer bicarbonate buffer, and then added to the incubation media containing duck pituitary. Then the incubation medium was added to the other incubation media with slices of the duck adrenal gland. After incubation, corticosterone liberated in the media was determined by fluorometry. This CRF activity was also present in extracts of the cerebrum and spinal cord. ACTH or ACTH-like activity was not detected in the duck hypothalamus, unlike the pigeon. More recently, Salem *et al.* (1970a,b) injected hypothalamic extracts of chickens into dexamethasone-treated rats. To ascertain whether the extracts induced ACTH release from the rat pituitary, depletion of adrenal ascorbic acid (AAAD) was examined in rat adrenal glands. Using this method, CRF was found in the hypothalamus of the chicken, but not in the cortex. In addition to corticotropin-releasing activity, chicken hypothalamus contained ACTH-like activity. The ACTH-like activity was demonstrated by measuring AAAD of the adrenals of hypophysectomized rats injected with the hypothalamic extracts. They calculated that ACTH-like activity contributed to the total AAA depleting activity of the chicken hypothalamus by 27–34% (Salem *et al.*, 1970b). The ACTH-like activity was not due to ACTH contamination, because the substance showing ACTH-like activity was heat resistant and thought to be a substance similar to α-MSH. CRF activity of the hypothalamus was not due to vasotocin, although vasotocin has ACTH-releasing activity, when used in large amounts. They calculated the relative AAAD activity in the chicken hypothalamus as follows: CRF, 73.5%; vasotocin, 5.0%; ACTH-like substance, 21.5%. Boiling for 1 hour completely destroyed the CRF activity. This is different from the CRF found in mammals (Salem *et al.*, 1970b). Using an intrapituitary microinjection technique, Sato and George (1972) detected CRF activity in the pigeon hypothalamic median eminence and showed a diurnal rhythm in the CRF activity.

4. Growth Hormone-Releasing Factor

Muller *et al.* (1967) have demonstrated growth hormone-releasing activity in the pigeon hypothalamus by measuring the decrease of growth hormone content in the rat pituitary after systemic injection of pigeon hypothalamic extracts.

5. Monoamines

As mentioned earlier (Section II,A), many papers have appeared dealing with monoamine distribution in the hypothalamus. The high concentration of monoamines in the nucleus infundibularis and median eminence (Sharp and Follett, 1968, 1970; Oksche *et al.*, 1970) suggests that monoamines could perform some role in regulating adenohypophyseal functions. Systemic injection of reserpine to reduce the monoamine level in the hypothalamus inhibits testicular development in the duck, pigeon, and chicken (Khazan *et al.*, 1960; Assenmacher *et al.*, 1961; Hagen and Wallace, 1961), but not in the White-crowned Sparrow (Brown and Mewaldt, 1967) and Japanese Quail (Sharp and Follett, 1969b). Implantation of reserpine into the nucleus tuberis exerts no effect on photoperiodic testicular response in the Japanese Quail (Sharp and Follett, 1969b). Neither noradrenaline nor dopamine induces gonadotropin release *in vitro* (Follett, 1970a). The physiological meaning of monoamines in the hypothalamus awaits further experimental elucidation.

VIII. Concluding Remarks

Considering the data collected so far, it may be stated that a great deal of interest has been uncovered, but there is a great deal more to be done. Significant problems in avian neuroendocrinology still to be resolved include the following.

1. There appear to be functional differences among the clusters of neurosecretory cells in the neurosecretory nuclei. The functions of these individual clusters of cells have not yet been determined.

2. There are monoaminergic fibers and strong monoamine oxidase activity around the perikarya of the neurosecretory cells. Electron microscopy has revealed that they are innervated by monoaminergic fibers. Is the monoaminergic fiber the only one that communicates with the neurosecretory cell?

3. Do the neurosecretory cells function as osmoreceptors?

4. The median eminence and the neural lobe function independently, although they receive neurosecretory fibers of apparently common origins. What is the mechanism involved in this mutual independence?

5. What are the physiological significances of neurohypophyseal hormones and monoamines in the median eminence?

6. The physiological significance of the ependymal secretion into the third ventricle and the ependymal absorption of the ventricular fluid is not known at the present time.

7. What is the role of the thick perivascular space of the primary plexus? Does it function as a reservoir for releasing factors?

8. In birds, nothing is known about the chemistry of releasing factors. Where are the sites of production of the factors in the hypo-

thalamus? How are the producing sites related to the regulating centers found by lesioning and by steroid implantation?

9. How are the production and the secretion of releasing factors regulated?

10. Are there regionally differentiated areas in the median eminence, each of which, containing a different releasing factor, corresponds to the differentiated areas of different cell types in the adenohypophysis?

11. Are all the releasing factors associated with granules or vesicles in the median eminence? If so, which types of granule contain the various releasing factors?

IX. General Summary

In the avian hypothalamus, there are two neurosecretory systems — fuchsinophilic (aldehyde fuchsin-positive) and nonfuchsinophilic. The former consists of the following three parts: (1) the cells of the nucleus supraopticus and the nucleus paraventricularis, in which neurohormones such as arginine vasotocin and oxytocin are produced; (2) their axons through which the neurohormones are transported; and (3) their axon terminals in two neurohemal regions — the anterior median eminence and the pars nervosa. Arginine vasotocin and oxytocin occur in both neurohemal regions. They are released from the pars nervosa into the systemic blood, and from the median eminence into the portal vessels. The release of these hormones seems to be controlled by monoaminergic fibers, since many fibers revealed by monoamine oxidase histochemistry and fluorescence microscopy are in contact with perikarya of the AF-positive neurosecretory cells. Arginine vasotocin is antidiuretic and vasopressor, and oxytocin is diuretic and vasodepressor. Arginine vasotocin is more potent in contractile actions on the fowl oviduct than oxytocin. Oxytocin induces increases of fatty acids and blood sugar in at least some species of birds. The biological significance of the neurohypophyseal hormones in the median eminence is not clearly known in relation to functions of the adenohypophysis. Although evidence is lacking, it is possible that this system produces some adenohypophysiotropic neurohormones that reach the pars distalis via the anterior median eminence and the anterior portal vessels.

The AF-negative (nonfuchsinophilic) neurosecretory system consists of cells whose localizations are not entirely known, but has axons and the axon terminals in the median eminence. The neuro-hormones produced in this system are adenohypophysiotropic (re-leasing factors) and reach the pars distalis and pars tuberalis via the adenohypophyseal portal system.

The avian median eminence is clearly divided into two regions — the anterior and posterior divisions. Both divisions are composed of five layers: (1) ependymal, (2) hypendymal, (3) fiber, (4) reticular, and (5) palisade. Some AF-positive fibers arising both from the nucleus supraopticus and the nucleus paraventricularis pass into the fiber layer toward the pars nervosa, forming the supraopticohypo-physeal tract. Other AF-positive fibers descend to the palisade layer of the anterior median eminence. Aldehyde fuchsin-negative fibers, as revealed by electron microscopy, terminate in both the anterior and posterior median eminences. Their terminals are located near the capillaries of the primary plexus of the hypophyseal portal vessels. Electron microscopy shows several types of granules and vesicles with different diameters in the AF-positive and AF-negative fibers of the median eminence. Large granules of 1500–2000 Å may be carriers of the neurohypophyseal hormones, and the granules of 1000 Å seem to be carriers of monoamines. The large granules are abundant in the anterior median eminence but are rare in the posterior median eminence. The granules of 1000 Å are present both in the anterior and posterior median eminences. Other granules are possible carriers of releasing factors. The perivascular space of the capillaries of the primary plexus sends more digitations into the parenchyma of the posterior median eminence than into that of the anterior median eminence.

The ependymal cells of the median eminence seem to secrete some substance into the third ventricle and absorb some material from the third ventricle. These ependymal cells extend their processes to the capillaries of the primary plexus. The presence of the ventriculo-hypophyseal system is suggested in relation to hypothalamic control of the adenohypophysis.

In the adenohypophyseal hormone-releasing factor system, the tubero-infundibular component seems to have an important role in gonadotropin secretion. Some of the cells of the nucleus infundibularis are monoaminergic and others are not. Both neurons send their axons to the palisade layer of the median eminence. These fibers form the tuberoinfundibular tract. The roles of the monoamine neurons are not known. The nonmonoaminergic neurons may produce gonadotropin-releasing factors. Destruction of the basal portion of the nucleus infundibularis interrupts gonadotropin secretion. Testosterone implanted in this nucleus interferes with gonadotropin secretion, resulting in atrophic testes. The preoptic area seems to be involved in ovulation. There is now some evidence for the occurrence of gonadotropin-releasing factors, a prolactin-releasing factor, an adrenocorticotropin-releasing factor, and a growth-hormone-releasing factor in the avian hypothalamus. These are released into the portal vessels and conveyed thereby to the adenohypophysis. It is not known, however, which neural components are responsible for the production of respective releasing factors; nor is the chemical nature of the releasing factors known.

REFERENCES

Abrahams, V. C., Koelle, G. B., and Smart, P. (1957). Histochemical demonstration of cholinesterases in the hypothalamus of the dog. *J. Physiol. (London)* **139**, 137–144.

Aghajanian, G. K., and Bloom, F. E. (1966). Electron-microscopic autoradiography of rat hypothalamus after intraventricular H^3-norepinephrine. *Science* **153**, 308–310.

Arai, Y. (1963). Diencephalic neurosecretory centers in the passerine bird, *Zosterops palpebrosa japonica*. *J. Fac. Sci., Univ. Tokyo, Sect.* 4 **10**, 249–268.

Assenmacher, I. (1952). La vascularisation du complexe hypophysaire chez le canard domestique. *Arch. Anat. Microsc. Morphol. Exp.* **47**, 448–572.

Assenmacher, I. (1953). Etude anatomique du système artériel cervicocéphalique chez l'oiseau. *Arch. Anat., Histol. Embryol.* **35**, 181–202.

Assenmacher, I. (1957a). Répercussions de lésions hypothalamiques sur le conditionnement génital du canard domestique. *C. R. Acad. Sci.* **245**, 210–213.

Assenmacher, I. (1957b). Nouvelles données sur le rôle de l'hypothalamus dans les régulations hypophysaires gonadotropes chez le canard domestique. *C. R. Acad. Sci.* **245**, 2388–2390.

Assenmacher, I. (1958). Recherches sur le contrôle hypothalamique de la fonction gonadotrope préhypophysaire chez le canard. *Arch. Anat. Microsc. Morphol. Exp.* **47**, 447–572.

Assenmacher, I. (1959). Regulations thyréotropes après section des veines portes hypophysaires chez le canard Pékin mâle. *Gumma J. Med. Sci.* **8**, 199–206.

Assenmacher, I., and Tixier-Vidal, A. (1959). Action de la section des veines portes hypophysaires sur le fonctionnement thyroïdien étudié à l'aide du radioiode ^{131}I chez le canard Pékin. *J. Physiol. (Paris)* **51**, 391–392.

Assenmacher, I., and Tixier-Vidal, A. (1963). Etude physiologique de l'activité thyroïdienne du canard mâle aprés section des veines portes hypophysaires. *Ann. Endocrinol.* **24**, 509–523.

Assenmacher, I., and Tixier-Vidal, A. (1964). Repercussions de la section des veines portes hypophysaires sur la préhypophyse du canard Pékin mâle, entier ou castré. *Arch. Anat. Microsc. Morphol. Exp.* **53**, 83–108.

Assenmacher, I., Tixier-Vidal, A., and Baylé, J. D. (1961). Inhibition du réflexe photosexual par la réserpine chez le canard mâle. *C. R. Soc. Biol.* **155**, 2235–2240.

Baylé, J. D. (1968). Evolution post-operatoire de la corticostéron mie après hypophysectomie ou déconnexion hypothalamo-hypophysaire chez le canard. *J. Physiol. (Paris)* **60**, 397–398.

Baylé, J. D., Astier, H., and Assenmacher, I. (1966). Activité thyroïdiene du pigeon après hypophysectomie ou autogreffe hypophysaire. *J. Physiol. (Paris)* **58**, 459.

Baylé, J. D., Boissin, J., and Assenmacher, I. (1967). Le taux plasmatique de la corticostérone libre chez l'Oiseau, après hypophysectomie ou déconnexion hypothalamo-hypophysaire. *C. R. Acad. Sci.* **265**, 1524–1526.

Baylé, J. D., Boissin, J., Daniel, J. Y., and Assenmacher, I. (1971). Hypothalamic-hypophyseal control of adrenal cortical function in birds. *Neuroendocrinology* **7**, 308–321.

Benoit, J., and Assenmacher, I. (1953a). Rapport entre la stimulation sexuelle préhypophysaire et la neurosécrétion chez l'oiseau. *Arch. Anat. Microsc. Morphol. Exp.* **42**, 334–386.

Benoit, J., and Assenmacher, I. (1953b). Action des facteures externes et plus particulierement du facteur lumineux sur l'activité sexuelle des oiseaux. *Reunion Endocrinol. Lang. Fr., 2nd, 1953* pp. 33–80.

Benoit, J., and Assenmacher, I. (1955). Le contrôle hypothalamique de l'activité préhypophysaire gonadotrope. *J. Physiol. (Paris)* **47**, 427–567.

Berde, B., and Boissonnas, R. M. (1966). Synthetic analogues and homologues of the posterior pituitary hormones. *In* "The Pituitary Gland" (G. W. Harris and B. T. Donovan, eds.), Vol. 3, pp. 624–661. Butterworth, London.

Bern, H. A., and Knowles, F. G. W. (1966). Neurosecretion. *In* "Neuroendocrinology" (L. Martini and W. F. Ganong, eds.), Vol. 1, pp. 139–186. Academic Press, New York.

Bern, H. A., and Nishioka, R. S. (1965). Fine structure of the median eminence of some passerine birds. *Proc. Zool. Soc. Calcutta* **18**, 107–119.

Bern, H. A., Nishioka, R. S., Mewaldt, L. R., and Farner, D. S. (1966). Photoperiodic and osmotic influences on the ultrastructure of the hypothalamic neurosecretory system of the White-crowned Sparrow, *Zonotrichia leucophrys gambelii*. *Z. Zellforsch. Mikrosk. Anat.* **69**, 198–227.

Bixler, E. J., Negro-Villar, A., Ringer, R. K., and Meites, J. (1968). Evidence for hypothalamic FSH-releasing factor in adult male turkeys. *Poultry Sci.* **47**, 1656.

Björklund, A., Falck, B., and Ljunggren, L. (1968). Monoamines in the bird median eminence. *Z. Zellforsch. Mikrosk. Anat.* **89**, 193–200.

Boissin, J. (1967). Le contrôle hypothalamo-hypophysaire de la fonction cortico-surrénalienne chez le canard. *J. Physiol. (Paris)* **59**, 423–444.

Boissin, J., Baylé, J., and Assenmacher, I. (1966). Le fonctionnement cortico-surrénalien du canard mâle après préhypophysectomie ou autogreffe hypophysaire ectopique *C. R. Acad. Sci.* **263**, 1127–1129.

Bradley, E. L., and Holmes, W. N. (1971). The effects of hypophysectomy on adrenocortical function in the duck (*Anas platyrhynchos*). *J. Endocrinol.* **49**, 437–457.

Brasch, M., and Betz, T. W. (1971). The hormonal activities associated with the cephalic and caudal regions of cockerel pars distalis. *Gen. Comp. Endocrinol.* **16**, 241–256.

Brown, I. L., and Mewaldt, L. R. (1967). Effects of reserpine on the White-crowned Sparrow (*Zonotrichia leucophrys gambelii*). *Brit. J. Pharmacol. Chemother.* **30**, 251–257.

Calas, A., and Assenmacher, I. (1970). Ultrastructure de l'éminence médiane du canard (*Anas platyrhynchos*). *Z. Zellforsch. Mikrosk. Anat.* **109**, 64–82.

Chen, C. Bixler, E. J., Weber, A., and Meites, J. (1968). Hypothalamic stimulation of prolactin release from the pituitary of turkey hens and poults. *Gen. Comp. Endocrinol.* **11**, 489–494.

Coon, J. M. (1939). A new method for the assay of posterior pituitary extracts. *Arch. Int. Pharmacodyn. Ther.* **62**, 79–99.

Dantzler, W. H. (1966). Renal response of chickens to infusion of hypertonic sodium chloride solution. *Amer. J. Physiol.* **210**, 640–646.

de Robertis, E. (1962). Ultrastructure and function in some neurosecretory system. *Mem. Soc. Endocrinol.* **12**, 3–20.

de Robertis, E. (1964). "Histophysiology of Synapses and Neurosecretion." Pergamon, Oxford.

de Robertis, E., Salganicoff, L., Zieher, L. M., and Rodriguez de Lores Arnaiz, A. (1963). Acetylcholine and cholinesterase content of synaptic vesicles. *Science* **140**, 300–301.

Dominic, C. J., and Singh, R. M. (1969). Anterior and posterior groups of portal vessels in the avian pituitary. *Gen. Comp. Endocrinol.* **13**, 22–26.

Duncan, D. (1956). An electron microscope study of the neurohypophysis of a bird, *Gallus domesticus. Anat. Rec.* **125**, 457–471.

Duvernoy, H., Gaint, F., and Koritke, J. G. (1969). Sur la vascularisation de l'hypophyse des oiseaux. *J. Neuro-Visceral Relat.* **31**, 109–127.

Egge, A. S., and Chiasson, R. B. (1963). Endocrine effect of diencephalic lesions in the White Leghorn hen. *Gen. Comp. Endocrinol.* **3**, 346–361.

Enemer, A., and Ljunggren, L. (1968). The appearance of monoamines in the adult and developing neurohypophysis of *Gallus gallus. Z. Zellforsch. Mikrosk. Anat.* **91**, 496–506.

Farner, D. S., and Oksche, A. (1962). Neurosecretion in birds. *Gen. Comp. Endocrinol.* **2**, 113–147.

Farner, D. S., Wilson, F. E., and Oksche, A. (1967). Avian neuroendocrine mechanism. *In* "Neuroendocrinology" (L. Martini and W. F. Ganong, eds.), Vol. 2, pp. 529–582. Academic Press, New York.

Follett, B. K. (1969). Diurnal rhythms of monoamine oxidase activity in the quail hypothalamus during photoperiodic stimulation. *Comp. Biochem. Physiol.* **29**, 591–600.

Follett, B. K. (1970a). Gonadotropin-releasing activity in the quail hypothalamus. *Gen. Comp. Endocrinol.* **15**, 165–179.

Follett, B. K. (1970b). Effects of neurohypophysial hormones and their synthetic analogues on lower vertebrates. *In* "International Encyclopedia of Pharmacology and Therapeutics," (H. Heller and B. T. Pickering, eds.), Sect. 41, Vol. I, pp. 321–350. Pergamon, Oxford.

Follett, B. K., and Farner, D. S. (1966). The effect of daily photoperiod on gonadal growth, neurohypophysial hormone content, and neurosecretion in the hypothalamo-hypophysial system of the Japanese Quail (*Coturnix coturnix japonica*). *Gen. Comp. Endocrinol.* **7**, 111–124.

Follett, B. K., Kobayashi, H., and Farner, D. S. (1966). The distribution of monoamine oxidase and acetylcholinesterase in the hypothalamus and its relation to the hypothalamo-hypophysial neurosecretory system in the White-crowned Sparrow, *Zonotrichia leucophrys gambelii. Z. Zellforsch. Mikrosk. Anat.* **75**, 57–65.

Frankel, A. I. (1970). Neurohumoral control of the avian adrenal: A review. *Poultry Sci.* **49**, 869–921.

Frankel, A. I., Graber, J. W., and Nalbandov, A. V. (1967a). The effect of hypothalamic lesions on adrenal function in intact and adenohypophysectomized cockerels. *Gen. Comp. Endocrinol.* **8,** 387–396.

Frankel, A. I., Graber, J. W., and Nalbandov, A. V. (1967b). Adrenal function in cockerels. *Endocrinology* **80,** 1013–1019.

Fuxe, K., and Ljunggren, L. (1965). Cellular localization of monoamines in the upper brain stem of the pigeon. *J. Comp. Neurol.* **125,** 355–382.

Gerschenfeld, H. M., Tramezzani, J., and de Robertis, E. (1960). Ultrastructure and function in neurohypophysis of the toad. *Endocrinology* **66,** 741–762.

Geschwind, I. I. (1969). Mechanism of action of releasing factors. *In* "Frontiers in Neuroendocrinology, 1969" (W. F. Ganong and L. Martini, eds.), pp. 389–431. Oxford Univ. Press, London and New York.

Gogan, F. (1968). Sensibilité hypothalamique à la testostérone chez le canard. *Gen. Comp. Endocrinol.* **11,** 316–327.

Gourdji, D., and Tixier-Vidal, A. (1966). Mise en évidence d'un contrôle hypothalamique stimulant de la prolactine hypophysaire chez le canard. *C. R. Acad. Sci.* **263,** 162–165.

Graber, J. W., and Nalbandov, A. V. (1965). Neurosecretion in the White Leghorn cockerel. *Gen. Comp. Endocrinol.* **5,** 485–492.

Graber, J. W., Frankel, A. I., and Nalbandov, A. V. (1967). Hypothalamic center influencing the release of LH in the cockerel. *Gen. Comp. Endocrinol.* **9,** 187–192.

Guillemin, R. (1964). Hypothalamic factors releasing pituitary hormone. *Recent Progr. Horm. Res.* **20,** 89–130.

Hagen, P., and Wallace, A. C. (1961). Effects of reserpine on growth and sexual development of chickens. *Brit. J. Pharmacol. Chemother.* **17,** 267–275.

Halász, B., Pupp, L., and Uhlarik, S. (1962). Hypophysiotrophic area in the hypothalamus. *J. Endocrinol.* **25,** 147–154.

Hartmann, J. F. (1958). Electron microscopy of the neurohypophysis in normal and histamine-treated rats. *Z. Zellforsch. Mikrosk. Anat.* **48,** 291–308.

Hartree, A. S., and Cunningham, F. J. (1969). Purification of chicken pituitary follicle-stimulating hormone and lutenizing hormone. *J. Endocrinol.* **43,** 606–616.

Herlant, M. (1967). Mode de libération des produits de neurosécrétion. *In* "Neurosecretion" (F. Stutinsky, ed.), pp. 20–35. Springer-Verlag, Berlin and New York.

Hirano, T. (1964). Further studies on the neurohypophyseal hormones in the avian median eminence. *Endocrinol. Jap.* **11,** 87–95.

Hirano, T. (1966). Neurohypophysial hormones in the pigeon median eminence in relation to reproductive cycle and formalin stress. *J. Fac. Sci., Univ. Tokyo, Sect. 4* **11,** 43–48.

Hirano, T., Ishii, Su., and Kobayashi, H. (1962). Effects of prolongation of daily photoperiod on gonadal development and neurohypophyseal hormone activity in the median eminence and the pars nervosa of the passerine bird, *Zosterops palpebrosa japonica. Annot. Zool. Jap.* **35,** 64–71.

Holmes, R. L. (1964). Comparative observation on inclusions in nerve fibers of the mammalian neurohypophysis. *Z. Zellforsch. Mikrosk. Anat.* **64,** 64–71.

Holmes, R. L., and Knowles, G. W. (1960). "Synaptic vesicles" in the neurohypophysis. *Nature (London)* **185,** 710–711.

Ishii, Se. (1967). Morphological studies of the distribution and properties of the granulated vesicles in the brain. *Arch. Histol. Jap.* **28,** 355–376.

Ishii, Su. (1970). Association of luteinizing hormone-releasing factor with granules separated from equine hypophysial stalk. *Endocrinology* **86,** 207–216.

Ishii, Su., Hirano, T., and Kobayashi, H. (1962). Preliminary report on the neurohypophysial hormone activity in the avian median eminence. *Zool. Mag. (Dobutsugaku Zasshi)* **71,** 206–211 (in Japanese with English summary).

Ishii, Su., Iwata, T., and Kobayashi, H. (1969). Intergranular localization of corticotropin-releasing factor and vasopressin in the horse median eminence. *Endocrinol. Jap.* **16,** 171–177.

Ishii, Su., Sarkar, A. K., and Kobayashi, H. (1970). Ovarian ascorbic acid-depleting factor in pigeon median eminence extracts. *Gen. Comp. Endocrinol.* **14,** 461–466.

Jackson, G. L., and Nalbandov, A. V. (1969a). Luteinizing hormone releasing activity in the chicken hypothalamus. *Endocrinology* **84,** 1262–1265.

Jackson, G. L., and Nalbandov, A. V. (1969b). Ovarian ascorbic acid depleting factors in the chicken hypothalamus. *Endocrinology* **85,** 113–120.

Jackson, G. L. (1971a). Avian luteinizing hormone-releasing factor. *Endocrinology* **89,** 1454–1459.

Jackson, G. L. (1971b). Comparison of rat and chicken luteinizing hormone-releasing factors. *Endocrinology* **89,** 1460–1463.

John, T. M., and George, J. C. (1967). Cyclic histochemical changes in the hypothalamo-hypophysial neurosecretory system of the migratory Wagtails, *Motacilla alba* and *Motacilla flava*. *J. Anim. Morphol. Physiol.* **14,** 216–222.

Kamiyoshi, M., Tagami, M., and Tanaka, K. (1969). *In vitro* demonstration of follicle-stimulating hormone releasing activity in the chicken hypothalamus. *Poultry Sci.* **48,** 1977–1978.

Kanematsu, S. (1968a). Hypothalamic control of the function of the anterior hypophysis. *J. Physiol. Anim. Nutri. (Tokyo)* **12,** 11–20.

Kanematsu, S. (1968b). Hypothalamic control of the secretion of prolactin and gonadotropins. *In* "Brain Function and Reproduction," Vol. I, pp. 99–124. Kyodo-Isho, Tokyo (in Japanese).

Kanematsu, S., and Mikami, S. (1966). Effect of hypothalamic lesions on the pituitary-thyroid system in the chicken. *Jap. J. Zootech. Sci.* **37,** Suppl., 28–29.

Kanematsu, S., and Mikami, S. (1970). Effects of hypothalamic lesions on protein-bound ^{131}Iodine and thyroidal ^{131}I uptake in the chicken. *Gen. Comp. Endorcinol.* **14,** 25–34.

Kanematsu, S., Sonoda, T., Kii, M., and Kato, Y. (1966). Effects of hypothalamic lesions on gonad in the hen. *Jap. J. Vet. Sci.* **28,** Suppl., 451.

Kanematsu, S., Daimon, T., and Mikami, S. (1967). Effect of hypothalamic lesions on PBI^{131} and ^{131}I uptake in the chicken. *Jap. J. Zootech. Sci.* **38,** Suppl., 61.

Kawashima, S., Farner, D. S., Kobayashi, H., Oksche, A., and Lorenzen, L. (1964). The effect of dehydration on acid-phosphatase activity, catheptic-proteinase activity, and neurosecretion in the hypothalamo-hypophyseal system of the White-crowned Sparrow, *Zonotrichia leucophrys gambelii*. *Z. Zellforsch. Mikrosk. Anat.* **62,** 149–181.

Khazan, N., Sulman, F. G., and Winnik, H. Z. (1960). Effect of reserpine on pituitary-gonadal axis. *Proc. Soc. Exp. Biol. Med.* **105,** 201–204.

Knowles, F. G. W. (1963). Techniques in the study of neurosecretion. *In* "Techniques in Endocrine Research" (P. Eckstein and F. Knowles, eds.), pp. 57–65. Academic Press, New York.

Kobayashi, H. (1963). Median eminence of birds. *Proc. Int. Ornithol. Congr., 13th, 1962* pp. 1069–1084.

Kobayashi, H. (1965). Histochemical, electron microscopic and pharmacological studies of the median eminence. *Proc. Int. Congr. Endocrinol., 2nd, 1964* Int. Congr. Ser. No. 83, Part 1, pp. 570–576.

Kobayashi, H., and Farner, D. S. (1960). The effect of photoperiodic stimulation of phosphatase activity in the hypothalamo-hypophyseal system of the White-crowned Sparrow, *Zonotrichia leucophrys gambelii. Z. Zellforsch. Mikrosk. Anat.* **53**, 1–24.

Kobayashi, H., and Farner, D. S. (1964). Cholinesterase in the hypothalamo-hypophyseal neurosecretory system of the White-crowned Sparrow, *Zonotrichia leucophrys gambelii. Z. Zellforsch. Mikrosk. Anat.* **63**, 965–973.

Kobayashi, H., and Farner, D. S. (1966). Evidence of a negative feedback on photoperiodically induced gonadal development in the White-crowned Sparrow, *Zonotrichia leucophrys gambelii. Gen. Comp. Endocrinol.* **6**, 443–452.

Kobayashi, H., and Ishii, Su. (1969). The median eminence as storage site for releasing factors and other biological active substances. *Proc. Int. Congr. Endocrinol., 3rd, 1968* Int. Congr. Ser. No. 184, pp. 548–554.

Kobayashi, H., Bern, H. A., Nishioka, R. S., and Hyodo, Y. (1961). The hypothalamo-hypophysial neurosecretory system of the parakeet, *Melopsittacus undulatus. Gen. Comp. Endocrinol.* **1**, 545–564.

Kobayashi, H., Oota, Y., and Hirano, T. (1962). Acid phosphatase activity of the hypothalamo-hypophyseal system of dehydrated rats and pigeons in relation to neurosecretion. *Gen. Comp. Endocrinol.* **2**, 495–498.

Kobayashi, H., Matsui, T., and Ishii, Su. (1970). Functional electron microscopy of the hypothalamic median eminence. *Int. Rev. Cytol.* **29**, 281–381.

Kobayashi, H., Wada, M., Uemura, H., and Ueck, M. (1972). Uptake of peroxidase from the third ventricle by ependymal cells of the median eminence. *Z. Zellforsch. Mikrosk. Anat.* **127**, 545–551.

Koelle, G. B. (1962). A new general concept of the neurohumoral function of acetylcholine and acetylcholinesterase. *J. Pharm. Pharmacol.* **14**, 65–90.

Koelle, G. B., and Geesey, C. (1961). Localisation of acetylcholinesterase in the neurohypophysis and its functional implications. *Proc. Soc. Exp. Biol. Med.* **106**, 625–628.

Konishi, T. (1965). A method for the quantitative analysis of the neurosecretory materials in the avian median eminence. *Zool. Mag. (Dobutsugaku Zasshi)* **74**, 313–328 (in Japanese with English summary).

Konishi, T. (1967). Neurosecretory activities in the anterior median eminence in relation to photoperiodic testicular response in young Japanese Quail (*Coturnix coturnix japonica*). *Endocrinol. Jap.* **14**, 60–68.

Konishi, T., and Kato, M. (1967). Light-induced rhythmic changes in the hypothalamic neurosecretory activity in Japanese Quail, *Coturnix coturnix japonica. Endocrinol. Jap.* **14**, 239–245.

Kook, Y., Cho, K. B., and Yun, L. O. (1964). Metabolic effects of oxytocin in the chicken. *Nature (London)* **204**, 385–386.

Kordon, C., and Gogan, F. (1964). Localisation par une technique de microimplantation de structures hypothalamiques responsables du feed-back par la testostérone chez le canard. *C. R. Soc. Biol.* **158**, 1795–1798.

Kragt, C. L., and Meites, J. (1965). Stimulation of pigeon pituitary prolactin release by pigeon hypothalamic extracts *in vitro*. *Endocrinology* 78, 1169–1176.

Laws, D. F. (1961). Hypothalamic neurosecretion in the refractory and postrefractory periods and its relationship to the rate of photoperiodically induced testicular growth in *Zonotrichia leucophrys gambelii*. Z. *Zellforsch. Mikrosk. Anat.* 54, 275–306.

Lederis, K. (1965). An electron microscopical study of the human neurohypophysis. Z. *Zellforsch. Mikrosk. Anat.* 65, 847–868.

Legait, E., and Legait, H. (1959). Recherches sur l'ultrastructure de la neurohypophyse. C. R. Ass. Anat. 45, 514–518.

Legait, H. (1959). Contribution a l'étude morphologique et expérimentale du système hypothalamo-neurohypophysaire de la poule Rhode-Island. Thesis, University of Louvain-Nancy, Société d'impressions typographiques.

Lepkovsky, S., and Yasuda, M. (1966). Hypothalamic lesions, growth and body composition of male chickens. *Poultry Sci.* 45, 582–588.

McCann, S. M., and Porter, J. C. (1969). Hypothalamic pituitary stimulating and inhibiting hormones. *Physiol. Rev.* 49, 240–284.

McFarland, L. Z., Yousef, M. K., and Wilson, W. O. (1966). The influence of ambient temperature and hypothalamic lesions of the disappearance rates of thyroxine-^{131}I in the Japanese Quail. *Life Sci.* 5, 309–315.

Martini, L. (1966). Neurohypophysis and anterior pituitary activity. *In* "The Pituitary Gland" (G. W. Harris and B. T. Donovan, eds.), Vol. 3, pp. 535–577. Butterworth, London.

Matsui, T. (1964). Effect of water deprivation on the hypothalamic neurosecretory system of the Tree Sparrow, *Passer montanus saturatus*. *J. Fac. Sci., Univ. Tokyo, Sect. 4* 10, 355–368.

Matsui, T. (1966a). Effect of prolonged daily photoperiods on the hypothalamic neurosecretory system of the Tree Sparrow (*Passer montanus saturatus*). *Endocrinol. Jap.* 13, 23–38.

Matsui, T. (1966b). Fine structure of the posterior median eminence of the pigeon, *Columbia livia domestica*. *J. Fac. Sci., Univ. Tokyo, Sect. 4* 11, 49–70.

Matsui, T., and Kobayashi, H. (1965). Histochemical demonstration of monoamine oxidase in the hypothalamo-hypophyseal system of the Tree Sparrow and the rat. Z. *Zellforsch. Mikrosk. Anat.* 68, 172–182.

Matsui, T., and Kobayashi, H. (1968). Surface protrusion from the ependymal cells of the median eminence. *Arch. Anat., Histol. Embryol.* 51, 431–436.

Matsuo, H., Baba, Y., Nair, R. M. G., Arimura, A., and Schally, A. V. (1971). Structure of the porcine LH- and FSH-releasing hormone. I. The proposed amino acid sequence. *Biochem. Biophys. Res. Commun.* 43, 1334–1339.

Matsuo, S., Vitums, A., King, J. R., and Farner, D. S. (1969). Light-microscope studies of the cytology of the adenohypophysis of the White-crowned Sparrow, *Zonotrichia leucophrys gambelii*. Z. *Zellforsch. Mikrosk. Anat.* 95, 143–176.

Meites, J., ed. (1970). "Hypophysiotropic Hormones of the Hypothalamus: Assay and Chemistry." Williams & Wilkins, Baltimore, Maryland.

Meites, J., and Nicoll, C. S. (1966). Adenohypophysis: Prolactin. *Annu. Rev. Physiol.* 28, 57–88.

Mikami, S. (1960). The structure of the hypothalamo-hypophyseal neurosecretory system in the fowl and its morphological changes following adrenalectomy, thyroidectomy and castration. *J. Fac. Agr., Iwate Univ.* 4, 359–379.

Mikami, S., Kanematsu, S., and Daimon, T. (1967). Effects of hypothalamic lesions of the appearance of thyroidectomy cells in the chicken. *Jap. J. Vet. Sci.* **29**, Suppl., 7.

Mikami, S., Vitums, A., and Farner, D. S. (1969). Electron microscopic studies on the adenohypophysis of the White-crowned Sparrow, *Zonotrichia leucophrys gambelii*. *Z. Zellforsch. Mikrosk. Anat.* **97**, 1–29.

Miller, R. A. (1961). Hypertrophic adrenals and their response to stress after lesions in the median eminence of totally hypophysectomized pigeons. *Acta Endocrinol. (Copenhagen)* **37**, 565–576.

Miller, R. A., and Riddle, O. (1942). The cytology of the adrenal cortex of normal pigeons and in experimentally induced atrophy and hypertrophy. *Amer. J. Anat.* **71**, 311–341.

Muller, E. E., Sawano, S., and Schally, A. V. (1967). Growth hormone-releasing activity in the hypothalamus of animals of different species. *Gen. Comp. Endocrinol.* **9**, 349–352.

Munsick, R. A., Sawyer, W. H., and Van Dyke, H. B. (1960). Avian neurohypophyseal hormones: Pharmacal properties and tentative identification. *Endocrinology* **66**, 860–871.

Nicoll, C. S. (1965). Neural regulation of adenohypophyseal prolactin secretion in tetrapods: Identification from *in vitro* studies. *J. Exp. Zool.* **158**, 203–210.

Nicoll, C. S., and Bern, H. A. (1965). Pigeon crop-stimulating activity (prolactin) in the adenohypophysis of lungfish and tetrapods. *Endocrinology* **76**, 156–160.

Nicoll, C. S., and Meites, J. (1962). Prolactin secretion *in vitro*: Comparative aspects. *Nature (London)* **195**, 606–607.

Nishioka, R. S., Bern, H. A., and Mewaldt, L. R. (1964). Ultrastructural aspects of the neurohypophysis of the White-crowned Sparrow, *Zonotrichia leucophrys gambelii*, with special reference to the relation of neurosecretory axons to ependyma in the pars nervosa. *Gen. Comp. Endocrinol.* **4**, 304–313.

Oehmke, H.-J. (1968). Regionale Strukturunterscheide im Nucleus infundibularis der Vögel (Passeriformes). *Z. Zellforsch. Mikrosk. Anat.* **92**, 406–421.

Oehmke, H.-J. (1969). Topographische Verteilung der Monoaminfluoreszenz im Zwischenhirn-Hypophysensystem von *Carduelis chloris* und *Anas platyrhynchos*. *Z. Zellforsch. Mikrostk. Anat.* **101**, 266–284.

Oehmke, H.-J. (1971). Vergleichende neurohistologische Studien am Nucleus infundibularis einiger australischer Vögel. *Z. Zellforsch. Mikrosk. Anat.* **122**, 122–138.

Oehmke, H.-J., Priedkalns, J., Vaupel-von Harnack, M., and Oksche, A. (1969). Fluoreszenz- und electronen- microscopische Untersuchungen am Zwischenhirn-Hypophysensystem von *Passer domesticus*. *Z. Zellforsch. Mikrosk. Anat.* **95**, 109–133.

Oksche, A., Laws, D. F., Kamemoto, F. I., and Farner, D. S. (1959). The hypothalamo-hypophyseal neurosecretory system of the White-crowned Sparrow, *Zonotrichia leucophrys gambelii*. *Z. Zellforsch. Mikrosk. Anat.* **51**, 1–42.

Oksche, A., Farner, D. S., Serventy, D. L., Wolff, F., and Nicholls, C. A. (1963). The hypothalamo-hypophyseal neurosecretory system of the Zebra Finch, *Taeniopygia castanotis*. *Z. Zellforsch. Mikrosk. Anat.* **58**, 846–914.

Oksche, A., Oehmke, H.-J., and Farner, D. S. (1970). Weitere Befunde zur Struktur und Funktion des Zwischenhirn-Hypophysensystems der Vögel. *In* "Aspects in Neuroendocrinology" (W. Bargmann and B. Scharrer, eds.), pp. 261–273. Springer-Verlag, Berlin and New York.

Oota, Y. (1970). An electron microscope study of ependymal cells of the pigeon, *Columbia livia domestica*. *Rep. Fac. Sci., Shizuoka Univ.* **5**, 69–86.

Oota, Y., and Kobayashi, H. (1962). Fine structure of the median eminence and pars nervosa of the pigeon. *Annot. Zool. Jap.* **36**, 128–138.

Oota, Y., and Kobayashi, H. (1966). On the synaptic vesicle-like structures in the neuro-secretory axon of the mouse neural lobe. *Annot. Zool. Jap.* **39**, 193–201.

Opel, H. (1963). Delay in ovulation in the hen following stimulation of the preoptic brain. *Proc. Soc. Exp. Biol. Med.* **113**, 488–492.

Opel, H. (1964). Premature oviposition following operative interference with the brain of the chicken. *Endocrinology* **74**, 193–200.

Opel, H. (1965). Oviposition in chickens after removal of the posterior lobe of the pituitary by an improved method. *Endocrinology* **76**, 673–677.

Opel, H., and Fraps, R. M. (1961). Blockade of gonadotrophin release for ovulation in the hen following stimulation with stainless steel electrode. *Proc. Soc. Exp. Biol. Med.* **108**, 291–296.

Opel, H., and Leopore, P. D. (1967). Ovulating hormone-releasing factor in chicken hypothalamus. *Poultry Sci.* **46**, 1302.

Payne, F. (1959). Cytologic evidence of secretory activity in neurohypophysis of the fowl. *Anat. Rec.* **134**, 433–453.

Péczely, P. (1969). Effects of the median eminence of the pigeon (*Columba livia domestica L.*) on the regulation of adenohypophyseal corticotropin secretion. *Acta Physiol.* **35**, 47–57.

Péczely, P., and Calas, A. (1970). Ultrastructure de l'éminence médiane du pigeon (*Columba livia domesticus*) dans diverses conditions experimentales. *Z. Zell-forsch. Mikrosk. Anat.* **111**, 316–345.

Péczely, P., and Zboray, G. (1967). CRF and ACTH activity in the median eminence of the pigeon. *Acta Physiol.* **32**, 229–239.

Péczely, P., Baylé, J. D., Boissin, J., and Assenmacher, I. (1970). Activité corticotrope et C.R.F. dans l'éminence médiane, et activité corticotrope de greffes hypophy-saires chez le pigeon. *C. R. Acad. Sci.* **270**, 3264–3267.

Pellegrino de Iraldi, A., Duggan, H. F., and de Robertis, E. (1963). Adrenergic synaptic vesicles in the anterior hypothalamus of the rat. *Anat. Rec.* **145**, 521–531.

Priedkalns, J., and Oksche, A. (1969). Ultrastructure of synaptic terminals in the nucleus infundibularis and nucleus supraopticus of *Passer domesticus*. *Z. Zellforsch. Mi-krosk. Anat.* **98**, 135–147.

Ralph, C. L. (1959). Some effects of hypothalamic lesions on gonadotropin release in the hen. *Anat. Rec.* **134**, 411–431.

Ralph, C. L., and Fraps, R. M. (1959). Effect of hypothalamic lesions on progesterone induced ovulation in the hen. *Endocrinology* **65**, 819–824.

Ralph, C. L., and Fraps, R. M. (1960). Induction of ovulation in the hen by injection of progesterone into the brain. *Endocrinology* **66**, 269–272.

Resko, J. A., Norton, H. M., and Nalbandov, A. V. (1964). Endocrine control of the adrenal in chickens. *Endocrinology* **75**, 192–200.

Rosenberg, L. L., Astier, H., Roche, G. L., Baylé, J. D., Tixier-Vidal, A., and Assen-macher, I. (1967). The thyroid function of the drake after hypophysectomy or hypo-thalamic pituitary disconnection. *Neuroendocrinology* **2**, 113–125.

Rossbach, R. (1966). Das Neurosekretorisch-Zwischenhirnsystem der Amsel (*Turdus merula L.*) im Jahresablauf und nach Wasserentzug. *Z. Zellforsch. Mikrosk. Anat.* **71**, 118–145.

Russell, D. H. (1968). Asetylcholinesterase in the hypothalamo-hypophyseal axis of the

White-crowned Sparrow, *Zonotrichia leucophrys gambelii*. *Gen. Comp. Endocrinol.* **11**, 51–63.

Russell, D. H., and Farner, D. S. (1968). Acetylcholinesterase and gonadotropin activity in the anterior pituitary. *Life Sci.* **7**, 1217–1221.

Salem, M. H. M., Norton, H. M., and Nalbandov, A. V. (1970a). A study of ACTH and CRF in chickens. *Gen. Comp. Endocrinol.* **14**, 270–280.

Salem, M. H. M., Norton, H. M., and Nalbandov, A. V. (1970b). The role of vasotocin and CRF in ACTH release in the chicken. *Gen. Comp. Endocrinol.* **14**, 281–289.

Sato, T., and George, J. C. (1972). Personal communication.

Sawyer, W. H. (1963). Neurohypophyseal peptides and water excretion in the vertebrates. *Mem. Soc. Endocrinol.* **13**, 45–59.

Schally, A. V., Arimura, A., Baba, Y., Nair, R. M. G., Matsuo, H., Pedding, T. W., Debeljuk, L., and White, W. F. (1971). Isolation and properties of the FSH and LH-releasing hormone. *Biochem. Biophys. Res. Commun.* **43**, 393–399.

Schally, A. V., Arimura, A., and Kastin, A. J. (1973). Hypothalamic regulatory hormones. *Science* **179**, 341–350.

Sharp, P. J., and Follett, B. K. (1968). The distribution of monoamines in the hypothalamus of the Japanese Quail, *Coturnix coturnix japonica*. *Z. Zellforsch. Mikrosk. Anat.* **90**, 245–262.

Sharp, P. J., and Follett, B. K. (1969a). The blood supply to the pituitary and basal hypothalamus in the Japanese Quail (*Coturnix coturnix japonica*). *J. Anat.* **104**, 227–232.

Sharp, P. J., and Follett, B. K. (1969b). The effect of reserpine on the pituitary-gonadal axis in quail. *In* "Seminar on Hypothalamic and Endocrine Function in Birds," (H. Kobayashi and D. S. Farner, eds.). p. 43.

Sharp, P. J., and Follett, B. K. (1969c). The effect of hypothalamic lesions on gonadotrophin release in Japanese Quail (*Coturnix coturnix japonica*). *Neuroendocrinology* **5**, 205–218.

Sharp, P. J., and Follett, B. K. (1970). The adrenergic supply within the avian hypothalamus. *In* "Aspects in Neuroendocrinology" (W. Bargmann and B. Scharrer, eds.), pp. 95–103. Springer-Verlag, Berlin and New York.

Shirley, H. V., and Nalbandov, A. V. (1956). Effects of neurohypophysectomy in domestic chickens. *Endocrinology* **58**, 477–483.

Skadhauge, E. (1964). Effects of unilateral infusion of arginine vasotocin into the portal circulation of the avian kidney. *Acta Endocrinol. (Copenhagen)* **47**, 321–330.

Snapir, N., Nir, I., Furuta, F., and Lepkovsky, S. (1969). Effect of administrative testosterone propionate on cocks functionally castrated by hypothalamic lesion. *Endocrinology* **84**, 611–618.

Stainer, I. M., and Holmes, W. N. (1969). Some evidence for the presence of a corticotropin-releasing factor (CRF) in the duck (*Anas platyrhynchos*). *Gen. Comp. Endocrinol.* **12**, 350–359.

Stetson, M. H. (1969a). The role of the median eminence in control of photoperiodically induced testicular growth in the White-crowned Sparrow, *Zonotrichia leucophrys gambelii*. *Z. Zellforsch. Mikrosk. Anat.* **93**, 369–394.

Stetson, M. H. (1969b). Formation of secondary neurohemal organs in the median eminence of the White-crowned Sparrow and Japanese Quail. *Gen. Comp. Endocrinol.* **13**, 392–398.

Stetson, M. H. (1969c). Hypothalamic regulation of FSH and LH secretion in male and female Japanese Quail. *Amer. Zool.* **7**, 1078–1079.

Sturkie, P. D., and Lin, Y. (1966). Release of vasotocin and oviposition in the hen. *J. Endocrinol.* **35**, 325–326.

Szenthágothai, J., Flerkó, B., Mess, B., and Halász, B. (1968). "Hypothalamic Control of the Anterior Pituitary," pp. 118–134. Akadémiai Kiad , Budapest.

Taguchi, S., Kobayashi, H., and Farner, D. S. (1966). Observation on the uptake of [35]Sulfur by the hypothalamo-hypophyseal system of the White-crowned Sparrow *Zonotrichia leucophrys gambelii*) following intraventricular injection of ^{35}S DL-cystein. *Z. Zellforsch. Mikrosk. Anat.* **69**, 228–245.

Takahara, H., Sonoda, T., and Kato, Y. (1967). Correlation between the center of the autonomic nervous system and the anterior hypophysis. IV. Cytological change of the hypophysis and the gonad after hypothalamic lesions in the cockerel. *Jap. J. Zootech. Sci. Suppl.* **38**, 60.

Tanaka, K., and Nakajo, S. (1959). Cholinesterase in the diencephalon of the hen in relation to egg laying. *Poultry Sci.* **38**, 991–995.

Tanaka, K., and Nakajo, S. (1962). Participation of neurohypophyseal hormone in oviposition in the hen. *Endocrinology* **70**, 453–458.

Tanaka, K., Kamiyoshi, M., and Tagami, M. (1969). *In vitro* demonstration of LH releasing activity in the hypothalamus of the hen. *Poultry Sci.* **48**, 1985–1987.

Tixier-Vidal, A. (1965). Caracters ultrastructuraux des types cellulaires de l'adenohypophyse du canard mâle. *Arch. Anat. Microsc. Morphol. Exp.* **54**, 719–780.

Tixier-Vidal, A., and Assenmacher, I. (1966). Etude cytologique de la préhypophyse du pigeon pendant la couvaison et la lactation. *Z. Zellforsch. Mikrosk. Anat.* **69**, 489–519.

Tixier-Vidal, A., Baylé, J. D., and Assenmacher, I. (1966). Etude cytologique ultrastructurale de l'hypophyse du pigeon après autogreffe ectopique. Absence de stimulation des cellules à prolactine. *C. R. Acad. Sci.* **262**, 675–678.

Tixier-Vidal, A., Follett, B. K., and Farner, D. S. (1968). The anterior pituitary of the Japanese Quail, *Coturnix coturnix japonica*. *Z. Zellforsch. Microsk. Anat.* **92**, 610–635.

Uemura, H. (1964a). Effects of gonadoectomy and sex steroid on the acid phosphatase activity of the hypothalamo-hypophyseal system in the bird. *Emberiza rustica latifascia. Endocrinol. Jap.* **11**, 185–203.

Uemura, H. (1964b). Effects of water deprivation the hypothalamo-hypophyseal neurosecretory system of the grass parakeet, *Melopsittacus undulatus*. *Gen. Comp. Endocrinol.* **4**, 193–198.

Uemura, H. (1964c). Cholinesterase in the hypothalamo-hypophysial neurosecretory system of the bird. *Zool. Mag. (Dobutsugaku Zasshi)* **73**, 118–126 (in Japanese with English summary).

Uemura, H. (1965). Histochemical studies on the distribution of cholinesterase and alkaline phosphatase in the vertebrate neurosecretory system. *Annot. Zool. Jap.* **38**, 79–96.

Uemura, H., and Kobayashi, H. (1963). Effects of prolonged daily photoperiods and estrogen on hypothalamic neurosecretory system of the passerene bird, *Zosterops palpebrosa japonica. Gen. Comp. Endocrinol.* **3**, 253–264.

Urano, A. (1968). Monoamine oxidase in the hypothalamic neurosecretory system and the adenohypophysis of the Japanese Quail and the mouse. *J. Fac. Sci., Univ. Tokyo, Sect. 4* **11**, 437–451.

Vitums, A., Mikami, S., Oksche, A., and Farner, D. S. (1964). Vascularization of the hypothalamo-hypophyseal complex in the White-crowned Sparrow, *Zonotrichia leucophrys gambelii. Z. Zellforsch. Mikrosk. Anat.* **64,** 541–569.

Vitums, A., Ono, K., Oksche, A., and Farner, D. S. (1966). The development of the hypophysial portal system in the White-crowned Sparrow, *Zonotrichia leucophrys gambelii. Z. Zellforsch. Mikrosk. Anat.* **73,** 355–366.

Wada, M. (1972). Effect of hypothalamic implantation of testosterone on photostimulated testicular growth in Japanese Quail (*Coturnix coturnix japonica*). *Z. Zellforsch. Mikrosk. Anat.* **124,** 507–519.

Wada, M. (1972). Effect of hypothalamic implantation of puromycin on photostimulated testicular growth in Japanese Quail (*Coturnix coturnis japonica*). *Gen. Comp. Endocrinol.* (in press).

Warren, S. P. (1968). Primary catecholamine fibers in the ventral hypothalamus of the White-crowned Sparrow, *Zonotrichia leucophrys gambelii.* Master's Thesis, University of Washington, Seattle.

Wilson, F. E. (1967). The tubero-infundibular system: A component of the photoperiodic control mechanism of the White-crowned Sparrow, *Zonotrichia leucophrys gambelii. Z. Zellforsch. Mikrosk. Anat.* **82,** 1–24.

Wilson, F. E. (1970). The tubero-infundibular region of the hypothalamus: A focus of testosterone sensitivity in male Tree Sparrow (*Spizella arborea*). *In* "Aspects in Neuroendocrinology" (W. Bargmann and B. Scharrer, eds.), pp. 274–286. Springer-Verlag, Berlin and New York.

Winget, C. M., Wilson, W. O., and McFarland, L. Z. (1967). Response of certain diencephalic, pituitary and plasma enzymes to light in *Gallus domesticus. Comp. Biochem. Physiol.* **22,** 141–147.

Wingstrand, K. G. (1951). "The Structure and Development of the Avian Pituitary from a Comparative and Functional Viewpoint." Gleerup, Lund.

Wolfson, A., and Kobayashi, H. (1962). Phosphatase activity and neurosecretion in the hypothalamo-hypophysial system in relation to the photoperiodic gonadal response in *Zonotrichia albicollis. Gen. Comp. Endocrinol., Suppl.* **1,** 168–179.

Wooley, P. (1959). The effect of posterior lobe pituitary extracts on blood pressure in several vertebrate classes. *J. Exp. Biol.* **36,** 453–458.

Chapter 5

AVIAN VISION

Arnold J. Sillman

I. Introduction

There can be very little doubt that birds, in general, are highly visual animals. One would expect, as Walls (1942) has suggested, that animals that travel at great speeds would have need for very good visual acuity so as to enable them better to perceive movement and to avoid collisions. Even a casual observation of their behavior—a hawk swooping down on its prey from great height, a swallow catching insects on the wing—indicates immediately that the bird depends on its visual system for its survival. Certainly, the great importance of vision to the bird is implied by the large size of the eyes. Some hawks and owls have eyes at least as large as those of man, and the ostrich eye, at 50 mm in diameter, is the largest eye of any land vertebrate (Walls, 1942). Even more striking than the absolute size of the avian eye is the relative size (Fig. 1). In some cases the fundi actually touch each other in the skull, and in some birds the combined weight of both eyes is greater than the weight of the brain.

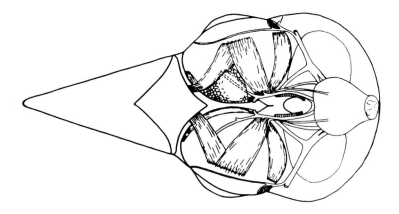

FIG. 1. Eyes and brain of *Passer domesticus*, viewed ventrally. (Redrawn from Slonaker, 1918.)

Thus, no treatment of the biology of birds could approach completeness without a consideration of avian vision. General information on this topic has already been compiled by several authors (Walls, 1942; Rochon-Duvigneaud, 1943; Duke-Elder, 1958; Pumphrey, 1961); this chapter leans heavily on these treatises. However, since there is not much point in belaboring material that has already been well presented, the current treatment of avian vision will concentrate on areas that either have not been stressed before in a general treatise, are at least highly characteristic of birds, or are still open to speculation and, therefore, active areas of investigation.

II. Aspects of Structure and Refraction

A. THE GLOBE

In terms of the general morphology of the avian eye, one cannot say that it differs drastically from that of any other class of vertebrates. However, there are differences in the shape of the eye among the birds and, in this respect, the globe can be classified into one of three broad categories. The great majority of birds possess what Walls refers to as the flat eye. This eye, illustrated in Fig. 2a, is similar to that found in the lizards and is characterized by an anterior–posterior axis that is much shorter than the other diameters. The Columbiformes and Galliformes are examples of birds with eyes that fall into this class. Birds that have greater need for visual acuity, including some of the Passeriformes and most of the Falconiformes, have eyes in which the ratio of the vertical and horizontal axes more closely approach unity so that the eye becomes globular in shape (Fig. 2b). The idea here is that the size of the image thrown onto the retina will be larger if the axial length of the eye is greater. Hence an increased axial length is a means of increasing the visual acuity. According to Walls some eagles have eyes with axial lengths so great that the eye is almost tubular in shape (Fig. 2c). Consequently, they have greater visual acuity. The owls also exhibit a tubular eye, but in this case some of the advantage gained in terms of acuity has been lost, since the lens of the eye has receded thereby decreasing the size of the retinal image. Most owls, of course, are either nocturnal or crepuscular

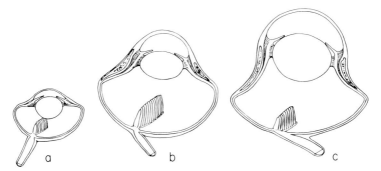

FIG. 2. The three characteristic shapes of the avian eye: (a) flat type of majority of birds; (b) globular type of most Falconiformes; (c) tubular type of owls and some eagles. (Redrawn from Walls, 1942.)

and, therefore, would not benefit so much from increased acuity as they would from increased sensitivity.

Duke-Elder (1958) suggests that the maintenance of the nonspherical shape of the avian eye requires special structures of support not found in the eyes of other vertebrates. One of these supportive devices is the cartilaginous cup present in the posterior portion of the globe. Another is a series of bony plates, the scleral ossicles, that completely encircles the eye just behind the cornea in the region of the concavity which is so clearly illustrated in Fig. 2c. There are usually fifteen of these plates, although they are known to vary in number from ten to eighteen. In the smaller eyes the ossicles are actually compact bone, but in the case of the larger eyes, such as those of the eagles, the plates are hollow, as is typical of much of the avian skeleton. In addition to their role as supportive structures, the scleral ossicles may serve to prevent the globe from changing shape during accommodation and, therefore, to facilitate the distortion of the refractive apparatus.

B. The Refractive Apparatus and Accommodation

In the avian eye, as in the eyes of all other terrestrial vertebrates, the major structure of refraction is the cornea. This is so because the greatest change in the index of refraction occurs at the cornea, as light passes from the air through the eye. The lens does play some part in refraction, but generally serves only as a fine adjustment during accommodation. Structurally, the cornea and lens do not vary much among vertebrate classes. However, there are interesting differences to be found in the mechanisms of accommodation.

Accommodation, of course, can be viewed as the alteration of the refractive apparatus, as an object moves closer to or farther away from the eye, so that the image on the retina is always properly focused. This is accomplished in various ways by different types of animals. In teleost fishes, for example, the lens is moved backward toward the retina as an object moves closer. In mammals, on the other hand, the lens changes shape and increases in power as an object approaches. In birds, however, it appears that the cornea plays a major role in the accommodation process (Slonaker, 1918; Walls, 1942; Gundlach et al., 1945; Duke-Elder, 1958). The contraction of Crampton's muscle, which extends from the cornea to the sclera, causes the cornea to increase in curvature and, therefore, to increase its refractive power. The lizards also have some ability to deform the cornea during accommodation, but it is not so well marked as it is in birds. The lens also functions in accommodation of the avian eye, its refractive power being increased by contraction of Brucke's muscle, which causes the ciliary body that surrounds the lens to press against it and deform it. In fact, the lens must be the major structure involved in accommodation in birds that dive and feed under water, for once submerged the cornea becomes optically inoperative, since its index of refraction is very similar to that of water. It is interesting, in this respect, that in diving birds the cornea is very much thicker than in other birds, and Crampton's muscle is very poorly developed. This is in contrast to the situation in hawks and owls in which the cornea is observed to alter greatly in curvature during accommodation and in which Crampton's muscle is very well developed.

III. Aspects of the Retina and Visual Perception

A. RETINAL ORGANIZATION

The retina is the true *raison d'être* of the eye. It is the organ that first senses the light, integrates the information, and then passes this information onto the brain in the form of nerve impulses. The other structures of the eye are designed to serve only to present the image to the retina. Accordingly, much of the discussion in this chapter will necessarily involve this complex receptor system.

The retina of the bird has been said to represent the ultimate in retinal organization (Walls, 1942), but in general it is not unlike the retinas of the other vertebrate classes. As illustrated in Fig. 3, the vertebrate retina can be divided up into several layers. Although the first of these, the pigment epithelium, is an interesting and incompletely understood layer, it does not appear to have a direct role in perception per se, and therefore we will say nothing more about it here.

The receptor cells actually span three retinal layers. They have their outer segments, which contain the visual pigment, in what might be referred to as the visual cell layer. Their nuclei are located in the outer nuclear layer, and their synaptic connections to more proximal cells are made in the region of the outer plexiform layer. The avian retina, as is true of that of most animals, is duplex in nature and thus contains both rods, responsible for dim light or scotopic vision, and cones, responsible for acute, bright light or photopic vision. The cones also serve to mediate color vision where this capability occurs. Not surprisingly, therefore, birds such as the diurnal predators and the passerines have retinas that are heavily dominated by cones, the rods being few in number and usually relegated to the periphery. The owl

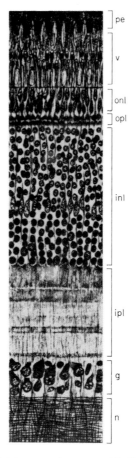

FIG. 3. The retina of the chick; pe=pigment epithelium; v=visual cell layer; onl =outer nuclear layer; opl=outer plexiform layer; inl=inner nuclear layer; ipl=inner plexiform layer; g=ganglion cell layer; and n=nerve fiber layer. (Relabeled from Walls, 1942.)

retina, on the other hand, has some cones but mostly rods. In addition to the usual single cone, the avian retina also contains double cones (Fig. 4). Such double cones have been described in the retinas of all classes of vertebrates with the exception of the placental mammals (Schultze, 1867), but their function is not at all clear, nor is it known

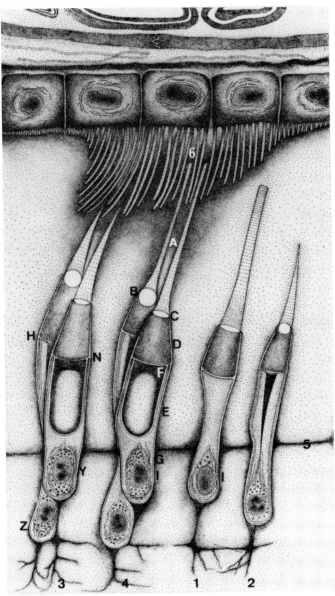

FIG. 4. Avian photoreceptors (*Zonotrichia leucophrys gambelii*); 1=rod; 2=single cone; 3=double cone; 4=double cone with rodlike outer segment; 5=external limiting membrane; 6=pigment epithelium; A=outer segments; B=oil droplet; C=connecting cilium; D=ellipsoid; E=paraboloid; F=basophilic inclusion bodies; G=basophilic perinuclear granules; H=chief cone; N=accessory cone; I, Z, Y=nuclei. (After Hartwig, 1968.)

how they differ functionally from the single cone. Recently, Hartwig (1968) reported that the retina of the White-crowned Sparrow (*Zonotrichia leucophrys gambelii*) contains, in addition to the normal double cone, a double cone with rodlike outer segments (Fig. 4). Hartwig suggested that this last type of cell is unique to the White-crowned Sparrow, but it is certainly possible that it will prove to be a more general cell type as the birds are studied more extensively.

In the inner nuclear layer are found the bipolar cells, the dendrites of which synapse with either rods or cones. The axon terminals of the bipolars synapse in the inner plexiform layer with the more proximally located ganglion cells whose axons leave the globe as the optic nerve. In accordance with the duplex nature of vision, several rods synapse with a single bipolar cell, thus increasing the sensitivity of the rod system, while generally only one cone synapses to a single bipolar cell, thus increasing the resolving power of the cone system. As a result, diurnal birds, whose retinas are dominated by cones, have a high concentration of bipolar cells and, consequently, a relatively thick inner nuclear layer. Also located in the inner nuclear layer of the vertebrate retina are the horizontal and amacrine cells. These cells do not have classical axons, but their processes do serve laterally to interconnect the other retinal neurons. They must, therefore, be responsible for some kind of integration of information. The inner nuclear layer of the avian retina is especially rich in these association cells, which, like the bipolars, add to the relative thickness of the layer (Detwiler, 1943). In fact, the entire retina of the bird is generally thicker than that of other vertebrates. However, whereas in diurnal birds the inner nuclear layer is substantially thicker than the outer nuclear layer, in nocturnal birds, where acute and complex vision is sacrificed for the sake of sensitivity, the outer nuclear layer is thicker than the inner nuclear layer.

With respect to the large complement of association cells in the retina of the diurnal bird, as well as to the large relative size of the eyes, it is interesting to note that many of the complex functions of the visual system that are relegated to the higher centers of the nervous system in the mammal are performed by the retina in the bird. For example, the pigeon retina contains five types of ganglion cells in addition to those capable of discriminating luminosities (Maturana and Frenk, 1963). These include cells that respond to verticality, horizontality, edges in general, moving edges, and convex edges. Thus, it would appear that the birds are similar to the amphibians and the reptiles, which have also been found to be capable of complex responses at the retinal level (Maturana, 1962).

B. AREAS AND FOVEAS – TYPE AND DISTRIBUTION

Animals, the birds among them, that have duplex retinas and that have any need at all for visual acuity during their activities tend to develop areas in their retinas. An area, and most animals have but one, can be defined as a site on the retina in which the cones become more densely packed at the expense of rods. Since visual acuity depends at least in part on the "grain" of the image, then any image falling upon a portion of the retina that has a much more dense population of cones will be perceived in greater detail than an image falling upon a portion of the retina not so specialized. Since, as mentioned just previously, the ratio of cones to their more proximal connecting neurons is about one to one, the retina in the region of the area tends to bulge out into the vitreous as more and more cones are incorporated. In very highly developed areas, such as are found in higher primates and diurnal lizards as well as the birds, this bulging would ordinarily be quite severe if not for the development of a pit within the area. This pit, or fovea as it is called, is formed by the shifting aside, out of the light path, of the more proximal neurons and their fibers. One must keep in mind that in the vertebrate eye the visual cells are closest to the sclera and are directed away from the light. In birds and diurnal lizards, which have the most highly developed areas, the foveal pit is also quite well developed, as illustrated in Fig. 5. In contrast, the higher primates, which have areas not quite so well developed as the birds and lizards, have foveal pits that are rather shallow and not so well formed (Fig. 6). The possible significance of the fovea will be discussed shortly. Suffice it to say here that the fovea is generally considered to be a point of very good vision.

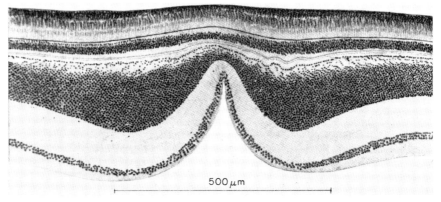

FIG. 5. Central fovea of a diurnal bird, the Least Tern (*Sterna albifrons*). (After Rochon-Duvigneaud, 1943.)

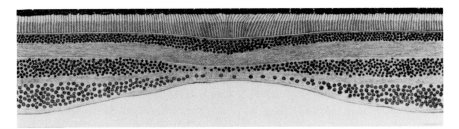

FIG. 6. Diagram of human fovea. (From Wolff, 1968.)

In terms of the fovea, the avian retina can be classified into several categories, as has been done in rather similar ways by both Wood (1917) and Duke-Elder (1958). First, it must be realized, that there are indeed birds that do not possess a fovea at all. Into this class fall such species as the domestic chicken and the California Quail (*Lophortyx californicus*). Duke-Elder places the turkey, the guinea hen* (*Numida pucherani*), and the pigeon also in the afoveal class, although he describes them as possessing foveas that are very shallow rather than nonexistent. Lockie (1952) has reported that the Northern Fulmar (*Fulmarus glacialis*) and the Manx Shearwater (*Puffinus puffinus*) also lack foveas, although their areas appear to be quite highly developed.

The vast majority of birds fall into Duke-Elder's second classification, termed central monofoveal. These have but one fovea, and as Wood (1917) and Slonaker (1897) have pointed out, most birds with a single fovea have it positioned near the center of the retina, slightly above and toward the nasal aspect of the optic disc where the optic nerve leaves the globe. Some examples of species having the single, central fovea include the Common Crow (*Corvus brachyrhynchos*) (Wood, 1917), the House Sparrow (*Passer domesticus*) (Lockie, 1952), and the American Dipper (*Cinclus mexicanus*) (Goodge, 1960).

The only species that possess a single fovea, but one that is located temporally and hence termed temporal monofoveal, are the owls. Walls (1942) points out that the European Swift (*Apus apus*) appears at first glance to have only a temporal fovea as do the owls, but upon closer examination one can observe some trace of a central fovea as well. He suggests that this may be an indication that at some time in their evolutionary history the owls may have had central foveas as well as temporal foveas and that the former eventually disappeared. Most owls, of course, are not diurnal animals, and therefore it is not unexpected that their foveas should be generally rather shallow and

*Presumably the Kenya Crested Guineafowl, *Guttera pucherani*. — ED.

poorly developed. It must be pointed out, however, that such is not always the case. For example, Oehme (1961) found the temporal foveas of *Strix aluco, Asio otus,* and *Tyto alba* to be bowl-shaped and shallow, whereas that of *Asio flammeus,* a diurnal hunter, is pitted and deep, as is typical of diurnal birds.

Mention above of the two foveas in swifts brings us to the next class of retina, the bifoveal retina. Birds that by nature of their feeding habits must be good judges of speed and distance possess temporal foveas in addition to central foveas. The hawks, eagles, terns, and swallows fall into this category. These birds all have eyes that are relatively frontal in their orientation, more so than, say, a pigeon or a crow. It is assumed that whereas the central foveas allow the animal to scan monocularly with acute vision, the temporal foveas permit the bird some degree of binocularity, a most desirable capability for birds that take their prey on the wing. It is not surprising to note, therefore, that the owls, which have completely lost the central fovea, are the most frontal of all birds. Man, who has excellent binocular vision, has the fovea located in much the same position as the temporal fovea of the owls (Wood, 1917). And the East Indian Long-nosed Tree-snake (*Ahaetulla mycterizans*), with "the sharpest sight and the most accurate judgment of distance of any [snake] in the world" has temporal foveas (Walls, 1942). Only birds, however, are known to have two foveas. In addition to those types of bird mentioned above, hummingbirds also have two foveas, as does the European Kingfisher (*Alcedo atthis*).

Some birds have an area that is not round, as is the usual case, but rather is ribbonlike and extends around the retina in a horizontal plane. This is illustrated quite nicely for the White-capped Albatross (*Diomedea cauta cauta*) in Fig. 7 (O'Day, 1940). It has been suggested that such an area might serve to increase sensitivity to objects on the horizon (Pumphrey, 1948a) or to aid in normal orientation (Duijm, 1958). Pumphrey points out that the ribbonlike area is usually found in birds that appear to lend importance to the horizon, and that such an area is never found in forest-dwelling birds. However, the ribbonlike structure is not limited to birds. Certain diurnal ground squirrels possess areas that can be described as horizontal bands (Walls, 1942), and such an area is present in the retina of the Red-eared Turtle (*Chrysemys scripta elegans*), in which it may be used for orientation (Brown, 1969). It should be noted that the ribbon can occur regardless of the type of fovea. For example, the Fulmar and the Manx Shear-water, both regarded as afoveal, have the ribbonlike area (Lockie,

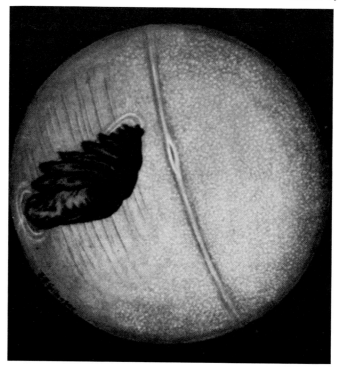

FIG. 7. Ribbonlike area in the retina of the White-capped Albatross (*Diomedea cauta cauta*). (After O'Day, 1940.)

1952). In some cases, as in the Common Tern (*Sterna hirundo*), the ribbon passes through the central fovea but not the temporal fovea. Some birds, such as the albatross mentioned above and the Giant Fulmar (*Macronectes giganteus*), have only a central fovea but also have a ribbonlike area passing through it in the horizontal plane. Finally, hawks, eagles, and swallows, which are bifoveal, have a ribbonlike area connecting the temporal and central foveas. Duke-Elder terms this last class of retina the infula–bifoveal type and suggests that it might represent the ultimate refinement in foveal development.

C. Visual Acuity and the Function of the Avian Fovea

Whereas it is rather easy to understand the value of an area in terms of increased visual acuity, it is somewhat difficult to see why a foveal pit should develop in the more highly refined areas. It is even less obvious why the avian fovea should be so much more striking in its depth than that of the mammals. A case can be made for the presence of the fovea in the mammals, for in these animals there are blood vessels in the retina that could be disadvantageous in an area designed for acute vision. As a result of the formation of the foveal depression, these blood vessels are removed from the light path. In birds, however, there is no retinal circulation, and, therefore, removal of blood vessels cannot be the reason for the development of the foveal pit. And, after all, it is in the birds in which the foveal depression is most striking.

Walls (1937, 1942) suggests that the function of the fovea is solely to increase the visual acuity of the animal. He points out that the relative refractive indices of the retinal tissue and the vitreous humor are such that the index of refraction of the retina is always substantially higher than that of the vitreous humor. Thus, he says, "if a light ray should strike the vitreo retinal boundary at anything but a right angle it will be refracted away from an imaginary perpendicular to the surface at the point of impact." According to Walls, the convex bulge into the vitreous humor as a result of the development of an area without a fovea would tend to converge the image and make it smaller, an undesirable condition in terms of visual acuity. Walls maintains, therefore, that the poorly developed foveas in the areas of species such as the typical teleost or the soft-shelled turtle serve merely to offset this disadvantage by making the area less convex. Supposedly, man's shallow fovea functions in a similar manner. Walls reasons, on the other hand, that a light ray incident on a very deep fovea, such as that found in birds and diurnal lizards, would be seriously diverged from its normal path (Fig. 8). The overall effect would be to magnify the size of the retinal image and, from the stand-

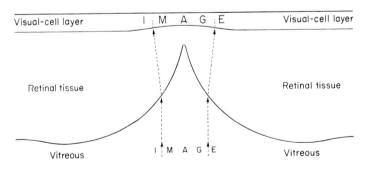

FIG. 8. The theoretical diverging effect of the well-developed avian fovea. (After Walls, 1942.)

point of visual acuity, would be most advantageous. Walls concludes that the shape of the fovea results in a 13% linear increase in the image size and a 30% increase in the area of the image in birds, more in lizards.

Pumphrey (1948a,b), however, takes issue with Walls' analysis. In fact, he takes issue with the entire idea, which is almost legendary, that the visual acuity of the bird is so very much greater than that of other animals. With respect to the claims made about the visual acuity of the bird, Pumphrey appears to be supported by the little experimental evidence that exists. For example, Hamilton and Goldstein (1933) report that the measured visual acuity of the pigeon is actually distinctly less than man and of about the same order as that of the monkey. Although this is somewhat in disagreement with Gundlach (1933), who concluded that pigeons have a visual acuity twice as good as that of man, both observations serve to illustrate the point that certainly not all diurnal birds possess extraordinary visual acuities as compared to man. On the other hand, the claim can be made that the pigeon is a semidomesticated animal, and as such it is to be expected that its visual acuity has declined over the years. In support of this idea, Donner (1951) has demonstrated that the domestic chicken has a visual acuity only one-twenty-fifth as good as that of man. Nevertheless, in the same series of experiments, Donner also demonstrated that the visual acuities of passeriform and falconiform birds, believed to have the very best resolving powers, are only about 2.5 times as great as that of man and about 2.8 times as great as that of the chimpanzee. Using a different approach, Oehme (1964) studied freshly enucleated eyes of European Buzzards (*Buteo buteo*) and Common Kestrels (*Falco tinnunculus*) and ascertained the nodal points and the intercenter distances between cones. Using the nodal point angles corresponding to the intercenter distances, he determined the visual acuity and concluded that the buzzard's maximum acuity would nearly correspond to that of man while the maximum acuity of the falcon would compare favorably with that of the rhesus macaque. All in all, therefore, it would seem that one must take seriously Pumphrey's assertion that the visual acuity of the diurnal bird is not so very much better than that of man.

Furthermore, in returning to our consideration of the fovea, Pumphrey suggests that the deep avian fovea would so distort an image, as a consequence of aberration, that if the fovea did anything at all in terms of visual acuity it would tend to decrease it rather than enhance it. Therefore, Pumphrey poses the question as to what value a fovea would be that not only fails to significantly increase visual acuity but might actually distort the image. He suggests that such a device would be of great help in fixation and detection of movement. Thus, as the image moved across the retina of the bird, it would be maintained in good focus because of the high density of cones of the avian extrafoveal retina, but as the image passed across the fovea, it would become slightly out of focus as a result of the aberration caused by the steep sides of the pit. By always maintaining the image slightly out of focus the bird would be able to "lock on" to the object and keep it in sight rather easily. Pumphrey relates his idea to the observation that, in general, the temporal foveas, which are used in binocular vision, are not so well developed or steep as the central foveas, which are used for scanning the sky or the ground at a distance. His point here is that after detecting distant movement of an object the bird approaches and must then have accurate measures of speed and distance. At that time, the bird is going to use its binocular capability and, therefore, its temporal rather than central foveas. Thus distortion, which could not be tolerated under such circumstances, is eliminated. The logic here is attractive. However, the argument suffers a bit when it is noted that the temporal foveas of eagles appear to be relatively well developed and very deep. Furthermore, a bird's visual acuity depends not only on the quality of the image but also on the nature of the organization of retinal elements receiving that image. It is quite conceivable that distortions in the image are compensated for by a specific arrangement of receptors and postreceptor elements so that the animal sees the distorted images clearly and sharply. Certainly if Walls is correct in his magnification theory, the deeper fovea could still give good acuity in spite of the distortion (Owen, 1971).

In summary then, Pumphrey is saying that birds appear to have much better vision than man because they do not tend to lose sight of a distant object as easily. He suggests that if man could find the object, he could see almost as well as a bird, but this ability is lacking. The better vision of the bird is a consequence also of the very large population of extrafoveal cones in the avian retina. In other words, although the actual acuity of the avian retina is probably not much greater than that of man, the bird can perceive a sharp image of an object falling even extrafoveally and then, if the object is moving, can keep it in sight much more easily than can man.

D. THE VISUAL PIGMENTS

The outer segments of the rods and cones contain the photosensitive material which is responsible for the absorption of light incident upon the visual cells and thus, in a way not yet fully understood, for the photoreceptor potential that ultimately initiates the nervous impulse that travels over the optic nerve and to the brain. Any complete discussion of the eye, regardless of the class of animals under consideration, must necessarily concern itself with these visual pigments; see Dartnall (1962) for an excellent review.

The pigment molecule is a conjugated protein that is classically pictured as consisting of an opsin prosthetic group attached by means of a Schiff base linkage to a vitamin A aldehyde chromophore. The chromophore, whose structure is illustrated in Fig. 9, may be based upon either the vitamin A_1 aldehyde (retinal) or the vitamin A_2 aldehyde (dehydroretinal), but it is always present in the native pigment in a specific isomeric structural configuration — the 11-*cis* form. The action of light is to isomerize the chromophore from the 11-*cis* form to the all-trans form, thus setting off a complex series of subsequent structural changes that eventually results in stimulation.

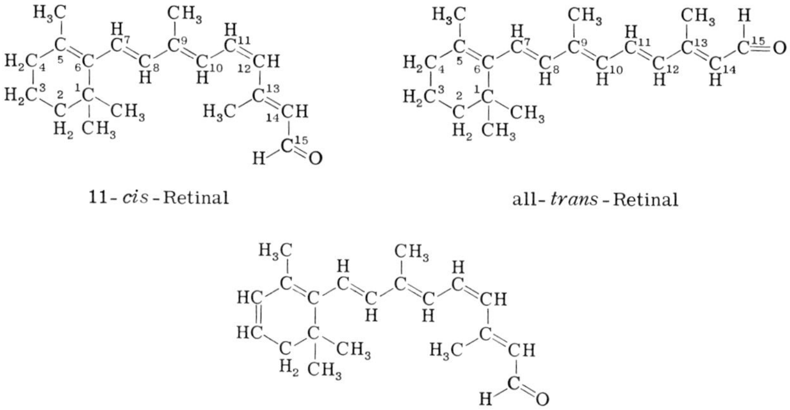

11-*cis*-Retinal   all-*trans*-Retinal

11-*cis*-Dehydroretinal

FIG. 9. Chromophores of the visual pigments.

Within a given class of visual pigments (i.e., either retinal or dehydroretinal) the chromophore has been shown not to vary and, therefore, any differences in pigment characteristics must be due to differences in their respective opsins. The pigments do indeed differ greatly in their characteristics, varying in their sensitivities to pH and various chemical treatments, as well as in their light-absorbing capabilities. This last factor is especially important, for it has become traditional to characterize a visual pigment not only by the nature of its chromophore, but also by the nature of its absorbance spectrum.

A typical visual pigment absorbance curve, this one determined for an extract prepared from the retina of the Burrowing Owl (*Speotyto cunicularia*), is illustrated by the open squares in Fig. 10. The peak absorbance, or λ_{max} as it is usually referred to, is at 502 nm, which means that this particular pigment is most efficient in absorbing light

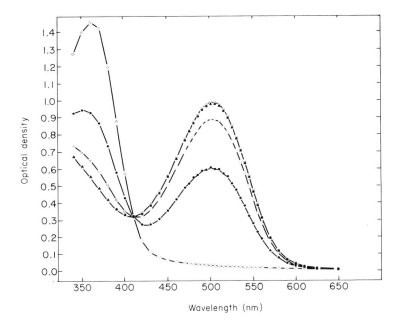

FIG. 10. The effect of light on the visual pigment of the Burrowing Owl (*Speotyto cunicularia*). Open squares represent the absorbance of the unexposed extract. Filled triangles represent absorbance after exposure to 650 nm light. Open circles after additional exposure to 625 nm light for 60 minutes; filled circles after additional 60 minutes' exposure to 600 nm light; open diamonds after another 60 minutes' exposure to 500 nm light. Decreases in optical density represent bleaching of visual pigment. Increases represent formation of product.

of 502 nm wavelength. One can see in Fig. 10 that exposing the visual pigment to the proper light causes it to bleach, proving that the material is photosensitive. Although all visual pigments have been found to have similarly shaped absorbance curves, their maxima may vary considerably. Most, but certainly not all, vertebrate visual pigments, which are generally accepted or known to be from the rods, absorb maximally close to 500 nm. Only one cone pigment, that of the accessory cone of the frog retina, has been shown, microspectrophotometrically, to absorb maximally in this region (Liebman and Entine, 1968). For this reason, and also because cone pigments, in general, have proven very elusive to extraction techniques, the avian visual pigments that have maxima from 500 nm to 506 nm and that constitute almost all of those analyzed (Table I) are most probably rod or scotopic visual pigments. However, one cannot be absolutely certain of this, since there is not yet an effective method of separating the rods from the cones before extraction, and no avian cones have been examined *in situ* by means of single-cell microspectrophotometry. Neverthe-

TABLE I
Visual Pigments of Certain Species of Birds

Species	λ_{max} (nm)	Reference
Galliformes		
Chicken (*Gallus gallus*)	562 (40%)	Bliss (1946; Wald
	502 (60%)	*et al.* (1955)
Turkey (*Meleagris gallopavo*)	562 (28%)	Crescitelli *et al.*
	503 (72%)	(1964)
Bobwhite Quail (*Colinus virginianus*)	500	Sillman (1969)
California Quail (*Lophortyx californicus*)	500	Sillman (1969)
Strigiformes		
Barn Owl (*Tyto alba*)	500	Köttgen and
		Abelsdorff (1896)
Great Horned Owl (*Bubo virginianus*)	502	Crescitelli (1958a,b)
Screech Owl (*Otus asio*)	503	Crescitelli (1958a,b)
Burrowing Owl (*Speotyto cunicularia*)	503	Sillman (1969)
Caprimulgiformes		
Poorwill (*Phalaenoptilus nuttallii*)	506	Sillman (1969)
Columbiformes		
Pigeon (*Columba livia*)	502 (80%)	Bridges (1962);
	544 (20%)	Sillman (1969)
Ring Dove (*Streptopelia "risoria"*)	502	Sillman (1969)
Mourning Dove (*Zenaida macroura*)	502	Sillman (1969)
Charadriiformes		
Western Gull (*Larus occidentalis*)	501	Crescitelli (1958a,b)
Pelecaniformes		
Brown Pelican (*Pelecanus occidentalis*)	502	Crescitelli (1958a,b)
Anseriformes		
Duck (*Anas platyrhynchos*)	502	Bridges (1962)
Psittaciformes		
Budgerigar (*Melopsittacus undulatus*)	505	Sillman (1969)
Falconiformes		
Roadside Hawk (*Buteo magnirostris*)	500	Sillman (1969)

TABLE I (*continued*)

Species	λ_{max} (nm)	Reference
Passeriformes		
Brown-headed Cowbird (*Molothrus ater*)	501	Sillman (1969)
House Finch (*Carpodacus mexicanus*)	502	Sillman (1969)
Brown Towhee (*Pipilo fuscus*)	500	Sillman (1969)
White-crowned Sparrow (*Zonotrichia leucophrys*)	502	Sillman (1969)
Zebra Finch (*Poephila guttata*)	502	Sillman (1969)
Bengalese Finch (*Lonchura striata*)	502 (94%) 490 (6%)	Sillman (1969)
Black-rumped Waxbill (*Estrilda troglodytes*)	502 (94%) 490 (6%)	Sillman (1969)
Chestnut Mannikin (*Lonchura malacca*)	503 (90%) ~480 (10%)	Sillman (1969)
Common Silverbill (*Lonchura malabarica*)	502 (94%) 490 (6%)	Sillman (1969)
White-browed Sparrow-Weaver (*Plocepasser mahali*)	504 (93%) ~485 (7%)	Sillman (1969)

less, there is little doubt that the pigments described for the four species of owls (*Bubo virginianus, Otus asio, Tyto alba* and *Speotyto cunicularia*) and the Poorwill (*Phalaenoptilus nuttallii*) are scotopic in nature, for these animals are either crepuscular or nocturnal and have retinas with very high concentrations of rods. The other, essentially diurnal, birds are somewhat more problematical, since their retinas contain almost entirely cones, but have some small percentage of rods as well. In a survey carried out by the author (Sillman, 1969), low yields of visual pigment were extracted from diurnal birds as compared to those from the Burrowing Owl and the Poorwill, as would be expected if the diurnal animals had but few rods in their retinas and the pigment extracted was from those cells. If one compares the Burrowing Owl with the Roadside Hawk (*Buteo magnirostris*), the former yielded at least twenty-five times as much visual pigment as the latter, although the eyes of the two species are of comparable size and the pigments have similar maxima. It is reasonable to conclude that the owls probably have adapted to dim-light environments by developing large rods (Kühne, 1878) with consequently much scotopic visual pigment, while hawks have adapted to bright-light environments by developing many cones but few and small rods with, consequently, very little scotopic pigment. Somewhat analogous to the case of the owls is the situation in some deep-sea fishes in which the rods are very long and yield very high concentrations of scotopic visual pigment (Munz, 1958a).

Exactly why the cone pigments should prove so difficult to extract is not yet understood. However, whatever the reasons, it must be said that no unequivocal extraction of a cone pigment has been made, although there have been numerous attempts. Nevertheless, it is interesting for the present discussion that the birds have yielded the two visual pigments that are most likely to prove to be true cone pigments. These are the 562 nm pigment found in the chicken (Wald, 1937; Bliss, 1946; Wald et al., 1955) and the turkey (Crescitelli et al., 1964), and the 544 nm pigment found in the pigeon (Bridges, 1962). The four major reasons for suspecting that these are pigments of the cones are (1) their maxima are obviously greatly displaced from the region of 500 nm, which is typical of rod pigments, (2) they were found in addition to another photopigment which in each case absorbed maximally near 500 nm, (3) they each constituted a major portion of the total pigment content of an extract prepared from cone-dominated retinas, and (4) their absorbance curves are in somewhat good agreement with spectral sensitivity data believed to represent cones (e.g., see Fig. 12). In the author's survey, five species of passerines also yielded second photopigments in addition to pigments absorbing maximally in the region of 500 nm. However, in these birds the maxima ranged from about 480 nm to 490 nm and, therefore, were not significantly displaced from the 500 nm region characteristic of scotopic pigments. Furthermore, these pigments usually constituted only 6–10% of the total pigment complement of the extract. Finally, no data are available as to the spectral sensitivities of the birds in question that might allow one to make a comparison with the visual pigment absorbance curves. Therefore, while the possibility that the minor components are cone pigments cannot be totally excluded from consideration, a conservative approach would force the conclusion that they are probably scotopic.

That visual pigments are indeed present in vertebrate cones has been shown by in situ microspectrophotometry on the retinas of the goldfish (Marks, 1965), monkey (Brown and Wald, 1963; Marks et al., 1964), tadpole and frog (Liebman and Entine, 1968). A microspectrophotometric analysis of the avian visual pigments would most certainly be a valuable contribution, but this may not yet be technically feasible because of the very small size of the avian cones.

Except in the case of Tyto alba, in which the chromophore type has not been determined, all the extracted bird visual pigments are known to be based upon retinal. This is in contrast to the teleost fishes and the amphibians, which have either retinal or dehydroretinal pigments and often have both in the same retina, but is similar to the reptiles and mammals, in which dehydroretinals also have never been found although, as in birds, multipigment systems are known to exist. In this respect it would be very interesting to analyze the visual pigments of a bird, such as the Anhinga (*Anhinga* sp.) or a kingfisher, that eats freshwater fish as a major component of the diet. Freshwater fish often have high concentrations of Vitamin A_2 in their systems, and it is conceivable that the retina of a bird feeding upon them would contain some dehydroretinal-based pigment. However, it is much more likely that only a retinal-based pigment would be found as it is known that mechanisms for converting vitamin A_2 to vitamin A_1 do exist in nature in amphibians (Crescitelli, 1958a,b) and fishes (Dartnall *et al.*, 1961), and can be observed in rats under the proper experimental conditions (Yoshikami *et al.*, 1969). The physiological importance relative to the constancy of the avian chromophore lies in the fact that visual pigments that are based upon dehydroretinal tend to absorb maximally at wavelengths that are somewhat longer than do visual pigments with similar opsins but with the retinal chromophore. Thus, the spectral sensitivity of the animal is shifted toward the red end of the spectrum.

The nature of the predominant avian scotopic pigment certainly displays a marked constancy in spectral location. This constancy is reminiscent of that of scotopic pigments in mammals and adult frogs and toads (Crescitelli, 1958a,b), but is quite unlike the great variability found in marine and freshwater teleosts, where it has been possible to draw some correlation between the pigment composition and the photic environment (Wald *et al.*, 1957; Munz, 1958a,b; Schwanzara, 1967). Birds such as pelicans and ducks, which presumably see under water, have also been found to have the standard scotopic pigment with maximum near 500 nm. And it is of interest that diet, which we have alluded to above, actually does not seem to influence the nature of the avian scotopic system. Among the birds examined so far there are seed eaters, fish eaters, insect eaters, and those that feed on small mammals. The diets of the five passerines that yielded a second pigment are not significantly different from those of several other passerines that did not yield such a pigment. Indeed, there is nothing obviously unique in the diet, coloration, or ecology of these five species that might indicate the biological significance of the second, presumably scotopic, photopigment. On the other hand, it is interesting to conjecture whether the extractable 562 nm pigment of the chicken and the turkey is not in some way related to factors, such as very specialized diet, that accompany a high degree of domestication.

E. COLOR VISION

A major reason for concern that cone pigments have not been positively identified in the avian retina stems from the fact that cones are recognized generally to be the photoreceptors mediating information about color (e.g., Pirenne, 1967) and the retina of birds is unique among vertebrates in that the cone population of most avian retinas is relatively high.

Nevertheless, despite the unfortunate lack of information regarding avian cone pigments, there is every reason to believe that birds employ color vision in their activities. Most birds are strictly diurnal and, therefore, active at a time when color perception would be of importance. Many birds are themselves very highly colored, and often display a striking sexual dimorphism which indicates that color vision might be of importance in mating behavior. And some birds have been observed to employ color—as in flowers—in their feeding. But aside from this somewhat indirect and more or less intuitive information implying that birds see colors, a fair amount of evidence of this capability has been gathered from experimental observations, both behavioral and physiological.

For example, Porter (1904, 1906) trained House Sparrows (*Passer domesticus*) and Brown-headed Cowbirds (*Molothrus ater*) to choose a glass covered with a particular colored paper. Both species appeared capable of discriminating between blue, green, yellow, and red papers, with the sparrow demonstrating somewhat better ability in the blue. Another behavioral study led Wood (1925) to conclude that the Satin Bowerbird (*Ptilonorhynchus violaceus*) can discriminate colors. Here animals were provided with various and sundry, small, colored items at the time that the male was building its bower, an elaborate structure into which the bird incorporates various objects as construction material. The bird was observed to choose objects of several shades of blue while avoiding objects colored orange or red as well as those that were all white or all black. In a more recent study on the chick of the Laughing Gull (*Larus atricilla*) Hailman (1966) found that the chicks react best to an object simulating the beak of the parent if the "beak" is painted either blue or red rather than yellow or green. Perhaps it is not merely coincidence that the adult gull does indeed have a red beak which the chick pecks to elicit feeding. Also, in yet another behavioral study, Klopfer (1968) showed that neonatal Pekin ducklings can distinguish red decoy adults from white decoy adults.

One thing that these four studies have in common is their apparent failure to control brightness. Unfortunately, if the brightness for each color tested is not identical or, even more important, does not appear identical to the animal under study, one can never be certain that the animal is discriminating between colors rather than between relative brightnesses. Such studies, therefore, are compromised in terms of yielding definitive answers with respect to color vision. Nevertheless, they do provide additional indications that birds are able to discriminate colors.

Fortunately, in the case of the pigeon, which probably has been studied more intensively with respect to vision than any other bird, we do not have to rely on evidence gathered in experiments in which brightness was not fully considered. In one study, Hamilton and Coleman (1933) trained pigeons to jump toward a screen upon which was projected a patch of light of a specific wavelength. Care was taken to vary the lights in intensity in such a way as to make certain that any discrimination must have been on the basis of hue rather than brightness. The authors found that pigeons could very easily discriminate hues in much the same fashion as man. In fact, they conclude that the pigeon has a trireceptor visual apparatus, which would be in accordance with the current concept of color vision at the receptor level.

In an experiment designed actually to test the importance of the nucleus rotundus to color vision in the pigeon, Hodos (1969) controlled the luminosity with a rather ingenious method whereby the birds themselves calibrated the stimulating lights by means of heterochromatic brightness matching. In this technique the birds were first trained to discriminate between two different spectral lights on the basis of brightness rather than color. The intensity of the brighter light was then gradually reduced until the pigeon could no longer make the distinction; i.e., until the lights appeared to be of similar brightness to the bird. Other pigeons were then successfully trained to discriminate between these lights in which color was now the only possible difference. The birds were found to be capable of distinguishing green from blue, yellow, or red, other conditions not being tested.

Besides behavioral experiments, there are physiological studies that indicate that the pigeon has the capacity to discriminate colors. Using a microelectrode, Donner (1953) recorded spike activity from

the ganglion cell layer of the pigeon retina in response to various
wavelengths of light of equal energy. He then plotted the sensitivity
of the retina, in terms of percent of maximum, against the wavelength.
Three different curves were obtained, each with a different maximum,
implying the presence of three different cellular networks, each
mediating a different region of the spectrum (Fig. 11). Although the
three curves are not comparable in form to visual pigment curves with
similar maxima, the data serve as an excellent indication that the
pigeon has a color-coded visual system.

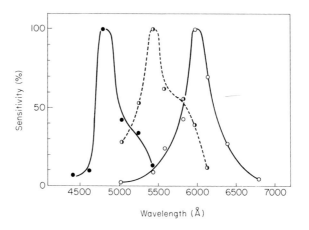

FIG. 11. Spectral sensitivity curves obtained from the ganglion cell layer of the
pigeon retina. See text for explanation. (Relabeled from Donner, 1953.)

Also taking an electrophysiological approach, Ikeda (1965) used the
electroretinogram (the mass electrical response of the retina to light)
to monitor the sensitivity of the pigeon retina to spectral colors.
Ikeda employed a high state of light adaptation and a flicker technique to ensure isolation of the cone response from the rod response.
She obtained a broad spectral sensitivity curve, illustrated in Fig. 12,
which displays a maximum near 547 nm. This curve agrees reasonably
well with the absorbance spectrum of the visual pigment that Bridges
(1962) extracted from the pigeon and interpreted as a cone pigment.
Most significantly, after adapting the retina to light of 547 nm wavelength, Ikeda found that a new spectral sensitivity curve was obtained,
this one peaking near 605 nm (Fig. 13). Although the agreement between this curve and that for a hypothetical visual pigment absorbing
maximally at 605 nm is not very good, the data are consistent with the

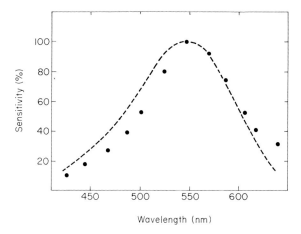

FIG. 12. Spectral sensitivity of the light-adapted pigeon retina, as obtained with the electroretinogram. (--) pigment 544; (●) sensitivity. (Replotted from Ikeda, 1965.)

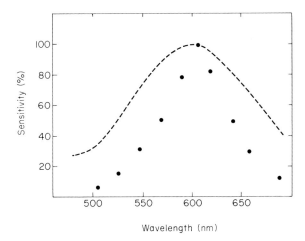

FIG. 13. Photopic spectral sensitivity of the pigeon retina after exposing the retina to a 547 nm adapting light; (--) hypothetical pigment 605; (●) sensitivity during selective adaption to 547 nm. (Replotted from Ikeda, 1965.)

idea that the pigeon retina contains a red-sensitive cone. The reasoning here is that the 547 nm adapting light bleached out the 544 nm visual pigment, and therefore, since only the red-sensitive pigment remained, the spectral sensitivity curve moved closer to that of the red-sensitive pigment. It is of interest to note that two of Donner's curves shown in Fig. 11 peaked at 540–550 nm and 590–610 nm, respectively, and that just recently Granda and Yazulla (1971) observed electrophysiologically that some cells of the pigeon's nucleus rotundus exhibit peak sensitivities in the 540 nm and 600–620 nm regions.

All things considered, it becomes clear that the pigeon does see colors, and that birds in general are most probably capable of color discrimination. But, in reality, surprisingly little work has been done to prove the point conclusively.

F. THE OIL DROPLETS

Among the most interesting and controversial structures found in the visual cells of many vertebrates, and sometimes associated with color vision, are the oil droplets. These spherical globules, reported first by Valentin (1840) and Hannover (1840), may be either colored or colorless, with the latter having been observed in vertebrates as advanced as the monotremes and marsupials, and as primitive as the sturgeons and lungfishes. The frogs, whose retinas contain yellow as well as colorless droplets, are considered to be the most primitive vertebrates containing colored oil globules. On the other hand, no oil droplets at all are reported for cyclostomes, elasmobranchs, holosteans, teleosts, urodele and coecilian amphibians, crocodilians, Pygopodidae, geckoes, snakes, and placental mammals (Walls, 1942). The diurnal birds do possess retinas with oil droplets, and it is in these animals (and the turtles) where the globules exhibit the brightest colors and are most varied, the pigeon retina containing red, orange, and yellow globules. Nocturnal or crepuscular birds, on the other hand, have only pale yellow or colorless droplets.

For the most part, the oil droplets are associated with the cones and are located in the inner segment (Fig. 4). Although Walls mentions that *Sphenodon,* a primitive reptile, and some lungfishes as well, contain oil droplets in their rods, this must be taken as the exceptional condition—the rods are usually devoid of oil globules. There is no doubt that both the single cone and the chief component of the double cone of the avian retina contain oil droplets, but it has yet to be determined definitely whether or not this is true for the accessory component of the double cone. Early workers (Dobrowolsky, 1871; Hoffman, 1877; Krause, 1894) did indeed describe a lipid globule in the accessory member, but later workers (Walls, 1942; Rochon-Duvigneaud, 1943) denied its existence. Still more recent investigations on the Great Tit (*Parus major*) (Engström, 1958), the pigeon (Cohen, 1963), the chicken (Morris and Shorey, 1967), and the White-crowned

Sparrow (*Zonotrichia leucophrys gambelii*) (Hartwig, 1968) fail to mention accessory cone oil droplets. But Meyer and Cooper (1966) produced some highly convincing evidence that the accessory component of the double cone of the domestic chicken does indeed contain an oil droplet, in this case yellowish green in color. These authors suggest that the relationship between the chief and accessory components of the double cone is such that the oil droplet of the accessory member is often, if not usually, obscured by the oil droplet of the chief member, and that this is what led many investigators to the conclusion that only the chief component contains an oil droplet.

The first extractions of the globules were performed by Capranica (1877) and by Kühne and Ayres (1878), but it was not until Wald and Zussman (1937, 1938) succeeded in purifying and crystallizing each of the droplet components from the chicken retina that some insight was provided as to the carotenoid nature of the material. Although the exact chemical composition of all the different droplets has yet to be determined, it appears likely that the red droplets of the chicken and the pigeon are astaxanthin (Wald and Zussman, 1937, 1938; Bridges, 1962). Sarcinin or sarcinaxanthin and a xanthophyll have also been described in avian oil droplet extracts (Wald and Zussman, 1937, 1938; Goodwin, 1952; Bridges, 1962). It is probable that most, if not all, retinas of diurnal species contain oil droplets composed of similar carotenoids, although the relative concentration of each may vary from species to species (Sillman, 1969).

Of course, the most interesting aspect of the oil droplets is their function. By virtue of their position between the source of light and the photosensitive material of the outer segments, it has been natural to reason, even from the earliest days, that the oil globules somehow serve as light filters. Schultze (1866) was the first to suggest the possibility that the globules serve a function in color vision, and this idea has been restated since by later investigators (Wald and Zussman, 1938; Hailman, 1964; King-Smith, 1969). Certainly each of the globules does have a distinct absorbance spectrum and, therefore, would allow only specific wavelengths of light to be transmitted through to the outer segment of its respective cone. In this respect, it is theoretically true that the differently colored globules, together with but a single visual pigment such as the 562 nm pigment extracted from the chicken and the turkey, could serve as a basis for color discrimination.

However, it is fairly certain now that whereas all the oil droplets absorb to one extent or another in the blue-green region of the spectrum, none of them has any significant absorbance in the red (Strother,

1963; King-Smith, 1969). In view of this, it becomes a little difficult to understand how the complete visible spectrum, including the red region, could be discriminated. With regard to this objection, King-Smith points out that, according to Hamilton and Coleman (1933), pigeons cannot discriminate hues in the region above 620 nm, whereas man can discriminate wavelengths up to 650 nm. Thus, the oil droplets cannot be positively eliminated as a factor in color discrimination solely on the basis of their failure to absorb in the red.

Walls (1942) argues against a color vision role for the oil droplets on the basis of the fact that the avian fovea is devoid of any red oil droplets. He reasons that it is not logical to conclude that the ability to discriminate color would be diminished in the fovea, which is traditionally considered to be the point of most acute vision. Therefore, Walls suggests that the globules do not function in color discrimination and, moreover, that the differently colored oil droplets each serves a different function. For example, to the yellow droplets he ascribes the role of reducing chromatic aberration, which is due mainly to the shorter wavelengths. The yellow droplets would filter out enough shorter wavelength light to allow a sharp image while transmitting enough of the visible spectrum to allow complete color vision. The yellow droplets could also serve to reduce glare that results from the scattering of light, since the shorter wavelengths are scattered to a greater degree than the longer. Thus, Walls believes that the yellow droplets of birds serve in much the same capacity as the yellow filters found in so many other vertebrates, such as the yellow cornea in some fishes and the yellow lens in diurnal squirrels, treeshrews, snakes, geckoes, and some lampreys as well as man. On the other hand, both the red and the yellow droplets could play a role in enhancing contrast. Thus, according to Walls, the yellow globules would enhance contrast between an object and the sky against which it is seen by removing much of the blue from the background. Similarly, the red droplets would increase the contrast between a green forest or field and an object seen against it by removing much of the green from that background. Enhanced contrast, of course, would greatly increase acuity — obviously of value to a highly visual animal. In this respect, it is significant that the retina of the pigeon has predominantly yellow droplets in its ventronasal quadrants (those portions of the retina with which the sky would be viewed), but predominantly red globules in the remaining dorsotemporal quadrant (that portion of the retina which would be used to view the land areas beneath the bird in flight).

At the moment, none of these suggested roles of the oil droplets has been proven experimentally. Thus, despite the fact that the theory describing decreased chromatic aberration and glare is most tempting, it can only be said in conclusion that the significance of these provocative globules is still open to question.

G. THE PECTEN

Anyone who is at all familiar with the eye of the bird is at least somewhat intrigued by the presence of the pecten. The first report of this structure, which is unique to the avian eye, is generally attributed to Perrault (1676) but, according to Wingstrand and Munk (1965), should actually be credited to Steno in 1673. Since those early reports, numerous studies have been carried out and many speculations put forth as to the exact function of the pecten. In fact, Wingstrand and Munk, in their most informative review, point out that thirty different functional theories have been published since the pecten was discovered.

The pecten, in one form or another, is present in the retinas of all birds and is always located with its base positioned directly over the optic disc, the point where the nerve fibers from the ganglion cells join together to form the optic nerve (Fig. 14). Since the optic disc already constitutes a blind spot on the retina, it is of great advantage, and most probably not merely fortuitous, that the nonsensory pecten is positioned where it is. Although the location of the pecten is constant in all birds, specific characteristics of the organ, such as its size, complexity, and degree of penetration into the vitreous humor, are subject to great variation. In some species, such as the Common Kestrel (*Falco tinnunculus*), the pecten is relatively small, is located almost entirely on the area of the optic disc, and extends out into the vitreous body to but a very slight degree (Fig. 15a). In other species, such as the Herring Gull (*Larus argentatus*), or the Barn Swallow, (*Hirundo rustica*), the pecten is a more elaborate structure that extends from its base ventrally and anteriorly so that it reaches far out into the vitreous humor and approaches the lens (Fig. 15b). In fact, in some species such as the Blue Jay (*Cyanocitta cristata*), the pecten almost touches the lens (Fig. 15c). Yet another distinct type of pecten is found in the kiwi (*Apteryx*) and is unique to that bird. Here the organ extends out into the vitreous humor as a narrow, conical structure that does not direct itself ventrally, but rather is situated along the axis between the center of the lens and the optic disc (Fig. 15d).

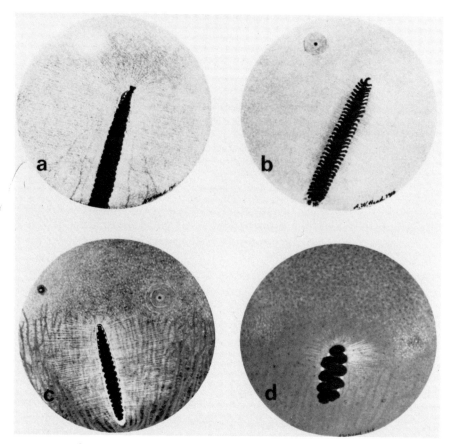

FIG. 14. The ocular fundi of some birds; (a) Blue Jay (*Cyanocitta cristata*); (b) Laughing Kookaburra (*Dacelo gigas*); (c) Kestrel (*Falco tinnunculus*); (d) European Nightjar (*Caprimulgus europaeus*). (After Wood, 1917.)

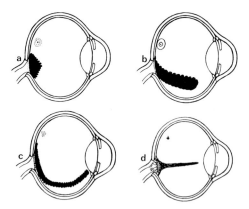

FIG. 15. Diagrams of the pectens of some birds; (a) Common Kestrel (*Falco tinnunculus*); (b) Barn Swallow (*Hirundo rustica*); (c) Blue Jay (*Cyanocitta cristata*); (d) Kiwi (*Apteryx*). (After Wood, 1917.)

In its finer structure, the pecten is basically of two morphological types. The primitive paleognathous birds, such as the Ostrich (*Struthio camelus*), usually have a pecten consisting of a central vertical panel that is surrounded by vane-like structures (Fig. 16a,c). On the other hand, the neognathous birds all have the pecten organized as an accordian-like pleated fin, as illustrated for the chicken and the Red-tailed Hawk (*Buteo jamaicensis*) in Fig. 16b,d. It is of interest that the pecten of the primitive kiwi has no vanes or pleats at all and, therefore, closely resembles the conus papillaris of some reptiles.

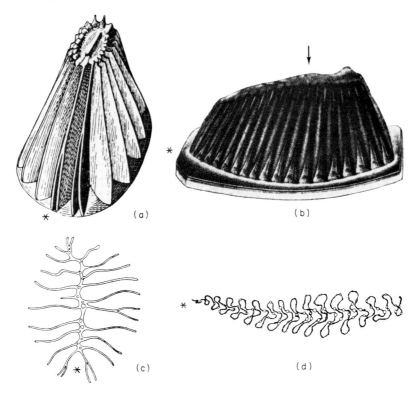

FIG. 16. Views of the pecten; (a) organ typical of the paleognathous birds; (b) organ typical of neognathous birds; (c) section of a near and parallel to its base, showing vane-like structures; (d) section of type of pecten illustrated by b, showing pleated structures characteristic of most pectens; (∘) dorsal end of pecten; (→) bridge of pecten. (After Walls, 1942; Franz, 1909; von Szily, 1922.)

Walls (1942) cautions, however, that the similarity in form is not an argument for linking the "ancestral" conus papillaris with its pecten "descendant," but is due probably to the fact that the eye of the Kiwi is quite degenerate and the degeneration applies also to its pecten.

The fact that the pecten is extremely vascular (Mann, 1924) has led to a consensus, even among the earliest investigators, that the organ functions in the nutrition of the retina. Relative to this view it should be recalled that the avian retina is devoid of blood vessels such as those that nourish the mammalian retina. The pecten, therefore, would certainly be extremely valuable as a nutritive device, and at the same time it would not offer the problem of having blood vessels interfering with acute vision. The optical interference of the retinal vessels presents no great problem in the mammal because, in contrast to birds, only the fovea is used in acute vision and the mammalian fovea is free of blood vessels. The idea that the pecten serves a metabolic function gains support from Wingstrand and Munk's (1965) finding that an oxygen diffusion gradient exists across the vitreous body between the pecten and the retina, and that this gradient is destroyed if the pecten blood supply is disturbed. These same authors also observed that in experimental animals where the pecten probably was not functional there was a marked retinal degeneration. The observation of Raviola and Raviola (1967) that the pecten has a high carbonic anhydrase activity is also consistent with the idea that the organ functions in the transport of oxygen. It is significant here that many of the surface capillaries of the pecten are separated from the vitreous humor by only a distance of about 1 μm (Dunn, 1968). Thus, the evidence is overwhelming in favor of the idea that the pecten functions in the metabolic support of the avian retina. However, the tremendous variation in its size and complexity indicates to many that the organ must have a function in addition to that of nutrition.

For example, several investigators have put forth the idea that the pecten is directly involved in the process of accommodation (e.g., Beauregard, 1876; Rabl, 1898). One suggestion with regard to this hypothesis is that the pecten acts as an erectile organ and thereby increases or decreases its size relative to the amount of blood present. It is reasoned that this change in pecten size would result in a corresponding alteration in the refractive apparatus by means of a pressure change transmitted through the vitreous humor. However, as Duke-Elder (1958) points out, although the size and complexity of the pecten do vary from species to species in direct relation to the accommodative capability of the animal, that capability itself varies with the metabolic level or visual effectivity of the retina, and, therefore, there is no evidence that would indicate that the two phenomena are not merely parallel. Furthermore, after performing some simple experiments, Mann (1924) concludes that changes in turgidity do not cause any change in the refraction of the avian eye. However, she does support the suggestion of Franz (1908) that the turgidity changes of the pecten could function in regulating the intraocular pressure of the eye, and that these changes could be important during changes in the bird's altitude while in flight.

Another theory on the function of the pecten suggests that the organ serves to maintain a high ocular temperature, particularly at great altitudes (Kajikawa, 1923). However, Walls (1942) points out that there seems to be rather little relationship between the flight habits of the various species of birds and the size or complexity of their respective pectens.

Thomson (1928, 1929) believes that the nonnutritive importance of the pecten lies in its capacity to function as a dark mirror. He suggests that the darkly pigmented pecten decreases glare and, at the same time, reflects an image onto the retina of an object approaching from the direction of the sun. Since the pecten is always found in the ventral portion of the retina (i.e., in the upper field of vision), Thomson reasons that it would allow birds such as the passerines to more easily detect the approach of high flying predatory birds such as hawks and eagles. The dark mirror would also function in helping fast flying birds to avoid obstacles above them, such as limbs of trees. But Thomson's theory does not explain why the hawks and eagles themselves, who fly at such great heights and are not in danger from avian predators, also have rather complex and large pectens. Furthermore, although Thomson maintains that it is optically feasible for the pecten to function as a dark mirror, Walls (1942) insists that it is inconceivable that the pecten could reflect an image good enough to be of any value.

The idea that the pecten reduces glare (Thomson, 1929; Beauregard, 1876) has been tested just recently with the pigeon by Barlow and Ostwald (1972). These authors conclude that the position of the organ relative to the light path makes it improbable that the pecten could reduce glare resulting from the direct incidence of light on its surface. However, they provide evidence that the pecten does serve to shade "one part of the retina from illumination by light scattered from an image on other parts of the retina." Such shading would reduce the threshold-elevating effect of scattered light and therefore would increase the sensitivity of the bird's eye. It is pointed out by Barlow and Ostwald that the pecten is in a good position to shade the area centralis from the sun's glare.

Yet another optical function of the pecten is advanced by Menner (1938), who suggests that the organ casts spokelike shadows by means of the vanes or pleats onto the retina in the region of the fovea. Thus, the image of an object moving across the bird's field of vision would be viewed intermittently and, therefore, the perception of that image would be enhanced—a sort of stroboscopic effect. Walls, however, feels that the shadows would be too far apart to be of any use in this respect and essentially discounts Menner's theory. Barlow and Ostwald believe that the orientation of the pecten with respect to the pupil would make such shadows rather small.

Finally, an interesting possibility, mentioned by Griffin (1953), is that the pecten might provide the bird with a fixed point which, along with the fovea, would be used as a navigational aid to determine the position of the sun during migration. This idea is attractive, as it would help explain how birds manage to find their way across vast distances, a problem yet unresolved. Unfortunately, however, the concept of the pecten as a sextant would be most difficult to test.

In contrast to those investigators who feel strongly that the pecten must have a major function other than that of retinal nutrition, Walls (1942) insists that the function of the structure is purely nutritional, and that the size and complexity of the pecten are related solely to the metabolic rate of the retina in question. Walls assumes that the cones have metabolic rates greater than rods, and therefore, that diurnal birds (i.e., birds with many cones in their retinas) would have greater need for a supplementary nutrient device than would nocturnal birds. Thus, diurnal birds should have more complex and larger pectens than nocturnal birds. Likewise, the degree of activity of the animal with regard to utilization of its eyes should also affect the metabolic rate of the retina in such a way that highly active birds should have more complex and larger pectens than less active birds. Although the assumptions that Walls makes have yet to be proven as fact, he can indeed state a pretty good case for his hypothesis. For example, nocturnal birds, such as owls and goatsuckers, have relatively small pectens with few folds. In fact, the pecten of the owl is incomplete in that it does not have a bridge joining the vane-like structures. In contrast, most passerine birds, which make good use of their eyes in acute vision, have relatively large pectens with many folds. The same is true of the hawks and eagles, birds also known for their great visual capability.

In summary, it seems clear today, as it did many years ago, that the avian pecten is related to the nutrition of the retina. The point in question is whether or not nutrition is the sole function of the organ. Walls insists quite strongly that it is, and certainly alternative theories have not been entirely satisfactory, nor have they met with wide acceptance. Nevertheless, this striking structure of the bird retina remains controversial; the riddle of the pecten remains as yet unsolved.

IV. Conclusion

An attempt has been made in the preceding pages to provide the reader with some insight into various aspects of avian vision. It should be apparent that a number of major problems have not yet been reconciled. The reasons for this are several. For example, the determination of the presence or absence of visual pigments in the avian cones will probably be delayed until the technique of *in situ* microspectrophotometry is refined to the point where it can be used to study cones as small as the bird's. On the other hand, it is entirely possible that the functions of the pecten and the well developed avian fovea may never be explained to the satisfaction of everyone, for it is extremely difficult to conceive of experiments that would provide direct evidence as to the respective roles of these structures. However, there appears to be no theoretical or technological reason preventing a study designed to determine whether or not color vision is a capability that most birds possess. The answer to this rather important question awaits only an investigator's time, effort, and interest.

REFERENCES

Barlow, H. B., and Ostwald, T. J. (1972). Pecten of the pigeon's eye as an inter-ocular eye shade. *Nature (New Biology)* **236**, 88–90.

Beauregard, H. (1876). Recherches sur les réseaux vasculaires de la chambre postérieure de l'oeil des Vertébrés. *Ann. Sci. Natur.: Zool. Paleontol.* **4**, 1–158.

Bliss, A. F. (1946). The chemistry of daylight vision. *J. Gen. Physiol.* **29**, 277–297.

Bridges, C. D. B. (1962). Visual pigments of the pigeon (*Columba livia*). *Vision Res.* **2**, 125–137.

Brown, K. T. (1969). A linear area centralis extending across the turtle retina and stabilized to the horizon by non-visual cues. *Vision Res.* **9**, 1053–1062.

Brown, P. K., and Wald, G. (1963). Visual pigments in human and monkey retinas. *Nature (London)* **200**, 37–43.

Capranica, S. (1877). Physiologisch-chemische Untersuchungen über die farbigen Substanzen der Retina. *Arch. Anat. Physiol., Physiol. Abt.* **1**, 283–296.

Cohen, A. I. (1963). The fine structure of the visual receptors of the pigeon. *Exp. Eye Res.* **2**, 88–97.

Crescitelli, F. (1958a). The natural history of visual pigments. *Ann. N.Y. Acad. Sci.* **74**, 230–255.

Crescitelli, F. (1958b). The natural history of visual pigments. *Proc. Annu. Biol. Colloq.* [*Oreg. State Univ.*] **19**, 30–51.

Crescitelli, F., Wilson, B. W., and Lilyblade, A. L. (1964). The visual pigments of birds. I. The turkey. *Vision Res.* **4**, 275–280.

Dartnall, H. J. A. (1962). The photobiology of visual processes. *In* "The Eye" (H. Davson, ed.), Vol. 2, 1st ed., pp. 321–533. Academic Press, New York.

Dartnall, H. J. A., Lander, M. R., and Munz, F. W. (1961). Periodic changes in the visual pigment of a fish. *Prog. Photobiol., Proc. Inst. Congr., 3rd, 1960* pp. 203–213.

Detwiler, S. R. (1943). "Vertebrate Photoreceptors." Macmillan, New York.

Dobrowolsky, W. (1871). Die Doppelzapfen. *Arch. Anat. Physiol. Wiss. Med.* **38**, 208–220.

Donner, K. O. (1951). The visual acuity of some passerine birds. *Acta Zool. Fenn.* **66**, 1–40.

Donner, K. O. (1953). The spectral sensitivity of the pigeon's retinal elements. *J. Physiol. (London)* **122**, 524–537.

Duijm, M. (1958). On the position of a ribbon-like central area in the eyes of some birds. *Arch. Neer. Zool.* **13**, 128–145.

Duke-Elder, S. (1958). The eyes of birds. *In* "System of Ophthalmology: The Eye in Evolution," Vol. 1, pp. 397–427. Kimpton, London.

Dunn, R. F. (1968). The morphology of the pecten oculi and the conus papillaris. *In* "Electron Microscopy 1968, Proceedings of the Electron Microscopy Society of America" (C. J. Arceneaux, ed.), pp. 170–171. Claitor's Publ. Div., Baton Rouge, Louisiana.

Engström, K. (1958). On the cone mosaic in the retina of *Parus major. Acta Zool. (Stockholm)* **39**, 65–69.

Franz, V. (1908). Das Pecten der Fächer, im Auge der Vögel. *Biol. Zentralbl.* **28**, 449–468.

Franz, V. (1909). Das Vogelauge. *Zool. Jahrb., Abt. Anat. Ontog. Tierre* **28**, 73–282.

Goodge, W. R. (1960). Adaptations for amphibious vision in the Dipper (*Cinclus mexicanus*). *J. Morphol.* **107**, 79–91.

Goodwin, T. W. (1952). "The Comparative Biochemistry of the Carotenoids." Chapman & Hall, London.

Granda, A. M., and Yazulla, S. (1971). The spectral sensitivity of single units in the nucleus rotundus of pigeon, *Columba livia. J. Gen. Physiol.* **57**, 363–384.

Griffin, D. R. (1953). Sensory physiology and the orientation of animals. *Amer. Sci.* **41**, 209–244.

Gundlach, R. H. (1933). the visual acuity of homing pigeons. *J. Comp. Psychol.* **16**, 327–342.

Gundlach, R. H., Chard, R. D., and Skahen, J. R. (1945). The mechanism of accommodation in pigeons. *J. Comp. Psychol.* **38**, 27–42.

Hailman, J. P. (1964). Coding of the colour preference of the gull chick. *Nature (London)* **204**, 710.

Hailman, J. P. (1966). Mirror-image color-preferences for background and stimulus-object in the gull chick (*Larus atricilla*). *Experientia* **22**, 257–258.

Hamilton, W. F., and Coleman, T. B. (1933). Trichromatic vision in the pigeon as illustrated by the spectral hue discrimination curve. *J. Comp. Psychol.* **15**, 183–191.

Hamilton, W. F., and Goldstein, J. L. (1933). Visual acuity and accommodation in the pigeon. *J. Comp. Psychol.* **15**, 193–197.

Hannover, A. (1840). Ueber die Netzhaut und ihre Gehirnsubstanz bei Wirbelthieren, mit Ausnahme des Menschen. *Arch. Anat. Physiol. Wiss. Med.* 320–345.

Hartwig, H.-G. (1968). Über Rezeptorentypen in der Retina von *Zonotrichia leucophrys gambelii*. *Z. Zellforsch. Mikrosk. Anat.* **91**, 411–428.

Hodos, W. (1969). Color discrimination deficits after lesions of the nucleus rotundus in pigeons. *Brain, Behav. Evolution* **2**, 185–200.

Hoffman, C. K. (1877). Zur Anatomie der Retina. III. Über den Bau der Retina bei den Vögeln. *Niederlaend. Arch. Zool.* **3**, 217–233.

Ikeda, H. (1965). The spectral sensitivity of the pigeon (*Columba livia*). *Vision Res.* **5**, 19–36.

Kajikawa, J. (1923). Beiträge zur Anatomie und Physiologie des Vogelauges. *Albrecht von Graefes Arch. Ophthalmol.* **112**, 260–346.

King-Smith, P. E. (1969). Absorption spectra and function of the coloured oil drops in the pigeon retina. *Vision Res.* **9**, 1391–1399.

Klopfer, P. H. (1968). Stimulus preferences and discrimination in neonatal ducklings. *Behaviour* **32**, 309–314.

Köttgen, E., and Abelsdorff, G. (1896). Absorption und Zersetzung des Sehpurpurs bei den Wirbeltieren. *Z. Psychol. Physiol. Sinnesorgane* **12**, 161–184.

Krause, W. (1894). Die Retina. V. Die Retina der Vögel. *Int. Monatsschr. Anat. Physiol.* **11**, 1–66.

Kühne, W. (1878). "On the Photochemistry of the Retina and on Visual Purple" (translated from the German by M. Foster). Macmillan, New York.

Kühne, W., and Ayres, W. C. (1878). On the stable colours of the retina. *J. Physiol. (London)* **1**, 109–130.

Liebman, P. A., and Entine, G. (1968). Visual pigments of frog and tadpole (*Rana pipiens*). *Vision Res.* **8**, 761–775.

Lockie, J. D. (1952). A comparison of some aspects of the retinae of the Manx Shearwater, Fulmar Petrel and House Sparrow. *Quart. J. Microsc. Sci.* **93**, 347–356.

Mann, I. C. (1924). The function of the pecten. *Brit. J. Ophthalmol.* **8**, 209–226.

Marks, W. B. (1965). Visual pigments of single goldfish cones. *J. Physiol. (London)* **178**, 14–32.

Marks, W. B., Dobelle, W. H., and MacNichol, E. F. (1964). Visual pigments of single primate cones. *Science* **143**, 1181–1183.

Maturana, H. R. (1962). Functional organization of the pigeon retina. *Int. Congr. Physiol. Sci. [Proc.], 22nd, 1962* Vol. 3, pp. 170–178.

Maturana, H. R., and Frenk, S. (1963). Directional movement and horizontal edge detectors in the pigeon retina. *Science* **142**, 977–979.

Menner, E. (1938). Die Bedeutung des Pecten im Auge des Vogels für die Wahrnehmung von Bewegungen, nebst Bemerkungen über seine Ontogenie und Histologie. *Zool. Jahrb., Abt. Allg. Zool. Physiol. Tierre* **58**, 481–538.

Meyer, D. B., and Cooper, T. G. (1966). The visual cells of the chicken as revealed by phase contrast microscopy. *Amer. J. Anat.* **118**, 723–734.

Morris, V. B., and Shorey, C. D. (1967). An electron microscope study of types of receptor in the chick retina. *J. Comp. Neurol.* **129**, 313–340.

Munz, F. W. (1958a). Photosensitive pigments from the retinae of certain deep-sea fishes. *J. Physiol. (London)* **140**, 220–235.

Munz, F. W. (1958b). The photosensitive retinal pigments of fishes from relatively turbid coastal waters. *J. Gen. Physiol.* **42**, 445–459.

O'Day, K. (1940). The fundus and fovea centralis of the albatross (*Diomedea cauta cauta* – Gould). *Brit. J. Ophthalmol.* **24**, 201–207.

Oehme, H. (1961). Vergleichend-histologische Untersuchungen an der Retina von Eulen. *Zool. Jahrb., Abt. Anat. Ontog. Tierre* **79**, 439–478.

Oehme, H. (1964). Vergleichende Untersuchungen an Greifvogelaugen Z. Morphol. Oekol. Tierre **53**, 618–635.

Owen, W. G. (1971). Personal communication.

Perrault, C. (1676). Mémoires pour servir à l'histoire naturelle des Animaux. *Mem. Acad. Roy Sci., Depuis 1666 jusq'a 1699, 3rd, 1729* As cited by Wingstrand and Munk (1965).

Pirenne, M. H. (1967). "Vision and the Eye." Chapman & Hall, London.

Porter, J. P. (1904). A preliminary study of the psychology of the English Sparrow. *Amer. J. Psychol.* **5**, 313–346.

Porter, J. P. (1906). Further study of the English Sparrow and other birds. *Amer. J. Psychol.* **17**, 248–271.

Pumphrey, R. J. (1948a). The sense organs of birds. *Ibis* **90**, 171–199.

Pumphrey, R. J. (1948b). The theory of the fovea. *J. Exp. Biol.* **25**, 299–312.

Pumphrey, R. J. (1961). Sensory organs: Vision. *In* "Biology and Comparative Physiology of Birds" (A. J. Marshall, ed.), Vol. 2, pp. 55–68. Academic Press, New York.

Rabl, C. (1898). Über den Bau und die Entwicklung der Linse. *Z. Wiss. Zool., Abt. A* **63**, 496–572.

Raviola, E., and Raviola, G. (1967). A light and electron microscopic study of the pecten of the pigeon eye. *Amer. J. Anat.* **120**, 427–462.

Rochon-Duvigneaud, A. (1943). "Les yeux et la vision des vertébrés." Masson, Paris.

Schultze, M. (1866). Zur Anatomie und Physiologie der Retina. *Arch. Mikrosk. Anat.* **2**, 175–286.

Schultze, M. (1867). Ueber Stäbchen und Zapfen der Retina. *Arch. Mikrosk. Anat.* **3**, 215–244.

Schwanzara, S. A. (1967). The visual pigments of freshwater fishes. *Vision Res.* **7**, 121–148.

Sillman, A. J. (1969). The visual pigments of several species of birds. *Vision Res.* **9**, 1063–1077.

Slonaker, J. R. (1897). A comparative study of the area of acute vision in vertebrates. *J. Morphol.* **13**, 445–502.

Slonaker, J. R. (1918). A physiological study of the anatomy of the eye and its accessory parts of the English Sparrow (*Passer domesticus*). *J. Morphol.* **31**, 351–459.

Strother, G. K. (1963). Absorption spectra of retinal oil globules in turkey, turtle and pigeon. *Exp. Cell Res.* **29**, 349–355.

Thomson, A. (1928). The riddle of the pecten, with suggestions as to its use. *Trans. Ophthalmol. Soc. U.K.* **48**, 293–331.

Thomson, A. (1929). The pecten, considered from an environmental point of view. *Ibis* **5**, 608–639.

Valentin, G. (1840). Die Fortschritte der Physiologie im Jahre 1839. e. Sinnesorgane. *Reper. Anat. Physiol.* **5**, 138–159.

von Szily, A. (1922). Vergleichende Entwicklungsgeschichte der Papilla nervi optici und der sog. axialen Gebilde. I. Morphogenese des Sehnerveneintrittes und des "Fächers" beim Hühnchen, als Beispiel für den Typus "Vögel." *Albrecht von Graefes Arch. Ophthalmol.* **107**, 317–431.

Wald, G. (1937). Photo-labile pigments of the chicken retina. *Nature (London)* **140**, 545–546.

Wald, G., and Zussman, H. (1937). Carotenoids of the chicken retina. *Nature (London)* **140**, 197.

Wald, G., and Zussman, H. (1938). Carotenoids of the chicken retina. *J. Biol. Chem.* **122**, 449–460.

Wald, G., Brown, P. K., and Smith, P. H. (1955). Iodopsin. *J. Gen. Physiol.* **38**, 623–681.

Wald, G., Brown, P. K., and Brown, P. S. (1957). Visual pigments and depths of habitat of marine fishes. *Nature (London)* **180**, 969–971.

Walls, G. (1937). Significance of the foveal depression. *Arch. Ophthalmol.* **18**, 912–919.

Walls, G. (1942). "The Vertebrate Eye and its Adaptive Radiation." Cranbrooke, Bloomfield Hills, Michigan.

Wingstrand, K. G., and Munk, O. (1965). The pecten oculi of the pigeon with particular regard to its function. *Biol. Skr. Dan. Videnskab. Selskab* **14**, 1–64.

Wolff, E. (1968). "Anatomy of the Eye and Orbit," (revised by R. J. Last), 6th ed. Lewis, London.

Wood, C. A. (1917). "The Fundus Oculi of Birds Especially as Viewed by the Ophthalmoscope." Lakeside Press, Chicago, Illinois.

Wood, C. A. (1925). Color sense of the satin bower bird. *Amer. J. Ophthalmol.* **8**, 120–122.

Yoshikami, S., Pearlman, J. T., and Crescitelli, F. (1969). Visual pigments of the vitamin A-deficient rat following vitamin A_2 administration. *Vision Res.* **9**, 633–646.

Chapter 6

CHEMORECEPTION

Bernice M. Wenzel

I. Introduction

Specialized chemoreception in birds has always been suspect, and to some extent it continues to be. The vocal prowess and bright plumage of so many birds, coupled with a style of life that seems independent of taste and smell judgments, emphasize auditory and visual functions and leave little reason to consider chemical sensitivity. Evidence accumulating since about 1960, however, gives increasing support to the idea of chemoreception while simultaneously raising a new set of questions. Anatomical evidence, especially impressive in the case of olfaction, has existed for some time, and convincing electrophysiological and behavioral observations have been more recent contributions.

Perception of chemical stimuli occurs by three neural routes in vertebrates: (1) the olfactory nerve with receptor endings in the posterodorsal reaches of the nasal cavity, (2) the taste fibers variously distributed in the facial and glossopharyngeal nerves with sense cells on the tongue and buccal lining, and (3) free nerve endings distributed widely over the body surface in the neural network that subserves the several qualities of cutaneous sensation. The last is usually called the common chemical sense and has been little studied. Investigators of chemoreception are almost exclusively concerned with the two special senses, and investigators of cutaneous sensation have never singled out this aspect for concentrated research. It must be kept in mind in connection with the controversy over avian chemoreception because some writers have argued that apparent olfactory or gustatory sensitivity in the bird may actually be due to common chemical sensitivity. This argument need no longer be taken seriously because the modalities of taste and smell do exist, but cutaneous chemoreception may also contribute. The evidence for specialized chemoreception forms the content of this chapter. A handbook on chemoreception is available and should be consulted for background information (Beidler, 1971).

II. Anatomy

The peripheral anatomy in both modalities is reasonably well understood at the levels of gross dissection and light microscopy, and a little work has been done on fine structure of the olfactory epithelium. Central pathways, however, remain almost entirely conjectural.

A. SENSE ORGANS

1. Taste

Although the numbers are relatively small, taste buds have been described in pigeon (Moore and Elliott, 1946), chicken (Lindenmaier and Kare, 1959), European Bullfinch (*Pyrrhula pyrrhula*) (Duncan, 1960b), and Japanese Quail (Warner *et al.*, 1967). Moore and Elliott divided the tongue into anterior and posterior portions at the dorsal fold about four-fifths of the way from the tip to the caudal limits, where there is a V-shaped line of papillae-like structures evidently very

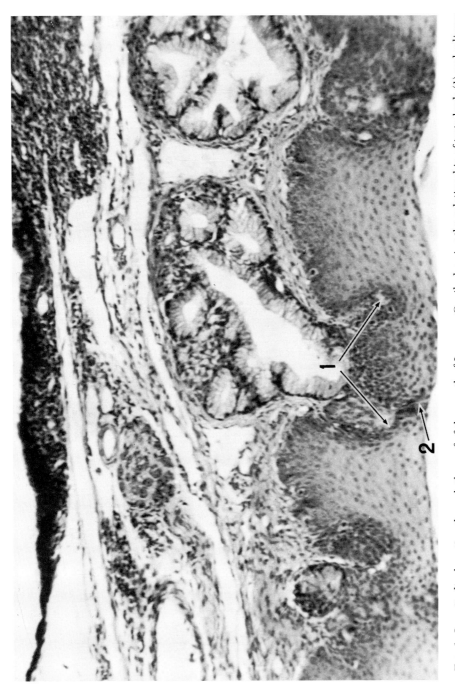

FIG. 1. Longitudinal section through the roof of the mouth of Japanese Quail, showing the relationship of taste buds (1) and salivary gland ducts (2). Hematoxylin and triosin stain; magnification × 240. (From Warner *et al.*, 1967.)

similar to the arrangement of circumvallate papillae on mammalian tongues. In birds, however, the taste buds are not located in the papillae. There is agreement that the greatest concentration of taste buds is in the posterior portion of the tongue and the floor of the pharynx and that they are totally absent from the cornified anterior tongue. Distribution of buds outside these regions probably varies somewhat with species. In the Japanese Quail, for example, the highest density is on the anterior palate and floor, but in the chicken they are only on the base of the tongue and floor. Also in the quail, the buds are conspicuously associated with salivary gland ducts, usually with two per gland (Fig. 1). The total numbers of buds reported for individual birds are very small compared with commonly studied mammals – 24 in the chicken, 37 in the pigeon, 46 in the European Bullfinch, and 62 in the Japanese Quail. By contrast, 10,000 have been reported in man (Pfaffmann, 1959).

In its gross structure, the avian taste bud bears a close resemblance to the mammalian form. It is a collection of cells extending about 70 μm in length and 30 μm in width reaching from the tunica propria through the epithelium. A duct communicates between the oral cavity and the lumen of the taste bud so that solutions can reach the sensory cells. No studies of ultrastructure have been reported nor is it known whether taste is the exclusive function of these buds or how many nerve fibers are associated with each one.

2. Smell

The receptor cells for olfaction in birds, as well as in other vertebrates, are simply the peripheral terminals of the olfactory nerve fibers. They are located in the olfactory epithelium which extends along the surface of the posterior concha. Bang (1960, 1964, 1965, 1966, 1968) has given extensive descriptions of the olfactory cavities of many avian species and pointed out a relationship between adaptations for effective airflow in the nasal cavity and a relatively large amount of olfactory epithelium. The complexity of the nasal cavity varies among different avian species, but in all forms it apparently provides at least some access to the olfactory region. Nasal glands in the anterior nasal cavity supply moisture for the inspired air, and an anterior valve in some tube-nosed forms apparently closes during diving and protects the interior of the cavity from seawater (Bang, 1960).

The structure of the olfactory epithelium is consistent for all vertebrates, including the two avian species that have been studied in detail (Brown and Beidler, 1966; Graziadei and Bannister, 1967). It consists of the receptor cells and, in addition, supporting sustentacular cells and basal cells (Fig. 2). The receptor cells are bipolar ciliated neurons with the cell body situated within the olfactory epithelium.

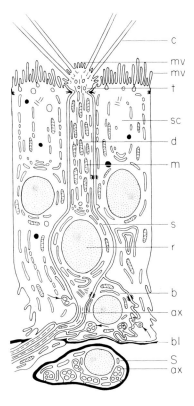

FIG. 2. Diagram of the olfactory epithelium of the domestic duck. The receptor cell (r) consists of a dendrite (d), soma (s), and axon (ax). The terminal swelling (t) bears cilia (c) and microvilli (mv). Each basal body of the cilia possesses several basal feet. Many mitochondria are seen in the dendrite. Supporting cells (sc) bear microvilli (mv). Basal cells (b) are observed near the basal lamina (bl). Olfactory axons can be observed enveloped in shallow pockets of basal cells and of supporting cells as well (arrows). Below the epithelium olfactory axons (ax) are entirely or partially enwrapped by Schwann cells (S). (From Graziadei and Bannister, 1967.)

The short dendritic process and terminal cilia extend through the mucous covering, while the longer axonal process is a component of the olfactory nerve. The only nonmammalian characteristic reported for these cells in birds is an increased number of microvilli on the terminal dendritic knob. No related functional property has been described, but the contributions of the separate parts of the receptor cell to sensory transduction are not yet known, so that it is impossible at present to assign significance to this interclass anatomical difference.

B. CRANIAL NERVES

1. Taste

According to functional studies (Kitchell *et al.*, 1959; Halpern, 1962; Landolt, 1970), taste impulses in birds are carried only in the IX cranial (glossopharyngeal) nerve and, unlike mammals, not at all in the VII (facial) nerve. Branches of the IX nerve also subserve cutaneous sensation in the tongue where there is no innervation from the V (trigeminal) nerve. As shown in Fig. 3, the lingual branch of the glossopharyngeal nerve divides into the lingual and the laryngolingual nerves, both of which innervate the tongue. Taste impulses, as well as those resulting from touch stimulation, are carried in the laryngolingual nerve, which innervates the posterior and pharyngeal areas. The lingual nerve is anterior and carries only cutaneous information. This situation, somewhat simplified in comparison with mammals, has presumably come about because of the bird's relatively small number of taste buds and their grouping in the posterior portion of the oral cavity. The corresponding area in the mammal is similarly innervated by a branch of the glossopharyngeal nerve.

2. Smell

The I cranial (olfactory) nerve is composed of the axons of the bipolar neurons whose cell bodies lie in the olfactory epithelium. They constitute two well-defined bundles of unmyelinated fibers that enter the olfactory bulbs directly and terminate in the mitral layer. The first synapse, therefore, occurs in the primary sensory area.

C. CENTRAL NUCLEI AND PATHWAYS

Almost no experimental studies have been reported for either modality. Virtually all of the information available is based on gross or microscopic observation of normal material.

1. Taste

Sensory glossopharyngeal fibers enter the medulla and join fibers from the VII (facial) and X (vagus) nerves to form the fasciculus soli-

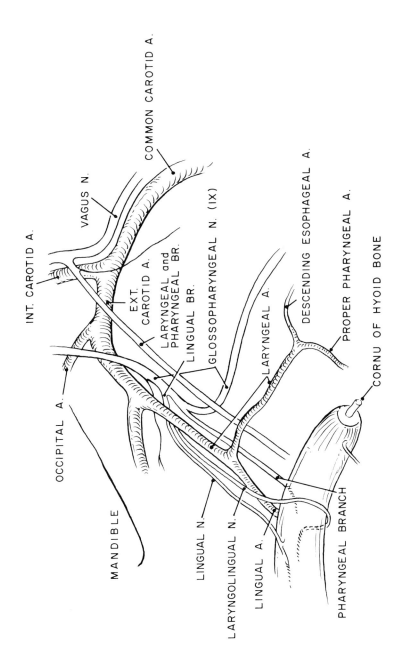

FIG. 3. Composite representation of the IX and X cranial nerves with their accompanying major arteries. Note that laryngeal artery is shown stretched as a result of retraction of cornu of hyoid bone. (From Landolt, 1970.)

tarius. This tract is said to be large in birds and is assumed to be related primarily to visceral sensitivity rather than gustatory (Sanders, 1929). The very small number of taste buds makes this assumption reasonable. The dorsomedial nucleus, which lies between the nucleus of the XII (hypoglossal) nerve and the dorsal efferent nucleus of the X nerve, was proposed as a gustatory center (Ariëns Kappers *et al.*, 1936) although no direct evidence was provided. It seems more likely that the nucleus solitarius would fill this role as it does in mammals. Further steps in the pathway to the thalamus and forebrain are completely unknown.

2. Smell

The size of the olfactory bulb varies widely among birds. In an effort to quantify the obvious range, Cobb (1960a,b) expressed the diameter of one olfactory bulb in a given species as a percent of the largest diameter of the forebrain for that same species. This index has been calculated for a diverse group of more than 100 species (Bang, 1971; Bang and Cobb, 1968) and has been found to vary from 37.0 for the Snow Petrel (*Pagodroma nivea*) to 3.0 for the Black-capped Chickadee (*Parus atricapillus*). In the latter case, the bulbs are actually fused and represent a minute structure. The largest values occur among Apterygiformes (34.0), Procellariiformes (29.0), Podicipediformes (24.5), Caprimulgiformes (23.8), and Gruiformes (22.2), and the smallest among Pelecaniformes (12.1), Piciformes (10.0), Passeriformes (9.7), and Psittaciformes (8.0). The earlier literature on microscopic anatomy of the olfactory system commented on the correlation between bulb size and the size of downstream components (Huber and Crosby, 1929).

Internally, the olfactory bulb is a layered structure around an olfactory ventricle. The outermost layer contains the entering fibers of the olfactory nerve, which terminate in the next layer in dense glomeruli made up also of dendritic endings of mitral and tufted cells. The cell bodies of the latter neurons lie internal to the glomerular layer in the mitral cell layer. Their axonal processes form the outflow of the bulb along with axons of the internal granular cells that compose the innermost layer of the bulb. The absence of gross differences between the avian and mammalian bulbs makes it reasonable to apply descriptions of mammalian bulbar ultrastructure (cf. Beidler, 1971) to birds, for no direct knowledge is available.

The central olfactory pathway has been described in a few species on the basis of microscopic examination of tissue from normal birds

(Craigie, 1930, 1932, 1940, 1941; Crosby and Humphrey, 1939; Huber and Crosby, 1929; Jones and Levi-Montalcini, 1958). The lack of correspondence between avian and mammalian forebrains precludes an equation between the two projection systems, however. One experimental study on the pigeon (Rieke and Wenzel, 1973) confirms earlier reports that the olfactory bulb projects to the prepiriform cortex, and suggests in addition that secondary fibers reach the parolfactory lobe and ventral hyperstriatum. The evidence also suggests that olfaction is represented in the neostriatum by way of a multisynaptic pathway.

III. Functional Capacity

A. NEUROPHYSIOLOGICAL EVIDENCE

The existence of basic anatomical structures provides only presumptive evidence for sensory function; thus, the demonstration of neural activity in response to application of natural physiological stimuli to the sense organ is powerful additional evidence. Such information is available for both taste and smell in birds, although it is still extremely sketchy for both senses and does little more than establish the existence of functional pathways. Especially needed are experiments on species for which behavioral data have been provided, more results from single units, and recordings from central points.

1. Taste

The sensitivity of the laryngolingual nerve to sapid solutions applied to the posterior and pharyngeal areas of the tongue has been well documented in the pigeon and chicken, the only forms studied (Kitchell et al., 1959; Halpern, 1962; Landolt, 1970). Both multi-unit and single-fiber activity have been recorded, and the results justify certain generalizations (Fig. 4). Salts and acids are generally effective stimuli in eliciting neural responses, but substances that are sweet to man are ineffective or minimally effective. Responses to saccharine occur in only about 60% of pigeons and are questionable in chickens. The bitter substance quinine hydrochloride elicits a reaction in chickens but not in pigeons. Individual differences within species are commonly observed with all substances.

A notable feature is the occurrence of responses of large magnitude to the application of distilled water at room temperature to the

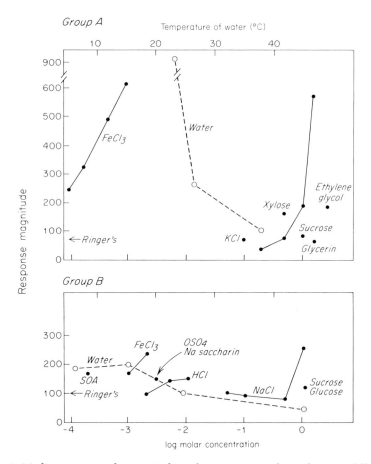

FIG. 4. Median summated magnitudes of responses in lingual nerve following chemical and thermal stimulation of the posterior tongue in 10 chickens. The ordinates represent the magnitude of the summated responses in arbitrary units, adjusted to 100 units for the following responses: Group A—distilled water at approximately 38°C; for these chickens (50% of sample), solutions were used at approximately 38°C, and except for sodium chloride and potassium chloride, were made up in Ringer's solution. Group B—distilled water at approximately 24°C; for these chickens, solutions were used at approximately 24°C (room temperature) and were made up in distilled water. (From Halpern, 1962.)

posterior tongue in many chickens (Fig. 4) (Kitchell *et al.*, 1959; Halpern, 1962). Kitchell *et al.* studied temperature responses as well as gustatory ones, and Landolt avoided the issue by presenting solutions only at tongue temperature. The so-called water response in mammalian taste nerves is a subject of active inquiry (cf. Beidler,

1971). In the case of the chicken, it seems parsimonious to consider that neuron populations with different functions are grouped together. Resolution of this problem awaits further studies of individual units. It is already known that tactile and thermal stimuli excite nerve impulses when applied to the anterior tongue and that taste stimuli do not (Halpern, 1962; Landolt, 1970).

In the case of such stimuli as salt and acid, which elicit good responses in the pigeon, there is a clear positive relationship between stimulus concentration and response magnitude (Landolt, 1970). On the basis of his single fiber studies, Landolt has proposed that the coding of stimulus intensity is accomplished by altering both rate of discharge in individual fibers and the number of fibers discharging. He could offer no suggestion about coding for stimulus quality. After looking for differences in temporal waveforms for different sapid solutions, he concluded that the overall shapes were roughly comparable. He also noted that the waveform after a given stimulus varied somewhat upon repeated application of the same stimulus.

The relation between these data and sensitivity in taste preferences will be discussed in Section III,B,1. Impulse traffic in the taste nerves appears adequate to account for some of the behavioral data presently available but not all. No recording has been done beyond the peripheral fibers.

2. Smell

Electrical activity in response to odorous stimuli has been recorded from both peripheral and central sites in the olfactory system. Tucker (1965) found that the action potential recorded from the olfactory nerve of 14 different avian species, listed in Table I, was indistinguishable from that recorded in macrosmatic reptilian and mammalian species. As the stimulus intensity was increased, the magnitude of the action potential also increased. Shibuya et al. (1970) have reported the same result with the Black-tailed Gull (Larus crassirostris). Tucker found no obvious differences in the recordings from the whole nerve when different odorous compounds were used as stimuli. In addition to the neural discharge, he also recorded the epithelial potential known as the electro-olfactogram, and it, too, resembled that already seen in other vertebrates from fish to monkey.

The mechanism of stimulus coding in the avian olfactory system is understood no better than for any other form at present. Only one study of single units in the olfactory nerve of a bird (the Black Vulture, Coragyps atratus) has been reported (Shibuya and Tucker,

TABLE I

SPECIES USED IN ELECTROPHYSIOLOGICAL EXPERIMENTS ON OLFACTION

Nerve recording (Tucker, 1965)	Bulb recording (Sieck, 1967; Sieck and Wenzel, 1969; Wenzel and Sieck, 1972)
Chicken (White Leghorn)	Black-footed Albatross
Common Crow (*Corvus brachyrhynchos*)	(*Diomedea nigripes*)
Muscovy Duck (*Cairina moschata*)	Chicken (White Leghorn)
Domestic (Embden) goose	Mallard Duck (*Anas*
Ring-billed Gull (*Larus delawarensis*)	*platyrhynchos*)
American Sparrow Hawk (*Falco sparverius*)	Domestic pigeon
Blue Jay (*Cyanocitta cristata*)	Manx (Black-vented)Shearwater
Common Nighthawk (*Chordeiles minor*)	(*Puffinus puffinus*
Domestic pigeon	*opisthomelas*)
Bobwhite Quail (*Colinus virginianus*)	
House Sparrow (*Passer domesticus*)	
Black Vulture (*Coragyps atratus*)	
Turkey Vulture (*Cathartes aura*)	
Common Yellowthroat (*Geothlypis trichas*)	

1967), and the data are characterized by the diversity of response patterns found. Very few units fired spontaneously during the recordings, so that the overall amount of background activity between stimuli was extremely low. Those units that did fire in the absence of stimulation either increased or decreased their activity when a stimulus was presented. As stimulus concentration increased, some fibers increased their firing rates while others decreased. Understanding of this topic awaits much more extensive research than has yet been completed. With extracellular recording, Shibuya and Tonosaki (1972) have found that unit responses from olfactory receptor cells of the chicken and shearwater (*Puffinus leucomelas*) are comparable to those from amphibians and reptiles although cell size is somewhat smaller.

Bulbar electrical activity has been recorded from a number of species (see Table I) and has been found to be generally comparable to that of mammals (Sieck, 1967; Sieck and Wenzel, 1969; Wenzel and Sieck, 1972). Figure 5 presents traces recorded through chronic macroelectrodes in the olfactory bulbs of a pigeon and a Manx (Black-vented) Shearwater (*Puffinus puffinus opisthomelas*). The wave bursts are very typical of recordings from any vertebrate olfactory system. Changes occurred with different stimulus compounds and with different concentrations of the same compound as well as from one bird to

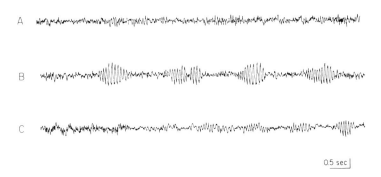

FIG. 5. Electrical activity from the olfactory bulb of a pigeon and a shearwater. (A) Intrinsic activity from the left bulb of a pigeon in purified air. (B) Activity from the same site as in A in response to 7.6×10^{-8} moles/ml pyridine. (C) Activity recorded from shearwater in response to 9.6×10^{-8} moles/ml trimethylpentane. In this trace, the point at which background activity was replaced by evoked activity is clearly apparent. Vertical calibration mark is 100 μV in A and B, 50 μV in C.

another of the same species. It is not possible at the present time to relate these changes to identifiable characteristics of the stimulus or of the individual birds. The fact that the recordings were genuine reflections of olfactory stimulation was established by showing that the activity disappeared following bilateral section of the olfactory nerve but continued with little alteration if the trigeminal nerve supply of the olfactory cavity was interrupted bilaterally. Nothing is known of the fate of electrical activity transmitted beyond the olfactory bulbs other than that some stimulus-related records have been obtained from the prepiriform area of the pigeon (Sieck, 1967; Sieck and Wenzel, 1969). Records have not yet been reported from individual units anywhere in the central olfactory pathway of a bird.

B. BEHAVIORAL EVIDENCE

Even before all of the above information was known, some studies were showing behavior that could justifiably be interpreted as taste or smell dependent. The potential for use of these senses is well established. The range and level of sensitivity are more difficult to measure and such research has been scarce.

1. Taste

Virtually all of the behavioral evidence for perception of taste stimuli by birds has been obtained in experiments that made use of some form of preference testing. If a bird consistently prefers one flavored solution over another, it can be concluded that it must have been able to taste the difference. It is obviously essential for the solutions to be equated in qualities other than taste to permit such a conclusion. Inasmuch as the crucial question in this work is whether birds are capable of perceiving taste, not whether they make important use of such perceptions under ordinary circumstances, the inferential nature of the conclusion is fully justified.

References to older work on avian gustation can be found in an earlier discussion (Kare, 1965). The modern work on preference taste testing in birds began in 1957 with studies of chickens (Kare *et al.*, 1957; Jacobs and Scott, 1957). The two-bottle testing technique was used in which the birds have access to two waterers, one of which typically contains a test solution and the other a control substance such as tap water or distilled water. Jacobs and Scott, in addition, sometimes paired two test solutions (e.g., sucrose and saccharine) and even used three bottles at times with two test solutions and one control. The amount of fluid consumed from each bottle was expressed as a percent of the day's total consumption. Conclusions of preference for and rejection of given solutions depended upon the relative amounts consumed. If consumption of control and test solutions did not differ, the behavior was identified as acceptance or indifference.

Kare *et al.* included representatives of all four taste qualities and found that sour and salt substances were substantially rejected, bitter solutions were either rejected or accepted, and sucrose was sometimes preferred but saccharine was rejected. Jacobs and Scott studied only sweet solutions and reported a pronounced preference for sucrose and an aversion for saccharine. They also noticed that the chickens did not drink the sucrose solutions in their customary way of drinking water. Instead of dipping the beak into the water and swallowing immediately, they dipped the beak frequently and shook it between dips so that more of the sucrose solutions was wasted. Regardless of this wastage, they still concluded that sucrose was preferred because so much more of those solutions disappeared. Later data (Kare and Ficken, 1963; Kare and Medway, 1959) showed that chickens are indifferent to a large group of sweet (according to man) carbohydrates except xylose, which they reject. Gentle (1972) reported a preference for 5% sucrose by Brown Leghorn chickens as well as a reduction in aversion to acid solutions if glucose was added. He also found that EEG arousal was a more sensitive indicator of taste perception than the preference data were.

Subsequent research of this general nature has been reported by several different investigators, and it all supports the general contention that birds are capable of distinguishing certain tastes. Disagreements have arisen over the question of acceptance or rejection of some substances, but it is important to remember that this type of issue is irrelevant to the question of basic sensory capacity. It has more to do with a debate over the possible usefulness of sensory information and will be discussed in Section IV,A.

Having established that birds do have the ability to taste, their level of acuity becomes of interest. Supposedly, it is possible to measure the absolute threshold for a given stimulus by determining the concentration at which a reaction of indifference changes to one of preference or aversion. Actually, however, this point is so strongly influenced by other factors, such as body fluid levels, metabolic condition, and the details of the testing procedure itself, that threshold estimates so obtained should be considered tenuous until they have been verified by experiments using other techniques of measurements. With this precaution in mind, it can be stated that the rejection threshold for inorganic acids was found by Duncan (1964) to be 0.0075 M in the wild pigeon, which is within the range of absolute thresholds reported for man (Pfaffmann, 1959) as is the value of 0.004 M for acetic acid (Duncan, 1960a). The behavior of the Great Tit (*Parus major*) changes from indifference to rejection as a solution of sodium chloride is altered from 0.08 M to 0.17 M (Warren and Vince, 1963). This value agrees in a general way with data obtained from pigeons showing preference at 0.085 M and rejection at 0.14 M and above (Duncan, 1962) and from House Finches (*Carpodacus mexicanus*) (Bartholomew and Cade, 1958), and shows less acuity than that of mammals. It should be stressed that no research has utilized more sensitive measuring techniques with birds.

A comparison between the behavioral and neurophysiological results shows both consistency and puzzling discrepancies. Neither Kitchell *et al.* (1959) nor Landolt (1970) found a response to quinine hydrochloride (0.02 M) in the taste nerves of pigeons, yet Duncan's (1960a) pigeons strongly rejected it at all concentrations from 0.003 M to 0.055 M, even in one-bottle testing. The lack of neural activity to sweet stimuli in pigeons agrees with their general indifference to such compounds. Taste nerves of chickens barely responded, if at all, to saccharine in Ringer's solution (Kitchell *et al.*, 1959), although both Kare *et al.* (1957) and Jacobs and Scott (1957) reported distinct aversion in two-bottle tests. Landolt's finding of no response to sucrose octaacetate in the pigeon nerve is also unusual when one considers the intense aversion to it shown by Bobwhite Quail (*Colinus virginianus*) and Japanese Quail (Brindley, 1965; Brindley and Prior, 1968; Cane and Vince, 1968). The neural and behavioral data are still so limited with respect to comparability of stimuli (chemical substance, concentration, diluent) and to number of replications that differences should not be labored. Landolt (1970), for example, calculated the relative stimulating efficiency of many of his stimuli, and Duncan (1962, 1964) has ranked sets of acid and salt stimuli from preference data. The two lists do not include the same chemicals, however, and most of Landolt's stimuli were at very high concentrations (2.0 *M*). The continued occurrence of behavioral responses to stimuli that regularly fail to arouse responses in apparent taste nerves, however, would certainly challenge either the validity of taste as the foundation for the behavior or the exclusion of other nerves for transmitting taste impulses.

2. Smell

The behavioral evidence for a functional olfactory system comes from two types of experiments, one in which birds have been trained to discriminate between odorous and odorless air, and another in which a change in a physiological index is interpreted as indicative of perception. In the latter procedure, heart rate, respiration, or both are continuously monitored while the bird is inhaling odorless air. Odor is added to the air for brief periods without providing any extraneous cue, and the heart and respiratory records are examined for consistent changes associated with introduction of the odorous stimulus. A variation on this procedure used "restlessness" in tame Mallard Ducks, whose eyes and ears were covered to reduce visual and auditory cues, as the index of perception (Whitten, 1971).

According to either method, the pigeon has been found to possess excellent olfactory perceptual ability (Henton *et al.*, 1966; Michelsen, 1959; Shumake *et al.*, 1970; Wenzel, 1967). The heart rates of pigeons increase over resting levels when odor is added to the air stream, and the heart rate change increases as stimulus intensity increases. These responses are greatly diminished following bilateral section of the olfactory nerve and are also affected, although not as much, by bilateral section of the ophthalmic branch of the trigeminal nerve which innervates the lining of the nasal cavity (Wenzel, 1971). It is probably fair to say that the total perceptual impact of an odorous stimulus is a combination of olfactory and trigeminal stimulation.

In operant conditioning experiments, pigeons have been trained to peck on a lighted disk at different rates depending on the presence or absence of odor in an entering air stream. Their success in learning this task when other possible cues are rigidly controlled is excellent evidence for their ability to perceive odors (Henton et al., 1966; Michelsen, 1959). The absolute threshold for amyl acetate measured in this way was 0.16% to 0.73% of vapor saturation (Henton, 1969) and the difference threshold was between 0.57 and 0.71 (Shumake et al., 1970).

The only species studied in addition to the pigeon have all been tested by the indirect procedure of cardiac and respiratory monitoring, as well as motor restlessness in the case of the response of ducks to skatol. They include Bobwhite Quail (Colinus virginianus), Canary (Serinus canarius), Manx Shearwater, Common Raven (Corvus corax), Mallard (Anas platyrhynchos), domestic chicken, Turkey Vulture (Cathartes aura), Humboldt Penguin (Spheniscus humboldti), and Brown Kiwi (Apteryx australis) (Wenzel, 1968; Wenzel, 1972; Wenzel and Sieck, 1972; Wenzel and Stonehouse, 1970). In all of these forms except the raven, positive evidence was obtained for perception of at least some odors. In most of the research, standard laboratory chemicals were used as stimuli in order to make the data comparable to results obtained in olfactory experimentation with mammals. Perception typically occurs for amyl acetate, pyridine, and trimethylpentane. The inclusion of more naturally occurring stimuli, such as fish or decayed meat, may or may not result in a perceptual response. On the basis of respiratory changes alone, Neuhaus (1963) concluded that the Greylag Goose (Anser anser) perceived the odor of skatol.

IV. Functional Significance

Although a persuasive amount of evidence has been presented for the ability of a number of birds to perceive tastes and odors, little is known about how, or even whether, such perceptions influence them in daily life. It is logically possible, if physiologically unusual, for functional capacity to be present without any appreciable reliance being placed on it. Some attempts have been made to solve this puzzle.

A. TASTE

The ability to taste is at least potentially useful in diet selection and in avoidance of toxic materials. Inasmuch as all of the behavioral data collected on avian taste expresses preference and rejection, it is natural to try to relate such results to intake patterns.

A few forms have been tested often enough to permit some generalizations about their taste preferences. The domestic chicken, feral pigeon, and Bobwhite and Japanese Quail have all been studied quite extensively. Red Junglefowl (*Gallus gallus*), House Finches, Great Tits, European Starlings (*Sturnus vulgaris*), Common Grackles (*Quiscalus quiscula*), and certain gulls have been subjects only in a few experiments. There has been a pronounced tendency for each set of investigators to conduct most of their experiments on a given species and always to use a single technique, which may differ in significant ways from the technique used almost exclusively in a different laboratory. Comparative statements, therefore, should be interpreted with caution until the effects of procedural differences have been shown to be negligible. Caution is also advisable about the representativeness of group data for individual birds, because individual differences in selection thresholds range from preference to total rejection for a given concentration of a single substance (Ficken and Kare, 1961). Age is also a variable. Bobwhite Quail, for example, are unable to discriminate sucrose octaacetate, a very bitter harmless substance, from water before 6 weeks of age, after which their discriminative capacity increases gradually during the next 2–4 months (Cane and Vince, 1968).

With regard to the basic taste qualities, present evidence suggests that there may be species differences for at least some compounds. Bitter tastes are generally rejected regardless of species or testing procedure, as has been shown for the pigeon (Duncan, 1960a), Bobwhite and Japanese Quail (Brindley, 1965; Brindley and Prior, 1968), chicken (Kare *et al.*, 1957), and the Great Tit (Cane and Vince, 1968). Salt is commonly rejected also, although feral pigeons (Duncan, 1962) showed indifference at 0.008 M and 0.017 M, which changed to preference at slightly higher concentrations of 0.043 M and 0.085 M and to total rejection for concentrations at 0.14 M and above. Others have argued that these data are not indicative of preference–aversion alone because the testing used only one bottle of fluid at a time and was done after 16½ hours of water deprivation so that thirst factors might have overridden taste factors. This argument has some cogency and is yet to be resolved. All other investigators have offered more than one bottle of fluid simultaneously. Pigeons, chickens, and Great Tits reject sour stimuli, but quail show a preference for them. Sweet solutions produce the most variable results. Quail and Great Tits show a preference for glucose, but pigeons and chickens are less predictable. Pigeons show a weak preference for a 3% glucose solution but otherwise are indifferent. With sucrose, however, they reject a 3% solution slightly, show strong preference for a 14% solution, and total rejection in a number of instances for a 28% solution, which is highly viscous (Duncan, 1960a). Chickens differ widely in their re-actions to sugars but reject saccharine consistently (Kare *et al.,* 1957). Kare's (1961; Kare and Ficken, 1963) argument that tastes do not follow the same qualitative categories for different species should not be ignored. Experiments are needed in which birds are tested for generalization between stimuli so that we can obtain some idea of the perceptual categories for a certain species. Such information could reduce a seeming jumble of contradictions to consistency.

Characteristics other than the taste of solutions contribute to the results in some cases. Japanese Quail are typically affected by such things as the colors of the drinkers and the relative positions of other items in and around cages (Brindley and Prior, 1968). Pick and Kare (1962) showed that visual and positional cues regularly associated with specific tastes could become influential over ingestion patterns of chickens. The pH values of solutions are of little importance for chickens (Fuerst and Kare, 1962). They accept solutions ranging from 2 to 10 and reject only at pH 1 and 13.

The influence of dietary factors may also be important, but the existing data do not provide strong support for this idea. Domestic chickens drink twice as much of a 10% sucrose solution than of water when their maintenance diet has lower caloric value, but they drink the same amount as water when maintained on a normal diet or on one of high caloric value. Red Junglefowl, on the other hand, show a moderate preference for sucrose solutions at all times, and they regu-late their caloric intake by adjusting the amounts of fluid and dry food ingested (Kare and Maller, 1967). The possible contribution of domestication to a blunting of taste impact may be considerable. Sodium-deprived chickens remain indifferent to dry food with sodium supplement and to 0.7% saline solution whether their deprivation resulted from dietary deficiency or subcutaneous injection of formalin. The induction of thiamine deficiency, however, results in an in-creased intake of thiamine-supplemented diet (Hughes and Wood-Gush, 1971). Birds that feed in seawater—the Herring Gull (*Larus argentatus smithsonianus*) and the Laughing Gull (*Larus atricilla*)— show the same general rejection threshold for solutions of sodium chloride or of seawater as do terrestrial forms (Harriman and Kare, 1966a,b; Harriman, 1967). When Herring Gulls were compared with European Starlings and Common Grackles, the gulls showed an aversion for 0.20 M sodium chloride solution while the grackles avoided 0.50 *M* and the Starlings 0.15 *M*. The Laughing Gull rejected 0.20 *M*, which was the lowest hypertonic solution used in the experiment, and also rejected seawater. The same results were obtained with gull chicks that were tested in a one-bottle procedure. House Finches, which can survive in an arid environment, are indifferent to sodium chloride solutions at 0.15 *M* or less, and at 0.2 *M* or higher drink only half as much as water (Bartholomew and Cade, 1958). In a more typically desert species, the Zebra Finch (*Poephila guttata*), sodium chloride solutions up to 0.5–0.6 *M* are readily accepted for drinking (Oksche *et al.*, 1963). Because of internal osmotic effects, however, the interpretation of the role of taste per se in experiments with hypertonic salt solutions must be cautious.

All of this research has told us that many birds do react to many taste stimuli, but there is absolutely no information about what, if anything, their taste sensitivity contributes to their normal behavior. It has been found that intake of both dry feed and fluid can be affected by flavor, but the intensity must be very strong, especially in the case of dry food (Duncan, 1960b; Kare and Pick, 1960). This proves nothing, however, about any possible influence of taste on normal selection. Yang and Kare (1968) showed that taste could be a factor in ingestion of insects by Red-winged Blackbirds (*Agelaius phoeniceus*). They presented 12 known insect secretions in a two-bottle procedure and found that salicylaldehyde and *p*-benzoquinone were rejected in very low concentrations indicating that taste, as well as other effects such as caustic action on the eyes could be involved in predation. On the basis of his discovery (Brower *et al.*, 1967) that Blue Jays (*Cyanocitta cristata bromia*) do not attack Monarch Butterflies (*Danaus plexippus*) after catching and eating only one or part of one with resultant emesis, Brower (1969) has proposed that birds are not finicky about tastes initially but may develop such habits on the basis of unpleasant experiences with certain substances. His research has not established that taste is a critical cue or is even necessary, however. By the same token, the fact that rice is heavily preferred by certain geese (McFarland and George, 1966) or that some West Indian birds developed and sustained a new habit of eating sugar as a result of food shortage after a hurricane (Hundley and Mason, 1965) cannot be taken as proof that taste factors are involved in a more significant way than metabolic effects. Wilcoxon *et al.* (1971) found that Bobwhite Quail are more responsive to the color added to water than the taste. The birds drank slightly sour blue water 30 minutes before injection of an illness-inducing compound. In a test 3 days later, they avoided blue water but drank uncolored sour water normally. These results might have been affected by the fact that Bobwhite Quail have a taste preference for sour (Brindley, 1965). A great deal of ingenious experimentation is necessary to increase our understanding of taste's role in avian behavior.

B. SMELL

Several instances of the use of olfaction in normal behavior have been documented, as described below, and others have been suggested on the basis of less controlled observations (Stager, 1967).

Extensive studies of vultures led Stager (1964) to conclude that the Turkey Vulture, and possibly the King Vulture (*Sarcoramphus papa*), locate the general area of carrion by means of odor cues. His procedure was to release ethyl mercaptan fumes from a hidden generator at the base of a canyon into still air in the early morning so that the fumes rose straight up the canyon. Shortly thereafter, Turkey Vultures were seen to congregate in the area and circle repeatedly. Stager suggested that this vulture is likely to use odor to find the general vicinity of food and vision to guide it to the specific location.

Kiwis have often been credited with olfactory function without any adequate evidence having been provided. Wenzel (1968, 1972) found that these birds were completely successful in identifying which of their three feeding stations contained food on a given night during experiments in a native bird reserve in New Zealand. The combination of nocturnal activity, poor eyesight, placement of external nares at the tip of the beak, and very large olfactory nerves and bulbs argues for their demonstrable olfactory ability to be valuable under completely natural conditions of foraging.

A variety of evidence has been presented by Grubb (1971, 1972) that Leach's Storm-Petrels (*Oceanodroma leucorrhoa*), Wilson's Storm-Petrels (*Oceanites oceanicus*), Greater Shearwaters (*Puffinus gravis*), and possibly Sooty Shearwaters (*P. griseus*) accomplish at least some navigation by reliance on olfactory cues. His most extensive studies were done on breeding Leach's Storm-Petrels on an island in the Bay of Fundy. He found that they reached the island at night by flying upwind, that nesting material served as an effective lure in darkness, that the birds landed through dense trees a short distance downwind of their burrows, and that they chose the arm of a Y maze that contained the odor of their own nest material rather than one odorized by control material taken from the ground. Finally, birds with plugs in their nostrils or with sectioned olfactory nerves had not returned to their burrows after 1 week, in contrast to handled or sham-operated controls. Grubb concluded that the evidence supported the hypothesis that Leach's Storm-Petrel depends on olfaction for many aspects of burrow location. Shallenberger (1973), on the other hand, found relatively little evidence for a reliance on odor cues by the Wedge-tailed Shearwater (*P. pacificus*) in Hawaii. He argued that the open terrain and relatively good lighting at the nesting site made it unnecessary in comparison with the conditions faced by the Leach's Storm-Petrels.

Stager (1967) has proposed that African honeyguides, which feed on beeswax, locate hives by the odor of the wax. He found that the birds were attracted to a concealed burning candle. Honeyguides can be netted repeatedly in the immediate vicinity of beehives even 10–12 days after the bees have abandoned them when visual and auditory cues would no longer be available (Archer and Glen, 1969). Because they were found at the hive itself and not in general circulation in the vicinity, it was proposed that they were able to pinpoint the location by means of a distance cue such as odor.

Recent experiments by Papi *et al.* (1971, 1972) have demonstrated a reliance on olfaction by homing pigeons. Birds with bilateral olfactory nerve section, with both nostrils plugged with cotton tampons, or with one nerve resected and the other nostril plugged all were profoundly disoriented as shown by the very low incidence of returns to the home loft. The few that returned arrived later than the sham-operated control birds. The authors consider the possibility of a nonspecific behavioral effect but think this is unlikely on the basis of their standardized observations of behavior patterns on 6 days before operation and 7 days after. They suggest that pigeons in the home loft learn to recognize odors from surrounding areas and to associate them with wind directions. When released at a distant point, they orient toward home, after finding a familiar odor at the release point, by flying opposite to the direction associated with that odor at the home site. This research needs much confirmation, since it represents a very unusual hypothesis.

Beyond these instances, avian reliance on olfaction cannot now be described. It may be of direct use to some species that are especially well equipped and virtually ignored by others that have minimal, though functional, equipment, such as songbirds and parrots. It may provide what we would think of as an esthetic experience with little or no regulatory function included, or it may even contribute to the normal operation of physiological systems in ways that are only beginning to be understood for other vertebrates (Johnston *et al.*, 1970). Not only does olfactory input influence many facets of reproduction in other forms but it has even been shown to affect certain aspects of general behavior in both rats and birds (e.g., Douglas and Isaacson, 1969; Heimer and Larsson, 1967; Papi *et al.*, 1971; Phillips and Martin, 1972; Sieck, 1972; Wenzel and Salzman, 1968; Wenzel *et al.*, 1969; Hutton and Wenzel, 1971). Perhaps its most significant contribution to many avian forms lies in this sphere of influence rather than in the transmission of specific information about odors.

ACKNOWLEDGMENT

Bibliographic assistance from the Brain Information Service under Contract NIH 70-2063 is gratefully acknowledged.

REFERENCES

Archer, A. L., and Glen, R. M. (1969). Observations on the behaviour of two species of honey-guides *Indicator variegatus* (Lesson) and *Indicator exilis* (Cassin). *Los Angeles County Mus. Contrib. Sci.* No. 160, pp. 1–6.

Ariëns Kappers, C. U. A., Huber, G. C., and Crosby, E. C. (1936). "The Comparative Anatomy of the Nervous System of Vertebrates, Including Man." Macmillan, New York.

Bang, B. G. (1960). Anatomical evidence for olfactory function in some species of birds. *Nature (London)* **188**, 547–549.

Bang, B. G. (1964). The nasal organs of the Black and Turkey Vultures: a comparative study of the cathartid species *Coragyps atratus atratus* and *Carthartes aura septentrionalis* (with notes on *Cathartes aura falklandica, Pseudogyps, bengalensis,* and *Neophron percnopterus*). *J. Morphol.* **115**, 153–184.

Bang, B. G. (1965). Anatomical adaptations for olfaction in the Snow Petrel. *Nature (London)* **205**, 513–515.

Bang, B. G. (1966). The olfactory apparatus of tube-nosed birds (Procellariiformes). *Acta Anat.* **65**, 391–415.

Bang, B. G. (1968). Olfaction in Rallidae (Gruiformes), a morphological study of thirteen species. *J. Zool.* **156**, 97–107.

Bang, B. G. (1971). Functional anatomy of the olfactory system in 23 orders of birds. *Acta Anat., Suppl.* **58**, 1–76

Bang, B. G., and Cobb, S. (1968). The size of the olfactory bulb in 108 species of birds. *Auk* **85**, 55–61.

Bartholomew, G. A., and Cade, T. J. (1958). Effects of sodium chloride on the water consumption of House Finches. *Physiol. Zool.* **31**, 304–310.

Beidler, L. M., ed. (1971). "Handbook of Sensory Physiology," Vol. 4, Pt. 1. Springer-Verlag, Berlin and New York.

Brindley, L. D. (1965). Taste discrimination in Bobwhite and Japanese Quail. *Anim. Behav.* **13**, 507–512.

Brindley, L. D., and Prior, S. (1968). Effects of age on taste discrimination in the Bobwhite Quail. *Anim. Behav.* **16**, 304–307.

Brower, L. P. (1969). Ecological chemistry. *Sci. Amer.* **220** (No. 2), 22–29.

Brower, L. P., Brower, J. von Z., and Corvino, J. M. (1967). Plant poisons in a terrestrial food chain. *Proc. Nat. Acad. Sci. U.S.* **57**, 893–898.

Brown, H. E., and Beidler, L. M. (1966). The fine structure of the olfactory tissue in the Black Vulture. *Fed. Proc. Fed. Amer. Soc. Exp. Biol.* **25**, 329.

Cane, V. R., and Vince, M. A. (1968). Age and learning in quail. *Brit. J. Psychol.* **59**, 37–46.

Cobb, S. (1960a). Observations on the comparative anatomy of the avian brain. *Perspect. Biol. Med.* **3**, 383–408.

Cobb, S. (1960b). A note on the size of the avian olfactory bulb. *Epilepsia* **1**, 394–402.

Craigie, E. H. (1930). Studies on the brain of the Kiwi (*Apteryx australis*). *J. Comp. Neurol.* **49**, 223–357.

Craigie, E. H. (1932). The cell structure of the cerebral hemisphere of the humming bird. *J. Comp. Neurol.* **56**, 135–168.

Craigie, E. H. (1940). The cerebral cortex in Palaeognathine and Neognathine birds. *J. Comp. Neurol.* **73**, 179–234.

Craigie, E. H. (1941). The cerebral cortex of the penguin. *J. Comp. Neurol.* **74**, 353–366.

Crosby, E. C., and Humphrey, T. (1939). Studies of the vertebrate telencephalon. I. The nuclear configuration of the olfactory and accessory olfactory formations and of the nucleus olfactorius anterior of certain reptiles, birds, and mammals. *J. Comp. Neurol.* **71**, 121–213.

Douglas, R. J., and Isaacson, R. L. (1969). Olfactory lesions, emotionality and activity. *Physiol. Behav.* **4**, 379–381.

Duncan, C. J. (1960a). Preference tests and the sense of taste in the feral pigeon. *Anim. Behav.* **8**, 54–60.

Duncan, C. J. (1960b). The sense of taste in birds. *Ann. Appl. Biol.* **48**, 409–414.

Duncan, C. J. (1962). Salt preferences of birds and mammals. *Physiol. Zool.* **35**, 120–132.

Duncan, C. J. (1964). The sense of taste in the feral pigeon. The response to acids. *Anim. Behav.* **12**, 77–83.

Ficken, M. S., and Kare, M. R. (1961). Individual variation in ability to taste. *Poultry Sci.* **40**, 1402.

Fuerst, W. J., Jr., and Kare, M. R. (1962). The influence of pH on fluid tolerance and preferences. *Poultry Sci.* **41**, 71–77.

Gentle, M. J. (1972). Taste preference in the chicken (*Gallus domesticus* L.). *Brit. Poultry Sci.* **13**, 141–155.

Graziadei, P., and Bannister, L. H. (1967). Some observations on the fine structure of the olfactory epithelium in the domestic duck. *Z. Zellforsch. Mikrosk. Anat.* **80**, 220–228.

Grubb, T. C., Jr. (1971). Olactory navigation by Leach's Petrel and other procellariiform birds. Doctoral Dissertation, University of Wisconsin, Madison.

Grubb, T. C., Jr. (1972). Smell and foraging in shearwaters and petrels. *Nature (London)* **237**, 404–405.

Halpern, B. P. (1962). Gustatory nerve responses in the chicken. *Amer. J. Physiol.* **203**, 541–544.

Harriman, G. E. (1967). Laughing Gull offered saline in preference and survival tests. *Physiol. Zool.* **40**, 273–279.

Harriman, G. E., and Kare, M. R. (1966a). Tolerance for hypertonic saline solutions in Herring Gulls, Starlings, and Purple Grackles. *Physiol. Zool.* **39**, 117–122.

Harriman, G. E., and Kare, M. R. (1966b). Aversion to saline solutions in Starlings, Purple Grackles, and Herring Gulls. *Physiol. Zool.* **39**, 123–126.

Heimer, L., and Larsson, K. (1967). Mating behavior of male rats after olfactory bulb lesions. *Physiol. & Behav.* **2**, 207–209.

Henton, W. W. (1969). Conditioned suppression to odorous stimuli in pigeons. *J. Exp. Anal. Behav.* **12**, 175–185.

Henton, W. W., Smith, J. C., and Tucker, D. (1966). Odor discrimination in pigeons. *Science* **153**, 1138–1139.

Huber, G. C., and Crosby, E. C. (1929). The nuclei and fiber paths of the avian diencephalon with consideration of telencephalic and certain mesencephalic centers and connections. *J. Comp. Neurol.* **48**, 1–225.

Hughes, B. O., and Wood-Gush, D. G. M. (1971). Investigations into specific appetites for sodium and thiamine in domestic fowls. *Physiol. Behav.* **6**, 331–339.

Hundley, M. H., and Mason, C. R. (1965). Birds develop a taste for sugar. *Wilson Bull.* **77**, 408.

Hutton, R. S., and Wenzel, B. M. (1971). Active avoidance responding in olfactory nerve sectioned pigeons. *4th Annu. Winter Conf. Brain Res., 1971.* Abstract.

Jacobs, H. L., and Scott, M. L. (1957). Factors mediating food and liquid intake in chickens. I. Studies on the preference for sucrose or saccharine solutions. *Poultry Sci.* **36**, 8–15.

Johnston, J. W., Jr., Moulton, D. G., and Turk, A., eds. (1970). "Advances in Chemoreception. I. Communication by Chemical Signals." Appleton, New York.

Jones, A. W., and Levi-Montalcini, R. (1958). Patterns of differentiation of the nerve centers and fiber tracts in the avian cerebral hemispheres. *Arch. Ital. Biol.* **96**, 231–284.

Kare, M. R. (1961). Comparative aspects of the sense of taste. *In* "Physiological and Behavioral Aspects of Taste" (M. R. Kare and B. P. Halpern, eds.), pp. 6–15. Univ. of Chicago Press, Chicago, Illinois.

Kare, M. R. (1965). The special senses. *In* "Avian Physiology" (P. D. Sturkie, ed.), 2nd ed., pp. 406–446. Cornell Univ. Press, Ithaca, New York.

Kare, M. R., and Ficken, M. S. (1963). Comparative studies on the sense of taste. *In* "Olfaction and Taste" (Y. Zotterman, ed.), pp. 285–297. Macmillan, New York.

Kare, M. R., and Maller, O. (1967). Taste and food intake in domesticated and jungle fowl. *J. Nutr.* **92**, 191–196.

Kare, M. R., and Medway, W. (1959). Discrimination between carbohydrates by the fowl. *Poultry Sci.* **38**, 1119–1127.

Kare, M. R., and Pick, H. L., Jr. (1960). The influence of the sense of taste on feed and fluid consumption. *Poultry Sci.* **39**, 697–706.

Kare, M. R., Black, R., and Allison, E. G. (1957). The sense of taste in the fowl. *Poultry Sci.* **36**, 129–138.

Kitchell, R. L., Ström, L., and Zotterman, Y. (1959). Electrophysiological studies of thermal and taste reception in chickens and pigeons. *Acta Physiol. Scand.* **46**, 133–151.

Landolt, J. P. (1970). Neural properties of pigeon lingual chemoreceptors. *Physiol. Behav.* **5**, 1151–1160.

Lindenmaier, P., and Kare, M. R. (1959). The taste end-organs of the chicken. *Poultry Sci.* **38**, 545–550.

McFarland, L. Z., and George, H. (1966). Preference of selected grains by geese. *J. Wildl. Manage.* **30**, 9–13.

Michelsen, W. J. (1959). Procedure for studying olfactory discrimination in pigeons. *Science* **130**, 630–631.

Moore, C. A., and Elliott, R. (1946). Numerical and regional distribution of taste buds on the tongue of the bird. *J. Comp. Neurol.* **84**, 119–131.

Neuhaus, W. (1963). On the olfactory sense of birds. *In* "Olfaction and Taste" (Y. Zotterman, ed.), pp. 111–124. Macmillan, New York.

Oksche, A., Farner, D. S., Serventy, D. L., Wolff, S., and Nicholls, C. A. (1963). The hypothalamo-hypophysial neurosecretory system of the Zebra Finch, *Taeniopygia castanotis. Z. Zellforsch. Mikrosk. Anat.* **58**, 846–914.

Papi, F., Fiore, L., Fiaschi, V., and Benvenuti, S. (1971). The influence of olfactory nerve section on the homing capacity of carrier pigeons. *Monit. Zool. Ital.* [N.S.] **5**, 265–267.

Papi, F., Fiore, L., Fiaschi, V., and Benvenuti, S. (1972). Olfaction and homing in pigeons. *Monit. Zool. Ital.* [N.S.] **6**, 85–95.

Pfaffmann, C. (1959). The sense of taste. *In* "Handbook of Physiology" (Amer. Physiol. Soc., J. Field, ed.), Sect. 1, Vol. I, pp. 507–533. Williams & Wilkins, Baltimore, Maryland.

Phillips, D. S., and Martin, G. K. (1972). Heart rate conditioning of anosmic rats. *Physiol. Behav.* **8**, 33–36.

Pick, H. L., Jr., and Kare, M. R. (1962). The effect of artificial cues on the measurement of taste preference in the chicken. *J. Comp. Physiol. Psychol.* **55**, 342–345.

Rieke, G. K., and Wenzel, B. M. (1973). Responses evoked in centers of the pigeon telencephalon following stimulation of the olfactory bulb. *Anat. Rec.* **175**, 424.

Sanders, E. B. (1929). A consideration of certain bulbar, midbrain and cerebellar centers and fiber tracts in birds. *J. Comp. Neurol.* **49**, 155–223.

Shallenberger, R. J. (1973). Breeding biology, homing behavior, and communication patterns of the Wedge-tailed Shearwater, *Puffinus pacificus.* Doctoral Dissertation, University of California, Los Angeles.

Shibuya, T., and Tonosaki, K. (1972). Electrical responses of single olfactory receptor cells in some vertebrates. *In* "Olfaction and Taste 4" (D. Schneider, ed.), pp. 102–108. Wissenschaftliche Verlagsgesellschaft MBH, Stuttgart.

Shibuya, T., and Tucker, D. (1967). Single unit response of olfactory receptors in vulture. *In* "Olfaction and Taste 2" (T. Hayashi, ed.), pp. 219–233. Pergamon, Oxford.

Shibuya, T., Ijima, M., and Tonosaki, K. (1970). Responses of the olfactory nerve in the seagull, *Larus crassirotris. Zool. Mag. (Tokyo)* **79**, 237–239.

Shumake, S. A., Smith, J. C., and Tucker, D. (1970). Olfactory intensity difference thresholds in the pigeon. *J. Comp. Physiol. Psychol.* **67**, 64–69.

Sieck, M. H. (1967). Electrophysiology of the avian olfactory system. Doctoral Dissertation, University of California, Los Angeles.

Sieck, M. H. (1972). The role of the olfactory system in avoidance learning and activity. *Physiol. & Behav.* **8**, 705–710.

Sieck, M. H., and Wenzel, B. M. (1969). Electrical activity of the olfactory bulb of the pigeon. *Electroencephalogr. Clin. Neurophysiol.* **26**, 62–69.

Stager, K. E. (1964). The role of olfaction in food location by the Turkey Vulture (Cathartes aura). *Los Angeles County Mus. Contrib. Sci.* No. 81, pp. 3–63.

Stager, K. E. (1967). Avian olfaction. *Amer. Zool.* **7**, 415–420.

Tucker, D. (1965). Electrophysiological evidence for olfactory function in birds. *Nature (London)* **207**, 34–36.

Warner, R. L., McFarland, L. Z., and Wilson, W. O. (1967). Microanatomy of the upper digestive tract of the Japanese Quail. *Amer. J. Vet. Res.* **28**, 1537–1548.

Warren, R. P., and Vince, M. A. (1963). Taste discrimination in the Great Tit (*Parus major*). *J. Comp. Physiol. Psychol.* **56**, 910–913.

Wenzel, B. M. (1967). Olfactory perception in birds. *In* "Olfaction and Taste 2" (T. Hayashi, ed.), pp. 203–217. Pergamon, Oxford.

Wenzel, B. M. (1968). The olfactory prowess of the Kiwi. *Nature (London)* **220**, 1133–1134.

Wenzel, B. M. (1971). Olfaction in birds. *In* "Handbook of Sensory Physiology" (L. M. Beidler, ed.), Vol. 4, Pt. 1, pp. 432–448. Springer-Verlag, Berlin and New York.

Wenzel, B. M. (1972). Olfactory sensation in the Kiwi and other birds. *Ann. N.Y. Acad. Sci.* **188**, 183–193.

Wenzel, B. M., and Salzman, A. (1968). Olfactory bulb ablation or nerve section and pigeons' behavior in non-olfactory learning. *Exp. Neurol.* **22**, 472–479.

Wenzel, B. M., and Sieck, M. (1972). Olfactory perception and bulbar electrical activity in several avian species. *Physiol. Behav.* **9**, 287–294.

Wenzel, B. M., Albritton, P. F., Salzman, A., and Oberjat, T. E. (1969). Behavioral changes in pigeons following olfactory nerve section or bulb ablation. *In* "Olfaction and Taste 3" (C. Pfaffmann, ed.), pp. 278–287. Rockefeller Univ. Press, New York.

Wenzel, B. M., and Stonehouse, B. J. (1970). Unpublished observations.

Whitten, A. J. (1971). A new behavioural method for further determination of olfaction in Mallard (*Anas platyrhynchos*). *J. Biol. Educ.* **5**, 291–294.

Wilcoxon, H. C., Dragoin, W. B., and Kral, P. A. (1971). Illness-induced aversions in rat and quail: Relative salience of visual and gustatory cues. *Science* **171**, 826–828.

Yang, R. S. H., and Kare, M. R. (1968). Taste response of a bird to constituents of arthropod defense secretions. *Ann. Entomol. Soc. Amer.* **61**, 781–782.

Chapter 7

MECHANORECEPTION

J. Schwartzkopff

I. Introduction

1. Differentiation and Biological Significance of the Various Receptive Organs

The development of senses that respond to mechanical stimulation is closely related to that of active movement. Therefore, the corresponding reactions are especially sensitive and manifold in birds, which are particularly mobile. The differentiation of mechanoreceptive sense organs has its starting point in the general sensitivity of

417

the skin in responding to contacts with other objects. Then receptors develop that are specialized for touch, air or water flow, and vibration.

The contact sensitivity responds as well to stimuli generated by external forces as to the effects of the animal's own movement. Sense organs located within muscles or tendons and sensitive nerve structures of the digestive tract or the circulatory system serve as proprioceptors. The related organs of equilibrium overlap functionally with the proprioceptors. The hearing organ, at the highest level of mechanoreception, evaluates sound waves, which are transferred from distant sources through air or water.

The inventory of acoustic signals has developed in phylogeny as a byproduct of various movements and especially of respiration. Eventually these signals are produced by specialized organs. Acoustic signals are, by their general physical properties, peculiarly useful for biological communication. Birds, among all other vertebrates, have achieved the most efficient bioacoustic information system (Thorpe, 1961; Brémond, 1963; Greenewalt, 1968; Il'ichev, 1968). This, however, must not be compared with the human language upon which the distinguished position of man is based.

2. Primary Process of Mechanoreception

The mechanosensitive organs originate from very different structures; therefore, the respective physiological processes need not necessarily be uniform. Mechanical stimulation is primarily supposed to affect sensitive molecular configurations. Very delicate and reversible changes of membrane texture might thus influence directly the ion permeability (local potential) from which excitation generates. Loewenstein and Rathkamp (1958) have identified the local potential of the Pacinian corpuscles generated by mechanical stimulation of segments of the sensitive axon membrane as a primary process (Fig. 4). Other experiments, however, seem to indicate, as an alternative, that intracellular plasmatic structures (e.g., of sensory cilia) are affected first, followed by permeability changes of the neighboring membrane segments (Vinnikov, 1965; Flock, 1967; Thurm, 1970).

The primary process of mechanical excitation, in any case, controls the conversion of energy by secondary processes made available through cellular metabolism. This is not an energy transformation in the sense of physics, but rather a translation of information or sensory transduction based on amplifying mechanisms. Correspondingly, enzymes are found in the sensitive region of tactile corpuscles (Loewenstein and Molins, 1958; Winkelmann and Myers, 1961), as well as at the ciliated surface of the labyrinthine receptors (Vinnikov *et al.*, 1965b), by which the metabolism responsible for the generation of local potentials is enhanced.

II. Sensitivity of the Skin and Proprioception

A. ONTOGENY AND ANATOMY

1. *General*

The overall somatosensory performance is certainly at a high level in birds, but information about the corresponding sense organs and nerve connections is rather fragmentary. Reptiles provide supplemental information, since the phylogenetic relations are close. The comparison with mammals is also based upon the similar functional level. In so far as sensory skin functions are mediated through specialized though elementary receptors, nerve fibers are involved; these originate from neurons of the spinal ganglia or the corresponding ganglia of the cranial nerves. In the beak, it is the ophthalmic branch of the trigeminal nerve, through which the Herbst corpuscles are supplied (Bailey, 1969; Quilliam and Armstrong, 1961).

The differentiation of the end organs in ontogeny is induced by neural processes that develop outward from the neural crest. Thus, the Herbst corpuscles in the beak of chickens or ducklings start developing after the sixteenth and twentieth day of incubation, respectively, when the nerve fibers have contacted the subepidermal tissues. However, transplantation experiments show that the properties of the subepidermal tissues determine the particular differentiations of the various organs or species (Dijkstra, 1933; Malinovský, 1967; Saxod, 1967, 1970b; Saxod and Sengel, 1968). In general, behavior in ontogeny in the Herbst corpuscles is similar to that in the Merkel and Grandry corpuscles. After dissection of the nerve, while the main axon degenerates and regenerates, the Herbst lamellae undergo only plasmatic changes, as do the tactile cells.

The degree of participation of elements of the autonomic nervous system in the sensitivity of the skin is not yet clear. Various types of tactile corpuscles (Fig. 1) are provided with additional fine nerve fibers, apparently of autonomic origin in birds as well as in other vertebrates (see Clara, 1925). Furthermore, simple nerve endings, loops, and networks have been identified within the skin and internal organs that are partially ascribed to the cerebrospinal system because of fiber diameter and myelin sheath. But others probably belong to the autonomic system, especially very fine fibers and peculiar ganglionic cells of the subepidermal layers (Schartau, 1938). In most cases, it cannot yet be decided whether these structures participate in trophic and other auxiliary functions of the receptors or actually mediate tactile, temperature, or pain sensations.

2. Free Nerve Endings

Myelinated nerve fibers that are dispersed among others within the skin of birds lose their sheaths near the epidermis, branch, and end in terminal boutons at the epidermal cells. These endings are obliterated gradually during the keratinization of the epidermis (Schartau, 1938). The feather papillae are especially richly supplied by such fibers. Plexiform networks of finer myelinated or unmyelinated fibers appear within the plain epidermis, ending to some extent intraepithelially. The bird's tongue and palate are supplied with more specialized endings (Botezat, 1906, 1909; Malinovský and Zemánek, 1971).

3. Corpuscles of Merkel and Grandry

Tactile corpuscles, named after two of the older anatomists, occur within the subepithelial connective tissue. In both types, a primarily myelinated nerve fiber approaches auxiliary tactile cell and ends with a delicate network or disk in close contact with or between two of the cells (Fig. 1). The electron microscope reveals synapselike contacts between the ramifications of nerve fibers and the tactile (or satellite) cell where cholinesterase is also found (Saxod, 1967, 1970a). These substructures suggest the possibility that the satellite cell

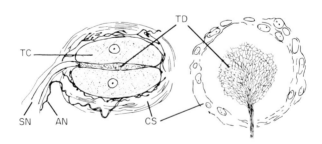

FIG. 1. Duck, tactile corpuscle of Grandry. AN = small (autonomous) nerve fiber; CS = connective tissue sheaths (capsule); SN = sensory nerve fiber; TC = tactile cell; TD = tactile disc (nerve plexus). (From von Buddenbrock, 1952.)

may have functions of a secondary sensory cell beyond that of mere mechanical transfer of the stimulus (Andres, 1969).

The Merkel corpuscles are smaller in general than the Grandry type and are located more superficially; the latter lie deeper within the corium. Accordingly, the Grandry corpuscles are enclosed by a mesodermic capsule which is lacking or very thin around the Merkel corpuscles (Dijkstra, 1933); but no major differences seem to exist between them. The satellite cells are homologous in both types of corpuscles. They are derived, according to some authors, from epithelial cells (other anatomists consider them to be modified Schwann cells), they are enveloped by (ectodermal) Schwann cells and the (mesodermal) perineurium of the nerve fiber (Botezat, 1906; Andres, 1969; Saxod, 1970a). The typical Grandry corpuscles, composed of two tactile cells and a nerve disk between them, have been found exclusively in the bill of ducks, while the Merkel corpuscles are widespread in various regions of the skin, the tongue, and the bill of all species (Malinovský, 1967; Malinovský and Zemánek, 1971).

4. Corpuscles of Herbst

The lamellated corpuscles of Herbst differ only slightly from mammalian corpuscles of Pacini (see Schildmacher, 1931; Poláček, 1969). The avian corpuscles are between 0.05 and 0.2 mm in length, about one third the length of the Pacinian corpuscles. Both types, however, consist of a central axon surrounded by an inner and an outer core. Between the cores are additional, nonmyelinated nerve fiber branches (Clara, 1925; Nafstad and Andersen, 1970) derived from the prevertebral ganglia of the sympathetic system, according to dissection experiments in dogs (Jurjewa, quoted by Schildmacher, 1931). The outer core of ellipsoid shape is enveloped by several layers of connective tissue (epineurium). Underneath connective tissue cells of perineural origin are combined to build up an onionlike system of capsules filled by some kind of intracellular jelly or fluid and interconnected by collagen fibers. The internal part of the outer core corresponds to the endoneurium; in birds, it rarely contains fibrocytes (Fig. 2). So the outer core as a whole has a cushionlike appearance (Clara, 1925; Saxod, 1970b).

The inner core is more cylindrical in shape, and it derives from a sensory nerve fiber and its Schwann (palisade) cells, the myelin sheaths of the latter being converted into a compound system of half lamellae. The most conspicuous, although not fundamental, dif-

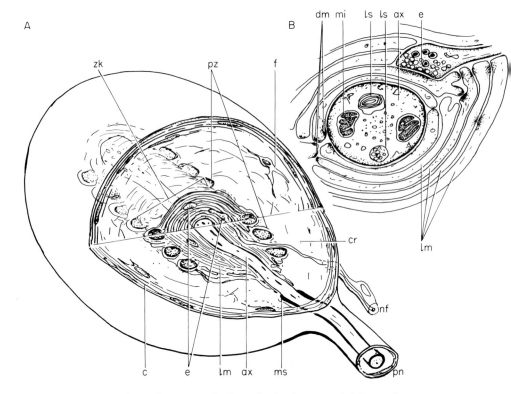

FIG. 2. Duck, Herbst corpuscle from the beak-region. (A) Semischematic presentation, the section planes showing the central core (zk), composed of the receptor axon(ax)
and the sheath of the palisade cells (pz) forming lamellae (lm). The myelin sheath (ms)
of the axon begins within the internal space of the capsule (cr). The capsule (c) of the
corpuscle originates from the perineural sheath (pn) of the receptive nerve fiber. The
capsule contains loose connective tissue, corresponding to the endoneurium, with
single fibrocytes (f). The unmyelinated nerve fiber (nf), intruding into the capsular
space, branches within the palisade lamellae, forming synaptic contacts (e). (B) Trans-
versal section across the axis of the central core (10,000-fold), showing axon (ax), palisade cell lamellae (lm) and the synaptic contacts of a vegetative nerve fiber (e). The
lamellae are connected with each other and with the axon membrane by desmosomes
(dm). Within the axon, mitochondria (mi) and lysosomes (ls) of various shape are to be
found. Centrally, neurofilaments and neurotubuli are visible. (By courtesy of Andres;
original drawing after electronmicroscopic pictures.)

ference between the Herbst and the Pacini corpuscles is the configuration of these lamellae and, in particular, the location of the
corresponding cell nuclei. In the bird, 10–20 nuclei at a time are
arranged in an orderly manner in two opposing rows at the flat sides

of the laterally depressed axon. Accordingly, the system of the lamellae is also more clearly bilaterally organized in birds than it is in mammals (Pease and Quilliam, 1957; Andres, 1969; Saxod, 1970b). Further, a peculiar vesicle or knob is formed at the end of the Herbst axon. In mammals as well as in birds, the axon membrane shows particular zones of contact with the adjacent lamellae (Andres, 1969; Saxod, 1970b). On the other side, the branches of the thin nerve fiber contact, by specialized junctions, the lamellae, but not directly the central axon (Nafstad and Andersen, 1970).

The first indications as to the function of the organs described above are derived from their concentration in certain organs (Schildmacher, 1931; Malinovský and Zemánek, 1971). The Herbst corpuscles occur in great numbers in the beaks of all birds, but particularly of the Scolopacidae. The bony tip of the bill of snipes forms comblike lamellae. Within these there is an orderly arrangement of numerous corpuscles (Clara, 1925; Bolze, 1969). Also the tip of the tongue of woodpeckers is richly supplied with lamellated corpuscles. These are to be found further in various parts of the subcutaneous connective tissue, particularly in regions that are mechanically exposed. For instance, the papillae of the contour feathers and their musculi arrectores are supplied with the organs as noted by Herbst himself (cf. Schwartzkopff, 1949; Poláček et al., 1966; Poláček, 1969). The occurrence in different regions of the periost is noteworthy, especially the grouping of Herbst corpuscles between tibia and fibula or radius and ulna (Schildmacher, 1931). Here the sense organs are rather well protected against external disturbances. They are contacted only by the muscles that deflect the toes or digits. Finally, the rich supply of the joint capsules, particularly in the wing, represents a significant difference in distribution from that of the Pacinian corpuscles in mammals (Poláček et al., 1966).

5. Mechanoproprioceptive Organs

Proprioceptors have developed in all vertebrates as part of the fast-contracting muscles. They are arranged in different ways, either to measure the active development of tension or the change of muscle length. Studies on amphibians, reptiles, and mammals emphasize their importance in the reflex adaptation of muscular activity (Huber and DeWitt, 1898; Hoffmann, 1964; Matthews, 1964; Barker, 1968; Fukami and Hunt, 1970; Hunt and Wylie, 1970). Birds have received little attention in this respect within recent decades (cf. Ariens Kappers *et al.*, 1936; Manni *et al.*, 1965), but certainly the avian scheme cannot be inferior to that of mammals.

Similarly, no detailed information concerning the proprioceptive supply of the internal organs in birds is available. Respiration, blood pressure, peristaltic movements of the intestine, defecation, and other internal functions are controlled by reflex action, mediated through specific sensory structures, which are preferentially modified nerve endings of the vegetative (autonomic) system. This, as such, is well developed in birds (Hirt, 1934; Stresemann, 1934; Ariens Kappers *et al.*, 1936).

B. PHYSIOLOGY OF SOMATOSENSORY RECEPTORS

The somatosensory functions, such as pain and temperature sensation, general touch, and specialized tactile or vibrational sensitivity, are generally mediated through the skin. The ascription to defined nerve endings or sense organs, however, in birds is sometimes ambiguous, often based on inference from information on mammals. The same is true in proprioception, e.g., it can only be supposed by comparative deduction that there are two types of muscle spindles that are sensitive either for tonic or for phasic stimulation (Fukami, 1970). In general, the avian muscle spindles show low mechanical thresholds. They differ from the mammalian organs in that the conduction velocity of the nerve fibers shows an unimodal distribution. The tendon organs have higher thresholds than the muscle spindles (Dorward, 1970).

1. *Functions of Nervous Centers and Connections*

Statements in the older literature make it clear that freshly decapitated birds are still able to move regularly and to respond to tactile stimuli (see Schwartzkopff, 1949). Therefore, the reflexes from skin and joint receptors as well as from muscular sense organs can be processed independently by the spinal cord (Ten Cate, 1960). Corresponding with the distribution of nerves, the somatosensory functions are arranged in segments of the skin (Fig. 3) (Kaiser, 1924, quoted by Stresemann, 1934). The borders of the dermatomes for tactile stimuli are roughly valid also for pain and temperature reactions.

The sizes of the trigeminal sensory nuclei, which serve the beaks of birds, correlate well with their biological significance. These nuclei are much larger in ibises, snipes (*Gallinago*), and lovebirds (*Agapornis*) than in crows (*Corvus*) or the Tawny Owl (*Strix aluco*) (Stingelin, 1961). A specific forebrain area adjacent to the olfactory bulb

FIG. 3. Dermatome of the nineteenth segment, isolated by severing the dorsal roots of the seventeenth, eighteenth, and twentieth through twenty-second spinal nerves; corresponding black zones are rendered insensitive. (After Kaiser, 1924; from Stresemann, 1934.)

increases in correlation with the trigeminal centers, being connected therewith through the quintofrontal tract. It supposedly serves as association center for combined oral sensations.

The caudal part of the mesostriatum (palaeostriatum according to Ariens Kappers *et al.*, 1936) seems to represent the uppermost center of general somatosensitivity in parrots according to the sectioning experiments of Kalischer (1905), while the more rostral part serves as a somatomotor center. The adjacent arrangements of corresponding sensory and motor functions in the basal ganglia of birds is analogous to the organization of the mammalian cortex.

An area in the pigeon brain (caudal ectostriatum) which is located near the sensory center found by Kalischer receives the corresponding nervous information 18 milliseconds after tactile stimulation of the skin (Erulkar, 1955, by electophysiological methods). The divergence in location may be explained by the methodological differences.

Another kind of somatosensory representation is established also in birds within the cerebellum, mainly in the IV–VIII folia. Here ganglionic elements can be specifically activated or inhibited by proprioceptive stimuli (joints, muscles) or by touching the skin (Gross, 1970). This corresponds to the role of the cerebellum in coordinating body movements.

2. Sensitivity for Temperature and General Tactile Stimuli

Peripheral sensors for temperature have so far been identified in birds only in the region of the beak (Kitchell *et al.*, 1959). Fibers of the glossopharyngeal nerve, supplying the tongue, contain many cold elements that react upon cooling by generating nerve signals among the taste receptors. It seems that true warm fibers are not present in the tongue, since increasing temperatures are answered only above 45°C, which is considered pain reaction. Homeothermia is controlled in birds apparently largely by central sensors that are sensitive to cooling and warming. They are located in the central canal of the spinal cord and possibly also in the brain stem (Rautenberg, 1969).

The various tactile stimuli are received through a variety of morphologically different endings of myelinated nerve fibers. Within the skin of the duck, these endings respond to tactile stimulation of defined areas (receptive fields) at different thresholds and adaptive rates. Dorward (1970) further identified pain receptors that show virtually no adaptation. These fibers, on the application of pressure, show a spatially diffuse response with a high threshold. The relations in reptiles are similar; Proske (1969) found receptive fields of various sizes and differences in thresholds and in adaptation rates.

The manifold reactions to mechanical stimuli correspond with the biological significance of the beak. Trigeminal nerve fibers have been identified as proprioceptors of the jaw musculature of the pigeon, as joint receptors and as tactile elements with narrow receptive fields, adapting rapidly or slowly (Zeigler and Witkowsky, 1968). Here, corpuscles of Merkel as well as corpuscles of Herbst probably participate in the sensation.

3. Mechanisms and Efficiency of Lamellated Corpuscles

Electrophysiological research on Herbst corpuscles (Skoglund, 1960; Winkelmann and Myers, 1961; Dorward and McIntyre, 1971) emphasizes the functional conformity with the Pacinian corpuscles in which the principles of mechanoreception have been elucidated by Loewenstein and co-workers. A local (receptor, generator) potential is generated by mechanical deformation (Fig. 4) of axon segments. When the generator potential surpasses a threshold level, it releases a propagated nerve impulse at the first node of Ranvier, still inside the outer core (Loewenstein and Rathkamp, 1958).

The cushion of the outer core and the lamellae of the inner core transfer mechanical energy to the central axon. The question is open as to whether the lamellated structures participate directly in the excitatory processes, for instance by establishing a distinct ionic concentration. The desmosomes and the occluding zones shown by

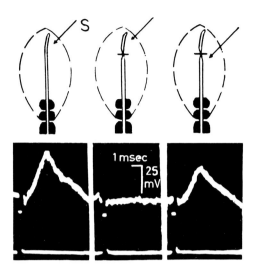

FIG. 4. Core of a decapsulated Pacinian corpuscle; generator potential elicited by tactile impulse (lower trace) to sensitive axon (S stimulating probe). Note that simultaneous compression of stimulated segment (middle) inhibits the potential, while compression peripheral to stimulus does not (right). No propagated nerve pulse is generated, since stimulus strength is constant at subthreshold level. (By courtesy of Loewenstein and Rathkamp, 1958.)

electron microscopy (Fig. 2) seem to tighten the cell borders and thus to endorse this suggestion (Andres, 1969; Saxod, 1970b). A nonspecific cholinesterase of high activity is present in the axon and in the surrounding lamellae, emphasizing the active role of these structures (Winkelmann and Myers, 1961; Loewenstein and Molins, 1958).

The additional fine fiber that extends between the inner and outer core conveys nerve signals at only 0.7 meters/second, while the main fiber has a conduction rate of about 35 meters/second outside the sheaths (Goto *et al.*, 1966). Electrical stimulation of the nonmyelinated fiber can enhance the sensitivity of the axon. But this does not prove unequivocally a direct regulation of the sensory functions, since the receptor metabolism may be influenced in a more general way.

The tactile organ is adapted, particularly by the structures of the outer core, for the elastic reception and transfer of rapid pressure changes, while permanent loads are compensated for (Loewenstein and Skalak, 1966). Thus, the corpuscle operates as a band pass, suppressing the lower frequencies of the stimulating force. Correspondingly, the corpuscles of Pacini and Herbst occur preferentially in body parts in which reactions to phasic stimuli or vibrations have been determined electrophysiologically in different mammals at about 200 Hz. The somewhat smaller corpuscles of birds are very sensitive to frequencies between 400 and 800 Hz (Skoglund, 1960; Dorward, 1970). Minor functional differences, e.g., of adaption rate, may be found within the organs of a single species.

Through conditioning experiments in birds, the thresholds for vibration have been determined (Schwartzkopff, 1949) (Fig. 5), which coincides with the above statements. The perception was ascribed to the Herbst corpuscles between tibia and fibula, since only the legs

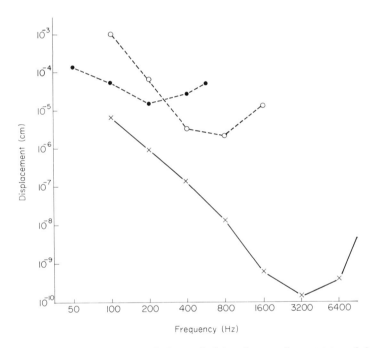

FIG. 5. European Bullfinch (*Pyrrhula pyrrhula*), vibrational sensitivity of the bird (O) compared with human (●) and with the bird's hearing (×). Sound intensity plotted as particle displacement. (From Schwartzkopff, 1955.)

are mobilized by the stimulus. Dorward and McIntyre (1971) confirmed the threshold sensitivity curve for vibrational stimuli using electrophysiological methods. The bending muscle of the toes transfers the vibrations to the sense organs.

In other parts of the body, the Herbst corpuscles seem to perform many control functions, from such as arranging the feather coat, its cleaning, and permanent adaptation to the varying requirements of thermoregulation, to breeding behavior, or to the aerodynamics in flight. A special type of control of this characteristic locomotor activity is suggested by the accumulation of corpuscles at the base of the wing feathers.

III. Labyrinth (Organ of Equilibrium)

The labyrinth is related phylogenetically to the lateral-line organ. It has developed from the pattern common to all vertebrates to become a "higher" sense organ, processing complex mechanical stimuli (Bethe *et al.*, 1926; De Burlet, 1934a; Vinnikov *et al.*, 1965a,b; Wersäll *et al.*, 1965).

A. ANATOMY, ONTOGENY, AND BIOCHEMISTRY

1. General

The earliest *Anlage* of the labyrinth appears in the developing chicken at the second day of incubation. An ectodermal indentation soon becomes the auditory vesicle, the simple epithelium of which differentiates into superior and inferior parts. Originating from the vesicle, a blind ending (ductus endolymphaticus) enters the subdural region of the brain. Later several sensory patches with auxiliary structures develop in relation to branches of the VIII cranial nerve (Held, 1926; De Burlet, 1934a). The nerve cells strongly influence the differentiation of sensory structures when auditory vesicles of chickens are cultivated in isolation, but the connection to the brain can be dispensed with (Friedman, 1968). Eventually, the sensory epithelia of the three cristae ampullares (Lüdtke and Schölzel, 1966), of the macula utriculi (Vinnikov *et al.*, 1965a) and of the papilla neglecta is derived from the pars superior (Fig. 6). The macula sacculi, papilla basilaris, and m. lagenae belong (in birds) to the pars inferior labyrinthi (De Burlet, 1934a). The basilar papilla, representing the organ of hearing, will be considered in detail in Section IV.

The labyrinth receives its blood supply rather uniformly through the labyrinthine artery and cochlear artery, while the venous drainage shows some variations in birds (Snapp, 1924). Its internal fluid is secreted by specialized regions of the wall that are richly supplied with blood vessels (e.g., tegmentum vasculosum). The ionic com-

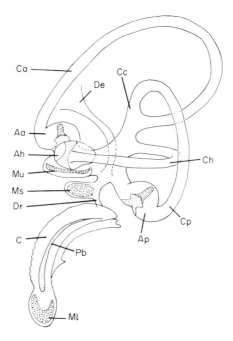

FIG. 6. Membranous labyrinth of birds (left side, lateral view). Aa, Ah, Ap = ampulla anterior, horizontalis, posterior, respectively; C = cochlear duct; Ca, Ch, Cp = canalis anterior, horizontalis, posterior; Cc = crus commune; De, Dr = ductus endolymphaticus, reuniens; Ml, Ms, Mu = macula lagenae, sacculi, utriculi; Pb = papilla basilaris. (Combined from Werner, 1939, and Lüdtke and Schölzel, 1966.)

position of the endolymph is similar to intracellular fluids, the potassium concentration being relatively high, the sodium relatively low (Dohlman *et al.*, 1959; Yashiki, 1968; Ishiyama *et al.*, 1969).

The delicate membranous labyrinth is encased by rather thin but solid osseous walls. In birds the osseus labyrinth conforms rather well in shape to the membranous labyrinth, protruding into the air-filled spaces of the skull. The adjacent midbrain, cerebellum, and medulla strongly influence its general shape (Werner, 1963).

A system of perilymphatic spaces separates the osseus and the membranous labyrinths, from which the perilymphatic duct (s. aquaeductus cochleae) provides an open connection to the cerebrospinal fluid system (Werner, 1958). The ionic composition of the perilymph is the same as that of the blood plasma from which it is derived by filtration. The specific gravity of avian perilymph (1.0006) is essentially identical with that of the endolymph (1.0017), whereas its viscosity is distinctly lower (0.76 against 1.19 cP at 44°C), becoming close to that of water (Money *et al.*, 1971). The comparatively high viscosity of the endolymph depends on its mucopolysaccharide content (Bélanger, 1961) rather than on protein. The latter is more highly concentrated in the perilymph (Dohlmann *et al.*, 1959).

2. Statolithic Organs—Maculae of the Utricle, Saccule, and Lagena

Several sensory patches with rather homogeneous epithelia covered by statolithic membranes have developed in birds. The macula utriculi has become more closely related to the pars superior than in lower vertebrates, according to its nerve supply. The macula sacculi and m. lagenae of the pars inferior are separated by the new basilar papilla (De Burlet, 1934a). The sensory epithelia of the utricle and the saccule extend roughly within one plane, the former horizontally, the latter obliquely. The macula lagenae, oriented vertically, has distinctly arching sides (Werner, 1938) (Fig. 7).

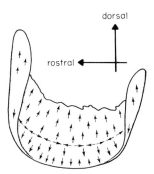

FIG. 7. Pigeon, lateral view of the left macula lagenae. Arrows indicate the polarization of hair cells (position of kinocilium in relation to the bundle of stereocilia). Note border line of the striola. (By courtesy of Jørgensen, 1970.)

The sense cells bear cilia (hair cells) that protrude into the gelatinous tectorial membrane, which is composed mainly of mucopolysaccharides and is secreted by the adjacent supporting cells. The membrane contains numerous statoconia, crystals of calcium carbonate, and to a minor extent, calcium phosphate crystals (Werner,

1939). The ultrastructure of the hair cells is the same in the sensory patches of maculae and cristae, the arrangement of the cilia being of special interest (Vinnikov, 1965; Vinnikov *et al.*, 1965a; Wersäll *et al.*, 1965). These are polarized in the sense that the single kinocilium, distinguished by its internal filaments, determines the structural (and physiological) axis of the cell through its position relative to the cluster of 40–50 homogeneous stereocilia; i.e., the hair cells of the lagena are subdivided into two opposing groups with averted kino-

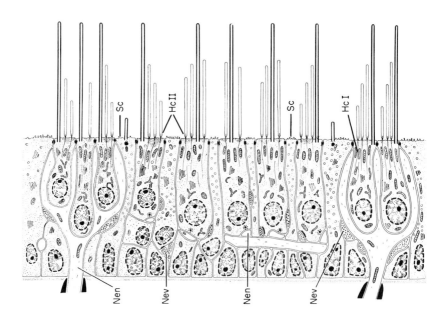

FIG. 8. Pigeon, cross section of lagenar sensory epithelium showing the striola with alternating polarization of hair cells. Note further the various types of hair cells and of nerve endings. Hc I, Hc II = hair cell type I, II; Nen = nonvesiculated nerve ending (chalices at type I Hc); Nev = vesiculated (efferent) ending. (By courtesy of Jørgensen, 1970.)

cilia (Jørgensen, 1970) (Figs. 7 and 8). The sensory epithelia of the utricle and the saccule are similarly organized according to their cellular polarization.

3. Cupola Organs—Ampullary Crests (and Papilla Neglecta)

The system of the semicircular canals is specialized for the perception of fluid movements and can therefore be compared with the lateral line organ. The endolymph can move circularly within the system, since the three canals communicate with the utricles at both ends. Each canal is oriented with respect to one spatial plane. This can be shown best in the horizontal canal (s. canalis lateralis, s.c. externus), which parallels the geophysical horizontal plane when the bird holds its head in the normal position (Duijm, 1951). Both of the other canals are oriented vertically, at about right angles to each other, and the anterior canal of the one side lies nearly parallel to the posterior canal of the other side. There is a general correlation between the size of the semicircular canals and the mobility of gallinaceous birds (Sagitov, 1964).

The bony walls of the horizontal and the posterior (vertical) canal fuse at their crossing in some species, permitting the perilymphatic clefts to communicate. Further, the membranous ends of both of the vertical canals combine to form the crus commune, a kind of utricular recess (De Burlet, 1934a). Each semicircular canal has a sensory crest at the bottom of the characteristic ampulla (Fig. 6). Further, the papilla neglecta is found in the utricle of some birds close to the ampulla posterior. Its shape is similar to the ampullary organs (De Burlet, 1934a). The cristae of both of the vertical canals develop a cross fold (septum cruciforme) without sensory cells. The hair cells show a homogenous polarization in each of the cristae insofar as studies in various vertebrates (but not yet in birds) have shown. The cilia are inserted into the gelatinous fan of the cupola, which corresponds with the tectorial membrane of the statolithic organs. It also is secreted by the supporting cells at very early stages of development (Schölzel and Lüdtke, 1967). The distal edge of the cupola reaches the roof of the ampulla. The epithelium adjacent to the sensory crest (transitional zone, planum semilunatum) has differentiated into light and dark cells, the ultrastructures of which indicate secretory and resorbing functions (Dohlmann et al., 1965; Lüdtke and Schözel, 1966; Ishiyama et al., 1969).

4. Nervous Pathways

The sensory patches of the vestibular (and auditory) labyrinth receive nervous connections through various branches of the VIII cranial nerve (n. stato-acusticus). The cell bodies of the (primary) elements combine to form the peripheral ganglia (ggl. vestibulare, s. Scarpae, and ggl. lagenare) (De Burlet, 1934a; Ariens Kappers *et al.*, 1936). However, the efferent fibers, that convey information from the brain to the labyrinth, originate from nerve cells in the medulla (Boord, 1961). The synapses at the junction of nerve endings with hair cells are ultrastructurally similar in all vertebrates (Wersäll *et al.*, 1965; Jørgensen, 1970). Within the nonauditory parts of the labyrinth, two types of hair cells have been established based on the synaptic appearance. The bottle-shaped type I cells are encased by one large (afferent) synaptic chalice (Fig. 8). The cylindrical type II cells receive several, smaller synaptic knobs, but only at their basal region. Here afferent and efferent endings can contact the hair cell membrane, whereas in the type I cells efferent endings can approach merely the outer membrane of the chalice.

The afferent knobs are recognized by some mitochondria and otherwise rather empty appearance. They are opposed by synaptic bars at the presynaptic (hair cell) side. Efferent synapses are identified by the high density of synaptic vesicles and the plasmatic duplications shown by the adjacent hair cell membrane (Fig. 9) (Cordier, 1964; Jahnke *et al.*, 1969).

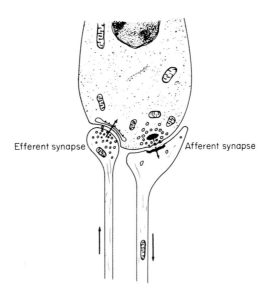

Efferent synapse   Afferent synapse

FIG. 9. Synaptic connections with hair cells of higher vertebrates. Note the synaptic vesicles (efferent) and the subsynaptic structures. (By courtesy of Flock, 1967.)

The neurites of the vestibular and lagenar ganglion enter the medulla and diverge on a complex system of second order centers, the so-called vestibular nuclei. Stingelin (1965) attempted to reconcile the somewhat confusing terminology of the earlier literature (see Ariens Kappers *et al.*, 1936); at least six major nuclear masses with further subdivisions can be discriminated. The passerine birds achieve the highest degree of differentiation, while in other groups the vestibular nuclei are more uniform and thus resemble those of the mammals (Bartels, 1925).

Nerve connections from the vestibular nuclei decussate to the contralateral centers, ascend to the cerebellum via the cerebellar nuclei, to the forebrain via the midbrain, and also to the oculomotor nuclei, and they also descend to the spinal cord via the inferior olivary nuclei. The complicated system of nervous connections and the considerable interspecific differences reflect the outstanding adaptive value of the labyrinth organ in birds in the control of manifold locomotory activities.

B. Sense Physiology of the Vestibular Labyrinth

1. Mechanism of Excitation

The tectorial membrane (cupola) of the labyrinthine sensory epithelia is shifted by the stimulating force, thus bending the cilia. Shearing (rather than pressure) represents the adequate stimulus for the hair cells (von Holst, 1950; Flock, 1967). The various receptors differ in their responses to the time course of the stimulation. The statolithic organs show tonic responses selectively; i.e., the nerve fibers discharge continuously if the displacement of the hairs is sustained. Other adjacent receptors may produce phasic responses to the same stimulus, signaling by on or off discharges only changes of the stimulating force. They are particularly sensitive for short impacts, acceleration, etc. In the cupola organs, the hair cells seem to react in a phasically selective manner. The sensory cells of the inner ear are sensitive to rapid changes in the direction of shearing.

Further, the hair cells indicate specifically the direction of the stimulus, corresponding to their structural polarization (see pp. 431–432). Bending the cilia toward the kinocilium excites the receptor, whereas the opposite displacement inhibits its activity (Lowenstein and Wersäll, 1959; Wersäll *et al.*, 1965; Harris *et al.*, 1970).

The transducer mechanism, located either at the shaft or, more probably, at the rootlet of the cilium, produces a receptor (or generator) potential across the hair-cell membrane whereupon a synaptic depolarization follows. The corresponding current may eventually elicit a propagated nerve impulse. The coding of information at the synapse depends, on one hand, on its shape (type I and II hair cells), and, on the other, on the synaptic functions as controlled through the activity of the efferent endings.

The processes of receiving and transmitting sensory information are sustained energetically by cellular metabolism. In the labyrinth, the endolymph has a distinctive function, since its ionic concentration and electric charge provide an extracellular energy store, apparently of greatest importance for the hearing organ (see p. 454). The endolymphatic, as well as the cellular mechanism of membrane polarization, are based eventually on the metabolism of glucose. The enzymes involved show greatest activities within the sensory and adjacent specialized epithelia; thus, for example, high activities of succinic dehydrogenase (Krebs cycle) occur within the parts of the ampullary crest of the pigeon (Ishiyama *et al.*, 1969). Prolonged sensory activity also increases protein metabolism markedly, as indicated by increasing RNA content (Abramyan, 1968; in the pigeon's saccule).

2. *Labyrinth Organ and Muscle Tonus*

The somatic muscle fibers maintain a constant slight tension. This is regulated by the continuous excitation generated by the labyrinth. The central nervous system modulates the labyrinthine tonus generators and can partially replace them after destruction of the vestibular nerve endings (Fischer, 1926; Groebbels, 1927a,b; Benjamins and Huizinga, 1927). Pigeons appear thoroughly languid soon after removal of both labyrinths. Several weeks or months later, they can again stand fairly normally by aid of central compensation, though the movements remain unsteady.

The impairment of the muscle tonus is even more significant after only one labyrinth has been destroyed, since bilateral asymmetry of the muscle tension results in this case. The experimental animal first shows a tendency to fall to the operated side. The musculature counteracts the imposed stretch by less resistance on the severed than on the intact side, even months later. Then, convulsive torsions of head and neck and finally rotation of the whole body develop, increasing through several weeks. The paroxysms eventually are reduced, and the bird may regain almost normal posture. If at this stage the remaining labyrinth is removed, the initial phenomena of unilateral operation appear, thus proving that the central nervous system has compensated for the loss of the first sense organ during the recovery period (see von Buddenbrock, 1952).

The maculae of the utricle and of the saccule are assumed to contribute mainly to the generation of muscle tonus in all vertebrates. In the birds, the lagena is also involved, but only in a supporting role (Huizinga, 1934; Schwartzkopff, 1949). As a peculiarity of birds, the loss of the semicircular canals alone seems to produce nearly the same behavioral impairment as destruction of the entire labyrinth.

3. Spatial Orientation and Regulation of Movement

a. General Cybernetic Relations. The state of activity within the bilaterally corresponding labyrinthine sensory patches shifts out of balance if these are stimulated externally by an asymmetric input. Changes in activity of somatic muscles follow, restoring the balance by negative feedback processes according to the laws of cybernetics. The out-of-balance state may also be created through internal stimuli, thus producing voluntary movements. Von Holst (1950) has emphasized the cybernetic aspects of vestibular functions by studies on fishes but including the other vertebrates by deduction. Many results of the older researchers, seemingly contradictory for so long, are better understood from the point of view of cybernetics (see Fischer, 1926; von Buddenbrock, 1952).

b. Static and Dynamic Functions of the Statolithic Maculae. The sensory patches of the utricle, saccule, and lagena recognize the direction of gravity through the statoconia, the weight of which produces different patterns of excitation within the epithelia corresponding to head position. The particular orientation of the single maculae (see p. 431) can be interpreted as functional optimization, since the absolute strength of shearing is maximal in the vertical epithelium, while its differential sensitivity is greatest in the horizontal position. Besides, single patches may be responsible in particular for certain postural reflexes; e.g., turning the eye vertically (*Raddrehung*) is controlled by the saccule (Benjamins and Huizinga, 1926).

Linear acceleration has the same influence on the statolithic organs as gravity (adequate stimulus). Birds demonstrate by reflectory movements (postural reflexes) that they can relate any body position to gravity, even when blindfolded. A duck, held by its body, adjusts its freely movable head so as to maintain its natural position in space

(Fig. 10). If translatory movements are imposed upon the animal, the neck muscles counteract. A simulated free fall elicits wing movements with its back down, steering movements of wings and tail turn it into normal position (Groebbels, 1928; von Buddenbrock, 1952).

FIG. 10. Postural reflex in the duck; note position of the head, independent of trunk. (After Huxley, 1913; from von Buddenbrock, 1952.)

c. Dynamic Functions of the Semicircular Canal System.   The ampullary crests with their fan-shaped cupolas are adapted especially for the perception of endolymph movements generated by circular acceleration (adequate stimulus) through the moment of inertia. Many experiments on birds confirm the direct observation by Steinhausen (1933) of the flow of endolymph in the ampullae of Pike. In these studies, the techniques of removing or plombing parts of the avian semicircular canal system (Fischer, 1926; Groebbels, 1927b; Huizinga, 1934, 1938) were applied. In other experiments, the fluid pressure on the cupola threshold was calculated as 0.0012 dyne cm^{-2}, which is very close to the pressure on the oval window at hearing threshold, 0.004 dyne cm^{-2} (Money *et al.*, 1971).

Any circular acceleration can be analyzed through one or several pairs of the semicircular canals, corresponding with their spatial arrangement. Stimulation of the sensory crest is answered by regulating reflexes, of which the eye nystagmus is the most easily accessible to experimentation. The line of sight is preserved against imposed turns of the head by reflexive counteractions of the oculomotor system. If the eye movement reaches its morphological limit, a fast back turn follows on the first, slow phase of the nystagmus. Another phenomenon, the post nystagmus, is produced by the retardation of a preceding rotation. It equals acceleration of the opposite sign (movement of inertia) and is suitable particularly for quantitative studies of vestibular functions in birds (Winget and Smith, 1962).

In the pigeon, the horizontal nystagmus cannot be produced after both horizontal canals have been incapacitated. Unilateral removal reduces the eye movements but does not eliminate them completely. Turning to the intact side is now more efficient than the opposite stimulation. Thus, the horizontal crest can process endolymphatic flow of both directions; shearing of the cilia is effective not only toward the kinocilium but also away from it. This probably is achieved by inhibition of the high spontaneous activity of the ampullary nerve in the latter. The double-acting control of ampullar reflexes distinguishes birds from amphibians and reptiles (von Buddenbrock, 1952). The consequences of removal of the vertical canals are similar to those of removal of the horizontal canals, in principle (Fischer, 1926). In this, the nonhomonymous canals of both sides cooperate in "over cross," according to their respective spatial orientations (see p. 433.

Interspecific differences among birds are up to now mainly based upon the comparison of morphologic data (cf. Werner, 1963), not upon functional tests. Thus, for example, Kimura (1931) and Hadžiselimović and Savković (1964) have shown that the size of the semicircular canals is correlated with the flight performance; pigeons, owls, thrushes, ravens, and falconiform birds have relatively larger canals than galliform birds and ducks.

4. Cooperation of the Labyrinth with Other Sense Organs

The vestibular organs, as instruments of spatial orientation, are connected with various motor centers and are also integrated with other sensory systems. The cerebellum plays a very important role, since vision, audition, and the somatosensations are represented within its cortical structures (Gross, 1970). It receives its most important information, however, from the vestibular organs.

There are especially close, direct relationships between the eye and the labyrinth. Not only are the nystagmic movements described above controlled through nerve connections between the rostrally located vestibular nuclei and the oculomotor centers of the midbrain, but also the information about the position of the resting head is translated into corresponding changes in eye direction (Benjamins and Huizinga, 1926).

The cooperation of the vestibular apparatus with the somatosensory system is shown by the *kipp* reflexes, which react to imposed movements around the longitudinal or transverse body axis. In these, the moment of inertia stimulates the labyrinth as well as the proprioceptors or the tactile receptors of the limbs. Correspondingly, the reflex activity of wings, legs, and tail is preserved at a reduced level after severing both vestibular nerves, and they can be eliminated only by cutting the dorsal roots of the spinal cord (Fischer, 1926).

Very little is known about common functions of the vestibular and the acoustic parts of the labyrinth, though in birds not only general phylogenetic considerations but particularly the highly developed lagena seem to favor such cooperation. As mentioned above, the lagenar nerve fibers are distributed partly to the vestibular and partly upon the acoustic nuclei of the medulla. In addition to the lagena, there is good reason to assume cooperation in the orientation mechanism of vestibular and auditory organs, but behavioral and physiological evidence is still lacking.

IV. Labyrinth (Auditory Organ)

A. ANATOMY

1. General Relations

During the phylogenetic and ontogenetic development, the labyrinth becomes isolated from direct contacts to the outside. Only the inner ear is connected with it secondarily. The spiracle of the earlier vertebrates becomes the eustachian tube and tympanic cavity (see De Burlet, 1934b; Werner, 1960). According to studies of the avian ontogeny, the auditory ossicle (columella) of birds develops from the *Anlage* of the hyomandibular cartilage (Smith, 1904; Schestakowa, quoted after Freye, 1952–1953). The otic capsule participates in the formation of its footplate, though only by inductive influence, according to Benoit (1964). The columella is coupled by a cartilaginous prolongation, the extracolumella, with the eardrum. The extracolumella develops three or four processes and originates from the lower segment of the hyoid arch (Fig. 11). Parallel to these structures, the membranous labyrinth differentiates the new sensory patch of the inner ear, the basilar papilla, which thus separates the older lagena from the saccule.

2. Outer Ear

Most of the higher vertebrates have developed a distinct outer ear that collects sound waves as the adequate stimulus, and protects the delicate eardrum against other disturbances. While the eardrum of birds and mammals forms one end of the auditory canal, the other end develops to form a variety of structures (Il'ichev, 1960); The outer ear of birds is not so close morphologically to that of mammals as it is functionally. In most birds, specialized feathers surround the auditory canal, being absent only in a few groups (e.g., the Struthioniformes, some of the gallinaceous birds, and the vultures). The feathers in front of the ear opening are adapted to minimize turbulence in flight and by this protect the hearing organ itself. Since their rami are not compact but isolated (without radioli), the permeation of sound waves is not obstructed (Stresemann, 1934; Schwartzkopff, 1949, 1955; Il'ichev, 1960, 1962, 1968). The caudal part of the orifice of the ear is formed by a muscular rim upon which specialized feathers combine into a tight margin. Thus, an enlarged ear funnel has been developed, for example, in song birds, parrots, and falconiform birds.

The shape of the outer ear varies adaptively among closely related species. For instance, the Osprey (*Pandion haliaetus*) has developed the slightest and the harriers (*Circus*) the most pronounced funnel structures among the birds of prey (Dement'ev and Il'ichev, 1963). In the social auks the ear serves for auditory communication, but diving requires special protection of the eardrum. The feathers that overlap the ear opening are stronger, while the opening itself is reduced in diameter. The caudal rim becomes enlarged in the species that dive into deeper waters, and can close the auditory canal as the water pressure increases. Thus, the outer (and middle) ear structures in birds of various diving habits are developed according to different ecological requirements (Kartashev and Il'ichev, 1964). In the ultimate case, the penguins have developed distinctive features of the outer ear which are almost exclusively protective, the functional role of the outer ear in auditory communication within the colony apparently having decreased (Il'ichev, 1961).

The most extensive adaptive variations of the ear funnel are found in the owls, many species of which have developed not only very large posterior ear flaps but also erectable anterior flaps. Both are furnished by specialized feathers and can serve either to close or enlarge the ear opening.

The structures of the funnels have become bilaterally asymmetrical in several genera of the Strigiformes (e.g., *Asio otus, A. flammeus, Strix aluco, Tyto alba*). The skull bones (Fig. 11) that constitute the auditory canal are also involved in some species (e.g., *Strix uralensis, S. nebulosa, Aegolius funereus;* Stresemann, 1934; Freye, 1952–1953; Norberg, 1968). This binaural asymmetry is unique among the freely moving vertebrates.

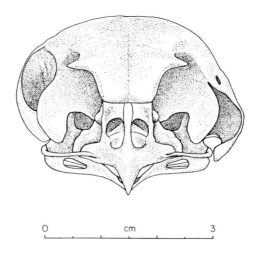

O   cm   3

FIG. 11. Boreal Owl (*Aegolius funereus*), anterior view of the skull. Axis through the center of the eardrum and the center of the ear aperture of the one side gives an angle of (vertical) divergence of about 40° to the corresponding axis of the other side. (By courtesy of Norberg, 1968.)

3. *Middle Ear*

The eardrum constitutes the medioventral wall of the auditory canal. Its membrane is composed of three layers. The epidermis of the canal continues upon it, becoming very delicate. It is but loosely attached to the underlying membrana propria of elastic connective tissue. The internal glabrous skin eventually borders the tympanic cavity. The eardrum of some birds is rimmed by the osseous annulus tympanicus — e.g., in the domestic fowl and in the owls (Stellbogen, 1930; Freye-Zumpfe, 1952–1953). In most species, at least a tendinous bridge spans the quadrate, isolating the eardrum from its movements as part of the mandibular joint. The eardrum of birds, in contrast to that of mammals, protrudes outward like a flat tent, supported mainly by the extracolumellar cartilage (Fig. 12; Pohlman, 1921; De Burlet, 1934b).

While the extracolumella supports the eardrum approximately in its center in the majority of birds, a distinctly eccentric support characterizes the owls. They are further distinguished by the extraordinary size and circular shape of the eardrum, whereas its size in other species does not show a close relation to the performance of the ear (Schwartzkopff, 1955, 1957a; Werner, 1962). The auditory ossicle, in addition to its attachment to the eardrum, is held in place by the ascending ligament, which runs from the tip of the extracolumella to

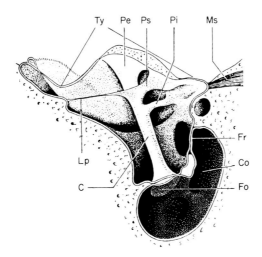

FIG. 12. Chicken, middle ear region; C = columella, Co = part of cochlea, Fo = fenestra ovalis, Fr = f. rotundis, Lp = ligamentum platneri, Ms = musculus stapedius, Pe = processus extracolumellaris, Pi = p. intracolumellaris, Ps = p. supracollumelaris, Ty = tympanum. (From Pohlman, 1921.)

the lateral wall of the eustachian tube, and through the ligament of Platner. The latter passes freely through the tympanic cavity, approximately in the opposite direction (Pohlman, 1921; Stellbogen, 1930; Freye-Zumpfe, 1952–1953).

The shape of the columella, and particularly that of its footplate, varies species-specifically, being more delicate in birds that are distinguished by keen auditory capacity (Krause, 1901). In some owls, the internal surface of the footplate forms a remarkable vesicular protrusion, apparently related to the eccentric insertion of the extracolumella in the eardrum. This arrangement, combined with an oblique angle of incidence, must produce rocking movements in sound conduction (Schwartzkopff, 1955; Pumphrey, 1961). The special form of the footplate supposedly produces less turbulence within the perilymph. Pistonlike displacement of the columella seems to predominate in the majority of bird species, and the columellar surface is correspondingly flat (Pohlman, 1921). Only one middle-ear muscle that acts upon parts of the eardrum and the extracolumella has developed in birds. It inserts at the exoccipital bone ventrally near the condyle (Stresemann, 1934). The middle-ear muscle of birds is probably homologous to the mammalian m. stapedius, since it is innervated by the facial nerve (Stellbogen, 1930), though it operates as a tensor

tympani (Wada, 1924). Several cranial bones participate in the formation of the tympanic cavity, the otic process of the quadrate and the
prootic bone being the most remarkable. They constitute an articular
crest that transverses the middle ear cavity. This communicates
through various openings with the air-filled spaces of the skull
(Stellbogen, 1930). Thus, the tympanic cavities of the two sides are not
truly isolated. Low-frequency changes of air pressure are transferred
directly from one eardrum to the other (Wada, 1924; Schwartzkopff,
1952), while high frequencies are attenuated by the osseous trabecula
within the connecting spaces. The tympanic cavity is coupled to the
inner ear by the round and the oval windows, the columella footplate
fitting with the latter by its fibrous rim. The round window of some
species (e.g., parrots, Denker, 1907) is covered by a stout membrane
supported by connective tissue, while it is very delicate and trans-
parent in others (e.g., passerine species, owls). The owls are further
distinguished by the size of the round window, while the oval window
is smaller than in comparable birds of other groups (Schwartzkopff,
1957b; Schwartzkopff and Winter, 1960).

4. Inner Ear—Cochlea with Lagena

The hearing organ of birds lies within the peripheral walls of the
skull, acoustically rather isolated by the surrounding air-filled cavities.
The osseous cochlea (including the lagena) forms a slightly bending
tube, the bilateral endings of which approach each other at the base of
the cranium. Internally, the cochlea is divided by a longitudinal
cartilageous frame. The basilar membrane stretches between its
shelves, constituting the down side of the cochlear duct, the blind
(apical) end of which contains the lagena (Fig. 7), while it communicates basally through the ductus reuniens with the remaining parts
of the labyrinth (Fig. 6; Held, 1926; De Burlet, 1934a).

a. Arrangement of Cells in the Cochlear Duct. The walls of the
cochlear duct, shown in Figs. 13 and 14, are composed of differentiated
epithelial cells. The sensory patch (papilla basilaris) covers the
basilar membrane and also partially the so called nervous cartilage.
In a single section, twenty to forty closely packed hair cells, with
interspersed supporting cells, can be seen (Figs. 13 and 14). The
inner hair cells (placed above the nervous or inner cartilage) are of
elongated and cylindrical shape, while the hair cells oriented toward
the outer cartilage are shorter and thicker. Therefore, according to
Takasaka and Smith (1971), tall, short, and intermediate types of hair

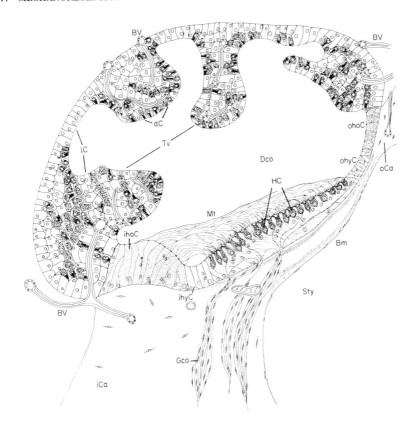

FIG. 13. Chicken, transverse section of cochlear duct; Bm = basilar membrane, BV = blood vessel, DC = dark cell, Dco = ductus cochlearis, Gco = ganglion cochleare, HC = hair cell, iCa = inner cartilage, ihoC = inner homogeneous cell, ihyC = inner hyaline cell, lC = light cell, Mt = membrana tectoria, oCa = outer cartilage, ohoC = outer homogeneous cell, ohyC = outer hyaline cell, Sty = scala tympani, Tv = tegmentum vasculosum. (Combined after Held, 1926.)

cells can be recognized. Also, the size of the cuticular plate at the base of the stereocilia and the distribution of the supplying nerve fibers show changes corresponding to the inner–outer gradient. The stereocilia of the avian auditory hair cells are particularly numerous (up to a hundred). The kinocilium (Fig. 15) is oriented transversally, i.e., from the nervous cartilage (Vinnikov, 1965; Jahnke et al., 1969; Takasaka and Smith, 1971; A. Flock, personal communication). The cilia are inserted into niches in the tectorial membrane, apparently more firmly than in mammals.

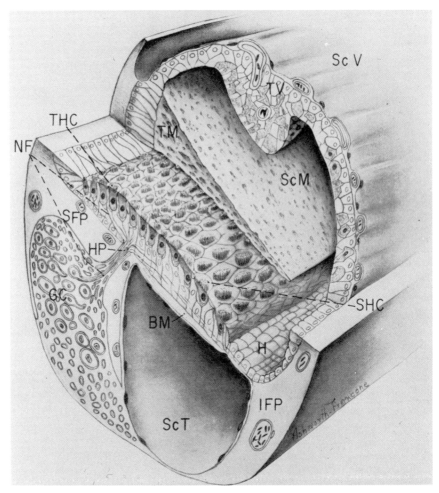

FIG. 14. Pigeon, semischematic reconstruction of cochlea; BM = basilar membrane, H = (outer) hyaline cell, HP = habenula perforata (pierced by auditory nerve fibers), IFP = outer cartilage (inferior fibrocartilaginous plate), NF = nerve fibers (peripheral process), ScM = cochlear duct (scala media), ScT = scala tympani, ScV = scala vestibuli, SFP = inner cartilage (superior fibrocartilaginous plate), SHC = short hair cell, THC = tall hair cell, TM = tectorial membrane. (By courtesy of Takasaka and Smith, 1971.)

The supporting cells are connected basally with the basilar membrane, while their apical surfaces bear microvilli and secrete the tectorial membrane. This is composed of acid mucopolysaccharide fibrils and proteins (Vinnikov *et al.*, 1965b). The adjacent sensory patch of the lagena shows a similar composition (Fig. 8) though the

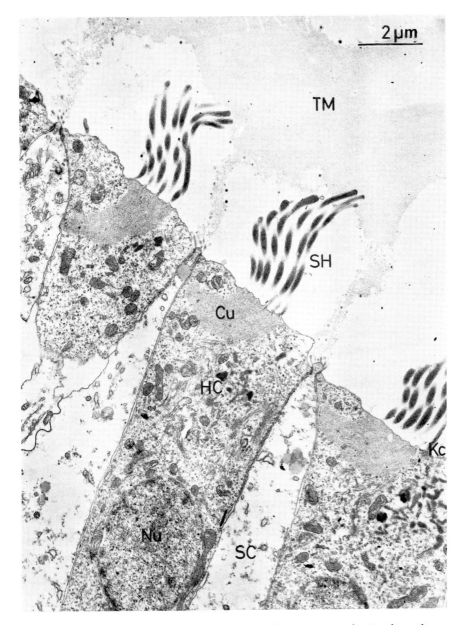

FIG. 15. Pigeon; transverse section of papilla basilaris. Cu = cuticle, Kc = kinocilium, Nu = nucleus, SC = secreting (supporting) cell, SH = sensory hairs (stereocilia), TM = membrana tectoria. (By courtesy of Jahnke *et al.*, 1969.)

hair cells are of type I and II as well and the kinocilia are oriented differently. The cells of the adjacent segments of the wall are closely related to the cochlear cells (Yashiki, 1968, 1969; Jørgensen, 1970).

The glandular epithelium of the transversely folded tegmentum vasculosum protrudes into the cochlear duct (Figs. 13–15), opposite to the basilary papilla. It is richly supplied with blood vessels (Amerlinck, 1923, 1931; Schmidt, 1964) and differentiates into two cell types. The light, cylindrical cells (osmiophobic, filled with vesicles) are located selectively in contact with the vessels. They are assumed to secrete the endolymphatic fluid. On the other hand, the dark, bottle-shaped cells (osmiophilic, with mitochondria) bear microvilli and a diplosome at their surface. These cells probably resorb sodium (Dohlman et al., 1965; Ishiyama et al., 1970).

The tegmental epithelium is bordered at both sides by the (inner or outer) homogeneous cells, upon which follow the respective hyaline cells, eventually joining the basilar papilla (Fig. 13). The orientation as inner or outer (e.g., hair cells) is sometimes confusing, since in birds as opposed to mammals the terms "anterior" or "posterior" are used synonymously with "inner" and "outer."

The sensory cells—modified type II haircells (see p. 434)—are approached by the afferent dendritic processes of neurons that combine to form the elongated cochlear ganglion (continued by the lagenar ganglion). The efferent endings belong to neurites of cells located in the medulla. The afferent nerve cells are embedded within the nervous cartilage shelf (s. anterior, s. quadrangular). The posterior cartilage is triangular in transverse section.

b. *Spatial Relations of the Inner Ear.* The perilymphatic space beneath the basilar membrane (scala tympani) widens basally near the round window, while a blind end of it extends under the apical part of the membrane (Fig. 16). Otherwise, the relations can be compared with the mammalian tympanic scala. However, a virtual vestibular scala has not developed in birds (von Békésy, 1944; Schwartzkopff and Winter, 1960), though it is shown misleadingly by histological sections. In live material (or before shrinking by histological treatment), the tegmentum vasculosum nearly touches the roof of the osseous cochlea (Figs. 14 and 16) being connected with it through a trabecular meshwork of connective tissue. Vestibular perilymphatic niches have developed only to some extent adjacent to the oval window and between cochlea and lagena. The vestibular system communicates twice with the tympanic scala, apically through the tube-shaped

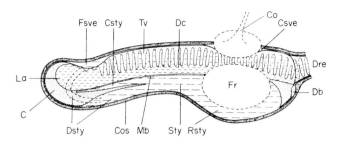

FIG. 16. Longitudinal section of the cochlea of a songbird; C = cartilageous shelf, Co = columella, Cos = osseous wall of cochlea, Csty = cavum scalae tympani, Csve = cisterna scalae vestibuli, Db = ductus brevis, Dc = d. cochlearis, Dre = d. reuniens, Dsty = d. scalae tympani (helicotrema), Fr = fenestra rotunda, Fsve = fossa scalae vestibuli, La = lagena, Mb = membrana basilaris, Rsty = recessus scalae tympani, Sty = scala tympani, Tv = tegmentum vasculosum. (From Schwartzkopff and Winter, 1960.)

helicotrema and basally through the ductus brevis of De Burlet (1934a; Schwartzkopff and Winter, 1960).

The basilar membrane of birds is short by comparison with that of mammals, but the sensory epithelium is broader (Denker, 1907). The total length of the inner ear increases but little with increasing body size (allometric exponent 0.25). Particularly "musical" birds do not show any special development. Only in the nocturnal owls, do the shape and size of the papilla basilaris approach that of the organ of Corti in mammals (Schwartzkopff, 1955, 1957a, 1963).

The oscillations of the inner ear structures are supposedly influenced by the tegmentum vasculosum. Though it may not impede the propagation of sound waves thoroughly, its soft, glandular tissue must exert a general damping influence. Older suggestions relating the regular tegmental folds to the mechanism of sound analysis in birds have been abandoned (Amerlinck, 1923, 1931).

5. *Auditory Pathways of the Avian Brain*

The neurites of the primary auditory nerve cells combine with lagenar fibers to form the auditory nerve which passes through the internal auditory canal to the medulla. Here the fibers are distributed on the magnocellular and angular nuclei (corresponding with the cochlear nuclei in mammals). Degeneration experiments show that the longitudinal organization of the sense organ is preserved by the arrangement of the respective nerve connections (Boord and Rasmussen, 1963). Only a few of the primary auditory fibers seem to ascend directly to the cerebellar nuclei; others diverge into the reticular formation (Ariens Kappers *et al.*, 1936; Stingelin, 1965).

Most of the ascending auditory fibers decussate within the medulla (Fig. 17). The dorsal striae connecting the magnocellular and laminar nuclei of both sides predominate in birds, while the ventral striae, partially originating from the superior olivary nuclei, appear more diffuse. The lateral lemniscus eventually combines all auditory fibers ascending from the medulla to the midbrain, forming two less conspicuous nuclei on its way (Winter, 1963; Stingelin, 1965; Boord, 1968).

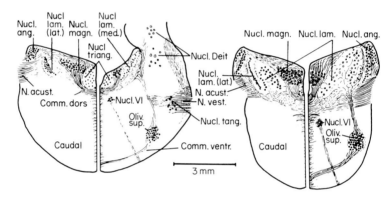

FIG. 17. Little Owl (*Athene noctua*) (left) and Barn Owl (*Tyto alba*) (right); transverse section of the medulla. Right half section of each from the level of nucleus of the VI cranial nerve, left half by 500 μm caudal to this. Note the second and third order auditory nuclei and the corresponding fiber connections, and the hypertrophic development in the nocturnal Barn Owl. N. Vest., Nucl. Deiters, Nucl. tangentialis, and Nucl. triangularis are parts of the vestibular nerve system. (From Winter and Schwartzkopff, 1961.)

The nucleus mesencephali lateralis pars dorsalis (MLD) and the isthmic and semilunar nuclei have been established as the respective midbrain centers (Ariens Kappers *et al.*, 1936; Boord, 1968; Potash, 1970; Newman, 1970). MLD, being the largest of these nuclear masses, has been studied by electrophysiological methods and experimental anatomy as well. A fiber path originating from it is relayed within the thalamic ovoid nucleus and ends eventually within the neostriatum, mainly within the mediocaudally situated "field L" (Figs. 18, 22, and 23) (Harmann and Philipps, 1967; Karten, 1967, 1968, 1970; Boord, 1969; Biedermann-Thorson, 1970a,b).

Delicate species-specific variations in the number and arrangement

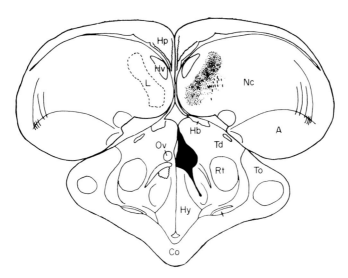

FIG. 18. Pigeon, transverse section through the brain at the level of the optic chiasma. Note lesion (black) in the area of the diencephalic ovoid nucleus and degenerating auditory fibers in the caudal neostriatum, field L; A = archistriatum, Co = chiasma nervi optici, Hb = ncl. habenularis, Hp = hippocampus, Hy = hypothalamus, L = auditory center of forebrain, Nc = neostriatum caudale, Ov = ncl. ovoidalis, Rt = ncl. rotundus, Td = thalamus dorsalis, To = tectum opticum. (By courtesy of Karten, 1968.)

of neurons have been revealed by meticulous comparison of the mesencephalic nucleus isthmi and nucleus semilunaris (Il'ichev and Dubinskaja, 1966; *Carduelis* and *Passer*). A coarser comparison of the development of the auditory nuclear mass in numerous birds does not show a correlation with the hearing capacities in most of the species. The number of neurons (counted within the medulla) increases slightly with body size (allometric exponent 0.15) (Winter, 1963), passerine species being by no means outstanding. The nocturnally hunting species of owls, distinguished also by the asymmetry of their outer ears (see p. 441), are provided with an extraordinary number of auditory neurons. Winter (1963) has counted 47,600 ganglionic cells within one-half of the medulla of Barn Owls (*Tyto alba*), but of only 13,600 for the Carrion Crow (*Corvus corone*) of about double the body weight. Also the Little Owl (*Athene noctua*) (11,200 auditory neurons), which hunts at dawn, does not exceed the normal bird. In the nocturnal owls, the binaurally activated laminar and superior olivary nuclei predominate in the gain of "computing units." This otherwise uniform laminar nucleus becomes folded and the neurons in it differentiate in shape (Fig. 17). Comparable relationships are reported from the midbrain, in which Cobb (1964) has studied the MLD of twenty-seven species. *Steatornis*, known for echo orientation, and various owls have also developed conspicuously compared to songbirds and pigeons.

B. PHYSIOLOGY OF HEARING

1. General Functions of the Ear and the Auditory Pathway

Any acoustic signal received by the avian ear starts a series of physiological processes, the particular knowledge of which is sometimes fragmentary but can be corroborated by comparison with other vertebrates (Pumphrey, 1949, 1961; Schwartzkopff, 1955, 1963, 1968; Vallancien, 1963; Grinnell, 1969; Il'ichev *et al.*, 1970).

a. Auxiliary Structures. The outer and the middle ear support the basic sense organ by collecting sound energy and transferring and adapting it to the inner ear. Removal of the ear funnels reduces the sensitivity of the cochlear microphonics of owls by more than fivefold. Similar experiments show that the outer ear of diving birds does not improve the loudness of an acoustic signal, while the improvement in songbirds is intermediate (Il'ichev and Isvekova, 1961). The middle ear transforms the aerial sound vibrations concentrated on the eardrum to match the particular impedance of the inner-ear fluids. This is achieved through several mechanisms that reduce the vibratory displacement and increase its force simultaneously. The plane quotient of the eardrum and columella footplate (Table I), the lever

TABLE I

IMPEDENCE MATCHING BY THE MIDDLE EAR

Species	Plane of eardrum/ oval window	Species	Plane of eardrum/ oval window
Mouse	24	European Coot (*Fulica atra*)	19
Man	27	Black-billed Magpie (*Pica pica*)	23
Guinea pig	29	Great Tit (*Parus major*)	25
Cat	34	Blackcap (*Sylvia atricapilla*)	29
Great Crested Grebe (*Podiceps cristatus*)	16	Budgerigar (*Melopsittacus undulatus*)	31
Pigeon	18	Long-eared Owl (*Asio otus*)	40

action of the auditory ossicle, and also the masses and elasticities of the various sound-conducting structures contribute to this (Schwartz-kopff, 1955; Pumphrey, 1961; Il'ichev, 1966a, 1968). As far as can be concluded from anatomy, the avian middle ear achieves the same level of performance as the mammalian ear. Physiologically, equivalence has already been stated by Bray and Thurlow (1942) from studies of cochlear distortion in the pigeon. Destruction of the middle ear or plugging the auditory canal reduces the sensitivity of cochlear microphonics in Bullfinches by 20–40 dB (Schwartzkopff, 1952). Bone conduction does not play an essential role, even with higher frequencies.

The functions of the single avian middle-ear muscle has not yet been elucidated thoroughly (Pohlman, 1921; Wada, 1924; Bray and Thurlow, 1942). But a function similar to that of the two muscles in mammals may be attributed to it, i.e., controlling the tension of the eardrum and the lever action of the columella and thus adapting the conduction of sound to various acoustic conditions.

b. Mechanics of the Inner Ear. The acoustic vibrations, on conduction to the oval window, mobilize the inner ear, including the round window membrane. Varying segments of the basilar membrane oscillate differentially depending on frequency, as has been observed directly by von Békésy (1944; in the fowl) and as has been concluded by Kimura (1924), Mazo (1955), and Gogniashvilii (1967) from studies of auditory damage, the cochlear microphonics after localized destruction of the inner ear, and biochemical activities, respectively. While the efficiency of peripheral frequency analysis is supposed to be less effective in birds than in mammals, the discrimination of sound intensities and of temporal patterns seems to prevail (Pumphrey, 1949, 1961; Griffin, 1953; Schwartzkopff, 1957b, 1968; Thorpe, 1961; Konishi, 1969a).

The oscillations of the cochlear fluids displace the hair cells in relation to the tectorial membrane, thus shearing the cilia and translating the acoustic stimuli into physiological changes (sensory coding). The mechanisms coincide in principle in birds and in mammals as shown by the similarity of the various bioelectrical and biochemical phenomena (Schmidt and Fernandez, 1962; Vinnikov *et al.*, 1965a,b,c; Gogniashvili, 1967; Necker, 1970).

c. Cochlear Potentials. The surface of the stimulated hair cells generates cochlear microphonics (CM) as the earliest event, almost without latency. CM reproduce the mechanical vibrations electrically. The upward dislocation of the basilar membrane corresponds with the depolarizing phase of the CM current that excites the auditory nerve fibers, while the alternating phase inhibits their activity (Schwartzkopff, 1957b, 1958). The inhibitory process is particularly influenced by metabolic impairment (Fig. 19; Necker, 1970). The

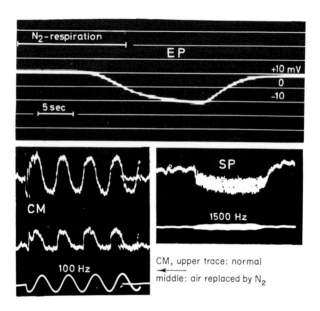

FIG. 19. Components of inner ear potentials; EP = endocochlear potential (House Sparrow, *Passer domesticus*) under short-time anoxia; CM = cochlear microphonics (Pigeon), note rectification by short time anoxia; SP = summation potential (European Starling, *Sturnus vulgaris*). (By courtesy of Necker, 1970.)

strict regime of excitatory and inhibitory events seems to be related to the uniform appearance and polarization of hair cells in the papilla basilaris. The cochlear duct further generates the positive endo-lymphatic DC potential (EP) (Fig. 19), which also depends very distinctly on metabolism. It is considerably lower (barely 20 mV) than in mammals (Schmidt and Fernandez, 1962) and seems to be generated by the electrogenic activity of the tegmentum vasculosum, in which Kuijpers *et al.* (1970) have found a correspondingly high enzyme activity (potassium- and sodium-sensitive ATPase). EP supposedly serves as an extracellular energy store, contributing to hair cell functions (Necker, 1970).

A summating potential (SP) (Fig. 19) similar to that of mammals has been described in the avian ear as a rectified derivation of CM (Stopp and Whitfield, 1964). The physiological role of SP is not clear in birds. It does not depend very much on metabolism and resembles in polarity and other attributes the depolarizing phase of CM. The action potential (AP) (Fig. 25) of the auditory nerve appears 0.5–1.2 msec after the excitatory phase of CM, depending on stimulus intensity. The nervous coding is rigidly correlated to the time course of the displacements of the basilar membrane (Schwartzkopff, 1957b, 1958). The physiological processes are very sensitive to metabolic impairment (Necker, 1970).

d. Central Nervous Processing. The auditory information, coded in nerve impulses and transferred to the brain through the fibers of the cochlear ganglion, receives its first treatment within the respective medullary nuclei and is then passed on to higher centers. The pattern of nerve discharges in the avian medulla reproduces the sound oscillations more closely than in mammals (volley principle). On the other hand, the avian neural elements respond preferentially to a confined frequency area (Figs. 20 and 21) (Schwartzkopff, 1957b; Stopp and Whitfield, 1961; Konishi, 1969a). The regular distribution of auditory nerve fibers to the secondary nuclei according to their cochlear connections is paralleled physiologically by a tonotopic arrangement of the best frequencies (Konishi, 1970). Also, the activation of the tertiary laminar nucleus corresponds with it binaural fiber connec-

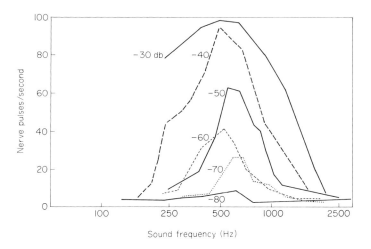

FIG. 20. Budgerigar (*Melopsittacus undulatus*); discharge rate of a secondary auditory nerve element as function of frequency and intensity. Voluntary intensity reference level (human threshold at about −120 dB).

tions. The direction of sound influences its discharge pattern, and more generally, the activity of the other medullary nuclei (Naumov and Il'ichev, 1964).

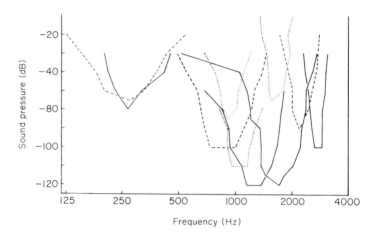

FIG. 21. Budgerigar (*Melopsittacus undulatus*), sensitivity of single auditory neurons in the medullary centers (frequency representation) (same intensity reference as in Fig. 19.) (From Schwartzkopff, 1957b.)

The processing of stimulus intensity through the lower stations of the auditory pathway is indicated generally by modulations of the nerve-pulse frequency and by different thresholds of the neural elements, varying up to 50 dB. While through the latter a coarse coding of loudness levels is achieved, the nerve discharges provide the base for the most sensitive discrimination (Schwartzkopff, 1957b; Stopp and Whitfield, 1961; Konishi, 1970). The tonal range of highest sensitivity as known from behavior has been corroborated by electrophysiological studies detecting preferentially neurons that represent this frequency area (Schwartzkopff, 1968). Moreover, Konishi (1969b, 1970) (Fig. 24) derived electrophysiological threshold curves from the combined capacities of single neurons. By this method, species-specific variations can be shown to correspond with the bioacoustic behavior, e.g., of House Sparrow (*Passer domesticus*) and Slate-colored Junco (*Junco hyemalis*).

A more complex treatment of the auditory information is applied by the nervous centers that follow the medullary nuclei. The auditory activity arrives here with a latency of 3–4 milliseconds. The adjacent semilunar nucleus discriminates between signals to either side. The neurons of the nucleus isthmi can be activated through the contralateral ear (Harman and Phillips, 1967). The spontaneous activity of the midbrain centers, including the main part of the torus semicircularis (MLD) is generally high; auditory stimuli may increase or diminish it (Stopp and Whitfield, 1961; Biedermann-Thorson, 1967). While the efficiency of sinusoidal stimuli is less pronounced, transitory or phasic parameters are processed preferentially and by high time resolution (2–4 milliseconds) (Erulkar, 1955; Harman and Phillips, 1967).

The roof of the avian forebrain (pallium, neopallium) seems not to be activated acoustically, according to various studies (Erulkar, 1955; Naumov and Il'ichev, 1964; Il'ichev, 1966b; Il'ichev et al., 1970; Harman and Phillips, 1967; Boord, 1969; Biederman-Thorson, 1970a,b; Leppelsack and Schwartzkopff, 1972). This corresponds with the brain capacities in the reptiles and discriminates the sauropsids from the mammals.

The uppermost hearing centers, located within the neostriatum, are activated through different fiber tracts and seem to operate differently (Figs. 22 and 23). Auditory information arrives at the caudal field L with latencies of 12–18 milliseconds. The fact that respective fiber connections originate from the mesencephalic MLD and are relayed in the diencephalic ovoid nucleus explain the rather high latency. Characteristic frequencies are not discriminated here, but time patterns are analyzed. Signals that are separated by 7 milliseconds

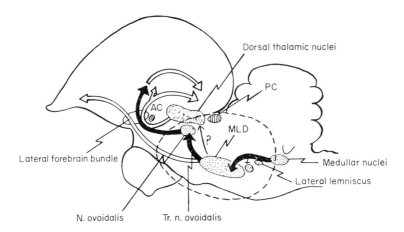

FIG. 22. Diagrammatic summary of the auditory pathway in birds. At all levels shown, both sides of the brain are activated by either ear. AC = commissura anterior, MLD = ncl. mesencephalicus lateralis pars dorsalis, PC = commissura posterior. (By courtesy of Harman and Phillips, 1967.)

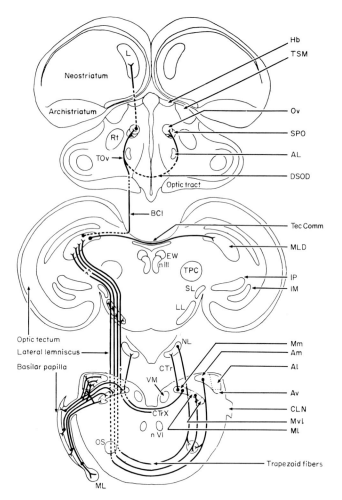

FIG. 23. Pigeon, diagram of composite transverse sections through selected levels of the brain showing the ascending auditory connections. Question marks indicate equivocal anatomical evidence. Al = nucleus angularis pars lateralis, Am = nucleus angularis pars medialis, Av = nucleus angularis pars ventralis, BCI = auditory fibers in brachium of inferior colliculus, CLN = cochlear and lagenar nerves, CTr = uncrossed dorsal cochlear tract, CTrX = crossed dorsal cochlear tract, DSOD = auditory fibers in dorsal supraoptic decussation, EW = Edinger–Westphal nucleus, Hb = habenular nuclei, IM = nucleus isthmi pars magnocellularis, IP = nucleus isthmi pars parvocellularis, L = auditory area of neostriatum, LL = nucleus of lateral lemniscus, ML = macula lagenae, Ml = nucleus magnocellularis pars lateralis, MLD = nucleus mesencephali lateralis pars dorsalis, Mm = nucleus magnocellularis pars medialis, Mvl = nucleus magnocellularis pars ventrolateralis, NL = nucleus laminaris, nIII = oculomotor nucleus, nIV = abducens nucleus, OS = superior olive, Ov = nucleus ovoidalis, Rt = nucleus rotundus, SL = nucleus semilunaris, SPO = nucleus semilunaris parovoidalis, Tee Comm = intertectal commissure, TSM = tractus septomesencephalicus, TOv = tractus nuclei ovoidalis, TPC = nuclei tegmenti pedenculopontinus pars compacta, VM = medial vestibular nucleus. (By courtesy of Boord, 1969).

can be distinguished. Sound stimuli to either ear may influence the same neural element, but differently, e.g., inhibiting or exciting it (Erulkar, 1955; Harman and Phillips, 1967; Biedermann-Thorson, 1970a,b). Also, the rostral region of the neostriatum processes auditory signals, although, no center has been identified morphologically. The latent period of this region is astonishingly short, only 5–11 milliseconds (Il'ichev, 1966b; Harmann and Phillips, 1967). Probably, the fibers that end here ascend directly from those nuclei of the midbrain (nucleus semilunaris) or the lateral lemniscus that have the shortest latencies by themselves. The rostral center differs further from the caudal center by its poor time resolution (40–50 milliseconds) (Il'ichev *et al.*, 1970).

Although the neostriatic auditory centers of birds differ from the respective cortical areas of mammals in general morphological relations, they seem to represent a comparable system of complicated nervous relays and interconnections. By analogy, areas of sensory acoustic functions have developed adjacent to corresponding motor centers that initiate or control sound production. Thus, Kalischer (1905) elicited calls from a parrot by electrical stimulation of a brain area located lateral to the caudal neostriatum. Similarly, but without parallels in the mammals, closely adjacent parts of the torus semi-circularis have differentiated in birds to control either sensory or motor acoustic functions (Murphey and Phillips, 1967; Newman, 1970; Potash, 1970).

Another type of sensomotor coordination combining auditory information with other modalities is represented by the cerebellum, in birds as well as in mammals. Most of its folia (II–XIII) are activated by auditory signals and similarly by visual stimuli, though different neurons may participate. In general, however, the processing of data provided by the lower mechanoreceptors predominates in the cerebellum (Gross, 1970).

In the mammal, all stations of the auditory pathways are interconnected by recurrent tracts that control the processing of nervous information. The fragmentary findings in birds have revealed so far an efferent connection that originates in the mesencephalic MLD, passes through the corresponding contralateral region and descends to the ventral striae (Karten, 1967). Another efferent bundle decussates at the floor of the fourth ventricle and ends at the hair cells of the papilla basilaris (and the other labyrinthine sensory patches). When this bundle is stimulated electrically (Figs. 23 and 25) the synaptic information transfer on the auditory nerve fibers (nervous coding) is inhibited (Boord, 1961; Desmedt and Delwaide, 1963). In the bird, the inhibition mediated through the efferent bundle is not as strong as in the mammal. This can be explained according to Smith (1968) by the finding that the efferent fibers within the avian ear contact almost exclusively the hair cell membrane directly, while in the mammal, contacts with axodendritic endings (to the afferent processes of the auditory nerve fibers) frequently occur.

2. General Performance of Hearing

a. Auditory Threshold. Findings based on very different methods indicate a rather consistent performance of the ear in various birds.

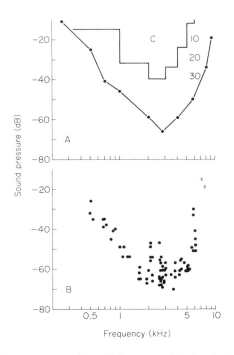

FIG. 24. Canary (*Serinus canarius*); audibility curve (A), thresholds of single auditory neurons at their characteristic frequencies in one individual (B), number of units classified according to their characteristic frequencies (C) from another bird. (By courtesy of Konishi, 1970.)

This is corroborated by the threshold curves (Figs. 5 and 24) that have been obtained from representatives of numerous families. The frequency range within which a single species can receive and process auditory signals is narrower in birds than in mammals. Further, birds as a whole do not hear ultrasonic vibrations (Table II). They also are

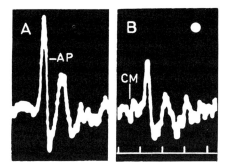

FIG. 25. Pigeon, efferent inhibition of auditory nerve action potential (AP). A, round-window response to click stimulation; B, additional activation of the efferent bundle at the bottom of the fourth ventricle, starting 10 milliseconds before the testing click. Note the reduction of AP while the cochlear microphonics (CM) increase slightly. (By courtesy of Desmedt and Delwaide, 1965.)

TABLE II
HEARING RANGE OF VARIOUS BIRDS[a]

Species	Hearing range (kHz)		
	Lower limit	Highest sensitivity	Upper limit
Anas platyrhynchos	0.3	2–3	8
Asio otus	0.1	6	18
Aythya valisineria	0.19	–	5.2
Bubo virginianus	0.06	1	8
Carduelis chloris	–	–	21
Columba livia	0.3	1–2	7.5
Corvus brachyrhynchos	0.3	1–2	8
Erithacus rubecula	–	–	21
Falco sparverius	0.3	2	8
Fringilla coelebs	0.2	3.2	–
Larus delawarensis	0.1	0.5–0.8	3
Loxia curvirostra	–	–	20
Melopsittacus undulatus	0.04	2	14
Eremophila alpestris	0.35	–	7.6
Passer domesticus	–	–	18
Phasianus colchicus	0.25	–	10.5
Pica pica	0.1	0.8–1.6	21
Plectrophenax nivalis	0.4	–	7.2
Pyrrhula pyrrhula	0.1	3.2	21
Serinus canarius	0.25	2.8	10
Spheniscus demersus	0.4	2–4	15
Strix aluco	0.1	3–6	21
Sturnus vulgaris	0.1	2	16

[a]Combined from data of several authors.

less sensitive to higher and lower tones within their hearing range than is man. In the intermediate range, however, the threshold sensitivity of most species studied equals that of mammals. The best performance is achieved between 1 and 6 kHz, varying interspecifically and in most cases in relation to the sound production (Trainer, 1946; Schwartzkopff, 1949; Konishi, 1962, 1969b, 1970).

b. *Difference Limen.* Behavioral studies show that birds can produce and discriminate numerous acoustic signals. It is an open question, however, which physical properties of the sound are evaluated. Differentiation of intensity and frequency especially can replace or supplement each other, as Pumphrey has emphasized (1949, 1961). Also, the transitory changes are of significance in the time pattern of species-specific sound production (Thorpe, 1961; Bremond, 1963; Schwartzkopff, 1962b; Greenewalt, 1968).

The significance of frequency discrimination for the human has, since the days of the early physiologists, involved considerations of the corresponding performance in birds (Ansley, 1954). Until recently, the statements were based essentially on the conditioning experiments of Knecht (1940) and of Wassiljew (1933). These data, however, do not provide satisfactory proof for the claimed equivalence of man and bird in frequency discrimination (ΔL of 0.003–0.005), according to a critical review by Greenewalt (1968). This author, by a new experimental approach, determines the discriminatory capabilities of the avian ear through the accuracy by which the frequency of distinct notes within a certain phrase of the song is reproduced. A reasonable mathematical treatment of the song of a Song Sparrow (*Melospiza melodia*) thus shows the variation coefficients in Table III. The pooled value (0.0049) is close to the human DL (Weber constant) of 0.0024 at the frequencies studied here (about 5.5 kHz). Other species apparently can maintain a constant intonation of distinct phrases with considerable precision in spite of relatively long intervals between the songs. This is comparable to the rare perfect pitch in the human. Again, the accuracy of intonation (calculated from the standard deviation) comes close to the human DL (Table IV).

The procedure of Greenewalt apparently opens a very interesting field of quantitative bioacoustic research. It is based on the crucial presupposition that the song studied is controlled by the bird's ear. This does not seem to be true for the "innate" calls, since young birds reared in incubators and held in isolation can produce these calls at a certain age (Sauer, 1954; Messmer and Messmer, 1956). Song,

TABLE III

COEFFICIENT OF VARIATION IN REPEATED NOTES OF AN INDIVIDUAL
SONG SPARROW (*Melospiza melodia*)[a]

Song: note	σ/f_m (standard deviation/mean frequency of phrase)				
	1	2	3	4	5
1	0.00477	0.00503	0.00422	0.00384	0.00533
2	0.00374	0.00456	0.00306	0.00365	0.00327
3	0.00518	0.00421	0.00242	0.00437	0.00334
4	0.00518	0.00557	0.00403	0.00418	0.00602
5	0.00526	0.00369	0.00367	0.00668	0.01038
6	0.00723	0.00691	0.00541	0.00451	0.00396
7	–	0.00784	–	–	0.00568

[a]Pooled coefficient of variation: 0.0049 (5.5 kHz). (From Greenewalt, 1968.)

TABLE IV

CONSTANCY OF INTONATION IN BIRDS[a]

Species	Mean frequency f_m (Hz)	Standard deviation σ (Hz)	Coefficient of variation σ/f_m	Human difference limen
Common Loon, *Gavia immer*	902	2.7	0.0029	0.0030
Varied Thrush, *Zoothera (Ixoreus) naevia*	2955	6.4	0.0022	0.0021
Wood Thrush, *Hylocichla mustelina*	2152	6.7	0.0031	0.0019
	2526	8.9	0.0035	0.0020
	3007	13.6	0.0045	0.0021
	3643	11.2	0.0031	0.0022
Carolina Chickadee, *Parus carolinensis*	3751	15.7	0.0042	0.0022
Song Sparrow, *Melospiza melodia*	2225	8.9	0.0040	0.0019
	2984	19.1	0.0064	0.0021
	4116	19.5	0.0047	0.0023
	7885	21.3	0.0027	0.0029

[a]From Greenewalt, 1968.

however, is probably innate in many songbirds by its general pattern only. Its actual composition can depend upon learning from other birds and the control of its production is through the ear. This was shown by the study of song development of birds reared in isolation or deafened surgically at early youth (Thielke-Poltz and Thielke, 1960; Thorpe, 1961; Konishi, 1964, 1965a,b). The situation becomes different after a song repertoire has been acquired. If the hearing organs of an adult songbird are removed, the motor program assembled up to this date can sustain the production of songs (and calls) almost unchanged. The control through the ear does not seem to be crucial any more (Schwartzkopff, 1949; Nottebohm, 1967; in the European Bullfinch (*Pyrrhula pyrrhula*) and the Chaffinch (*Fringilla coelebs*). Thus the conclusions drawn by Greenewalt refer to the process of song acquisition.

Nevertheless, the findings discussed above are corroborated indirectly by DL determinations (0.024) in the pigeon through conditioning methods (Price *et al.*, 1967). The performance of the pigeon is less sensitive by about one-sixth when compared with songbirds, according to Greenewalt. This corresponds with the general level of bioacoustic efficiency.

The auditory difference limen of intensity, unfortunately, has been barely studied in birds. Occasional observations on Bullfinches show that they discriminate intensity differences of 14% when the two ears are stimulated differentially (Schwartzkopff, 1952). This emphasizes the assumption that sound intensity DL of birds comes close to the human DL of about 10%.

Students of avian hearing and sound production almost unanimously emphasize the excellent time resolution of acoustic signals by birds (Thorpe, 1961; Pumphrey, 1961; Schwartzkopff, 1962a,b,c; Konishi 1969a, 1970). It seems from several estimates that the avian ear operates at least tenfold more precisely than the human ear. Greenewalt (1968), who studied the accuracy of time-interval reproduction within a song phrase, found standard deviations between 0.14 and 0.80 milliseconds. This is comparable with the time discrimination of bats in echolocation, but it is difficult to compare quantitatively with human performance. Without doubt, the human ear cannot analyze certain rapid changes within a bird's song, thus proving its inferiority in this respect. On the other hand, man (and probably also birds) can perceive differences of time or phase up to 10^{-5} or even 10^{-6} seconds (see Schwartzkopff, 1962a, 1967; Batteau, 1967), although, without separating the respective oscillations. Thus, the comparative interpretation becomes difficult, since the bird does not indicate through differential behavior whether he separates single events or discriminates pitches.

3. Auditory Localization

Spatial orientation in most birds depends primarily on the eye, which is only assisted by the ear. Some species of swiftlet (*Collocalia*), however, and various members of the owl family have achieved very efficient mechanisms of auditory localization, thus adapting to life in the dark. In general, all sound parameters that are available to the avian ear are utilized together for the determination of direction and distance to a source of sound. However, the binaural time difference seems to be less important in the smaller birds than it is in mammals (Schwartzkopff, 1952, 1962a; Il'ichev, 1967).

a. Localization in Diurnal Birds. According to the behavioral studies of Granit (1941) the spatial difference limen of song birds (22° in the horizontal plane) is rather poor. But it is sufficient to locate through repeated trials a fellow bird, calling from a hiding place. In principle, the performance can be improved almost *ad libitum* when probing is repeated and the information averaged by the brain. Thus, the seemingly much better performance of the clucking hen (DL 4°) in locating its chirping chick must not be greatly different from that of the song bird (Engelmann, 1928).

The directional characteristics of the Bullfinch ear (Fig. 26) show, that the luring call (mean frequency at 3 kHz) may produce considerable intensity differences between the two ears of a listening fellow, sufficient to explain the spatial DL (15% at 22°). In addition, directional differences of the frequency composition are generated by a complex sound and are probably evaluated by the bird as they are by man. But the social behavior does not depend on localization in a crucial way whereas it does depend on "special" communication (in raising the offspring) (Hüchtker and Schwartzkopff, 1958).

b. Localization in Owls. The most outstanding performance in auditory localization is shown by various owls, in which the behavior as well as the adapation of all parts of the hearing system emphasize these capacities (Stresemann, 1934; Freye, 1952–1953; Schwartzkopff, 1962a,c; Winter, 1963). A Barn Owl flying in a dark room can catch a mouse running on the floor (Payne and Drury, 1958). The owls differ from the bats by not utilizing echolocation; they further do not perceive ultrasound. Those owls that hunt at night are linked by intermediate forms to species that in daytime hunt invisible prey covered by vegetation (e.g., Short-eared Owl (*Asio flammeus*) (Dement'ev and Il'ichev, 1963).

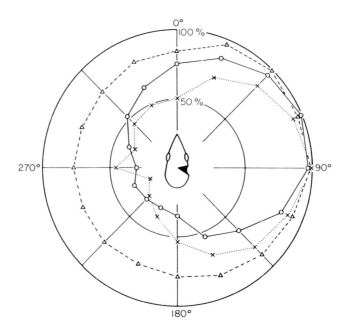

FIG. 26. Bullfinch (*Pyrrhula pyrrhula*), polar diagram of directional sensitivity (right ear) after measurements of cochlear microphonics; ($\Delta = 800$ Hz, $\mathbf{O} = 3200$ Hz, $\times = 12{,}800$ Hz. (From Schwartzkopff, 1952.)

The broad-headed owls apparently are more predisposed for the processing of binaural time difference than the other species. Long-eared Owls (*Asio otus*) answer click signals that arrive at the ear with a binaural difference of 3×10^{-5} seconds by turning the head to the leading side, thereby achieving the level of human performance. Clues derived from binaural time difference are, however, equivocal in three-dimensional orientation. Though the asymmetry of the outer ears does not disturb the evaluation of time (Norberg, 1968), binaural time-difference perception as known in the human could not possibly have required the development of the asymmetric structures.

The general directional sensitivity of the outer ear can be modified actively by the owls much more than in other birds (Schwartzkopff, 1962c, 1963). If the orifice of the outer ear is narrowed by the anterior and posterior ear flaps, the directionality is improved on one side, but on the other side the general sensitivity of the ear decreases. As shown by the polar diagram of Fig. 26 for a songbird, not only the intensity but also the frequency content of a signal depends on the direction of the sound shadow and thereby on the shapes of the head and the ear funnels.

The spatial information derived from intensity together with frequency is no longer ambivalent when processed differentially by asymmetric ear structures (Pumphrey, 1949; Schwartzkopff, 1962c, 1963; Payne, 1962, 1967; Norberg, 1968). The directional characteristics of the three species with asymmetric ears thus far studied (Long-eared Owl, Barn Owl, Boreal Owl (*Aegolius funereus*) show clearly that sound spectra close to the upper limit of hearing (beyond 12 kHz) can provide unequivocal clues for spatial orientation, but sound of lower frequencies is insufficient in this respect. Thus, any attempt to explain the undoubted localization performance of owls involves the difficulty that an extraordinarily sensitive discrimination of intensities and frequencies must be postulated for the highest tones. Nothing is known in this respect about owls in particular, but in the other birds or in man the respective capacities are poor. Payne (1962, 1971), studying intensively and by various methods the auditory localization of the Barn Owl, has found a spatial DL of less than 1°. According to the theory of this author, the owl locates its prey by adjusting the direction of the beak so as to receive at either ear maximum sound intensities of various higher frequencies.

A solution of the problem can be derived hypothetically from the studies of Batteau (1967) in man. Here monaural auditory localization (with fixed head) is based upon the contours of the pinna, which are poor enough in the human. The monaural detection of sound source is cancelled if the pinna is bent or filled with wax. Noise waves arriving at the ear normally produce a monaural time difference between the primary signal entering the auditory canal directly and the secondary one reflected by contours of the pinna. The delay of about 10^{-4} to 10^{-5} seconds generates at the basilar membrane a complex time pattern that depends on the direction of sound incidence on the ear structures. Comparable patterns must originate in the ears of owls, the dimensions being similar to the human, but the structures more distinct. Since the ears of the owls under discussion provide asymmetric patterns, the analysis of these patterns can replace the frequency discrimination in the above argument, which is difficult to explain.

The postulated analysis of monaural time or phase differences fits with the general predisposition of the avian ear for time rather than frequency discrimination (see p. 464). The peculiar length of the basilar membrane in owls is now understood to be instrumental in time-pattern analysis, serving in auditory localization.

c. Echo Orientation. Though sound production is at its most complex level of development in birds, only a few species utilize the echo of their calls for orientation. The most famous example is the Oilbird (*Steatornis caripensis*) which nests in long caves in tropical South America (Griffin, 1953). Some cave swiftlets of the genus *Collocalia* also use "bird sonar," while others do not (Novick, 1959; Medway and Wells, 1969). Birds flying in the dark can only avoid obstacles, but not hunt for moving targets as bats do. They produce short probing calls (1 millisecond) of 4–8 kHz, but no ultrasound. The silent period between the signals may be as short as 2–3 milliseconds, thus proving similar time resolution as in bats. *Collocalia vanikorensis granti*, which were forced to fly in complete darkness by Griffin and Suthers (1970), could avoid rods of 6 mm diameter, thus showing an ability about one-tenth of that of bats that were tested in comparable experiments. Recent bioacoustic findings (Poulter, 1969) indicate that penguins swimming without visual control locate their prey through sonar calls, but also in these highly specialized birds hearing does not reach the ultrasonic range (Wever *et al.*, 1969).

REFERENCES

Abramyan, R. A. (1968). Histochemical study of the avian saccule under conditions of relative rest and under adequate stimulation. *Zh. Evol. Biokhim. Fiziol.* **4,** 376–383.

Amerlinck, A. (1923). Contribution à l'étude de la membrane de Reissner et de l'épithélium de revément du canal cochléaire des oiseaux. *Arch. Biol.* **33,** 301–328.

Amerlinck, A. (1931). Nouvelles recherches sur l'histogenèse et la structure du labyrinthe membraneux de l'oreille des oiseaux. *Arch. Biol.* **40,** 19–56.

Andres, K.-H. (1969). Zur Ultrastruktur verschiedener Mechanorezeptoren von höheren Wirbeltieren. *Anat. Anz.* **124,** 551–565.

Ansley, H. (1954). Do birds hear their songs as we do? *Proc. Linn. Soc. N. Y.* **63,** 39–40.

Ariens Kappers, C. U., Huber, G. C., and Crosby, E. C. (1936). "The Comparative Anatomy of the Nervous System of Vertebrates, Including Man," Vols. I–III. Hafner, New York.

Bailey, S. E. R. (1969). The responses of sensory receptors in the skin of the green lizard, *Lacerta viridis,* to mechanical and thermal stimulation. *Comp. Biochem. Physiol.* **29,** 161–172.

Barker, D. (1968). L'innervation motrice du muscle strié des vertébrés. *Actual. Neurophysiol.* **8,** 23–71.

Bartels, M. (1925). Über die Gegend des Deiters- und Bechterewskernes bei Vögeln. *Z. Anat. Entwicklungsgesch.* **77,** 726–782.

Batteau, D. W. (1967). The role of the pinna in human localization. *Proc. Roy. Soc., Ser. B* **158,** 158–180.

Bélanger, L. F. (1961). Observations on the intimate structure and composition of the chick labyrinth. *Anat. Rec.* **139,** 539–546.

Benjamins, C. E., and Huizinga, E. (1926). Die Raddrehung wird bei den Tauben von Sacculusotolithen ausgelöst, *Z. Hals- Nasen- Ohrenheilk.* **15,** 228–230.

Benjamins, C. E., and Huizinga, E. (1927). Untersuchungen über die Funktion des Vestibularapparates bei der Taube. *Pfluegers Arch. Gesamte Physiol. Menschen Tiere* **217**, 105–123.

Benoit, J. A. A. (1964). Etude expérimentale de l'origine de la columelle auriculaire de l'embryon de poulet. *Arch. Anat. Microsc. Morphol. Exp.* **53**, 357–366.

Bethe, A., von Bergmann, G., Embden, G., and Ellinger, A. (1926). "Handbuch der normalen und pathologischen Physiologie," Vol. XI, Part 1. Springer-Verlag, Berlin and New York.

Biederman-Thorson, M. (1967). Auditory responses of neurons in the lateral mesencephalic nucleus (inferior colliculus) of the barbary dove. *J. Physiol. (London)* **193**, 695–705.

Biederman-Thorson, M. (1970a). Auditory evoked responses in the cerebrum (field L) and ovoid nucleus of the Ring Dove. *Brain Res.* **24**, 235–245.

Biederman-Thorson, M. (1970b). Auditory responses of units in the ovoid nucleus and cerebrum (field L) of the Ring Dove. *Brain Res.* **24**, 247–256.

Bolze, G. (1969). Anordnung und Bau der Herbst'schen Körperchen in Limicolenschnäbeln im Zusammenhang mit der Nahrungsfindung. *Zool. Anz.* **181**, 313–355.

Boord, R. L. (1961). The efferent cochlear bundle in the caiman and pigeon. *Exp. Neurol.* **3**, 225–239.

Boord, R. L. (1968). Ascending projections of the primary cochlear nuclei and Ncl. laminaris in the pigeon. *J. Comp. Neurol.* **133**, 523–542.

Boord, R. L. (1969). The anatomy of the avian auditory system. *Ann. N. Y. Acad. Sci.* **167**, 186–198.

Boord, R. L., and Rasmussen, G. L. (1963). Projection of the cochlear and lagenar nerves on the cochlear nuclei of the pigeon. *J. Comp. Neurol.* **120**, 463–475.

Botezat, E. (1906). Die Nervenapparate in den Mundteilen der Vögel und die einheitliche Endigungsweise der peripheren Nerven bei den Wirbeltieren. *Z. Wiss. Zool.* **84**, 205–360.

Botezat, E. (1909). Die sensiblen Nervenapparate in den Hornpapillen der Vögel. *Anat. Anz.* **34**, 449.

Bray, C. W., and Thurlow, W. R. (1942). Interference and distortion in the cochlear responses of the pigeon. *J. Comp. Psychol.* **33**, 279–289.

Brémond, J. C. (1963). Acoustic behavior of birds. *In* "Acoustic Behavior of Animals" (R.-G. Busnel, ed.), pp. 709–750. Elsevier, Amsterdam.

Clara, M. (1925). Über den Bau des Schnabels der Waldschnepfe. *Z. Mikrosk.-Anat. Forsch.* **3**, 1–108.

Cobb, S. (1964). A comparison of the size of an auditory nucleus (n. mesencephalicus lateralis, pars dorsalis) with the size of the optic lobe in twenty-seven species of birds. *J. Comp. Neurol.* **122**, 271–280.

Cordier, R. (1964). Sur la double innervation des cellules sensorielles dans l'organe de Corti du pigeon. *C. R. Acad. Sci.* **258**, 6238–6240.

De Burlet, H. M. (1934a). Vergleichende Anatomie des statoakustischen Organs. a) Die innere Ohrsphäre. *In* "Handbuch der vergleichenden Anatomie der Wirbeltiere" (L. Bolk *et al.*, eds.), Vol. 2, Part 2, pp. 1293–1380. Urban & Schwarzenberg, Berlin.

De Burlet, H. M. (1934b). Vergleichende Anatomie des statoakustischen Organs. b) Die mittlere Ohrsphäre. *In* "Handbuch der vergleichenden Anatomie der Wirbeltiere" (L. Bolk *et al.*, eds.), Vol. 2, Part 2, pp. 1381–1432. Urban & Schwarzenberg, Berlin.

Dement'ev, G. P., and Il'ichev, V. D. (1963). Das äussere Ohr der Greifvögel. *Falke* 10, 123-125, 158-164, and 187-191.

Denker, A. (1907). "Das Gehörorgan und die Sprechwerkzeuge der Papageien." Bergmann, Wiesbaden.

Desmedt, J. E., and Delwaide, P. J. (1965). Functional properties of the efferent cochlear bundle of the pigeon revealed by stereotaxic stimulation. *Exp. Neurol.* 11, 1-26.

Dijkstra, C. (1933). Die De- und Regeneration der sensiblen Endkörperchen des Entenschnabels (Grandry- und Herbst-Körperchen) nach Durchschneidung des Nerven, nach Fortnahme der ganzen Haut und nach Transplantation des Hautstückchens. *Z. Mikrosk.-Anat. Forsch.* 34, 75-158.

Dohlman, G., Ormerod, F. C., and McLay, K. (1959). The secretory epithelium of the internal ear. *Acta Oto-Laryngol.* 50, 244-249.

Dohlman, G., Ormerod, F. C., and McLay, K. (1965). The mechanism of secretion and absorption of endolymph in the vestibular apparatus. *Acta Oto-Laryngol.* 59, 275-285.

Dorward, P. K. (1970). Response patterns of cutaneous mechanoreceptors in the domestic duck. *Comp. Biochem. Physiol.* 35, 729-735.

Dorward, P. K., and McIntyre, A. K. (1971). Responses of vibration-sensitive receptors in the interosseous region of the duck's hind limb. *J. Physiol., London* 219, 77-87.

Duijm, M. (1951). On the head posture in birds and its relation to some anatomical features. *Proc. Kon. Ned. Akad. Wetensch., Ser. C* 54, 202-271.

Engelmann, W. (1928). Untersuchungen über Schallokalisation bei Tieren. *Z. Psychol.* 105, 317-370.

Erulkar, S. D. (1955). Tactile and auditory areas in the brain of the pigeon. *J. Comp. Neurol.* 103, 421-457.

Fischer, M. H. (1926). Die Funktion des Vestibularapparates (der Bogengänge und Otolithen) bei Fischen, Amphibien, Reptilien und Vögeln. *In* "Handbuch der normalen und pathologischen Physiologie" (A. Bethe, G. V. Bergmann, G. Embden, and A. Ellinger, eds.), Vol. XI, Part 1, pp. 791-867. Springer-Verlag, Berlin and New York.

Flock, A. (1967). Ultrastructure and function in the lateral line organs. *In* "Lateral Line Detectors" (P. Cahn, ed.), pp. 163-197. Indiana Univ. Press, Bloomington.

Freye, H.-A. (1952-1953). Das Gehörorgan der Vögel. *Wiss. Z. Univ. Halle-Wittenberg* 2, 267-297.

Freye-Zumpfe, H. (1952-1953). Befunde am Mittelohr der Vögel. *Wiss. Z. Univ. Halle-Wittenberg* 2, 445-461.

Friedman, I. (1968). The chick embryo otocyst in tissue culture: A model ear. *J. Laryngol. Otol.* 82, 185-201.

Fukami, Y. (1970). Tonic and phasic muscle spindles in snake. *J. Neurophysiol.* 33, 28-35.

Fukami, Y., and Hunt, C. C. (1970). Structure of snake muscle spindles. *J. Neurophysiol.* 33, 9-27.

Gogniasvili, O. S. (1967). Histochemische Untersuchungen des Cortiorgans der Vögel unter den Bedingungen einer Toneinwirkung. *Zh. Evol. Biokhim. Fiziol.* 3, 272-275.

Goto, K., Sorimachi, M., Shibazaki, S., and Loewenstein, W. (1966). A dual nerve supply of Pacinian corpuscle. *J. Physiol. Soc. Jap.* 28, 27-37.

Granit, O. (1941). Beiträge zur Kenntnis des Gehörsinns der Vögel. *Ornis Fenn.* 18, 49-71.

Greenewalt, C. H. (1968). "Bird song: Acoustics and Physiology." Random House (Smithsonian Inst. Press), New York.

Griffin, D. R. (1953). Acoustic orientation in the oil bird, *Steatornis. Proc. Nat. Acad. Sci. U.S.* **39**, 884–893.

Griffin, D. R. (1954). Bird sonar. *Sci. Amer.* **190**, 79–83.

Griffin, D. R., and Suthers, R. A. (1970). Sensitivity of echolocation in cave swiftlets. *Biol. Bull.* **139**, 495–501.

Grinnell, A. D. (1969). Comparative physiology of hearing. *Ann. Rev. Physiol.* **31**, 545–580.

Groebbels, F. (1927a). Die Lage- und Bewegungsreflexe der Vögel. VI. Mitteilung. Regenerationsbefunde im Zentralnervensystem der Taube nach Entfernung des Labyrinths und seiner Teile. *Pfluegers Arch. Gesamte Physiol. Menschen Tiere* **218**, 89–97.

Groebbels, F. (1927b). Die Lage- und Bewegungsreflexe der Vögel. VII. Mitteilung. Wirkung zweiseitiger Labyrinthoperationen auf die Lage- und Bewegungsreflexe der Haustaube. *Pfluegers Arch. Gesamte Physiol. Menschen Tiere* **218**, 408–417.

Groebbels, F. (1928). Die Lage- und Bewegungsreflexe der Vögel. XI. Mitteilung. Die Analyse der Stützreaktion. *Pfluegers Arch. Gesamte Physiol. Menschen Tiere* **221**, 50–65.

Gross, N. B. (1970). Sensory representation within the cerebellum of the pigeon. *Brain Res.* **21**, 280–283.

Hadžiselimović, H., and Savković, L. J. (1964). Appearance of semicircular canals in birds in relation to mode of life. *Acta Anat.* **57**, 306–315.

Harman, A. L., and Phillips, R. E. (1967). Responses in the avian midbrain, thalamus and forebrain evoked by click stimuli. *Exp. Neurol.* **18**, 276–286.

Harris, G. G., Frishkopf, L. S., and Flock, A. (1970). Receptor potentials from hair cells of the lateral line. *Science* **167**, 76–79.

Held, H. (1926). Die Cochlea der Säuger und der Vögel, ihre Entwicklung und ihr Bau. *In* "Handbuch der normalen und pathologischen Physiologie" (Bethe, v. Bergmann, Emden, Ellinger, eds.), Vol. XI, Part 1, pp. 466–534. Springer-Verlag, Berlin and New York.

Hirt, A. (1934). Die vergleichende Anatomie des sympathischen Nervensystems. *In* "Handbuch der vergleichenden Anatomie der Wirbeltiere" (L. Bolk *et al.*, eds.), pp. 685–776. Urban and Schwarzenberg, Berlin.

Hoffmann, C. (1964). Vergleichende Physiologie der mechanischen Sinne. *Fortschr. Zool.* **16**, 268–332.

Huber, G. C., and DeWitt, L. (1898). A contribution on the motor nerve endings and on the nerve endings in muscle spindles. *J. Comp. Neurol.* **7**, 169–230.

Hüchtker, R., and Schwartzkopff J. (1958). Soziale Verhaltensweisen bei hörenden und gehörlosen Dompfaffen (*Pyrrhula pyrrhula* L.). *Experientia* **14**, 106–107.

Huizinga, E. (1934). Experimentelle Untersuchungen am Bogengangsapparat der Taube. *Acta Oto-Laryngol.* **20**, 76–102.

Huizinga, E. (1938). Über Rollbewegungen bei der Taube. *Pfluegers Arch. Gesamte Physiol. Menschen Tiere* **240**, 713–717.

Hunt, C. C., and Wylie, R. M. (1970). Response of snake muscle spindles to stretch and intrafused fiber contraction. *J. Neurophysiol.* **33**, 1–9.

Il'ichev, V. D. (1960). External part of auditory analyser in birds. I. General morphology and functional peculiarities. *Zool. Zh.* **39**, 1871–1877.

Il'ichev, V. D. (1961). Einige Besonderheiten des äusseren Gehöranalysators bei Pinguinen. *Nauch. Dokl. Vyssh. Shk.* **2**, 51–54.

Il'ichev, V. D. (1962). Entwicklungsmorphologie. Zusätzliche Fächer in den Ohrfedern der Vögel, ihr Bau und ihre Funktion. *Dokl. Akad. Nauk SSSR* **144**, 1185–1188.

Il'ichev, V. D. (1966a). Functional peculiarities and evolution factors in the sound-transmitting system in birds. *Zool. Zh.* **14**, 1421–1435.

Il'ichev, V. D. (1966b). Electrophysical characteristics of the acoustic representation in the large hemisperes of birds. *Zh. Vyssh. Nerv. Deyatel. im. I. P. Pavlova* **16**, 480–488.

Il'ichev, V. D. (1967). Acoustic orientation of birds as a zoological problem (some problems of the bioacoustics of birds). *Zool. Zh.* **46**, 1741–1757.

Il'ichev, V. D. (1968). Adaptions of auditory system in birds and their role on evolution. *Zh. Obshch. Biol.* **29**, 31–47.

Il'ichev, V. D., and Dubinskaja, G. R. (1966). Anatomische Unterschiede der akustischen Kerne des Gehirns bei nahen Gattungen der Vögel. *Zool. Zh.* **45**, 1580–1582.

Il'ichev, V. D., and Isvekova, L. M. (1961). Some peculiarities of the function of the external portion of auditory analizer in birds. *Zool. Zh.* **40**, 1704–1714.

Il'ichev, V. D., Gurin, S. S., and Temcin, A. N. (1970). Elektrophysiologische Charakteristik des Gehörsystems der Vögel. *Nauch. Dokl. Vyssh. Shk.* **1**, 38–48.

Ishiyama, E., Cutt, R. A., and Keels, E. W. (1969). Succinic dehydrogenase in the pigeon ampulla. *Arch. Oto-Laryngol.* **90**, 574–580.

Ishiyama, E., Cutt, R. A., and Keels, E. W. (1970). Ultrastructure of the tegmentum vasculosum and transitional zone. *Ann. Otol., Rhinol. Laryngol.* **79**, 998–1009.

Jahnke, V., Lundquist, P. G., and Wersäll, J. (1969). Some morphological aspects of sound perception in birds. *Acta Oto-Laryngol.* **67**, 583–601.

Jørgensen, J. M. (1970). On the structure of the macula lagenae in birds with some notes on the avian maculae utriculi and sacculi. *Vidensk. Medd. Dan. Naturh. Foren.* **133**, 121–147.

Kalischer, O. (1905). Das Grosshirn der Papageien in anatomischer und physiologischer Beziehung. *Abh. Preuss. Akad. Wiss. Phys. Math. Kl.* **IV**, 1–105.

Kartashev, N. N., and Il'ichev, V. D. (1964). Über das Gehörorgan der Alkenvögel. *J. Ornithol.* **105**, 113–136.

Karten, H. J. (1967). The organization of the ascending auditory pathway in the pigeon (*Columba livia*). I. Diencephalic projections of the inferior colliculus (nucleus mesencephali lateralis, pars dorsalis). *Brain Res.* **6**, 409–427.

Karten, H. J. (1968). The ascending auditory pathway in the pigeon (*Columba livia*). II. Telencephalic projections of the nucleus ovoidalis thalami. *Brain Res.* **11**, 134–153.

Karten, H. J. (1970). Telencephalic projections of the nucleus rotundus in the pigeon (*Columba livia*). *J. Comp. Neurol.* **140**, 35–52.

Kimura, M. (1924). Beiträge zur experimentellen Schallschädigung. *Z. Hals- Nasen-Ohrenheilk.* **8**, 13–45.

Kimura, T. (1931). Morphologische Untersuchungen über das membranöse Gehörorgan der Vögel. *Folia Anat. Jap.* **9**, 91–142.

Kitchell, R. L., Ström, L., and Zottermann, Y. (1959). Electrophysiological studies of thermal and taste reception in chickens and pigeons. *Acta Physiol. Scand.* **46**, 133–151.

Knecht, S. (1940). Über den Gehörsinn und die Musikalität der Vögel. *Z. Vergl. Physiol.* **27**, 169–232.

Konishi, M. (1964). Effects of deafening on song development in two species of juncos. *Condor* **66**, 85–102.

Konishi, M. (1965a). Effects of deafening on song development in American Robins and Black-headed Grosbeaks. Z. Tierpsychol. 22, 584–599.

Konishi, M. (1965b). The role of auditory feedback in the control of vocalization in the White-crowned Sparrow. Z. Tierpsychol. 22, 770–783.

Konishi, M. (1969a). Time resolution by single auditory neurones in birds. Nature (London) 222, 566–567.

Konishi, M. (1969b). Hearing, single-unit analysis, and vocalizations in song-birds. Science 166, 1178–1181.

Konishi, M. (1970). Comparative neurophysiological studies of hearing and vocalizations in song-birds. Z. Vergl. Physiol. 66, 257–272.

Krause, G. (1901). "Die Columella der Vögel, ihr Bau und dessen Einfluss auf die Feinhörigkeit." Friedländer, Berlin.

Kuijpers, W., Houben, N. M. D., and Bonting, S. L. (1970). Distribution and properties of ATPase activities in the cochlea of the chicken. Comp. Biochem. Physiol. 36, 669–676.

Leppelsack, H.-J., and Schwartzkopff, J. (1972). Eigenschaften von akustischen Neuronen im kaudalen Neostriatum von Vögeln. J. Comp. Physiol. 80, 137–140.

Loewenstein, W. R., and Molins, D. (1958). Cholinesterase in a receptor. Science 128, 1284.

Loewenstein, W. R., and Rathkamp, R. (1958). The sites for mechano-electric conversion in a Pacinian corpuscle. J. Gen. Physiol. 41, 1245–1264.

Loewenstein, W. R., and Skalak, R. (1966). Mechanical transmission in a Pacinian corpuscle. An analysis and a theory. J. Physiol. (London) 182, 346–378.

Lowenstein, O., and Wersäll, J. (1959). A functional interpretation of the electron-microscopic structure of the sensory hairs in the cristae of the elasmobranch Raja clavata in terms of directional sensitivity. Nature (London) 184, 1807–1808.

Lüdtke, H., and Schölzel, H. (1966). Die Morphogenese der Cristae im Labyrinth des Hühnchens. Zool. Jahr., Abt. Allg. Zool. Physiol. Tiere 72, 291–308.

Malinovský, L. (1967). Die Nervenendkörperchen in der Haut von Vögeln und ihre Variabilität. Z. Mikrosk.-Anat. Forsch. 77, 279–303.

Malinovský, L., and Zemánek, R. (1971). Sensory innervation of the skin and mucosa of some parts of the head in the domestic fowl. Folia Morphol. 19, 18–23.

Manni, E., Bortolami, R., and Azzena, G. B. (1965). Jaw muscle proprioception and mesencephalic trigeminal cells in birds. Exp. Neurol. 12, 320–328.

Matthews, P. B. C. (1964). Muscle spindles and their motor control. Physiol. Rev. 44, 219–288.

Mazo, J. B. (1955). Der Mikrophoneffekt der Schnecke als Mittel zur Untersuchung der funktionellen Entwicklung des peripheren Gehörorgans. Probl. Fiziol. Akust. 3, 95–101.

Medway, Lord, and Wells, D. R. (1969). Dark orientation by the Giant Swiftlet Collocalia gigas. Ibis 111, 609–611.

Messmer, E., and Messmer, I. (1956). Die Entwicklung der Lautäusserungen und einiger Verhaltensweisen der Amsel (Turdus merula merula L.) unter natürlichen Bedingungen und nach Einzelaufzucht in schalldichten Räumen. Z. Tierphysiol. 13, 341–441.

Money, K. E., Bonen, L., Beatty, J. D., Kuehn, L. A., Sokoloff, M., and Weaver, R. S. (1971). Physical properties of fluids and structures of vestibular apparatus of the pigeon. Amer. J. Physiol. 220, 140–147.

Murphey, R. K., and Phillips, R. E. (1967). Central patterning of a vocalization in fowl. Nature (London) 216, 1125–1126.

Nafstad, P. H. J., and Andersen, A. E. (1970). Ultrastructural investigation on the inner-vation of the Herbst corpuscle. *Z. Zellforsch. Mikrosk. Anat.* **103**, 109–114.

Naumov, N. P., and Il'ichev, V. D. (1964). Klanganalyse im Grosshirn der Vögel. *Natur-wissenschaften* **51**, 644.

Necker, R. (1970). Zur Entstehung der Cochleapotentiale von Vögeln: Verhalten bei O₂-Mangel, Cyanidvergiftung und Unterkühlung sowie Beobachtungen über die räumliche Verteilung. *Z. Vergl. Physiol.* **69**, 367–425.

Newman, J. D. (1970). Midbrain regions relevant to auditory communication in song-birds. *Brain Res.* **22**, 259–261.

Norberg, A. (1968). Physical factors in directional hearing in *Aegolius funereus* (Linné) (Strigiformes), with special reference to the significance of the asymmetry of the external ears. *Ark. Zool.* **20**, 181–204.

Nottebohm, F. (1967). The role of sensory feedback in the development of avian vocali-zations. *Proc. Int. Congr. Ornithol., 14th, 1966* pp. 265–280.

Novick, A. (1959). Acoustic orientation in the Cave Swiftlet. *Biol. Bull.* **117**, 497–503.

Payne, R. (1962). How the Barn Owl locates prey by hearing. *Living Bird* **1**, 151–159.

Payne, R. (1971). Acoustic location of prey by Barn Owls (*Tyto alba*). *J. Exp. Biol.* **54**, 535–573.

Payne, R., and Drury, W. H. (1958). *Tyto alba*, marksman of the darkness. *Natur. Hist., N.Y.* **67**, 316–323.

Pease, D. C., and Quilliam, T. A. (1957). Electron microscopy of the pacinian corpuscle. *J. Biophys. Biochem. Cytol.* **3**, 331–342.

Pohlman, A. G. (1921). The position and functional interpretation of the elastic liga-ments in the middle-ear region of *Gallus*. *J. Morphol.* **35**, 229–269.

Polácek, P. (1969). Die Ultrastructur des Herbstschen Körperchens im Vergleich mit dem Vater-Pacinischen Körperchen. *Sb. Ved. Pr., Lek. Fak. Karlovy Univ. Hradci Kralove* **12**, 411–416.

Polácek, P., Sklenská, A., and Malinovský, L. (1966). Contribution to the problem of joint receptors in birds. *Folia Morphol.* **14**, 33–42.

Potash, L. M. (1970). Neuroanatomical regions relevant to production and analysis of vocalization within the avian torus semicircularis. *Experientia* **26**, 1104–1105.

Poulter, T. C. (1969). Sonar of penguins and fur seals. *Proc. Calif. Acad. Sci.* **36**, 363–380.

Price, L. L., Dalton, L. W., and Smith, J. C. (1967). Frequency DL in the pigeon as de-termined by conditioned suppression. *J. Audit. Res.* **7**, 229–239.

Proske, U. (1969). An electrophysiological analysis of cutaneous mechanoreceptors in a snake. *Comp. Biochem. Physiol.* **29**, 1039–1046.

Pumphrey, R. J. (1949). The sense organs of birds. *Smithson. Inst., Annu. Rep.* pp. 305–330.

Pumphrey, R. J. (1961). Sensory organs: Hearing. *In* "Biology and Comparative Physi-ology of Birds" (A. J. Marshall, ed.), Vol 2, pp. 69–86. Academic Press, New York.

Quilliam, T. A., and Armstrong, J. (1961). Structural and denervation studies of the Herbst corpuscle. *In* "Cytology of Nervous Tissue," pp. 33–38. Taylor & Francis, London.

Rautenberg, W. (1969). Die Bedeutung der zentralnervösen Thermosensitivität für die Temperaturregulation der Taube. *Z. Vergl. Physiol.* **62**, 235–266.

Sagitov, A. K. (1964). The vestibular analyzer and the degree of mobility of gallinaceous birds. *Tr. Samarkand. Gos. Univ.* **137**, 5–38.

Sauer, F. (1954). Die Entwicklung der Lautäusserungen vom Ei ab schalldicht gehaltener Dorngrasmücken (*Sylvia c. communis* Latham) im Vergleich mit später isolierten und mit wildlebenden Artgenossen. *Z. Tierpsychol.* 11, 10–93.

Saxod, R. (1967). Histogénèse des corpuscules sensoriels cutanès chez le poulet et le cánard. *Arch. Anat. Microsc. Morphol. Exp.* 56, 153–166.

Saxod, R. (1970a). Etude au microscope électronique de l'histogénèse du corpuscule sensoriel cutané de Grandry chez le canard. *J. Ultrastruct. Res.* 32, 477–496.

Saxod, R. (1970b). Etude au microscope électronique de l'histogénèse du corpuscule sensoriel cutané de Herbst chez le canard. *J. Ultrastruct. Res.* 33, 463–482.

Saxod, R., and Sengel, P. (1968). Sur les conditions de la différenciation des corpuscules sensoriels cutanées chez le poulet et le canard. *C. R. Acad. Sci., Ser. D* 267, 1149–1152.

Schartau, O. (1938). Die periphere Innervation der Vogelhaut. *Zoologica (Stuttgart)* 95, 1–17.

Schildmacher, H. (1931). Untersuchungen über die Funktion der Herbstschen Körperchen. *J. Ornithol.* 74, 374–415.

Schmidt, R. S., and Fernandez, C. (1962). Labyrinthine DC potentials in representative vertebrates. *J. Cell. Comp. Physiol.* 59, 311–322.

Schmidt, S. (1964). Blood supply of pigeon inner ear. *J. Comp. Neurol.* 123, 187–203.

Schölzel, H., and Lüdtke, H. (1967). Die Entwicklung der Cupula im Labyrinth des Hühnchens. *Zool. Jahrb. Abt. Allg. Physiol. Tiere* 74, 164–177.

Schwartzkopff, J. (1949). Über Sitz und Leistung von Gehör und Vibrationssinn bei Vögeln. *Z. Vergl. Physiol.* 31, 527–608.

Schwartzkopff, J. (1952). Untersuchungen über die Arbeitsweise des Mittelohres und das Richtungshören der Singvögel unter Verwendung von Cochlea-Potentialen. *Z. Vergl. Physiol.* 34, 46–68.

Schwartzkopff, J. (1955). Schallsinnesorgane, ihre Funktion und biologische Bedeutung bei Vögeln. *Acta Int. Congr. Ornithol., 11th, 1954* pp. 189–208.

Schwartzkopff, J. (1957a). Die Grössenverhältnisse von Trommelfell, Columella-Fussplatte und Schnecke bei Vögeln verschiedenen Gewichts. *Z. Morphol. Oekol. Tiere* 45, 365–378.

Schwartzkopff, J. (1957b). Untersuchung der akustischen Kerne in der Medulla von Wellensittichen mittels Mikroelektroden. *Verh. Deut. Zool. Ges. Graz* pp. 374–379.

Schwartzkopff, J. (1958). Über den Einfluss der Bewegungsrichtung der Basilarmembran auf die Ausbildung der Cochlea-Potentiale von *Strix varia* (Barton) und *Melopsittacus undulatus* (Shaw). *Z. Vergl. Physiol.* 41, 35–48.

Schwartzkopff, J. (1962a). Die akustische Lokalisation bei Tieren. *Ergeb. Biol.* 25, 136–176.

Schwartzkopff, J. (1962b). Vergleichende Physiologie des Gehörs und der Lautäusserungen. *Fortschr. Zool.* 15, 213–336.

Schwartzkopff, J. (1962c). Zur Frage des Richtungshörens von Eulen (Striges). *Z. Vergl. Physiol.* 45, 570–580.

Schwartzkopff, J. (1963). Morphological and physiological properties of the auditory system in birds. *Proc. Int. Ornithol. Congr., 13th, 1962* pp. 1059–1068.

Schwartzkopff, J. (1967). Hearing. *Annu. Rev. Physiol.* 29, 485–512.

Schwartzkopff, J. (1968). Structure and function of the ear and of the auditory brain area in birds. *In* "Hearing Mechanisms in Vertebrates" (A. V. S. De Reuck and J. Knight, eds.), pp. 41–59. Churchill, London.

Schwartzkopff, J., and Winter, P. (1960). Zur Anatomie der Vogel-Cochlea unter natürlichen Bedingungen. *Biol. Zentralbl.* 79, 607–625.

Skoglund, C. R. (1960). Properties of Pacinian corpuscles of ulnar and tibial location in cat and fowl. *Acta Physiol. Scand.* **50**, 385–386.

Smith, C. A. (1968). Morphological features of axo-dendritic relationships between efferent and cochlear nerves in the cochlea of mammal and pigeon. *Struct. Funct. Inhibitory Neuronal Mech. Proc. Int. Meet. Neurobiol., 4th, 1968* pp. 141–146.

Smith, G. (1904). The middle ear and columella in birds. *Quart. J. Microsc. Sci.* **48**, 11–22.

Snapp, C. F. (1924). Blood supply in the labyrinth of birds. *Anat. Rec.* **27**, 29–45.

Steinhausen, W. (1933). Über die Beobachtung der Cupula in den Bogengangsampullen des Labyrinthes des lebenden Hechtes. *Pfluegers Arch. Gesamte Physiol. Menschen Tiere* **232**, 500–512.

Stellbogen, E. (1930). Über das äussere und mittlere Ohr des Waldkauzes. *Z. Morphol. Oekol. Tiere* **19**, 686–731.

Stingelin, W. (1961). Grössenunterschiede des sensiblen Trigeminuskerns bei verschiedenen Vögeln. *Rev. Suisse Zool.* **68**, 247–251.

Stingelin, W. (1965). "Qualitative und quantitative Untersuchungen an Kerngebieten der Medulla oblongata bei Vögeln." Karger, Basel.

Stopp, P. E., and Whitfield, I. C. (1961). Unit responses from brain stem nuclei in the pigeon. *J. Physiol. (London)* **158**, 165–177.

Stopp, P. E., and Whitfield, I. C. (1964). Summating potentials in the avian cochlea. *J. Physiol. (London)* **175**, 45–46.

Stresemann, E. (1934). Sauropsida, Aves. *In* "Handbuch der Zoologie" (W. G. Kükenthal and Krumbach, eds.), Vol. VII, Part 2. de Gruyter, Berlin.

Takasaka, T., and Smith, C. A. (1971). The structure and innervation of the pigeon's basilar papilla. *J. Ultrastruct. Res.* **35**, 20–65.

Ten Cate, H. (1960). Locomotor movements in the spinal pigeon. *J. Exp. Biol.* **37**, 609–613.

Thielcke-Poltz, H., and Thielcke, G. (1960). Akustisches Lernen verschieden alter schallisolierter Amseln (*Turdus merula* L.) und die Entwicklung erlernter Motive ohne und mit künstlichem Einfluss von Testosteron. *Z. Tiersychol.* **17**, 211–244.

Thorpe, W. H. (1961). "Bird Song." Cambridge Univ. Press, London and New York.

Thurm, U. (1970). Mechanosensitivity of motile cilia. *Neurosci. Res. Program, Bull.* **8**, 496–498.

Trainer, J. E. (1946). The auditory acuity of certain birds. Thesis, Cornell University, Ithaca, New York.

Vallancien, B. (1963). Comparative anatomy and physiology of the auditory organ in vertebrates. *In* "Acoustic Behavior of Animals" (R.-G. Busnel, ed.), pp. 522–556. Elsevier, Amsterdam.

Vinnikov, J. A. (1965). Principles of structural, chemical, and functional organization of sensory receptors. *Cold Spring Harbor Symp. Quant. Biol.* **30**, 293–299.

Vinnikov, J. A., Govardovskii, V. A., and Osipova, I. V. (1965a). Substructural organization of the gravitation organ of the pigeon utriculus. *Biofizika* **10**, 641–644.

Vinnikov, J. A., Titova, L. K., and Aronova, M. Z. (1965b). Vergleichende histochemische Untersuchung der Cholinesterase in den rezeptorischen Strukturen des Labyrinths und in den Organen der Seitenlinien bei den Wirbeltieren. *Acta Histochem.* **22**, 120–154.

Vinnikov, J. A., Osipova, I. V., Titova, L. K., and Govardovskii, I. (1965c). Electron microscopy of Corti's organ of birds. *Zh. Obshch. Biol.* **26**, 138–150.

von Békésy, G. (1944). Über die mechanische Frequenzanalyse in der Schnecke verschiedener Tiere. *Akust. Z.* **9**, 3–11.

von Buddenbrock, W. (1952). "Vergleichende Physiologie," Vol. 1. Birkhaeuser, Basel.

von Holst, E. (1950). Die Arbeitsweise des Statolithenapparates bei Fischen. *Z. Vergl. Physiol.* **32**, 60–120.

Wada, Y. (1924). Beiträge zur vergleichenden Physiologie des Gehörorgans. *Pflügers Arch. Gesamte Physiol. Menschen Tiere* **202**, 46–69.

Wassiljew, M., Ph. (1933). Über das Tonunterscheidungsvermögen der Vögel für hohe Töne. *Z. Vergl. Physiol.* **19**, 424–438.

Werner, C. F. (1938). Funktionelle und vergleichende Anatomie des Otolithenapparates bei Vögeln. *Z. Anat. Entwicklungsgesch.* **108**, 775–791.

Werner, C. F. (1939). Die Otolithen im Labyrinth der Vögel, besonders beim Star und der Taube. *J. Ornithol.* **87**, 10–23.

Werner, C. F. (1958). Der Canaliculus (Aquaeductus) cochleae und seine Beziehungen zu den Kanälen des IX. und X. Hirnnerven bei den Vögeln. *Zool. Jahr. Abt. Anat. Ontog. Tiere* **77**, 1–8.

Werner, C. F. (1960). "Das Gehörorgan der Wirbeltiere und des Menschen." Thieme, Stuttgart.

Werner, C. F. (1962). Allometrische Grössenunterschiede und die Wechselbeziehung der Organe (Untersuchungen am Kopf der Vögel). *Acta Anat.* **50**, 135–157.

Werner, C. F. (1963). Schädel-, Gehirn- und Labyrinthtypen bei den Vögeln. *Morphol. Jahrb.* **104**, 54–87.

Wersäll, J., Flock, A., and Lunquist, P. G. (1965). Structural basis for directional sensitivity in cochlear and vestibular sensory receptors. *Cold Spring Harbor Symp. Quant. Biol.* **30**, 115–132.

Wever, E. G., Herman, P. N., Simmons, J. A., and Hertzler, D. R. (1969). Hearing in the Blackfooted Penguin, *Spheniscus demersus*, as represented by the cochlear potentials. *Proc. Nat. Acad. Sci. U.S.* **63**, 676–680.

Winget, C. M., and Smith, A. S. (1962). Quantitative measurement of labyrinthine function in the fowl by nystagmography. *J. Appl. Physiol.* **17**, 712–718.

Winkelmann, R. K., and Myers, Th. T. (1961). The histochemistry and morphology of the cutaneous sensory end organs of the chicken. *J. Comp. Neurol.* **117**, 27–35.

Winter, P. (1963). Vergleichende qualitative und quantitative Untersuchungen an der Hörbahn von Vögeln. *Z. Morphol. Oekol. Tiere* **52**, 365–400.

Yashiki, K. (1968). Fine structure of the sensory epithelium of the lagena of birds. *Yonago Acta Med.* **12**, 47–60.

Zeigler, H. P., and Witkovsky, P. (1968). The main sensory trigeminal nucleus in the pigeon: A single-unit analysis. *J. Comp. Neurol.* **134**, 255–264.

Chapter 8

BEHAVIOR

Robert A. Hinde

479

I. Introduction

Birds evolved from diapsid stock, the earliest known specimens having been found in Jurassic rocks. Like many other aspects of their anatomy, the brain structure provides evidence of their reptilian affinities; the large striatal regions and the pallial areas are similar in many respects to those of modern lizards and very different from those of mammals. Partly because plasticity of behavior is often held to be characteristic of the mammalian line and partly because avian brain structure suggests an accentuation of trends to be seen in lizards, the behavior of birds is often contrasted with that of mammals as "rigid" or "stereotyped." Such generalizations are overstatements. Although every species of bird has a large repertoire of more or less stereotyped responses, so does every mammal. The feeding behavior of the horse is no more plastic than that of the Great Tit (*Parus major*), and species-characteristic responses play their part in the sexual behavior of the rat just as they do in that of the Chaffinch (*Fringilla coelebs*). In fact, avian capabilities for learning are comparable in many respects with those of most mammals.

This marked learning ability, coupled with the large repertoire of striking and thus easily recognizable movement patterns possessed by each species, render birds a particularly interesting group for the study of behavior. Progress in this work has been facilitated by the similarities between their sensory capacities and those of man. Further, birds are numerous and widespread, and many species make excellent laboratory animals. They have undergone a marked adaptive radiation, so that many types of life history and many variations on behavioral themes are found within the group, and their aesthetic appeal has ensured that the necessary foundation of a broad knowledge of their natural history has been securely laid.

In view of these facts, it is not surprising that the literature on bird behavior already fills many thousands of scientific papers and innumerable volumes. A number of excellent reviews of parts of this field are already available. The earlier work was discussed by Maier and Schneirla (1935) and by Warden *et al.* (1936), and Armstrong (1947) has summarized much of the literature on reproductive behavior. More recently, Thorpe (1963) has reviewed the literature on avian learning, and studies of birds play a prominent part in recent texts on animal behavior (Marler and Hamilton, 1966; Hinde, 1970).

It is thus unnecessary, as well as impracticable, to attempt a broad survey of the study of bird behavior here. Rather, attention has been concentrated on some of the more recent investigations and on some current issues of interest and controversy likely to have general significance. This has inevitably meant that many aspects of natural history have been omitted, and it has not been possible to do justice to the diversity of behavior to be found within the group. Furthermore, no detailed exposition of theoretical ideas is presented, though it will be evident that this chapter draws much of its material from work stimulated by Lorenz (1935, 1937, 1950) and Tinbergen (1951).

II. Analysis of Bird Behavior

A. PROBLEMS AND GOALS

It is convenient to divide the problems that arise in the study of behavior into three groups, which, though theoretically separable, are in practice interrelated. First, the behavior selected for study must be related to events or conditions which immediately precede it: this is one form of causal analysis. Second, we must understand the ontogeny of the behavior: this also is a form of causal analysis, but involves delving further into the past. The third group of problems concerns questions extending both forward in time and even further backward into the past. These include such questions as what are the consequences of each pattern of behavior and which of them provides material for the action of natural selection, and how has behavior evolved. Each of these also involves causation, but in the first case the behavior itself is the cause and the object of study is the effect, while in the second we are looking at causes of the behavior, but in the very remote past.

The species of birds are so numerous, and their behavior so diverse, that to gain even a moderate understanding of the behavior of a handful of species is an enormous task. The most we can hope for is a series of principles or generalizations, each with a qualification defining its scope. The need for such qualifications is one reason why descriptive studies of the behavior of a wide range of species are essential to the more analytical studies with which this chapter is mostly concerned.

B. DESCRIPTION AND CLASSIFICATION

Description is in any case an essential preliminary to analysis. Two methods are available. One refers ultimately to the strength, degree, and patterning of muscular contraction, but is usually limited to patterns of limb or body movement. The other method involves reference to the consequences of the behavior. Terms such as "approaching" (without specification of the method of locomotion) or "carrying nest material" are of this type.

Once behavior has been described, we can recognize repeated instances of the same "type" of behavior. These types must then be classified into groups, the criteria used depending on the problem in which we are interested. Among the more useful groups are those involving types of behavior with similar immediate causation (e.g., activities influenced by the stimulus situation "rival male" can be described as "agonistic behavior"), those with a similar function (e.g., "reproductive behavior"), and those with a similar historical origin (e.g., homologous but different display movements in closely related species). It is of course essential to keep the categories distinct; "sex behavior" defined causally as all activities positively affected by male sex hormone will not be coextensive with "sex behavior" defined functionally as activities contributing to successful mating.

C. STEREOTYPED PATTERNS

Comparative study reveals that many motor patterns used by birds are characteristic of the species. The precise way in which the wings are moved during flight or the tail is flicked before takeoff, the manner in which food is caught, the postures used by the male in courting the female, and the call notes and song are usually similar in all members of a given species but may differ between even closely related ones (Lorenz, 1935, 1950).

A concept useful in the analysis of these species-characteristic movements has been the "fixed action pattern." This refers to movements that may themselves be quite complicated, involving the use of many different effectors and muscle groups, and consisting of a temporal pattern of muscular contractions, but that cannot be analyzed into successive responses depending on qualitatively different external stimuli: therefore, they are not chain reactions. Since fixed action patterns are conspicuous and easily recognizable, they form a convenient starting point for analysis (Lorenz, 1935, 1950; Tinbergen, 1942, 1951). They can be identified in nearly every aspect of a bird's life — the begging of the young, courtship and threat postures, sleeping, hunting, and so on. While some are characteristic of the species, others are to be found throughout a genus, family, or higher systematic category; they are thus useful to the systematist, and indeed it was for this reason that attention was first focused on them (e.g., Lorenz, 1950). An example of one fixed action pattern that differs slightly between closely related species is shown in Fig. 1.

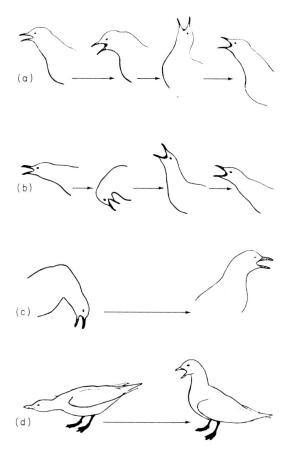

FIG. 1. The oblique long call of four species of gulls. Each sequence is to be read from left to right. Note the slight interspecies differences. (From Tinbergen, 1959.)

Since fixed action patterns depend on external factors only for their elicitation, they may vary in completeness but often vary little in the relationships among their parts. In many cases they show remarkable constancy. For example, the "head-throw" display of the Common Goldeneye (*Bucephala clangula*) has a mean duration of 1.29 seconds and a standard deviation of only ±0.08 seconds (Dane *et al.*, 1959). In other cases, however, they are more variable.

Many movement patterns once elicited are independent of further external stimuli except insofar as they are oriented with respect to the environment. In such cases the basic form of the movement remains recognizable, but there is superimposed an "orientation component." Where the orientation depends on the same stimuli and muscle groupings as the fixed action pattern itself, the two are inseparable: this is the case with many threat and courtship postures, such as the "head-up" threat of the Great Tit (Fig. 2). Sometimes,

(a)   (b)

FIG. 2. (a) Head-up threat posture used by Great Tits in disputes over territories. (b) Head-forward threat posture used by Blue Tits in disputes over territories and over food. The Great Tit displays its black ventral stripe, the Blue Tit its white cheeks. (From a drawing by Yvette Spencer-Booth in Hinde, 1970.)

however, the orientation depends on different stimuli. For instance, many ground-nesting birds retrieve an egg that has rolled out of the nest by placing the beak beyond it and drawing it back toward the breast, simultaneously making lateral balancing movements to prevent the egg rolling out to one side. These balancing movements cease if the egg is replaced by a cylinder or removed altogether, but the sagittal movement usually continues. Here, then, the latter can be regarded as the fixed action pattern and the former as an orientation or taxis component which is continuously affected by environmental stimuli (Lorenz and Tinbergen, 1938).

D. APPETITIVE AND CONSUMMATORY BEHAVIOR

It is sometimes convenient to divide the behavior of a species into "appetitive" and "consummatory" components (Sherrington, 1906; Craig, 1918). According to this, the more variable earlier phases in a behavior sequence, such as patrolling the territory or "looking for" food, are classed as appetitive, while the final stereotyped act which brings the sequence to an end, such as striking the rival or swallowing, is termed consummatory. This distinction is useful in the early stages of analysis, though it rests on several heterogeneous and uncorrelated characters (e.g., place in behavior sequence, degree of rigidity, effect of performance on motivation of whole sequence) and does not imply distinct types of underlying mechanisms. In practice, appetitive and consummatory behavior differ only in degree, and many behavior patterns have characteristics of both.

Appetitive behavior is usually labeled in terms of the behavior to which it leads — food seeking, fighting, sleeping, etc. Three characteristics may assist in its identification: (1) the motor pattern(s), (2) the orientation component, and (3) the stimuli to which the animal is particularly responsive while showing the behavior. Thus, a feeding Great Tit may hop (motor pattern) under a beech tree (orientation component) "looking for" (to use a convenient shorthand) beech mast. A nesting Great Tit may hop under the trees looking for moss. The responsiveness to stimuli is the most difficult characteristic to investigate, and yet often the most interesting: it is the changes in responsiveness that mark off the different phases of appetitive behavior. Since responsiveness is a relative matter (thus, passerines are nearly always responsive to a flying hawk, whatever else they are doing, or a bird may snatch up a particle of food while looking for nest material), the labeling of appetitive behavior is often neither easy nor precise.

For behavior that can more conveniently be characterized as consummatory, it is also usually possible to describe the motor pattern and the orientation, though some consummatory behavior lacks an orientation component. Often it is also possible to identify "consummatory stimuli" (e.g., "food in stomach") that are responsible for the subsequent fall in motivation and may be compared with the "stimuli to which the bird is particularly responsive" when showing appetitive behavior.

E. SIGN STIMULI AND RELEASERS

Much avian behavior can be evoked by relatively simple stimulus situations. For instance, Lack (1939), investigating the behavior of territory-holding European Robins (*Erithacus rubecula*), found that stuffed adult robins placed in the territory were nearly always threatened, while stuffed juvenile robins, which resemble the adult but lack the red breast, were ignored. Further, a bunch of red breast feathers was threatened more readily than a complete stuffed juvenile; the red breast is thus more effective in eliciting threat than all other characteristics together (Fig. 3).

FIG. 3. Threat display of European Robin (*Erithacus rubecula*). (Redrawn from Daanje, 1951.)

Similarly Tinbergen and Perdeck (1950) investigated the stimuli eliciting the begging response of young Herring Gulls (*Larus argentatus*) with cardboard models. The adult of this species has a red spot near the tip of its beak, and the experiments showed that "the object that releases the pecking . . . is characterized for the chick by (1) movement, (2) shape (elongate, not too short, thin), (3) lowness, (4) downward pointing position, (5) nearness, and (6) the bill patch, which must be (a) red and (b) differing, by contrast, from the ground colour of the bill" (Tinbergen, 1951). Head shape and color had no influence on the response. Thus, the chick selects from among the stimuli it receives (see also Hailman, 1967).

Similar principles apply in other sensory modalities. For example, the song of the White-throated Sparrow (*Zonotrichia albicollis*) consists of a series of clear notes, and Falls (1963, 1969) was able to make artificial songs differing in defined ways from normal ones. These were played to wild birds, and their effectiveness in eliciting responses was assessed. It was found that the song had to contain unvarying pure tones, without harmonics. The notes, and the intervals between them, had to be within certain limits of length. The pitch changes between successive notes were of less importance. Other characters, such as the broken nature of some notes, and the slurs that occur in the natural song, appeared to be of no significance. The frequency of the notes, and the length and loudness of the song, could also be varied considerably. Thus, again only certain features were effective in producing a response, and this implies selective responsiveness to those features.

In practice, selection takes place in stages. In the first instance, the sense organs themselves are more responsive to changes in some types of physical variable than in others, and the properties of the perceptual mechanisms ensure that some stimulus configurations are responded to more readily than others. For instance, circles are more conspicuous than other closed shapes to most vertebrates. In these cases, stimuli or configurations of stimuli that are especially conspicuous have this property for all types of behavior.

In other cases of selective responsiveness, however, the stimulus character in question is effective in eliciting one type of behavior only. Such responsiveness may be characteristic of the species and independent of any previous experience of the particular stimulus situation, but not explicable in terms of properties of the sense organs or perception. For example, a male Chaffinch in breeding condition may make a hovering copulatory approach to a female stuffed in the soliciting posture, but take little notice of one stuffed in a perching position.

These species-characteristic stimulus–response connections are, however, much modified by individual experience. Indeed, it is a necessary working assumption that every time a bird responds to a stimulus, learning occurs that modifies responsiveness on future occasions. When the stimulus–response relations do not depend on previous experience of the stimulus object, the essential stimulus characters are the same for all members of the species. Once learning has occurred, however, individuals may come to respond preferentially to different aspects of the naturally occurring stimulus situation.

Those characteristics of the stimulus situation that are especially important in eliciting responses are termed "sign stimuli" (Russell, 1943). Many responses depend initially on relatively few sign stimuli, but these are often themselves complex and have configurational as

well as purely quantitative characteristics. The stimuli that elicit
instinctive responses in nature are not necessarily the optimal ones.
Thus, in his experiments with young Herring Gulls, Tinbergen found
that a long, thin, red rod with three white rings near the end, which
thereby displayed in an exaggerated form just those characters which
are important in eliciting begging (see above), was more effective
than a real Herring Gull's head.

Few sign stimuli can be specified along a single physical scale,
since relational properties are often important. For example, the
gaping of 8-day-old European Blackbirds (*Turdus merula*) is both
elicited and directed by visual stimuli, the gape being directed toward
the head of the parent. Some models used by Tinbergen and Kuenen
(1939) to study this response are shown in Fig. 4. Gaping was directed

FIG. 4. Cardboard models used to elicit gaping responses from nestling thrushes
(From Tinbergen, 1951.)

toward the smaller of the two projections of the smaller model, but to
the larger one on the larger model. It thus seems that the stimulus for
gaping is characterized not by its absolute size, but by its size in relation to the whole.

Any one response may depend on a number of sign stimuli, which
may be in more than one sensory modality. Absence of any of them
results in a decrease in the intensity of the response, but not in a
change in its nature. The reduction in the intensity of the response
depends on the extent of the deficiency in the stimulus situation, but
not on which sign stimuli are absent (heterogeneous summation,
Seitz, 1940/1941).

Stimuli may affect behavior in a variety of ways. First, they may
elicit a response or lower its threshold. Sometimes the response is

immediate, as when the alarm call of a passerine elicits flying to cover. In other cases the stimulus acts slowly or continuously to produce a state of responsiveness, as when stimuli from the mate produce a change in endocrine state, or when stimuli from the territory produce readiness to attack an intruder, or the "discriminative" stimulus in a learning situation permits the bird to differentiate between situations in which a response will and will not be rewarded. There are further resemblances between the discriminative and the "incentive–motivational" effects of stimuli, but these will not be discussed here (see, e.g., Bacon and Bindra, 1967).

Second, stimuli may direct a response. When an alarm call elicits "flying to cover" from a passerine bird, the flight is directed by stimuli from a nearby bush (e.g., Marler, 1956b). Finally, stimuli may inhibit a response or raise its threshold. For example, the presence of a female often reduces the song of male passerines, and stimuli from the nest it has built reduce the nest-building behavior of a female canary (Hinde, 1958).

F. THE PROBLEM OF MOTIVATION

A particular stimulus does not always elicit the same response. If the external situation is constant, altered responsiveness must be ascribed to changes in the internal state. Conventionally, temporary and reversible changes in the internal state are labeled as motivational, though it is usual to exclude changes that last for less than a second or so and those that occur in the sense organs (e.g., adaptation) or effectors (e.g., muscular fatigue).

While for some purposes it is convenient to postulate changes in an "internal drive" or "tendency" to account for changes in responsiveness to a constant stimulus (see, e.g., Sections III, H and J below), full understanding involves analysis. This requires first the isolation of the various factors that influence responsiveness. Some of the more important of these are the following.

a. Stimuli. The various ways in which stimuli may affect responsiveness were referred to above: the effect may be either to enhance or to reduce responsiveness. While most cases that have been investigated involve stimuli external to the bird, internal stimuli also may be important. However, with certain exceptions (see factor b below), these have been little studied in birds.

b. Hormones. Although much of the work on the effects of hormones on bird behavior refers to the domestic chicken, a considerable volume of data on other species is now accumulating. Recent summaries have been given by Lehrman (1961) and van Tienhoven (1961). Hormones may affect behavior in a variety of ways. First, they may affect the central nervous system directly. This has been demonstrated by chronically implanting very small amounts of crystalline hormone in specific parts of the brain. For example, localized androgenic implants have been shown to induce male copulatory behavior in capons (Barfield, 1964) and courtship in castrated male Ring Doves (Hutchison, 1967), while progesterone induces incubation behavior in doves (Komisaruk, 1967).

Second, hormones may affect peripheral structures and thus indirectly the input to the central nervous system. For example, prolactin causes feeding of young by Ring Doves in part because it causes engorgement of the crop. If the crop region is anesthetized, injections of prolactin do not reduce parental feeding (Lehrman, 1955). Experiments by Klinghammer and Hess (1964) and by Hansen (1966), however, indicate that this is not a necessary factor in inducing feeding by regurgitation. Again, brood-patch development in canaries can be induced by exogenous estrogen and progesterone: it involves increased sensitivity to tactile stimulation, and this probably plays an important role in the control of nest-building (Hinde *et al.*, 1963; Hinde, 1965).

Third, hormones may produce changes in structures serving as social stimuli, such as the cock's comb. In addition to these, hormones may produce nonspecific central effects, mediated through a general activating system; they may produce indirect central effects via the pituitary and gonadotropin release; and by acting early in development, they may influence behavioral sex. Such effects have been little studied in birds.

Whatever their mode of action, the effect of hormones on behavior may be affected by experience (Lehrman and Wortis, 1960, 1967) and by other factors in the situation, such as the presence of a mate (Bruder and Lehrman, 1967). Such observations must be considered in relation to Beach's (1951) finding that the effects of forebrain ablations on the sex behavior of pigeons could be offset by androgen administration.

c. Diurnal Changes. Most activities show some sort of diurnal rhythm (see Volume V).

d. Other Internal Changes. In general, the threshold of stimulation necessary to elicit a response, or the strength of the response elicited by a constant stimulus, varies with the time since the response was last given. This is commonplace in the case of feeding and drinking. As another example, the copulatory behavior of a male Chaffinch to a stuffed female in the soliciting posture is reduced after ejaculation and recovers in about 1 hour. Again, Fig. 5a shows the

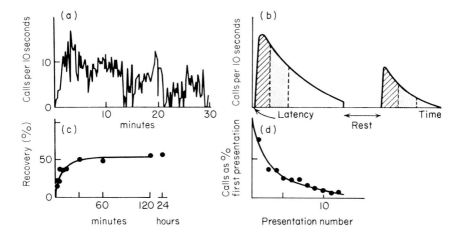

FIG. 5. Mobbing response given by Chaffinches to predators. (a) The course of the response to a stuffed owl in one individual. (b) Hypothetical habituation curves for two successive stimulus presentations, to show measures taken. Diagonally striped area indicates calls in first 6 minutes. (c) Recovery (number of calls in first 6 minutes of second presentation as percentage of number of calls in first 6 minutes of first presentation). (d) Response to live Little Owl (*Athene noctua*) presented for 20 minutes per day. (Redrawn from Hinde, 1954, in Horn and Hinde, 1970.)

number of calls given by a Chaffinch in successive 10-second periods while mobbing a stuffed owl. There is an initial warm-up phase, followed by a slow waning, but superimposed on these are shorter-term fluctuations in response strength. It thus seems that each response, or each moment of responding, must be associated with complex internal changes which affect subsequent responding. For the first few minutes the resultant of these is to augment response frequency, but subsequently the response wanes. If the owl is presented twice in succession, as shown in the idealized Fig. 5b, the recovery during a rest interval can be measured. It occurs rapidly during the first 30 minutes or so, but thereafter ceases, leaving a more or less permanent decrement (Fig. 5c). Successive daily presentations of the stimulus show that the permanent decrements are cumulative (Fig. 5d). As yet the internal changes responsible for these fluctuations in response strength cannot be studied directly, but they can be classified according to their recovery periods, whether they increase or decrease response strength, and the extent to which they are specific to the particular stimulus used (Hinde, 1960).

In the cases just mentioned, the occurrence of the behavior under study can be related to changes outside the nervous system. Sometimes, however, behavior appears to occur "spontaneously." Indeed, the occurrence of many hormonally influenced activities, such as singing, cannot be immediately related to hormonal fluctuations; the endocrine state merely regulates their probability. Of course spontaneity in the behavior of the whole animal does not necessarily imply spontaneity of the central nervous system, and spontaneity of the nervous system does not necessarily imply spontaneity of the cells within it. At the behavioral level, spontaneity of fixed action patterns is easily noticed and often dramatic. For example, Lorenz (1937) described the behavior of a well-fed European Starling (*Sturnus vulgaris*) which had had no opportunity to catch flies for some time, and suddenly went through all the movements of searching for a fly, catching and killing it, although no fly was discernible to the observer. Similarly, canaries deprived of nest material will go through the movements of weaving material into a nonexistent nest (Hinde, 1958). Spontaneity in appetitive behavior is less easily remarked, but nevertheless commonplace.

When the various factors that affect the occurrence of a particular pattern of behavior have been identified, the role of central nervous structures can be investigated. Although knowledge concerning the functioning of the avian brain is much less advanced than knowledge concerning the mammalian brain, the techniques of both ablation and electrical stimulation within the central nervous system have been used in birds (e.g., von Holst and von St. Paul, 1963; Åkerman *et al.*, 1960; Åkerman, 1966; Harwood and Vowles, 1966; Brown, 1969). Electrical stimulation can produce integrated sequences of behavior and long-lasting changes in responsiveness or mood. As in mammals, electrical stimulation of some parts of the avian brain produces positive reinforcing effects (Goodman and Brown, 1966).

G. AMBIVALENT BEHAVIOR

For most of the time, factors for more than one type of behavior are present. For example, there may be conflicting tendencies to feed and fly with the flock, and during the breeding season each member of a pair has conflicting tendencies to attack, flee from, and behave sexually toward its mate. It is often impossible to do more than one thing at a time, and such ambivalence may have various outcomes.

a. Inhibition of All but One Response. Sometimes only one of the alternative patterns is shown, the others being suppressed completely. Thus the appearance of a flying predator causes passerines to flee to cover and an immediate suppression of all other behavior.

b. Alternation. The bird may alternate between the two alternatives. A fighting bird, with conflicting tendencies to attack and flee from its rival, may edge a little toward him and then away again, and so on repeatedly. A bird that is uneasy while it is feeding will take a few beakfuls and then look around for predators and then feed again.

c. Intention Movements. Sometimes the bird will make a series of incomplete movements expressing one or other or both of the conflicting tendencies (Fig. 6) (see Daanje, 1951).

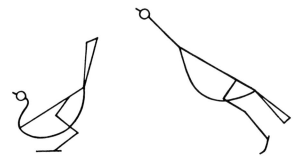

FIG. 6. Two phases of a takeoff leap. Redrawn after Daanje (1951).

d. Compromise Behavior. Sometimes a response is given that resembles responses associated with both the incompatible tendencies. Thus, the tail flicking of a hungry bird, which does not approach a food dish because of a frightening object there, expresses both hunger (flying to dish) and fear (flying away) (Andrew, 1956).

e. Ambivalent Posturing. Sometimes the bird adopts a posture that simultaneously expresses both tendencies. Many threat (e.g., Fig. 7) and courtship postures are of this type, the various components (e.g., wing raising and tail spreading) of the posture expressing one or other of the conflicting tendencies. This is discussed further below.

f. Redirection Activities. Here the motor pattern of one of the conflicting tendencies is shown, but is oriented to another object.

FIG. 7. Two forms of the upright threat posture of the Herring Gull (*Larus argentatus*). Attack involves pecking downward with the beak and striking with the wings. Intention movements for both of these can be seen in the aggressive form of the posture (left), but not in the anxiety form (right). (After Tinbergen, 1959.)

Thus, the male Black-headed Gull (*Larus ridibundus*) with a female on his territory often redirects onto other birds the aggressiveness that she elicits (Moynihan, 1955).

g. *Autonomic Activities.* A variety of autonomic responses occur in conflict situations, including defecation and thermoregulatory responses (Andrew, 1956; Morris, 1956).

h. *Displacement Activities.* Sometimes a bird under the influence of two or more conflicting tendencies shows an apparently irrelevant type of behavior. For instance, fighting Great Tits may suddenly break off the attack and peck vigorously at buds, or courting ducks may preen their wing feathers (Lorenz, 1941). Such cases are labeled "displacement activities" (Tinbergen, 1940, 1952; Kortlandt, 1940). This is undoubtedly a heterogeneous category, and little is yet known about their causation. In some cases, however, external stimuli for the irrelevant activity were already present. This is the case with the Great Tit example quoted above, which thus also has characters of a redirection activity. Similarly, Räber (1948) showed that when a turkey had conflicting tendencies to attack and flee, the appearance of displacement feeding or drinking depended on the availability of food or water. Internal causal factors for the irrelevant behavior also may already be present. For instance, the initial posture of the thwarted animal may produce proprioceptive stimuli already associated with the irrelevant activity (Tinbergen, 1952).

In addition to such factors, several suggestions concerning the mechanisms of displacement activities have been advanced:

1. As we have seen, autonomic activity resulting from the conflict situation may result in apparently irrelevant behavior such as warm-

ing or cooling responses. These may provide stimuli for somatic activities: for instance, feather movements may provide skin stimuli that elicit scratching or preening (Andrew, 1956; Morris, 1956).

2. When mutual incompatibility prevents the appearance of those patterns of behavior that would otherwise have had the highest priority, behavior that would otherwise have been suppressed may be permitted to appear: this is known as the disinhibition hypothesis. For example, the occurrence of displacement preening in incubating terns (*Sterna* spp.) has been ascribed to inhibitory relations between the mechanisms controlling escape behavior and incubation, and between incubation and preening. When incubating is partially inhibited by escape, its inhibitory effect on preening is reduced and preening appears (van Iersel and Bol, 1958; see also Andrew, 1956; Rowell, 1961). In some cases, the absolute and relative strengths of the two tendencies that are in conflict appear to influence which displacement activity appears (Kruijt, 1964).

3. Sometimes apparently irrelevant activities occur in situations where there is no evidence for conflict between two behavioral systems, for instance, when appetitive behavior is physically thwarted. McFarland (1966, 1967) has suggested that such cases, as well as those more usually ascribed to disinhibition, may be due to frustration causing a switch of attention to new stimuli.

These are only some of the suggestions for the mechanisms underlying the appearance of apparently irrelevant activities. Others include effects of general arousal (Hinde, 1959; Bindra, 1959) and the operation of a dearousal system (Delius, 1967). It is therefore necessary to stress that the category of displacement behavior is not homogeneous, and that it is often difficult to decide what should and should not be called displacement.

i. Regression. Sometimes intense conflict seems to result in regression to a more juvenile phase. This is well known in mammals but has been little studied in birds (but see Holzapfel, 1949). Neurosis, likewise, cannot be discussed here.

As we shall see later, some of these modes of behavior have become elaborated in evolution for a signal function. Most threat and courtship displays have been evolved from intention movements, redirection activities, or displacement activities.

H. INTEGRATION OF PARTIAL PATTERNS

The types of behavior shown by any species can be classified into functional groups — nest building, sexual behavior, feeding, and so on.

The means by which discrete activities are integrated to produce functional sequences must now be considered.

The simplest way is the chain response: each response brings the animal into the stimulus situation releasing the next one. Most chains of appetitive behavior–consummatory behavior are of this type. During each phase of the appetitive behavior the bird is particularly sensitive to the stimulus situation that elicits the next phase, and, apparently, the perception of the stimuli eliciting one phase can act as a reinforcement for the learning of the behavior performed in the preceding one. As an example, the chaining of the various responses involved in the copulation of turkeys is shown in Fig. 8 (Schein and Hale, 1965), although it must be emphasized that the sequence of responses is seldom so smooth as this idealized diagram suggests. Normally a given act by one individual may produce any of several (but not usually all) of the responses from the other. Variability in the sequence implies that whichever response appears depends in part on internal factors.

Often the elements of such a behavior sequence share common causal factors, and the later activities in the sequence require a higher intensity of those factors than do the earlier ones. Perhaps because it is more economical in terms of nervous mechanisms, functionally related activities are thus often also causally related. Thus, in many birds, the male sex hormone influences song, territorial behavior, and all male sexual activities. This sharing of common causal factors is thus a second method by which functional integration is brought about.

A special case of common causal factors concerns the consummatory stimuli which bring a sequence of behavior to an end. Insofar as they affect a number of different patterns of behavior, they play a role in their temporal integration. For example, stimuli from the nest that the female canary has built play a part in producing the reduction in nest-building behavior that occurs around the time the eggs are laid. Often this inhibitory effect also affects a number of other functionally related types of activity. Thus, in the Great Cormorant (*Phalacrocorax carbo*), the possession of a nest has an inhibitory effect on all other nesting activities, even though the causal factors for nest building (hormones, twigs, etc.) are still present (Kortlandt, 1940).

Since at any one moment causal factors for a variety of types of behavior are likely to be present, but a bird on the whole does only one thing at a time, there must be inhibitory relations between the

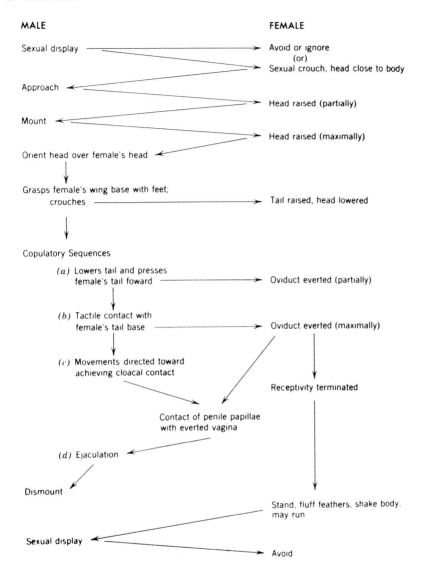

MALE   FEMALE

Sexual display ———————————————→ Avoid or ignore
                                                        (or)
                                                     Sexual crouch, head close to body

Approach ←
                                                  ←——— Head raised (partially)

Mount ←
                                                  ←——— Head raised (maximally)

Orient head over female's head ←

Grasps female's wing base with feet;
         crouches ———————————————→ Tail raised, head lowered

Copulatory Sequences

      (a) Lowers tail and presses
              female's tail foward ———————→ Oviduct everted (partially)

      (b) Tactile contact with
              female's tail base ———————————→ Oviduct everted (maximally)

      (c) Movements directed toward
              achieving cloacal contact

                          Contact of penile papillae
                            with everted vagina   Receptivity terminated

      (d) Ejaculation ←

Dismount

                                                     Stand, fluff feathers, shake body.
                                                     may run
Sexual display ←

                                                  ———→ Avoid

FIG. 8. The sequence of behavior leading to copulation in turkeys. (From Hale and Schein, 1962.)

different types of behavior. This is a statement in terms of behavior; the mechanisms underlying inhibition may be diverse.

We have seen that the performance of an activity may have positive or negative, short-term or long-term effects on the probability of that response occurring subsequently. If a number of activities are linked together either positively (e.g., by chaining) or negatively (e.g., by inhibitory relations), consequences of the performance of one may also have repercussions on the probability of the others. Such effects play a part, for instance, in the integration of the nest-building sequence of canaries (Hinde and Stevenson, 1970).

I. STUDY OF THE ONTOGENY OF BEHAVIOR

Development involves a virtually continuous interaction between the organism and its environment. Attempts to divide behavior into that which is innate, inborn, or instinctive, and that which is learned, are inevitably sterile. The environment may affect behavior in diverse ways, not all of which are usefully described as learning, and all behavior must depend on both genetic and environmental factors. Similarly, a dichotomy between processes or consequences of maturation and experience is of only limited usefulness. Processes of maturation occur in, and depend on, the intraorganismic environment: they are not influenced by extraorganismic factors only insofar as the environment relevant to them is maintained constant by homeostatic mechanisms (e.g., Hinde, 1968, 1970; Lehrman, 1970).

In practice, the analysis of development depends on the study of differences. Differences in the behavior of genetically identical birds reared in different environments can be ascribed to experience, while differences between genetically dissimilar birds reared in the same environment can be ascribed to the differences in their genetic constitution. But the finding that a difference in behavior is due to a genetic difference does not mean that experience does not affect the development of that behavior. For example, tits (*Parus* spp.) use the foot for holding large food objects, but Chaffinches do not, even when reared from the egg in tits' nests. However, observation and experiment show that learning enters into the development of the behavior in tits (Vince, 1964). Thus, tits are innately different from Chaffinches in using the foot in feeding, even though learning is essential for the development of the behavior.

Once it has been established that a difference is due to genetic or to experiential factors, the next stage is to see how the difference is produced. A genetic change, for instance, may affect behavior by producing changes in sense organs, effectors, body proportions, or general excitability, as well as in neural organization specific to the response. For example, the differences between the food preferences of European finches are related to differences in beak size and body size and proportions, and arise in large part through individuals learning to select those foods that they can take most efficiently (Hinde, 1959; Kear, 1962; Newton, 1967).

J. DEVELOPMENT OF BEHAVIOR—LEARNING

Among birds, the behavior characteristic of the species can be molded by individual experience to an extent greater than in any other group except mammals. Indeed, this learning ability is essential for many aspects of avian economy—their active, wide-ranging life and complex social behavior are possible only by virtue of the modifiability of their behavior. Furthermore, through these faculties they are able both to exploit the resources of their environment more fully and to avoid its dangers more successfully.

The literature on learning in birds has recently been thoroughly reviewed by Thorpe (1963), who has classified the types of learning into various categories. Here only a few selected topics will be considered.

1. Motor Patterns

As we have seen in the case of the use of the foot by tits, the role of experience in the development of motor patterns is not always immediately apparent. As another example, the ability of birds with grossly abnormal beaks to feed themselves shows that the feeding patterns are more labile than appears at first sight. With such labile patterns the direction of learning is normally controlled by structural or other species-characteristic features, so that the normal pattern nearly always develops. Unfortunately, except in the special case of song (see below), few experimental studies have been made. Grohmann (1939) showed that improvement in flying ability during a certain period of development is due to maturation and not practice, and it seems likely that maturation plays an important role in many simple activities (but see Section I,G). Conflicting views are held on the development of the pecking response in chicks. Kuo (1932) traces even the first simple head movements of the newly hatched chick to similar movements passively induced before hatching, although some authors consider that he overestimates the role of experience (e.g., Lorenz, 1961, 1965). After hatching, improvement in accuracy is due in part to practice and in part to increased postural stability (Cruze, 1935; Moseley, 1925).

There is, however, clear evidence that some responses appear, completely integrated, when the appropriate stimulus is first presented, even with birds that have been reared in isolation from others. This, of course, does not imply that individual learning plays no part in the development of the constituent parts of the pattern, but it does mean that the total behavior appears as an integrated whole in the absence of example or reward. To give but a few examples: Chaffinches reared in isolation give the same courtship and threat postures to the same stimuli as normal birds. Great Tits reared by hand crouch to a high-pitched whistle just as they would to the alarm call of their parents. Canaries, reared without nest material, will go through all the movements of nest building in an empty pan: further, when given material for the first time, they will carry it to the nest pan almost immediately (see also Heinroth and Heinroth, 1924–1933; Sauer, 1954, 1956). Observations of this type show, then, that each individual has the potentiality for developing certain motor patterns which are characteristic of the species and that these will appear the first time the appropriate stimulus is presented.

Some movement patterns are very resistant to modification by experience. For example, some lovebird species (*Agapornis*) carry nest material in the bill, others by tucking it into the rump feathers. Hybrids between the two often show tucking movements which are unsuccessful through inadequate orientation or selection of nest material, and they may continue to show these movements with considerable persistence (Dilger, 1960).

On the other hand, quite stereotyped movements may be produced by reinforcement techniques (e.g., Skinner, 1948) or under conditions of confinement (e.g., Sargent and Keiper, 1967).

2. *Modifications in Stimulus–Response Relations*

Although such responses often appear on the first occasion that the appropriate stimulus is presented, there are many cases in which the eliciting stimuli are wholly or partially learned. Thus, Craig (1912) showed that young doves, which had never drunk before, showed no tendency to drink at the sight or sound of water: only the presence of water inside the mouth elicited the reflex of swallowing. Probably the first drink usually results from the bird "accidentally" pecking at a drop or pool of water; subsequently, it learns very rapidly the characteristics of the container or other surroundings of the water and later still, presumably, the characteristics of water itself.

Often, responding on the first occasion that the appropriate stimulus is presented is due to an initial responsiveness to a very wide, though not unlimited, range of stimulus characters. Thus, naive young birds will peck at a wide range of objects, although size, solidity, roundness, brightness, and color are relevant variables, their optimum values varying between species (e.g., Rheingold and Hess, 1957; Fantz, 1957; Kear, 1964; Hailman, 1967; Dawkins, 1968). Similarly the young of many nidifugous species will flee or crouch when any moving shape passes overhead; will tend to approach any conspicuous object, especially if it moves; and are attracted by a wide range of repetitive sounds.

In many cases, therefore, development involves increasing specificity of responsiveness. Chicks rapidly come to peck only at potential food objects (e.g., Hailman, 1967), and young murres (*Uria*) to respond only to the calls of their own parents (Tschanz, 1968). In many cases, habituation plays an important role in this. The fleeing response of young nidifugous birds to shapes passing overhead is initially elicited by a wide range of stimuli, but shapes that are often seen lose their effectiveness (Schleidt, 1961; Davies, 1962). The obverse of habituation is the reinforcement associated with some stimuli and not with others. Chicks continue to peck at stimuli that subsequently release grasping and swallowing, and young finches come to prefer those seeds they can dehusk most efficiently (Kear, 1962); the processes involved are, however, complex (Hogan, 1973).

Another factor of importance in limiting responsiveness is the fear induced by strange situations. For example young Common Gallinules (*Gallinula chloropus*) hand-reared from birth will readily peck at food in forceps, but if they are taken from the nest when a few days old the forceps at first elicit fear responses, and can be used only after habituation. We shall meet this again in the study of imprinting.

While some processes restrict the range of effective stimuli, others extend it. To be included here are the various types of associative learning discussed by Thorpe (1963). Pigeons have been used more extensively than any other species in the study of learning in birds. The operant situation devised by Skinner (1938) contains a key which the pigeon can peck. A suitable reinforcer is delivered to the bird either with every response or on some previously arranged schedule. This "schedule of reinforcement" is of crucial importance in determining the subsequent rate and pattern of responding (Ferster and Skinner, 1957). Witholding reinforcement results in a decrease in the rate of pecking ("extinction"). Several studies have now shown that a variety of species-characteristic patterns in addition to pecking, such as the calls of Budgerigars (*Melopsittacus undulatus*) (Ginsburg, 1960) and chicks (Hoffmann *et al.*, 1966) and the preening of pigeons (Hogan, 1964), can be increased by appropriate positive reinforcement. In the great majority of experiments food has been used as a reinforcer, but stimuli known to elicit species-characteristic behavior patterns can also act as reinforcers for an operant response. The reinforcing properties may be either positive, as with the visual presentation of an adversary for fighting cocks (Thompson, 1964) or song for a male Chaffinch (Stevenson, 1969), or negative, as with an alarm call of the Chaffinch (Thompson, 1969). Even a conspicuous flashing light, which can elicit the approach response of a young duckling, can act as a reinforcer (Bateson and Reese, 1969).

Trial and error learning has been studied in the laboratory with puzzle boxes and mazes. In the former the bird must learn how to open the door of a box in order to reach a reward. The method by which the door is opened has no obvious connection with the door, and the first solution comes by chance. After that, however, the birds fairly rapidly learn to concentrate their attempts in a particular place, and later on to open the door. In mazes (Diebschlag, 1940) the performance of pigeons is in some ways comparable to that of mammals, but it soon becomes rather stereotyped and thus, having learned one maze, they change only slowly to another.

Learning abilities change considerably with both age and experience. The complexity of the age changes suggest the operation of multiple factors (Vince, 1960; Cane and Vince, 1968). There are also considerable species differences in learning ability, especially with more complex problems (e.g., Kruschinski *et al.*, 1963). Some of these may depend on differences in curiosity (Wünschmann, 1963). In "learning set" performance, pigeons and chickens are comparable to cats (Zeigler, 1961; Plotnik and Tallarico, 1966).

3. Orienting of Movement Patterns

The importance of modifications to the orienting stimuli again cannot be overestimated. Practically every type of appetitive behavior is subject to modification in this way, and it is unnecessary to enumerate instances. The most remarkable are concerned with the homing abilities of birds. This implies not only the ability to fly on a given course, but also to navigate, i.e., to fix present position and fly the correct course for home. The literature on bird navigation has been reviewed several times recently (Matthews, 1968; Schmidt-Koenig, 1965) (see, also, Volume V).

4. Social Learning

Several studies have shown how the behavior of one bird may be affected if it is given the opportunity to observe another. Birds that see another bird eat may learn to feed at the same source (e.g., Hinde and Fisher, 1951; see also Klopfer, 1959; Alcock, 1969), but if they see it eat distasteful or noxious food they may subsequently avoid similar food items (Rothschild and Ford, 1968). Although there is no evidence for the learning of new motor patterns by birds except in the case of song learning (see below), Dawson and Foss (1965) found that a Budgerigar learning to take the lid off a pot tends to select the same movement from its repertoire as a more experienced bird that it had been allowed to observe.

5. Learning in Complex Situations

To illustrate the complexity of the learning of which birds are capable, a few more heterogeneous studies from field and laboratory must be cited. The Eurasian Nutcracker (*Nucifraga caryocatactes*) stores hazelnuts during the autumn and relies on them for food in winter and spring. Although the caches may be hidden under 0.5 m of snow, the birds are remarkably adept at finding them again, over 80% of diggings being successful (Swanberg, 1951).

A number of laboratory experiments have attempted to investigate the extent of the "higher faculties" possessed by birds. For example, Kruschinski *et al.* (1963) tested whether birds could chose the right way to go around a screen after they had seen, through a slit in the screen, two boxes moving in opposite directions, one with food and the other without. Crows and Black-billed Magpies (*Pica pica*) were found to be very successful, hens and pigeons considerably less so. Several authors have assessed the ability of birds to learn sequences of left–right or go–no go decisions. Zeier (1966) found some pigeons to be successful in sequences up to eight decisions in length (see also Rahmann-Esser, 1964). Common Jackdaws (*Corvus monedula*) and Budgerigars can be trained to distinguish between short sequences of notes, recognition persisting after considerable changes of tempo, pitch, and timbre (Reinert and Reinert-Reetz, 1962; Reinert, 1965).

Of particular interest are experiments on the "number concept": Koehler (1943) has investigated two types of counting ability. The first is the ability to compare groups of units presented simultaneously by means of seen numbers of those units only. In a typical experiment of this type, the bird is taught to discriminate between 2, 3, 4, 5, and 6 black spots on the lids of small boxes, the "key" for its choice being a group of one of these numbers of objects lying in front. The second is the ability to estimate numbers of incidents following each other, and thus to "keep in mind" numbers presented successively in time, independent of rhythms or other clues. Thus, birds were taught to open boxes placed in a row until a certain number of baits had been secured and then to stop, the distribution of baits among the boxes being varied between experiments. In each case, the limit attainable is about 5 or 6. The role of "rhythm" in these experiments is still an open issue (Pastore, 1962b). Insofar as it is possible to perform comparable experiments in which counting named numbers is eliminated, the limit achieved by man is comparable to that attained by birds.

III. Bird Behavior Illustrated by Functional Groups of Activities

Birds have undergone an adaptive radiation that in recent geological time has been rivaled by no other vertebrates except the teleosts. The diversity of their ways of life and the small proportion of species that have been studied in any detail makes any attempt to catalog or review the behavior of birds out of the question. Therefore only a few functional aspects of bird behavior are considered here, and these very briefly. The aim is to provide not a review, but some indications of the diversity of behavior within the group, of how the principles discussed in Section II find application to these varied facets of the bird's life, and of some areas of current interest.

A. Prehatching Behavior

Kuo's development of a method for observing chick embryos through a window in the eggshell opened up exciting possibilities that have been insufficiently exploited. Kuo (1932) observed the changing action patterns of the embryo, their form being clearly affected by the intraegg environment. His descriptions indicate that the processes of development within the egg involve complex interactions between the changing organism and its changing environment. Some of his observations have been repeated more recently with duck eggs (Gottlieb and Kuo, 1965). They have been used as a basis for interesting suggestions about the role of the intraegg environment in the development of species-characteristic responses (e.g., Schneirla, 1965; Lehrman, 1953), and more experimental work is urgently needed.

While Kuo (see also 1967) emphasized the role of intraegg stimulation, Hamburger (1964; Hamburger et al., 1965, 1966; Decker and Hamburger, 1967; Hamburger and Oppenheim, 1967) has stressed the spontaneity of many of the movements of the chick embryo. Cyclic movements start at $3\frac{1}{2}$ days and continue throughout embryonic life. If the spinal cord is sectioned, the movements continue in each section. This implies that potential pacemakers are present at each level, although the more anterior ones are probably normally dominant. These movements are not substantially dependent on sensory input and probably arise independently.

Several authors have been concerned with the development of the sensory systems. The onset of responses seems always to occur in the order nonvisual photic, tactile, vestibular, proprioceptive, auditory, and visual (Gottlieb, 1968). Chicks can certainly respond to extraegg stimuli before hatching, e.g., to warning calls by remaining quiescent (Baeumer, 1955) and to light by a change in frequency of beak clapping (Oppenheim, 1968), and the responsiveness to auditory stimulation contributes to the development of the embryo and affects behavior after hatching (Gottlieb, 1971). Vince (e.g., 1969) has shown that signals produced by the embryos in the period before hatching promote synchronization of hatching in the clutch. She has also investigated the means by which the signals are produced (see also Driver, 1967).

B. DEVELOPMENT OF VISUOMOTOR ABILITIES

Relatively little work on the development of space perception and visuomotor abilities has been done with birds, and in some of the early work, retinal degeneration during dark-rearing may have confused the results. Doves reared for 2 months with their eyes covered by translucent hoods show normal optokinetic behavior (Siegel, 1953a,b). Chicks avoid a visual cliff 24 hours after hatching, even if they have been kept previously in darkness (Shinkman, 1963; see also Kear, 1967): a considerable number of experiments investigating the cues used here have been reviewed by Walk (1965). The readiness with which young birds venture out over a visual cliff is affected by experience (Tallarico and Farrell, 1964; Krames and Carr, 1968), and there are also species differences (Kear, 1967).

Some experiments on the development of the pecking response were cited in Section II,H. Although form perception, size constancy, and brightness constancy develop independently of visual experience in ducklings (Pastore, 1962a), Siegel (1953a,b) found that doves reared with translucent goggles took rather longer to learn to discriminate shapes than did normally reared controls, and Gottlieb (1968) found that early auditory stimulation played a part in the development of ducklings' ability to discriminate the species maternal call.

C. Preening, Locomotion, etc.

Although relatively little attention has been paid to maintenance activities other than feeding, a few recent studies have been concerned with their taxonomic variability or their relations to functional requirements; e.g., bipedal locomotion (Kunkel, 1962), bathing (Nicolai, 1962), scratching (Simmons, 1961), and various maintenance activities (Wickler, 1961).

D. Feeding

The feeding habits of birds are so diverse, and the structural and behavioral adaptations so numerous, that only a few points can be mentioned here. The responsiveness of naive chicks or nestlings to simple stimulus patterns has already been mentioned (Section II,H) and a detailed review is given by Hailman (1967). Even the means of grasping and swallowing the food brought by the parents may be controlled by relatively simple stimulus characters that are responded to appropriately by naive birds (Hunt and Smith, 1964; Oberholzer and Tschanz, 1968). At first, pecking in passerine birds (personal observation) and chicks (Hogan, 1965) shows little relationship to food deprivation. However, reinforcement affects the restriction of the stimuli eliciting pecking, although the way in which it operates is complex. From experiments with chicks, Sterritt and Smith (1965) concluded that feedback stimuli from pecking and the arrival of nutrients in the crop were both reinforcing, but only in interaction with one another (see also Hogan, 1973). A number of environmental factors, including stimulation from social companions, influence the development of feeding behavior (Tolman, 1967; May and Dorr, 1968), and the sort of food ingested may affect subsequent preferences (Rabinowitch, 1968). Wortis (1969) has studied the transition from dependent to independent feeding in the Ring Dove, and reviewed the literature on other species.

The adults of each species have a number of motor patterns used in feeding. As we have seen, learning may enter into the development of these patterns, even though they are species-characteristic (pp. 499–500). Such patterns are usually conservative in evolution, being similar in closely related species. In only relatively few cases, in which there has been marked divergence in the type of prey taken, do related species use markedly different motor patterns (contrast, for example, *Recurvirostra avosetta* with other waders).

In accordance with Gause's principle (1934), however, there is usually little competition between species over food (Lack, 1954). Thus the sign stimuli eliciting or guiding the behavior differ more than the motor patterns themselves. In some cases it is primarily the eliciting stimuli that differ. Thus, different species of finch eat seeds of different sizes. In other cases, the type of food taken is similar (although not necessarily the same), but the place where it is found differs between species. Thus when food is scarce the various *Parus* species feed in different places in the trees (Hartley, 1953; Gibb, 1954).

However, although there are characteristic species differences in feeding behavior, it is not very clear how these differences arise for, while there may be specific preferences for particular stimuli, learning plays a large part in the development of feeding behavior in the individual. Even in domestic chicks learning occurs at every peck and influences the course of subsequent feeding behavior. Learning affects not only where the bird searches, but also the nature of the stimuli to which it is responsive (Croze, 1970). As we have seen, specific preferences may depend partly on learning, the course of which is affected by specific structural characteristics (Section II,H), observation of the parents (e.g., Norton-Griffiths, 1967), and so on. The development of feeding preferences may thus be quite complex even in more or less monophagous species, and must be even more so in polyphagous ones. It may be more than coincidence that many polyphagous species also have a reputation for intelligence (e.g., Corvidae and Paridae).

Furthermore, although few physiological experiments on feeding in birds comparable with the work on mammals have been carried out, it is clear that the motivational basis of feeding behavior is far from simple. Among recent studies, Wood-Gush and Gower (1968) have studied the detailed behavioral consequences of food deprivation in domestic fowl; Delius (1968) has investigated the effects of hunger and thirst on color preferences in pigeons; and McFarland (1965a,b,c) has made a sophisticated analysis of eating, drinking, and temperature control and their interactions in the Ring Dove.

The motivation of hunting may be to some degree independent of that of feeding. For instance, a Little Owl (*Athene noctua*) that has been satiated with food may go through all the actions of catching and killing mice even though no mice are present (Thorpe, 1948). Blue Tits (*Parus caeruleus*) often learn to invade houses, employing the motor patterns that they normally use in feeding in tearing down wallpaper, etc., even though there is no food there and no shortage of food elsewhere (Hinde, 1953b). In Common Ravens (*Corvus corax*), hunger affects hoarding behavior as well as feeding (Gwinner, 1965).

E. TOOL USING

A number of cases of tool using are known in birds. The Galapagos Woodpecker-Finch (*Camarhynchus pallidus*) uses a spine to extract insects from crevices in bark (Lack, 1947). There is some evidence that this behavior does not develop if the bird is reared in isolation for the early months (Millikan and Bowman, 1967). The Egyptian Vulture (*Neophron percnopterus*) breaks open eggs by throwing stones at them (van Lawick-Goodall, 1970; Alcock, 1970). Other cases seem to be limited to occasional individuals (e.g., Morse, 1968). A number of species are able to learn to pull up food suspended on the end of a piece of string, which can perhaps be regarded as a form of tool using (Vince, 1961; Thorpe, 1963).

F. RESPONSES TO PREDATORS

Nearly all species are susceptible to predation. Among the adaptations promoting safety from predators can be mentioned the following.

a. Development of Cryptic Coloration. Sometimes this is extraordinarily effective: the females of many ground-nesting birds cannot be detected from a distance of a few yards (Cott, 1940).

b. Use of Postures that Enhance the Value of Cryptic Coloration. Thus, most ground-nesting birds remain very still when a predator approaches, and some frogmouths (*Podargus*) that nest on the branches of trees adopt a posture that makes them seem like a broken-off stump (Cott, 1940).

c. Responses to Stimuli Likely to Indicate Danger. These are of two types. First, responses to loud noises, quickly moving objects, unusual objects in a familiar environment, and so on. These are subject to rapid habituation if not reinforced by further stimuli indicative of danger. Second, each species has responses to stimuli indicative of particular predators — owls, hawks, snakes, and so on. Just how specific these responses actually are is uncertain: certainly it is possible to combine characters of an owl and a mammal to produce a composite object which is very effective at evoking a response. Such responses to specific predators are probably less liable to habituation than the general responses mentioned earlier, though even here habituation does occur (see p. 491). Many species have two or more distinct responses to predators. In passerines, flying predators produce an immediate dash to cover and crouching, while mammalian or perched avian predators induce mobbing, a form of ambivalent behavior compounded from fleeing, investigatory, and aggressive behavior. Actual attacks on predators by small birds are rare, but larger species, such as the European Bittern (*Botaurus stellaris*), may fight vigorously in defence of their nest.

d. Exploratory Behavior. Any bird finding itself in a new environment will actively explore it. This exploratory behavior undoubtedly helps to protect birds from predators, as well as to exploit the area's resources. Furthermore, "curiosity" is shown toward strange objects, especially those having some but not many of the characters of a predator, and also to a real predator by young birds whose predator responses have not yet matured (Hinde, 1954a,b).

e. Warning Calls. Gregarious species have a system of warning calls that increase the safety of individuals in the flock. These may have special properties that make them difficult to locate (Marler, 1959), and their effectiveness in warning genetically related individuals may, from the point of view of natural selection, counterbalance the additional danger to which the caller is exposed (Maynard Smith, 1965; see also Perrins, 1968; Driver and Humphries, 1969).

f. Distraction Display. Many species have a special display that appears to function by directing a predator away from the nest or young onto themselves. They achieve this by simulating another easily available type of prey—a sick or injured bird, a rodent, etc. Such displays apparently depend on conflicting tendencies to attack and to flee from the predator, and sometimes also to incubate or brood the young. The behavior may have evolved from the genuinely incapacitated condition shown by some birds when suddenly startled (e.g., Curio, 1964) and has been elaborated in evolution in accordance with its display function (Armstrong, 1949; Simmons, 1952).

g. Spacing Out. Spacing out of nests serves as a defence against predation (Tinbergen *et al.*, 1967; see also Kruuk, 1964). This is to be understood in terms of the formation of a "search image" by the predator (Croze, 1970).

G. Flocking and Social Behavior

Most birds are gregarious for at least part of the year. Flocking is more common outside the breeding season, many species that defend large breeding territories becoming gregarious for the rest of the year. On the other hand, some sea birds nest colonially but are solitary at other times, and some species (e.g., some weavers) are always gregarious (see Emlen, 1952).

Although most birds flock by preference with conspecific individuals, mixed flocks are common. In such cases the sign stimuli that elicit flocking can apparently be quite unspecific, another bird of about the same size being sufficient to induce the response. However, many species have evolved social signals—call notes, conspicuous rump and wing markings, and so on—which function in flock integration. Often these are quite similar in unrelated species—presumably by convergence, since the color patterns are often such that they make the movements of taking flight conspicuous, and the call notes have characteristics that make them easily locatable and are audible over a distance not greatly exceeding that necessary for this function (Marler, 1957).

When a member of a gregarious species becomes separated from its flock it shows a special type of appetitive behavior. In the Great Tit, for instance, this consists of hopping through the bushes, peering around (presumably looking for flock companions), and giving the flocking calls frequently. When it rejoins the flock, it resumes feeding quietly. Flocking behavior thus has several types of appetitive behavior—looking for the flock, flying in company, etc.—but no consummatory act, the appetitive behavior ceasing when other individuals are close. In view of such facts, social behavior must be regarded as a category in its own right and not merely as an adjunct of feeding, sexual and other types of behavior, as was formerly supposed (Tinbergen, 1951).

Among birds that flock closely there is usually a strong tendency for "social facilitation"; that is, the performance of a particular pattern of behavior by one member of the flock increases the likelihood that other individuals will behave similarly. In some species this is so marked that an individual that has been fed to satiation will start to feed again when it sees other individuals feeding (Katz and Révész, 1921), while an individual which is hungry will not feed for a while when placed with satiated individuals (Lorenz, 1935). No doubt this plays an important part in flock integration.

Flocks may be so closely integrated that the birds fly only a few centimeters apart and twist and turn synchronously, or they may be merely loose aggregations. The degree of flock integration depends on two groups of factors: those promoting integration, such as the various sign stimuli and appetitive behavior considered above, and opposing disruptive factors. Among the latter are the food-seeking movements of individuals, which in many species would tend to produce a random dispersal if not guided by the social patterns. There are also, however, factors tending to produce overdispersal — that is, to space the birds out more than randomly — in particular, the aggressive and individual-distance behavior of the birds composing the flock. The nature of the flock is thus the result of these conflicting tendencies (Emlen, 1952; Hinde, 1952).

Ultimately, the structure of the flock is a compromise between the conflicting advantages of gregarious and solitary life. Among the former are probably the decreased susceptibility to attack by predators enjoyed by birds in a flock and, in some species, the increased efficiency in the exploitation of food sources. By a process known as "local enhancement" the attention of the other members of a flock may be called to a new food source discovered by one individual. On the other hand, too great a proximity to other individuals involves too great a risk of food robbery. Clearly the balance between these factors will be determined by many other aspects of the species biology, such as predation pressure and the nature and availability of food (e.g., Goss-Custard, 1970). The few species that are never gregarious — some raptors and some birds that hold territory throughout the year — probably remain solitary because their feeding habits necessitate it (Lack, 1954).

H. FIGHTING AND TERRITORY

While in a flock, most birds attempt to maintain a space around themselves (the "individual distance") free from other birds, by threatening, attacking, or fleeing from other individuals that approach (Hediger, 1950; Conder, 1949). The primary functions of this individual distance are probably increased immunity from food-robbing by other birds, or adequate room for sudden takeoff, according to the species. Its extent varies between species and, within species, according to the occasion: sometimes it is zero (e.g., many estrildids).

Sometimes the area defended is not that around the bird, but around some object or situation that is important for survival or reproduction, such as a mate, song post, nest site, or territory. Such territorial behavior occurs in many other groups (e.g., fish, reptiles, and mammals) but it has been studied particularly in birds. Avian territories are defended mainly during the breeding season, but in some species at other times as well. Territory size varies from several square miles (some birds of prey) to the distance the sitting bird can peck (many cliff-nesting species). It is absent apparently only in the Emperor Penguin (*Aptenodytes forsteri*), for which the advantages of close huddling as a protection from the cold winds more than compensate for the lack of territorial seclusion (Stonehouse, 1953). The biological significance of territory has been much discussed; it undoubtedly varies among species (reviews by Hinde, 1956b; Tinbergen, 1957; Brown, 1964; Lack, 1966).

A territory owner normally attacks conspecific individuals on his territory, but flees from his neighbors if he is trespassing on their territory. The boundary is the line along which the tendencies to attack and to flee are more or less in balance in each of the two rivals. During a skirmish, each combatant is in an ambivalent condition, with tendencies to attack and flee from the rival. The precise nature of the behavior or threat posture shown depends on the nature of the balance. These threat postures are social signals – they serve to intimidate the rival (Figs. 2 and 3) (Tinbergen, 1952, 1953; Hinde, 1952, 1953a).

In addition to threat postures, submissive and appeasement postures play an important part in fighting. These serve to indicate "nonaggressiveness" and thus to reduce the aggressiveness of a dominant rival. They are given primarily in situations where a tendency to flee is in conflict with some other incompatible tendency. Many of them have evolved from intention movements of fleeing, or displacement activities, and depend for their effect on the nonpresentation of stimuli for aggression to the rival (e.g., compare Figs. 2 and 9).

The external stimulus for aggression is normally certain stimuli from a fellow member of the species (Marler, 1955, 1956a,b). Sometimes it is elicited by members of other species, e.g., if they approach the nest, and by predators. It ceases on the disappearance of the eliciting stimuli. There are no consummatory stimuli distinct from the absence of the eliciting ones (contrast feeding). It has been shown for a number of species that the male sex hormone increases aggressiveness (Shoemaker, 1939), but the generality of this effect remains to be assessed. The defensive aggressive behavior of doves at the nest can be enhanced by prolactin and progesterone, and progesterone increases the aggressiveness of males toward other males (Vowles and

FIG. 9. Submissive posture of Great Tit. From a drawing by Yvette Spencer-Booth.

Harwood, 1966). Luteinizing hormone is said to increase aggressiveness in European Starlings (Davis, 1957). In the Red-billed Quelea (*Quelea quelea*), Crook and Butterfield (1970) found that luteinizing hormone increases social rank in nonbreeding birds, but testosterone increases in encounters over a socially valent commodity such as nest material.

As the last example shows, the means by which the internal state affects aggressiveness may be quite subtle. It may involve giving significance to an external object or area, proximity to which precipitates the fight. Thus, a Great Tit in reproductive mood will attack a male who, though 100 m distant, is intruding on his territory, but tolerate him 10 m away across the boundary. Two minutes later, the same territory owner may, while feeding, ignore an intruder 1 m away from him inside the territory. Again, although caged finches and buntings fight more when food deprived, this is not due to a direct effect of hunger on the threshold for aggression, but to the increased frequency with which birds come into proximity (Marler, 1955, 1956a; Andrew, 1957).

In addition to proximity, the probability that a bird will behave aggressively is enhanced by frustration (e.g., Azrin *et al.*, 1966), and Caryl (1969) has shown that male Zebra Finches (*Poephila guttata*) show more aggressive behavior if another individual is present in an adjacent cage, especially if that individual is familiar.

I. PAIR FORMATION AND COURTSHIP

Birds may be monogamous or polygamous, polyandrous or promiscuous. The pair bond may last for life or for the duration of a mating attempt. With such diversity, few generalizations about sex relations can be given here: the subject has been reviewed by Armstrong (1947). There are, however, some principles about the mechanisms involved that have now been established for a number of groups and are probably widely applicable.

Each individual carries characters that may elicit attacking, fleeing, and/or sexual behavior in other individuals. Thus, whereas during fights the rivals have two conflicting tendencies (Section III,H), during courtship the mates have at least three. The nature of the courtship display shown at any time depends on the strengths and relative strengths of these tendencies. As the tendencies change through the season, so do the courtship displays (e.g., Tinbergen, 1952; Hinde, 1952).

The courtship of some cardueline and fringilline finches can be used to illustrate this (Figs. 10 and 11). The first response of the male in pair formation is an aggressive one given to all other members of

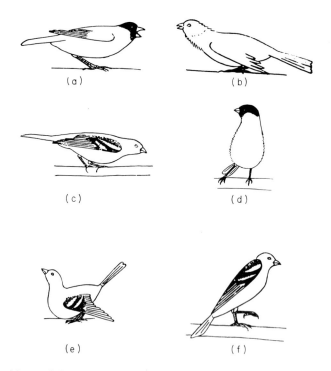

FIG. 10. (a) Head forward posture of European Bullfinch (*Pyrrhula pyrrhula*). A posture of this type is widespread among passerines. The body is oriented towards the rival. (b) Courtship posture of male canary. The body is oriented obliquely to the female. (c) Early courtship posture of Chaffinch. The body is oriented laterally to the female. (d) Courtship posture of Bullfinch. The body is oriented laterally to the female. (e) Soliciting posture of female Chaffinch. (f) Precopulatory posture of male Chaffinch. (Redrawn from Hinde, 1955, 1956a.)

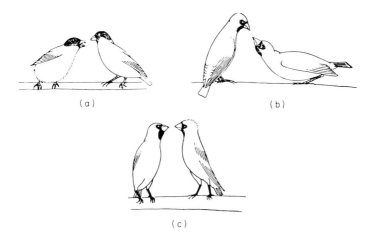

FIG. 11. (a) Courtship of Bullfinches. The female is on the left. (b) and (c) Billing ceremonies derived from courtship feeding in the Hawfinch (*Coccothraustes coccothraustes*). (Redrawn from Hinde, 1955, 1956a.)

the species which approach. At this stage, the male may already be established on a territory (Chaffinch) or the birds may still be in flocks. If the newcomer is a male or nonreceptive female, it may behave aggressively or flee. If it is a receptive female, she may adopt a submissive posture and show a passive resistance to the male's attack. After a while the male begins to behave sexually. As his sexual tendency increases, his tendency to attack the female decreases: she exploits this and becomes dominant to him. Correlated with this change, the head-forward threat posture used by the male becomes gradually replaced by courtship postures. These differ from the threat postures primarily by an increasingly oblique or lateral orientation of the body toward the female, and an increasingly upright position of the body and neck. At this stage, his tendency to attack the female may be relatively weak, but full sexual behavior is prevented by his tendency to flee. Eventually, sexual factors predominate, and copulation attempts occur (Hinde, 1955/1956).

This account is, of course, highly schematic, and there are great variations between species even among these finches. Thus, in the Chaffinch, the male's changeover from aggressive to courtship display is usually almost immediate, while in many carduelines the male's aggressiveness is much more persistent. In many species "courtship feeding" occurs; that is, the male feeds the female who begs like a young bird: in the Chaffinch this is absent. Furthermore, there are of course great interspecies differences in the displays.

Nevertheless, it is probable that the courtship of most birds consists of variations on this theme. At any rate, considerable understanding of the complexity of courtship behavior can be obtained once its essential ambivalence is recognized. Andrew (1961) has pointed out that displays may be associated with other tendencies in addition to the conventional attacking, fleeing, and sexual tendencies (for instance, nest building); and others have emphasized the probable importance of a purely social tendency (Kunkel, 1962; Fischer, 1965). Andrew (1956) has also argued that it is more useful to analyze displays in terms of incompatible responses, rather than in terms of incompatible tendencies each of which may affect a number of responses. A detailed analysis of the threat displays of the Great Tit by Blurton-Jones (1968) goes a long way toward defining the spheres of usefulness of each point of view.

It is impossible here to do justice to the diversity of courtship types to be found among birds. Many of the variations are correlated with the degree of sexual dimorphism—the more dissimilar the sexes are in color, the more rapidly sex recognition occurs and also the more different are the displays of male and female. But in each species the evolution of the courtship pattern has been influenced by many other aspects of the life history—the degree of gregariousness, territorial behavior, the nature of the habitat, availability of food, sex ratio, and so on. These principles are displayed to best advantage in those groups in which a large number of species have been studied, such as the gulls (Tinbergen, 1959, 1967; Beer, 1966; E. Cullen, 1957; Moynihan, 1955, 1959, 1962; N. G. Smith, 1966) and the weavers (Crook, 1964).

J. Evolution of Displays

Threat and courtship displays have been elaborated in evolution from the various types of behavior that appear in conflict situations — primarily intention movements, redirection activities, and displacement activities (p. 494). The intention movements are often those of locomotion (Fig. 6), but incipient movements of striking, copulating, etc., also occur. The displacement activities (i.e., apparently functionally irrelevant activities) are extremely diverse. Among the commonest are various types of feather raising or lowering and preening and cleaning movements.

These movements have changed in evolution, becoming more effective for their signal function. The changes involved consist largely of absolute and relative threshold changes in the various components of the movement. These result in changes in the relations between the components, so that, for instance, the withdrawal of the head in the first phase of a takeoff leap may be associated with the depressed tail of the second. Further, they may result in a decrease in the variability of the movement such that the same response is given to a larger range of eliciting factors (Morris, 1957). It has been suggested that in some cases the movement may become emancipated from the causal factors that originally controlled it, so that it is now evoked by quite different ones (Lorenz, 1951). This seems to be the case, for instance, in court-ship feeding: often a female will beg for food from her mate when she has just been feeding herself, and even when she is carrying food in her beak (but see Krebs, 1970). However, as yet too little is known about the possible existence of causal factors common to the original and display contexts for the generality of the principle of emancipa-tion to be assessed (Tinbergen, 1948, 1952, 1959; Moynihan, 1955; Morris, 1957; Cullen, 1966).

Among cardueline finches, many of the components of the display postures are loosely associated with one or other of the conflicting tendencies—wing raising with attacking and crest raising with fleeing, for instance (but see Stokes, 1962). The relationship between the display components and tendencies are similar in a wide range of species and interspecies hybrids.

As well as changes in the display movements, conspicuous struc-tures have been developed that make the movements more striking. Since, within groups of closely related species, the same display posture may exhibit diverse structural characters, it would seem that the movement is the more primitive and that the structural features were evolved later to make it more conspicuous. Nevertheless, there has probably usually been a parallel elaboration of both movement and structure (Figs. 2, 3, and 10c). All these changes are to be under-stood as adaptations to the signal functions of the displays.

K. COMMUNICATION

As we have seen, all species of birds are social for at least some of the time, and some are social for all the time. Social life demands com-munication with other members of the species, and this depends pri-marily on visual and auditory signals. The behavior of one individual may be affected by the mere presence of another, or by incipient in-tention movements that it has learned are associated with impending action. The most important elements in social communication, how-ever, are the sounds, postures, and structures that have been adapted in evolution for a signal function, as discussed in Section III,J. The evolution of such social releasers has usually also been associated with the evolution of appropriate responsiveness to the signal.

The precise number of signals in the repertoire of any species cannot be assessed. Many of the signals intergrade so that the total number is arbitrary. But as a rough guide estimates of the number of vocal signals in birds' repertoires are of the order of 5–20 (Thorpe, 1961). The number of visual signals is probably of a similar order, or rather more.

These totals are not large. One wonders how sufficient information to integrate a complex social life can be transmitted by such a small number of signals. Two points must be made here. First, many of the postures listed in descriptions of communicatory behavior are combinations of more elemental components. In a study of the agonistic behavior of Blue Tits at a winter feeding station, Stokes (1962) showed that the correlation between individual component and subsequent behavior was usually not large, and that this was due in part to interaction between the components. Combinations of components showed more reliable associations with subsequent behavior. Second, each signal is made in a context, and that context is essential for its interpretation. This has been demonstrated in detail by W. J. Smith's (1966) studies of tyrannid flycatchers (see also W. J. Smith, 1969).

The precise nature of each signal used in social communication is presumably a consequence of conflicting selection pressures for conspicuousness, crypticity, distinctiveness, and so on. For example, the form of call notes varies with their function. In particular, those that do not convey information about position (some alarm calls) share certain characteristics (long duration, no sudden changes in pitch, beginning and end gradual). Social calls, which must be easily locatable, have the opposite characteristics. Among other factors ultimately governing the forms of the calls is the need for distinctiveness from the calls of other sympatric species. This is especially the case with song. With the "see" alarm call, on the other hand, which serves in interspecific communication, there is a selective value in convergence on a common type.

The study of song development has produced particularly interesting data in recent years. In a few species, the species-characteristic song develops even in individuals auditorally isolated from others from hatching, e.g., Song Sparrow (*Melospiza melodia*) (Mulligan, 1966). In most cases, however, some such experience is necessary. In

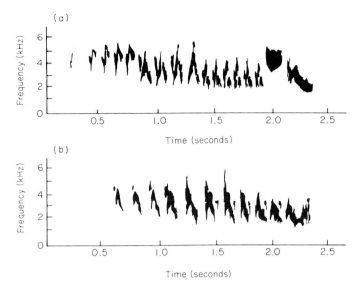

FIG. 12. Chaffinch song. (a) Example of normal song. (b) Song of an individual reared in isolation. (After Thorpe, 1961.)

the Chaffinch, birds reared in isolation produce only a relatively simple song (Fig. 12). However, exposure to the species-characteristic song during the autumn, long before the bird itself starts to sing, results in near-normal song. Apparently, the Chaffinch first learns what the song is like and later, during the period when the amorphous subsong gradually develops into the full song, how to sing it (e.g., Thorpe, 1961; Nottebohm, 1970). The latter seems to involve a monitoring of the bird's own output to conform to the model acquired previously. In harmony with this view, the full song can act as a reinforcer for an operant response in birds with appropriate experience and in the appropriate hormonal state (Stevenson, 1969), and deafening after experience of the song but before its production prevents song development (Nottebohm, 1970). However, the Chaffinch will not learn to sing any song that it hears. Apparently selectivity depends on a tendency to imitate songs with the species-characteristic note structure; although in other species there is a tendency to imitate the song produced by the (foster) parent (e.g., Immelmann, 1969).

A considerable volume of work on song development in this and other species is reviewed by Thorpe (1961), Konishi and Nottebohm (1969), Immelmann (1969), Marler (1970), and Nottebohm (1970). The ontogeny of visual displays has been studied in only a few species (e.g., Kruijt, 1964). Other aspects of bird vocalizations are discussed in various papers in Hinde (1969).

L. Nest Site and Nest

Although a few species, e.g., murres (Tschanz, 1968), build no nest, most birds lay their eggs in a more or less elaborate one. The form of these nests is extremely diverse, and the technique of building is often complex. The nest site is apparently recognized by a relatively small number of sign stimuli: in some (e.g., hole-nesting) species these may be partially visual, but usually tactile senses seem to be primary (Craig, 1918).

Nest building depends in the first instance on a relatively limited number of motor patterns. Even simple nests, however, pose a real problem to the ethologist. Surely, behavior of great complexity must be required for their construction. Thorpe (1963), following an earlier description by Tinbergen, has listed the distinct movements or com-

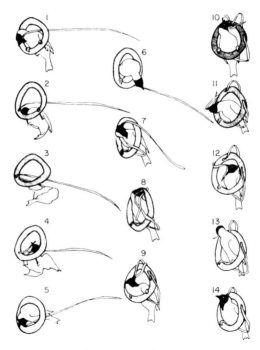

Fig. 13. A typical sequence of movements by a male Village Weaverbird (*Ploceus cucullatus*) as he weaves a single strip torn from a leafblade of elephant grass into his ring. (From Collias and Collias, 1962.)

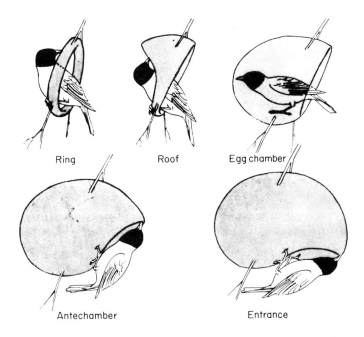

Ring   Roof   Egg chamber

Antechamber   Entrance

FIG. 14. Normal stages in nest-building by a male Village Weaverbird. (From Collias and Collias, 1962.)

binations of movements in the nest building of the Long-tailed Tit (*Aegithalos caudatus*), and has shown that responsiveness to at least eighteen distinct stimulus situations is also necessary.

Among the most complex nests are those built by the weavers, and their nest-building behavior has been described in some detail (Crook, 1960; Collias and Collias, 1962, 1964). As Figs. 13 and 14 show, the actual stitching behavior is quite complicated, but the production of the correct shape is largely due to the fact that at each stage in the building the bird perches in the same place but orients its building to the edge of the existing fabric.

In many species, individuals reared in isolation build normal nests when they come into breeding condition. Nevertheless, learning enters into the integration of the behavior. Canaries kept without nest material develop bizarre habits that enable them to perform only part of the nest-building sequence (Hinde, 1958). Furthermore, they learn to go to a perch where nest material is present and available to them rather than to one where it is present but not available. There is considerable evidence that opportunity to perform even small fragments of the nest-building sequence is reinforcing (Hinde and Steel, 1972).

The integration of the sequence in adult birds is discussed by Hinde and Stevenson (1969).

The finished nest of the canary consists of a cup of grass and a lining of feathers. The changeover from carrying grass to the nest to carrying feathers is a consequence of stimuli from the cup which the female herself has built, perceived through the hormone-sensitized brood patch (see Section II,F).

M. PARENTAL BEHAVIOR

Birds can be divided into two groups — nidicolous species, in which the young are completely helpless on hatching and entirely dependent on the parents for food and protection; and nidifugous species, which leave the nest almost at once and feed themselves, although they still depend on their parents for protection. Some groups (e.g., Laridae) are intermediate.

Parental behavior involves several functional groups of behavior patterns, including those of incubating the eggs and brooding the young, removing egg shells and feces, protecting the young from predators with warning calls (p. 509) and distraction displays (p. 509) as well as feeding them. Where both parents play a part, their duties often bring them into close proximity, with the consequent tendencies to attack, flee from, and behave sexually toward each other (pp. 514–516). Complicated ceremonies (e.g., nest-relief ceremonies) have been evolved which apparently smooth parental relations. While the primitive condition seems to have been for both sexes to share parental responsibilities, there has been an evolutionary trend in most orders for the female to take the major share: in a few orders, however, the reverse is the case.

In order for reproduction to be successful, the parent(s) must show an overlapping sequence of types of behavior — pair formation, courtship and copulation, nest site selection and nest-building, egg-laying, incubation, and feeding the young. Each activity must appear at the right time in relation to the others and to the physical and social environment. The processes involved have been studied especially in the Ring Dove (e.g., Lehrman, 1965), the domestic canary (e.g., Hinde, 1965; Hinde and Steel, 1966), and the Budgerigar (Brockway, 1969; R. E. Hutchison, 1971). Data on other species have been reviewed by Lehrman (1961, 1962) and Farner (1967). The integration of the sequence depends on a complex nexus of events involving changes in the environment, some of them, such as construction of the nest, resulting from the bird's behavior; the bird's changing endocrine state;

and the behavioral changes themselves. Some of the interactions involved are shown in Fig. 15, although it must be emphasized that the relationships indicated there represent only the very incomplete knowledge at present available for only one species.

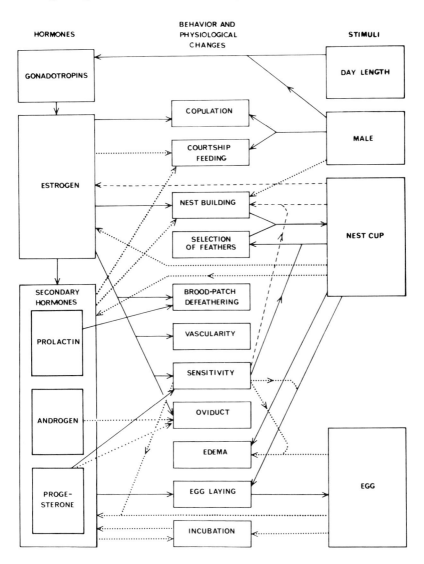

FIG. 15. Relations between external stimuli, hormonal conditions, and reproductive development in female canaries. Solid lines indicate positive effects, dashed lines indicate negative effects, and dotted lines probable effects not yet established with certainty.

N. Filial Behavior

Several aspects of filial behavior have been discussed in previous sections, and attention here is confined to one topic of special interest, the establishment of the parent–young relationship. As we have seen, most filial responses can at first be elicited by a wide range of stimulus objects. Quite soon, however, young birds of most species come to respond only to their own parents. Recognition may depend on either auditory or visual characteristics, or both (e.g., Beer, 1970; Hutchison *et al.*, 1968). The narrowing of the range of effective stimuli has been studied most extensively in the context of the following response of nidifugous species, under the rubric of "imprinting." This is not a special form of learning, as suggested earlier (Lorenz, 1935), but involves a learning about the properties of the environment comparable to that studied previously in the context of "perceptual learning" or "exposure learning." Once the environment becomes familiar, strange objects within it elicit fear responses: largely for this reason, imprinting is most likely to occur within a "sensitive period" which may be only a day or two in length (e.g., Bateson, 1966; Sluckin, 1964). This learning is superimposed on a tendency to approach conspicuous objects in the environment, proximity to which is reinforcing (Bateson and Reese, 1969), and a tendency to approach the maternal calls (Gottlieb, 1971). Although no conventional reinforcement is necessary for the following response to become directed toward a conspicuous moving object in the environment, under natural conditions learning in the course of feeding, brooding, etc., must also play a part.

Experience with, and the consequent familiarity of, individuals with particular characteristics may subsequently lead to the choice of mates with similar characteristics in adulthood. However, a responsiveness to features typical of the species that is independent of previous experience of them also plays a role (e.g., Nicolai, 1964; Schütz, 1965; Immelmann, 1969; de Lannoy, 1967; Klinghammer, 1967).

REFERENCES

Åkerman, B. (1966). Behavioural effects of electrical stimulation in the forebrain of the pigeon. I and II. *Behaviour* **26,** 323–338 and 339–350.

Åkerman, B., Andersson, B., Fabricius, E., and Svensson, L. (1960). Observations on central regulation of body temperature and of food and water intake in the pigeon (*Columba livia*). *Acta Physiol. Scand.* **50,** 328–336.

Alcock, J. (1969). Observational learning in three species of birds. *Ibis* **111,** 308–321.

Alcock, J. (1970). The origin of tool-using by Egyptian Vultures *Neophron percnopterus*. *Ibis* **112**, 542.

Andrew, R. J. (1956). Some remarks on behaviour in conflict situations, with special reference to *Emberiza* spp. *Brit. J. Anim. Behav.* **4**, 41–45.

Andrew, R. J. (1957). The aggressive and courtship behaviour of certain Emberizines. *Behaviour* **10**, 255–308.

Andrew, R. J. (1961). The displays given by passerines in courtship and reproductive fighting. A review. *Ibis* **103a**, 315–348.

Armstrong, E. A. (1947). "Bird Display and Behaviour." Lindsay Drummond, London.

Armstrong, E. A. (1949). Diversionary display. *Ibis* **91**, 88–97 and 179–188.

Azrin, N. H., Hutchinson, R. R., and Hake, D. F. (1966). Extinction-induced aggression. *J. Exp. Anal. Behav.* **9**, 191–204.

Bacon, W. E., and Bindra, D. (1967). The generality of the incentive-motivational effects of classically conditioned stimuli in instrumental learning. *Acta Biol. Exp. (Warsaw)* **27**, 185–197.

Baeumer, E. (1955). Lebensart des Haushuhns. *Z. Tierpsychologie* **12**, 387–401.

Barfield, R. J. (1964). Induction of copulatory behaviour by intracranial placement of androgen in capons. *Amer. Zool.* **4**, 133.

Bateson, P. P. G. (1966). The characteristics and context of imprinting. *Biol. Rev.* **41**, 177–220.

Bateson, P. P. G., and Reese, E. P. (1969). The reinforcing properties of conspicuous stimuli in the imprinting situation. *Anim. Behav.* **17**, 692–699.

Beach, F. A. (1951). Effects of forebrain injury upon mating behavior in male pigeons. *Behaviour* **4**, 36–59.

Beach, F. A., ed. (1965). "Sex and Behavior." Wiley, New York.

Beer, C. G. (1966). Adaptations to nesting habitat in the reproductive behaviour of the Black-billed Gull *Larus bulleri*. *Ibis* **108**, 394–410.

Beer, C. G. (1970). Individual recognition of voice in the social behavior of birds. *In* "Advances in the Study of Behavior" (D. S. Lehrman, R. A. Hinde, and E. Shaw, eds.), Vol. 3, pp. 27–74. Academic Press, New York.

Bindra, D. (1959). "Motivation." Ronald Press, New York.

Blurton-Jones, N. G. (1968). Observations and experiments on causation of threat displays of the Great Tit (*Parus major*). *Anim. Behav. Suppl.* **1**, 2.

Brockway, B. F. (1969). Roles of budgerigar vocalization in the integration of breeding behaviour. *In* "Bird Vocalizations" (R. A. Hinde, ed.), pp. 131–158. Cambridge Univ. Press, London and New York.

Brown, J. L. (1964). The evolution of diversity in avian territorial systems. *Wilson Bull.* **76**, 160–169.

Brown, J. L. (1969). The control of avian vocalizations by the central nervous system. *In* "Bird Vocalizations" (R. A. Hinde, ed.), pp. 79–96. Cambridge Univ. Press, London and New York.

Bruder, R. H., and Lehrman, D. S. (1967). Role of the mate in the elicitation of hormone-induced incubation behavior in the Ring Dove. *J. Comp. Physiol. Psychol.* **63**, 382–384.

Cane, V. R., and Vince, M. A. (1968). Age and learning in quail. *Brit. J. Psychol.* **59**, 37–46.

Caryl, P. G. (1969). Social control of fighting behaviour in the Zebra Finch, *Taeniopygia guttata* (Vieillot). Ph.D. Thesis, Cambridge University.

Collias, N. E., and Collias, E. C. (1962). An experimental study of the mechanisms of nest-building in a weaverbird. *Auk* **79**, 568–595.

Collias, N. E., and Collias, E. C. (1964). Evolution of nest-building in the weaverbirds (Ploceidae). *Univ. Calif., Berkeley, Publ. Zool.* **73**, 1–162.

Conder, P. (1949). Individual distance. *Ibis* **91**, 649–656.

Cott, H. B. (1940). "Adaptive Colouration in Animals." Methuen, London.

Craig, W. (1912). Observations on doves learning to drink. *J. Anim. Behav.* **2**, 273–279.

Craig, W. (1918). Appetites and aversions as constituents of instincts. *Biol. Bull.* **34**, 91–107.

Crook, J. H. (1960). Nest form and construction in certain West African weaverbirds. *Ibis* **102**, 1–25.

Crook, J. H. (1964). The evolution of social organisation and visual communication in the weaverbirds (Ploceinae). *Behaviour, Suppl.* **10**, 1–178.

Crook, J. H., and Butterfield, P. A. (1970). Gender role in the social system of *Quelea*. *In* "Social Behaviour in Birds and Mammals" (J. H. Crook, ed.), pp. 211–248. Academic Press, New York.

Croze, H. (1970). Searching image in Carrion Crows. *Z. Tierpsychol., Beit.* **5**, 1–85.

Cruze, W. W. (1935). Maturation and learning in chicks. *J. Comp. Psychol.* **19**, 371–409.

Cullen, E. (1957). Adaptations in the Kittiwake to cliff-nesting. *Ibis* **99**, 275–302.

Cullen, J. M. (1966). Reduction of ambiguity through ritualization. *Phil. Trans. Roy. Soc. London, Ser. B*, **251**, 363–374.

Curio, E. (1964). Fluchtmangel bei Galapagos-Töpeln. *J. Ornithol.* **105**, 334–339.

Daanje, A. (1951). On locomotory movements in birds and the intention movements derived from them. *Behaviour* **3**, 48–98.

Dane, B., Walcott, C., and Drury, W. H. (1959). The form and duration of the display actions of the Goldeneye *(Bucephala clangula)*. *Behaviour* **14**, 265–281.

Davies, S. J. J. F. (1962). The response of the Magpie Goose to aerial predators. *Emu* **62**, 51–55.

Davis, D. E. (1957). Aggressive behavior in castrated Starlings. *Science* **126** [3267], 253.

Davis, D. E. (1963). Hormonal control of aggressive behavior. *Science* **126**, 253.

Dawkins, R. (1968). The ontogeny of a pecking preference in domestic chicks. *Z. Tierpsychol.* **25**, 170–186.

Dawson, B. V., and Foss, B. M. (1965). Observational learning in Budgerigars. *Anim. Behav.* **13**, 470–474.

Decker, J. D., and Hamburger, V. (1967). The influence of different brain regions on periodic motility of the chick embryo. *J. Exp. Zool.* **165**, 371–384.

de Lannoy, J. (1967). Zur Prägung von Instinkthandlungen. *Z. Tierpsychol.* **24**, 162–200.

Delius, J. D. (1967). Displacement activities and arousal. *Nature (London)* **214**, 1259–1260.

Delius, J. D. (1968). Color preference shift in hungry and thirsty pigeons. *Psychol. Sci.* **13**, 273–274.

Diebschlag, E. (1940). Über den Lernvorgang bei der Haustaube. *Z. Vergl. Physiol.* **28**, 67–104.

Dilger, W. C. (1960). The comparative ethology of the African parrot genus *Agapornis*. *Z. Tierpsychol.* **17**, 649–685.

Driver, P. M. (1967). Notes on the clicking of avian egg-young, with comments on its mechanism and function. *Ibis* **109**, 434–437.

Driver, P. M., and Humphries, D. A. (1969). The significance of the high intensity alarm call in captured passerines. *Ibis* **111**, 243–244.

Emlen, J. T. (1952). Flocking behavior in birds. *Auk* **69**, 160–171.

Falls, J. B. (1963). Properties of bird song eliciting responses from territorial males. *Proc. Int. Ornithol. Congr., 13th, 1962* Vol. 1, pp. 259–271.

Falls, J. B. (1969). Functions of territorial song in the White-throated Sparrow. *In* "Bird Vocalizations" (R. A. Hinde, ed.), pp. 207–232. Cambridge Univ. Press, London and New York.

Fantz, R. L. (1957). Form preferences in newly hatched chicks. *J. Comp. Physiol. Psychol.* **50**, 422–430.

Farner, D. S. (1967). The control of avian reproductive cycles. *Proc. Int. Ornithol. Congr., 14th, 1966* pp. 107–134.

Ferster, C. B., and Skinner, B. F. (1957). "Schedules of Reinforcement." Appleton, New York.

Fischer, H. (1965). Das Triumphgeschrei der Graugans (*Anser anser*). *Z. Tierpsychol.* **22**, 247–304.

Gause, G. F. (1934). "The Struggle for Existence." Williams & Wilkins, Baltimore, Maryland.

Gibb, J. (1954). Feeding ecology of tits. *Ibis* **96**, 513–543.

Ginsburg, N. (1960). Conditioned vocalization in the Budgerigar. *J. Comp. Physiol. Psychol.* **53**, 183–186.

Goodman, I. J., and Brown, J. L. (1966). Stimulation of positively and negatively reinforcing sites in the avian brain. *Life Sci.* **5**, 693–704.

Goss-Custard, J. D. (1970). Feeding dispersion in some overwintering wading birds. *In* "Social Behaviour in Birds and Mammals," (J. H. Crook, ed.), pp. 3–36. Academic Press, New York.

Gottlieb, G. (1968). Prenatal behaviour of birds. *Quart. Rev. Biol.* **43**, 148–174.

Gottlieb, G. (1971). "Development of Species Identification in Birds." Univ. of Chicago Press, Chicago, Illinois.

Gottlieb, G., and Kuo, Z. Y. (1965). Development of behavior in the duck embryo. *J. Comp. Physiol. Psychol.* **59**, 183–188.

Grohmann, J. (1939). Modifikation oder Funktionsreifung? Ein Beitrag zur Klärung der wechselseitigen Beziehungen zwischen Instinkthandlung und Erfahrung. *Z. Tierpsychol.* **2**, 132–144.

Gwinner, E. (1965). Beobachtungen über Nestbau und Brutpflege des Kolkraben (*Corvus corax*) in Gefangenschaft. *J. Ornithol.* **106**, 145–178.

Hailman, J. P. (1967). The ontogeny of an instinct. *Behaviour, Suppl.* **15**, 1–59.

Hale, E. B., and Schein, M. W. (1962). The Behaviour of Turkeys. *In* "The Behaviour of Domestic Animals" (E. S. E. Hafez, ed.), pp. 531–564. Bailliere-Tindall, London.

Hamburger, V. (1964). Ontogeny of behaviour and its structural basis. *Comp. Neurochem., Proc. Int. Neurochem. Symp., 5th, 1962* pp. 21–34.

Hamburger, V., and Oppenheim, R. (1967). Prehatching motility and hatching behavior in the chick. *J. Exp. Zool.* **166**, 171–204.

Hamburger, V., Balaban, M., Oppenheim, R., and Wenger, E. (1965). Periodic motility of normal and spinal chick embryos between 8 and 17 days incubation. *J. Exp. Zool.* **159**, 1–14.

Hamburger, V., Wenger, E., and Oppenheim, R. (1966). Motility in the chick embryo in the absence of sensory input. *J. Exp. Zool.* **162**, 133–160.

Hartley, P. H. T. (1953). An ecological study of the feeding habits of the English titmice. *J. Anim. Ecol.* **22**, 261–288.

Harwood, D., and Vowles, D. M. (1966). Forebrain stimulation and feeding behavior in the Ring Dove (*Streptopelia risoria*). *J. Comp. Physiol. Psychol.* **62**, 388–396.

Hediger, H. (1950). "Wild Animals in Captivity." Butterworth, London.

Heinroth, O., and Heinroth, M. (1924–1933). "Die Vögel Mitteleuropas." Berlin.

Hinde, R. A. (1952). The behaviour of the Great Tit (*Parus major*) and some other related species. *Behaviour, Suppl.* **2**, 1–201.

Hinde, R. A. (1953a). The conflict between drives in the courtship and copulation of the Chaffinch. *Behaviour* **5**, 1–31.

Hinde, R. A. (1953b). A possible explanation of paper-tearing behaviour in birds. *Brit. Birds* **46**, 21–23.

Hinde, R. A. (1954a). Factors governing the changes in strength of a partially inborn response, as shown by the mobbing behaviour of the Chaffinch (*Fringilla coelebs*). I. The nature of the response, and an examination of its course. *Proc. Roy. Soc., Ser. B* **142**, 306–331.

Hinde, R. A. (1954b). II. The warning of the response. *Proc. Roy. Soc., Ser. B* **142**, 331–358.

Hinde, R. A. (1955/1956). A comparative study of the courtship of certain finches (Fringillidae). *Ibis* **97**, 706–745, **98**, 1–23.

Hinde, R. A. (1956). The biological significance of the territories of birds. *Ibis* **98**, 340–369.

Hinde, R. A. (1958). The nest-building behaviour of domesticated canaries. *Proc. Zool. Soc. London* **131**, 1–48.

Hinde, R. A. (1959). Behaviour and speciation in birds and lower vertebrates. *Biol. Rev.* **34**, 85–128.

Hinde, R. A. (1960). Factors governing the changes in strength of a partially inborn response, as shown by the mobbing behaviour of the Chaffinch (*Fringilla coelebs*). III. The interaction of short-term and long-term incremental and decremental effects. *Proc. Roy. Soc., Ser. B* **153**, 398–420.

Hinde, R. A. (1965). Interaction of internal and external factors in integration of canary reproduction. *In* "Sex and Behavior" (F. A. Beach, ed.), pp. 381–415. Wiley, New York.

Hinde, R. A. (1968). Dichotomies in the study of development. *In* "Genetic and Environmental Influences on Behavior" (J. M. Thoday and A. S. Parkes, eds.), pp. 1–14. Oliver & Boyd, Edinburgh.

Hinde, R. A., ed. (1969). "Bird Vocalizations." Cambridge Univ. Press, London and New York.

Hinde, R. A. (1970). "Animal Behaviour," 2nd ed. McGraw-Hill, New York.

Hinde, R. A., and Fisher, J. (1951). Further observations on the opening of milk bottles by birds. *Brit. Birds* **44**, 393–396.

Hinde, R. A., and Steel, E. A. (1966). Integration of the reproductive behaviour of female canaries. *Symp. Soc. Exp. Biol.* **20**, 401–426.

Hinde, R. A., and Steel, E. A. (1972). Reinforcing events in the integration of canary nest-building. *Anim. Behav.* **20**, 514–525.

Hinde, R. A., and Stevenson, J. G. (1969). Integration of response sequences. *In* "Advances in the Study of Behaviour" (D. S. Lehrman, R. A. Hinde, and E. Shaw, eds.), Vol. 2, pp. 267–296. Academic Press, New York.

Hinde, R. A., Bell, R. Q., and Steel, E. A. (1963). Changes in sensitivity of the canary brood patch during the natural breeding season. *Anim. Behav.* **11**, 553–560.

Hoffmann, H. S., Schiff, D., Adams, J., and Searle, J. J. (1966). Enhanced distress vocalization through selective reinforcement. *Science* **151**, 352–354.

Hogan, J. A. (1964). Operant control of preening in pigeons. *J. Exp. Anal. Behav.* **7**, 351–352.

Hogan, J. A. (1965). An experimental study of conflict and fear: An analysis of behaviour of young chicks toward a mealworm. I. The behaviour of chicks which do not eat the mealworm. *Behaviour* **25**, 45–97.

Hogan, J. A. (1973). How young chicks learn to recognize food. *In* "Constraints on Learning: Limitations and Predispositions" (R. A. Hinde and J. S. Hinde, eds.), pp. 119–140. Academic Press, London.

Holzapfel, M. (1949). Die Beziehung zwischen den Treiben junger und erwachsener Tiere. *Schweiz. Z. Psychol.* **8**, 32–60.

Horn, G., and Hinde, R. A. (1970). "Short-Term Changes in Neural Activity and Behaviour." Cambridge Univ. Press, London and New York.

Hunt, G. L., and Smith, W. J. (1964). The swallowing of fish by young Herring Gulls. *Ibis* **106**, 457–461.

Hutchison, J. B. (1967). Initiation of courtship by hypothalamic implants of testosterone propionate in castrated doves (*Streptopelia risoria*). *Nature (London)* **216**, 591–592.

Hutchison, R. E. (1971). The integration of reproductive behaviour in the female Budgerigar. Ph.D. Thesis, Cambridge University.

Hutchison, R. E., Stevenson, J. G., and Thorpe, W. H. (1968). The basis for individual recognition by voice in the Sandwich Tern (*Sterna sandvicensis*). *Behaviour* **32**, 150–157.

Immelmann, K. (1969). Song development in the Zebra Finch and other estrildid finches. *In* "Bird Vocalizations" (R. A. Hinde, ed.), pp. 61–74. Cambridge Univ. Press, London and New York.

Katz, D., and Révész, G. (1921). Experimentelle Studien zur vergleichenden Psychologie (Versuche mit Hühnern). *Z. Angew. Psychol.* **18**, 307–330.

Kear, J. (1962). Food selection in finches with special references to interspecific differences. *Proc. Zool. Soc. London* **138**, 163–204.

Kear, J. (1964). Colour preference in young Anatidae. *Ibis* **106**, 361–369.

Kear, J. (1967). *Wildfowl Trust 18th Annu. Rep.* pp. 122–124.

Klinghammer, E. (1967). Factors influencing choice of mate in altricial birds. *In* "Early Behavior: Comparative and Developmental Approaches" (H. W. Stevenson *et al.*, eds.), pp. 5–42. Wiley, New York.

Klinghammer, E., and Hess, E. H. (1964). Parental feeding in Ring Doves (*Streptopelia roseogrisea*): Innate or learned? *Z. Tierpsychol.* **21**, 338–347.

Klopfer, P. H. (1959). Social interactions in discrimination learning with special reference to feeding behavior in birds. *Behaviour* **14**, 282–299.

Koehler, O. (1943). "Zahl" – Versuche en einem Kohlraben und Vergleichsversuche an Menschen. *Z. Tierpsychol.* **5**, 575–712.

Komisaruk, B. R. (1967). Effects of local brain implants of progesterone on reproductive behavior in Ring Doves. *J. Comp. Physiol. Psychol.* **64**, 219–224.

Konishi, M., and Nottebohm, F. (1969). Experimental studies in the ontogeny of avian vocalizations. *In* "Bird Vocalizations" (R. A. Hinde, ed.), pp. 29–48. Cambridge Univ. Press, London and New York.

Kortlandt, A. (1940). Eine Übersicht der angeborenen Verhaltenweisen des Mitteleuropaischen Kormorans (*Phalacrocorax carbo sinensis*). *Arch. Neer. Zool.* **14**, 401–442.

Krames, L., and Carr, W. J. (1968). The effect of previous visual experience upon the response to depth in domestic chickens. *Psychon. Sci.* **10**, 249–250.

Krebs, J. R. (1970). The efficiency of courtship feeding in the Blue Tit (*Parus caeruleus*). *Ibis* **112**, 108–110.

Kruijt, J. P. (1964). Ontogeny of social behaviour in Burmese Red Junglefowl (*Gallus gallus spadiceus* Bonnaterre. *Behaviour, Suppl.* **12**, 1–201.

Kruschinski, L. W., Swetuchina, W. M., Molodkina, L. N., Popowa, N. P., and Maz, W. N. (1963). Vergleichende physiologisch-morphologische Erforschung komplizierter Verhaltensformen von Vögeln. *Z. Tierpsychol.* **20**, 474–486.

Kruuk, H. (1964). Predators and anti-predator behaviour of the Black-headed Gull (*Larus ridibundus* L.). *Behaviour, Suppl.* **11**, 1–129.

Kunkel, P. (1962). Zur Verbreitung des Hüpfens und Laufens unter Sperlingsvögeln (Passeres). *Z. Tierpsychol.* **19**, 417–439.

Kuo, Z. Y. (1932). Ontogeny of embryonic behavior in Aves. IV. The influence of embryonic movements upon the behavior after hatching. *J. Comp. Psychol.* **14**, 109–122.

Kuo, Z. Y. (1967). "The Dynamics of Behavior Development." Random House, New York.

Lack, D. (1939). The behaviour of the Robin. I and II. *Proc. Zool. Soc. London, Ser. A* **109**, 169–178.

Lack, D. (1947). "Darwin's Finches." Cambridge Univ. Press, London and New York.

Lack, D. (1954). "The Natural Regulation of Animal Numbers." Oxford Univ. Press, London and New York.

Lack, D. (1966). "Population Studies of Birds." Oxford Univ. Press (Clarendon), London and New York.

Lehrman, D. S. (1953). A critique of Konrad Lorenz's theory of instinctive behaviour. *Quart. Rev. Biol.* **28**, 337–363.

Lehrman, D. S. (1955). The physiological basis of parental feeding behaviour in the Ring Dove (*Streptopelia risoria*). *Behaviour* **7**, 241–286.

Lehrman, D. S. (1961). Gonadal hormones and parental behavior in birds and infra-human mammals. *In* "Sex and Internal Secretions" (W. C. Young, ed.), 3rd ed., pp. 1268–1382. Williams & Wilkins, Baltimore, Maryland.

Lehrman, D. S. (1962). Interaction of hormonal and experiential influences on development of behavior. *In* "Roots of Behavior" (E. L. Bliss, ed.), pp. 142–156. Harper (Hacher), New York.

Lehrman, D. S. (1965). Interaction between internal and external environments in the regulation of the reproductive cycle of the Ring Dove. *In* "Sex and Behavior (F. A. Beach, ed.), pp. 355–380. Wiley, New York.

Lehrman, D. S. (1970). Semantic and conceptual issues in the nature-nurture problem. *In* "Development and Evolution of Behavior" (L. R. Aronsen *et al.*, eds.), pp. 17–52. Freeman, San Francisco, California.

Lehrman, D. S., and Wortis, R. P. (1960). Previous breeding experience and hormone-induced incubation behavior in the Ring Dove. *Science* **132**, 1667–1668.

Lehrman, D. S., and Wortis, R. P. (1967). Breeding experience and breeding efficiency in the Ring Dove. *Anim. Behav.* **15**, 223–228.

Lorenz, K. (1935). Der Kumpan in der Umwelt des Vogels. *J. Ornithol.* **83**, 137–213, and 289–413.

Lorenz, K. (1937). Über die Bildung des Instinktbegriffes. *Naturwissenschaften* **25**, 289–300, 307–318, and 324–331.

Lorenz, K. (1941). Vergleichende Bewegungsstudien an Anatinen. *J. Ornithol.* **89,** Suppl., 194–294.

Lorenz, K. (1950). The comparative method in studying innate behaviour patterns. *Symp. Soc. Exp. Biol.* **4,** 221–268.

Lorenz, K. (1951). Über die Entstehung auslösender "Zeremonien." *Vogelwarte* **16,** 9–13.

Lorenz, K. (1961). Phylogenetische Anpassung und adaptive Modifikation des Verhaltens. *Z. Tierpsychol.* **18,** 139–187.

Lorenz, K. (1965). "Evolution and Modification of Behavior." Univ. of Chicago Press, Chicago, Illinois.

Lorenz, K., and Tinbergen, N. (1938). Taxis und Instinkthandlung in der Eirollbewegung der Graugans. I. *Z. Tierpsychol.* **2,** 1–29.

McFarland, D. J. (1965a). The effect of hunger on thirst motivated behaviour in the Barbary Dove. *Anim. Behav.* **13,** 286–292.

McFarland, D. J. (1965b). Hunger, thirst and displacement pecking in the Barbary Dove. *Anim. Behav.* **13,** 293–300.

McFarland, D. J. (1965c). Control theory applied to the control of drinking in the Barbary Dove. *Anim. Behav.* **13,** 478–492.

McFarland, D. J. (1966). The role of attention in the disinhibition of displacement activities. *Quart. J. Exp. Psychol.* **18,** 19–30.

McFarland, D. J. (1967). On the causal and functional significance of displacement activities. *Z. Tierpsychol.* **23,** 217–235.

Maier, N. R. F., and Schneirla, T. C. (1935). "Principles of Animal Psychology." McGraw-Hill, New York.

Marler, P. (1955). Studies of fighting in Chaffinches. *Brit. J. Anim. Behav.* **3,** 111–117 and 137–146.

Marler, P. (1956a). *Brit. J. Anim. Behav.* **4,** 23–30.

Marler, P. (1956b). The voice of the Chaffinch and its function as a language. *Ibis* **98,** 231–261.

Marler, P. (1957). Specific distinctiveness in the communication signals of birds. *Behaviour* **11,** 13–39.

Marler, P. (1959). Development in the study of animal communication. *In* "Darwin's Biological Work" (P. R. Bell, ed.), pp. 150–206. Cambridge Univ. Press, London and New York.

Marler, P. (1970). A comparative approach to vocal learning: song development in White-crowned Sparrows. *J. Comp. Physiol. Psychol.* **71,** Monogr., 1–25.

Marler, P., and Hamilton, W. J. (1966). "Mechanisms of Animal Behavior." Wiley, New York.

Matthews, G. V. T. (1968). "Bird Navigation." Cambridge Univ. Press, London and New York.

May, J. G., and Dorr, D. (1968). Initiative pecking in chicks as a function of early social experience. *Psychon. Sci.* **11,** 175–176.

Maynard Smith, J. (1965). The evolution of alarm calls. *Amer. Natur.* **99,** 59–63.

Millikan, G. C., and Bowman, R. I. (1967). Observations on Galapagos tool-using finches in captivity. *Living Bird* **6,** 23–41.

Morris, D. (1956). The feather postures of birds and the problem of the origin of social signals. *Behaviour* **9,** 75–113.

Morris, D. (1957). "Typical intensity" and its relation to the problem of ritualisation. *Behaviour* **11,** 1–12.

Morse, D. H. (1968). The use of tools by Brown-headed Nuthatches. *Wilson Bull.* **80**, 220–224.

Moseley, D. (1925). The accuracy of the pecking response in chicks. *J. Comp. Psychol.* **5**, 75–97.

Moynihan, M. (1955). Some aspects of reproductive behavior in the Blackheaded Gull (*Larus ridibundus ridibundus* L.) and related species. *Behaviour, Suppl.* **4**, 1–201.

Moynihan, M. (1959). Notes on the behavior of some North American gulls. IV. The ontogeny of hostile behavior and display patterns. *Behaviour* **14**, 214–239.

Moynihan, M. (1962). Hostile and sexual behavior patterns of South American and Pacific Laridae. *Behaviour, Suppl.* **8**, 1–365.

Newton, I. (1967). The adaptive radiation and feeding ecology of some British finches. *Ibis* **109**, 33–98.

Nicolai, J. (1962). Über Regen-, Sonnen- und Staubbaden bei Tauben (Columbidae). *J. Ornithol.* **103**, 125–139.

Nicolai, J. (1964). Der Brutparasitismus der Viduinae als ethologisches Problem: Prägungsphänomene als Faktoren der Rassen- und Artbildung. *Z. Tierpsychol.* **21**, 129–204.

Norton-Griffiths, M. (1967). Some ecological aspects of the feeding behaviour of the Oystercatcher *Haematopus ostralegus* on the edible mussel *Mytilus edulis*. *Ibis* **109**, 412–424.

Nottebohm, F. (1970). Ontogeny of bird song. *Science* **167**, 950–956.

Oberholzer, A., and Tschanz, B. (1968). Zur Verhalten der jungen Trottellumme (*Uria aalge*) gegenüber Fisch. *Rev. Suisse Zool.* **75**, 43–51.

Oppenheim, R. W. (1968). Light responsivity in chick and duck embryos just prior to hatching. *Anim. Behav.* **16**, 276–280.

Pastore, N. (1962a). Perceptual functioning in the duckling. *J. Psychol.* **54**, 293–298.

Pastore, N. (1962b). Further experiments in counting ability: canaries and Mynas. *Z. Tierpsychol.* **19**, 665–686.

Perrins, C. (1968). The purpose of the high-intensity alarm call in small passerines. *Ibis* **110**, 200–201.

Plotnik, R. J., and Tallarico, R. B. (1966). Object-quality learning-set formation in the young chicken. *Psychon. Sci.* **5**, 195–196.

Räber, H. (1948). Analyse des Balzverhaltens eines domestizierten Truthahns (*Meleagris*). *Behaviour* **1**, 237–266.

Rabinowitch, V. E. (1968). The role of experience in the development of food preferences in gull chicks. *Anim. Behav.* **16**, 425–428.

Rahmann-Esser, M. (1964). Erlernen rhythmischer Handlungsfolgen bei Hühnern. *Z. Tierpsychol.* **21**, 837–853.

Reinert, J. (1965). Takt und Rhythmusunterscheidung bei Dohlen. *Z. Tierpsychol.* **22**, 623–671.

Reinert, J., and Reinert-Reetz, W. (1962). Das Erkennen erlernter Tonfolgen in abgewandelter Form durch einen Wellinsittich. *Z. Tierpsychol.* **19**, 728–740.

Rheingold, H. L., and Hess, E. H. (1957). The chick's "preference" for some visual properties of water. *J. Comp. Physiol. Psychol.* **50**, 417–421.

Rothschild, M., and Ford, B. (1968). Warning signals from a Starling *Sturnus vulgaris* observing a bird rejecting unpalatable prey. *Ibis* **110**, 104–105.

Rowell, C. H. F. (1961). Displacement grooming in the Chaffinch. *Anim. Behav.* **9**, 38–63.

Russell, E. S. (1943). Perceptual and sensory signs in instinctive behaviour. *Proc. Linn. Soc. London* **154**, 195–216.

Sargent, T. D., and Keiper, R. R. (1967). Stereotypes in caged canaries. *Anim. Behav.* **15**, 62–66.

Sauer, F. (1954). Die Entwicklung der Lautäusserungen vom Ei Abschalldichtgehaltener Dorngrasmücken (*Sylvia c. communis* L.) im Vergleich mit später isolierten und mit wildlebenden Artgenossen. *Z. Tierpsychol.* **11**, 10–93.

Sauer, F. (1956). Über das Verhalten junger Gartengrasmücken *Sylvia borin. J. Ornithol.* **97**, 156–187.

Schein, M. W., and Hale, E. B. (1965). Stimuli eliciting sexual behavior. *In* "Sex and Behavior" (F. A. Beach, ed.), pp. 440–482. Wiley, New York.

Schleidt, W. M. (1961). Reaktionen von Truthühnern auf fliegende Raubvögel und Versuche zur Analyse ihrer AAM's. *Z. Tierpsychol.* **18**, 534–560.

Schmidt-Koenig, K. (1965). Current problems in bird orientation. *In* "Advances in the Study of Behavior" (D. S. Lehrman, R. A. Hinde, and E. Shaw, eds.), Vol. 1, pp. 217–278. Academic Press, New York.

Schneirla, T. C. (1965). Aspects of stimulation and organization in approach/withdrawal processes underlying vertebrate behavioral development. *In* "Advances in the Study of Behavior" (D. S. Lehrman, R. A. Hinde, and E. Shaw, eds.), Vol. 1, pp. 1–74. Academic Press, New York.

Schütz, F. (1965). Sexuelle Prägung bei Anatiden. *Z. Tierpsychol.* **22**, 50–103.

Seitz, A. (1940/1941). Die Paarbildung bei einigen Cichliden. I. *Z. Tierpsychol.* **4**, 40–84.

Sherrington, C. S. (1906). "The Integrative Action of the Nervous System." Yale Univ. Press, New Haven, Connecticut.

Shinkman, P. G. (1963). Visual depth-discrimination in day-old chicks. *J. Comp. Physiol. Psychol.* **56**, 410–414.

Shoemaker, H. H. (1939). Effect of testosterone propionate on behaviour of the female canary. *Proc. Soc. Exp. Biol. Med.* **41**, 299–302.

Siegel, A. I. (1953a). Deprivation of visual form definition in the Ring Dove. I. Discriminatory learning. *J. Comp. Physiol. Psychol.* **46**, 115–119.

Siegel, A. I. (1953b). Deprivation of visual form definition in the Ring Dove. II. Perceptual-motor transfer. *J. Comp. Physiol. Psychol.* **46**, 249–252.

Simmons, K. E. L. (1952). The nature of the predator-reactions of breeding birds. *Behaviour* **4**, 161–172.

Simmons, K. E. L. (1961). Problems of head-scratching in birds. *Ibis* **103a**, 37–49.

Skinner, B. F. (1938). "The Behavior of Organisms; An Experimental Analysis." Appleton, New York.

Skinner, B. F. (1948). "Superstition" in the pigeon. *J. Exp. Psychol.* **38**, 168–172.

Smith, N. G. (1966). Adaptations to cliff-nesting in some arctic gulls (*Larus*). *Ibis* **108**, 68–83.

Smith, W. J. (1966). Communication and relationships in the genus *Tyrannus. Nuttall Ornithol. Club., Publ.* **6**, 1–250.

Smith, W. J. (1969). Messages of vertebrate communication *Science* **165**, 145–150.

Sterritt, G. M., and Smith, M. P. (1965). Reinforcement effects of specific components of feeding in young leghorn chicks. *J. Comp. Physiol. Psychol.* **59**, 171–175.

Stevenson, J. G. (1969). Song as a reinforcer. *In* "Bird Vocalizations" (R. A. Hinde, ed.), pp. 49–60. Cambridge Univ. Press, London and New York.

Stokes, A. W. (1962). Agonistic behaviour among Blue Tits at a winter feeding station. *Behaviour* **19**, 118–138.

Stonehouse, B. (1953). The Emperor Penguin. I. *Falkland Isl. Depend. Surv., Sci. Rep.* **6**.

Swanberg, P. O. (1951). Food storage, territory and song in the Thick-billed Nutcracker. *Proc. Int. Congr. Ornithol., 10th, 1950* pp. 545–554.

Tallarico, R. B., and Farrell, W. M. (1964). Studies of visual depth perception: an effect of early experience on chicks on a visual cliff. *J. Comp. Physiol. Psychol.* **57**, 94–96.

Thompson, T. I. (1964). Visual reinforcement in fighting cocks. *J. Exp. Anal. Behav.* **7**, 45–49.

Thompson, T. I. (1969). Conditioned avoidance of the mobbing call by Chaffinches. *Anim. Behav.* **17**, 517–522.

Thorpe, W. H. (1948). The modern concept of instinctive behaviour. *Bull. Anim. Behav.* **7**, 2–12.

Thorpe, W. H. (1961). "Bird Song." Cambridge Univ. Press, London and New York.

Thorpe, W. H. (1963). "Learning and Instinct in Animals," 2nd ed. Methuen, London.

Tinbergen, N. (1940). Die Überspringbewegung. *Z. Tierpsychol.* **4**, 1–40.

Tinbergen, N. (1942). An objectivistic study of the innate behaviour of animals. *Bibl. Biotheor., Leiden* **1**, 39–98.

Tinbergen, N. (1948). Social releases and the experimental method required for their study. *Wilson Bull.* **60**, 6–51.

Tinbergen, N. (1951). "The Study of Instinct." Oxford Univ. Press, London and New York.

Tinbergen, N. (1952). Derived activities: their causation, biological significance, origin and emancipation during evolution. *Quart. Rev. Biol.* **27**, 1–32.

Tinbergen, N. (1953). "The Herring Gull's World." Collins, London.

Tinbergen, N. (1957). The functions of territory. *Bird Study* **4**, 14–28.

Tinbergen, N. (1959). Comparative studies of the behaviour of gulls (Laridae): A progress report. *Behaviour* **15**, 1–70.

Tinbergen, N. (1967). Adaptive features of the Black-headed Gull *Larus ridibundus* L. *Proc. Int. Ornithol. Congr., 14th, 1966* pp. 43–59.

Tinbergen, N., and Kuenen, D. J. (1939). Über die auslösenden und die richtunggebenden Reizsituationen der Sperrbewegung von jungen Drosseln (*Turdus m. merula* L. *und T. e. ericetorum* Turton). *Z. Tierpsychol.* **3**, 37–60.

Tinbergen, N., and Perdeck, A. C. (1950). On the stimulus situation releasing the begging response in the newly-hatched Herring Gull chick (*Larus argentatus* Pont.). *Behaviour* **3**, 1–38.

Tinbergen, N., Impekoven, M., and Franck, D. (1967). An experiment on spacing out as a defence against predation. *Behaviour* **28**, 307–321.

Tolman, C. W. (1967). The feeding behaviour of domestic chicks as a function of rate of pecking by a surrogate companion. *Behaviour* **29**, 57–62.

Tschanz, B. (1968). Trottellummen. *Z. Tierpsychol., Suppl.* **4**, 1–103.

van Iersel, J. J. A., and Bol, A. C. A. (1958). Preening of two tern species. A study on displacement activities. *Behaviour* **13**, 1–88.

van Lawick-Goodall, J. (1970). Tool-using. *In* "Advances in the Study of Behavior" (D. S. Lehrman, R. A. Hinde, and E. Shaw, eds.), Vol. 3, pp. 195–250. Academic Press, New York.

Vince, M. A. (1960). Developmental changes in responsiveness in the Great Tit (*Parus major*). *Behaviour* **15**, 219–243.

Vince, M. A. (1961). "String-pulling" in birds. III. The successful response in Green-finches and canaries. *Behaviour* **17**, 103–129.

Vince, M. A. (1964). Use of the feet in feeding by the Great Tit *Parus major. Ibis* **106**, 508–529.

Vince, M. A. (1969). Embryonic communication, respiration and the synchronisation of hatching. *In* "Bird Vocalizations" (R. A. Hinde, ed.), pp. 233–260. Cambridge Univ. Press, London and New York.

von Holst, E., and von St. Paul, U. (1963). On the functional organisation of drives. *Anim. Behav.* **11**, 1–20; translated from *Naturwissenschaften* **18**, 409–422 (1960).

Walk, R. D. (1965). The study of visual depth and distance perception in animals. *In* "Advances in the Study of Behavior" (D. S. Lehrman, R. A. Hinde, and E. Shaw, eds.), Vol. 1, pp. 100–154. Academic Press, New York.

Warden, C. J., Jenkins, T. N., and Warner, L. H. (1936). "Comparative Psychology." Ronald Press, New York.

Wickler, W. (1961). Über die Stammesgeschichte und den taxonomischen Wert einiger Verhaltensweisen der Vögel. *Z. Tierpsychol.* **18**, 320–342.

Wood-Gush, D. G. M., and Gower, D. M. (1968). Studies on motivation in the feeding behaviour of the domestic cock. *Anim. Behav.* **16**, 101–107.

Wünschmann, A. (1963). Quantitative Untersuchungen zum Neugierverhalten von Wirbeltieren. *Z. Tierpsychol.* **20**, 80–109.

Wortis, R. P. (1969). The transition from dependent to independent feeding in the young Ring Dove. *Anim. Behav., Suppl.* **2**, 1–54.

Zeier, H. (1966). Über sequentielles Lernen bei Tauben, mit spezieller Berücksichtigung des "Zähl"-Verhaltens. *Z. Tierpsychol.* **23**, 161–189.

Zeigler, H. P. (1961). Learning-set formation in pigeons. *J. Comp. Physiol. Psychol.* **54**, 252–254.

AUTHOR INDEX

Numbers in italics refer to the pages on which the complete references are listed.

537

Ranch, V. M., 193, *279*
Ranney, R. E., 56, 58, *103*
Rathkamp, R., 418, 426, *473*
Rautenberg, W., 426, *474*
Raviola, E., 380, *386*
Raviola, G., 380, *386*
Ray, I., 208, *279*
Reed, F. E., 220, *285*
Reese, E. P., 502, 524, *525*
Refetoff, S., 213, 214, 216, 217, *279*
Regan, N., *104*
Regaud, C., 25, *103*
Reineke, E. P., 216, 217, 218, 219, 221, 223, *276, 279, 281*
Reinert, J., 503, *532*
Reinert-Reetz, W., 503, *532*
Reinhart, W. H., 55, *103*
Reisfeld, R. A., 255, *278*
Resko, J. A., 197, 199, 201, *279*, 326, *344*
Révész, G., 510, *529*
Rheingold, H. L., 501, *532*
Richardson, K. C., 70, *103*
Riddle, O., 4, 6, 27, 33, 55, 58, 59, 63, 69, 86, *88, 96, 103*, 120, 145, 164, 166, *179*, 193, 197, 202, 204, 222, 223, 261, *276, 279*, 327, *343*
Rieke, G. K., 397, *414*
Riley, G. M., 148, *179*
Riley, J., 232, 233, *268*
Ringer, R. K., 218, 221, *281*, 329, *337*
Ringoen, A. R., 57, *103*
Robbins, J., 213, *268*
Robertis, J. M., 48, *101*
Roberts, J. D., 14, 15, *99*
Robin, N. I., 213, 214, 216, 217, *279*
Roche, G. L., 325, *344*
Roche, J., 212, *279*
Rochon-Duvigneaud, A., 350, 357, 374, *386*
Rodrigues de Lores Arnaiz, A., 309, *338*
Roepke, R. R., 55, *103*
Romanoff, A. J., 70, *103*
Romanoff, A. L., 48, 70, *103*, 186, 211, 258, *279*
Rookledge, K. A., 65, *94*, 152, 153, 161, *175*, 244, *271*
Roos, T. B., 54, 58, *103*
Roosen-Runge, E. C., 9, *103*
Rosati, P., 10, *90*

Rosenberg, L. L., 212, 213, 214, 216, 217, 220, 227, 228, 229, 230, *263, 269, 279, 280*, 325, *344*
Rossbach, R., 290, 312, 313, 314, *344*
Rothchild, I., 53, 59, *103, 104*
Rothschild, M., 503, *532*
Roudneva, L. M., 65, *101*
Rounds, D. E., 56, *92*
Rowan, W., 17, *104*
Rowell, C. H. F., 495, *532*
Rubin, B. L., 10, *91*
Rudolph, H. J., 170, *179*
Russell, D. H., 317, *344, 345*
Russell, E. S., 487, *533*
Russo, R. P., 114, 115, *180*
Rzasa, J., 71, *104*

S

Sadlier, R. M. F. S., 47, *99*
Saeki, Y., 86, *104*, 166, *179*
Sagitov, A. K., 433, *474*
St. Pierre, R. L., 260, *280*
Salem, M. H. M., 169, *179*, 329, 332, *345*
Salem, M. H. R., 199, 200, *280*
Salganicoff, L., 309, *338*
Salley, K. W., 196, *283*
Salvatore, G., 212, *279*
Salzman, A., 405, 411, *415*
Samuels, L. T., 10, *104*
Samols, E., 240, 243, 245, 246, 248, *280*
Sandberg, A. A., 190, 192, *280*
Sanders, E. B., 396, *414*
Sandor, T., 189, 190, *280*
Sargent, T. D., 500, *533*
Sarkar, A. K., 152, *176*, 202, 203, 206, 207, 261, *264, 272, 280*, 309, 329, *340*
Sarkar, M., 145, *171*
Sato, T., 237, *280*, 329, 332, *345*
Sato, Y., 120, 140, 145, 148, *176*
Sauer, F., 462, *474*, 500, *533*
Sauers, A. K., 194, *277*
Savard, K., 48, 49, 50, 51, 53, *90*
Savkovic, L. J., 439, *471*
Sawano, S., 323, *343*
Sawyer, W. H., 314, 315, *343, 345*
Saxena, R. N., 33, *106*
Ŝaxod, R., 419, 420, 421, 423, 427, *475*
Sayler, A., 48, 51, 54, *104*

INDEX TO BIRD NAMES

SUBJECT INDEX

A

Acetylcholine, 309, 310
Acetylcholinesterase (AChE), 120, 141, 145, 299, 317
Accessory sex organs, 65–78, *see also* specific organs
female, 69–72
male, 66–69
Acid phosphatase, 316, 317
Adenohypophyseal hormone-releasing factors, 288
Adenohypophysectomy, 315, 326, 327, 331
Adenohypophysis, 34, 38, 109–170, 288, 291, 299, 313, 316, 317, 320–322, 324, 327, 328, 330, 331, 334–336
cellular differentiation in, 115, 116
embryology of, 114–116
morphogenesis of, 114, 115
lobes of, 114
Adenosinetriphosphatase (ATPase), potassium- and sodium-sensitive, 454
Adipose tissue, 210, 244, 245
brown, 209
Adrenal-cortical function, 314
regulation of, 196–208
rhythm of, 201–203
annual, 202, 203
diurnal, 201, 202
Adrenal cortex, 185, 186, 188–208
carbohydrate metabolism, 193, 194
electrolyte metabolism, 195, 196
interrenal tissue, 186

lipid metabolism, 194
protein metabolism, 194, 195
Adrenal glands, 184–210, 261, 314, 326, 327, *see also* specific glands
testes and, 203–206
Adrenal hypertrophy, 204
Adrenal medulla, 185, 186, 208–210
carbohydrate metabolism, 210
chromaffin tissue, 185
lipid metabolism, 209, 210
Adrenalectomy, 134, 149, 150, 195, 196
Adrenaline, 208, 242
Adrenocorticotropin (ACTH), 37, 137, 149, 168, 169, 186, 198, 199, 202, 204, 242, 261, 262, 314, 322, 327, 331, 332
Adrenocorticotropic hormone-releasing factor, 291, 336
Aggressiveness, 513
Albumin, 213
Aldosterone, 185, 188–191, 195, 205
Alkaline phosphatase, 257, 317
Alternation, 493
Amines, biogenic, 119
Amino acids, 119
Ampulla anterior, 430
Ampulla horizontalis, 430
Ampulla posterior, 430
Ampullary crests, 433, 436
circular acceleration, 438
Androgen, 26–30, 57, 80, 81
Androstenedione, 10, 49
Archistriatum, 451
Ascorbic acid, 261

N

Nasal glands, 195, 196, 392
Navigation, 502
Neopallium, 457
Neostriatum, 397, 457, 458
  caudal, 451
Nerve(s), *see* specific types
Nerve endings, free, 420
Nests, 520–522
  building, 496, 498, 520, 522
  site, 520–522
Neuroendocrinology, 287–336
Neurohormones, hypophysiotropic, 291, 299
Neurohypophyseal hormones, 288, 307, 309, 310, 313–315, 333
  releasing factors, 328–333
Neurohypophysectomy, 315, 323
Neurohypophysis, 297, 308–310
Neurons
  catecholaminergic, 200
  serotoninergic, 200
Neurosecretory cells, 288, 299, 312, 318, 333
  releasing factors, 320
Neurosecretory system, 288–307
  enzymes in, 316–318
  Gomori-negative, 291, 320–334
  Gomori-positive, 289–291, 298, 299, 310–318
Node of Ranvier, 426
Noradrenaline, 333
Norepinephrine, 185, 208–210
Nucleus angularis, 458
Nucleus basilis, 299
Nucleus entopeduncularis, 290
Nucleus habenularis, 451, 458
Nucleus hypothalamicus posterior medialis, 297–299, 323, 325, 326
Nucleus isthmi, 458
Nucleus laminaris, 458
Nucleus lateralis externus hypothalami, 290
Nucleus lemniscus, 458
Nucleus magnocellularis, 458
Nucleus magnocellaris interstitialis dorsalis, 290, 314
Nucleus ovoidalis, 451, 458
Nucleus paraventricularis, 289, 290, 297–299, 311, 313, 321, 325, 326, 334, 335
Nucleus rotundus, 374, 451, 458
Nucleus semilunaris, 458
Nucleus tuberis, 294, 297–299, 321–324, 327, 333
Nystagmus, 438

O

Oculomotor nuclei, 435, 458
Olfaction, 389, 392, 409, 410, *see also* Smell
  use in homing, 410
Olfactory bulb, 396, 397, 424
Olfactory epithelium, 393
Olfactory nerve, 394, 404
Olive
  inferior, nuclei, 435
  superior, 458
Oocytes, 43
Optic chiasma, 322
Optic disc, 358, 377
Optic nerve, 451
Osmoreceptors, 333
Ovary, 40–46, 85, 322, *see also* Accessory sex organs
  effect of hormones on, 56–60, *see also* specific substances
  histophysiology of, 46–56
Oviduct, 42, 50, 59, 61, 86, 322
Oviposition, 323
Ovotestis, 4
Ovulation, 86, 322–324
Oxytocin, 313–315, 329, 330, 334

P

Pacinian corpuscles, 418, 421, 422, 426, 427
Pair formation, 513–516, 522
Palaeostriatum, 425
Pallium, 457
Palmic acid, 194
Pancreas, endocrine, 185, 236–248
  carbohydrate metabolism, 241–244
  cell types, 237
  lipid metabolism, 244, 245
Pancreatectomy, 241–243, 247
Pancreozymin, 247
Papilla basilaris, 429, 430, 444, 448, 449